THE INITIATION OF DNA REPLICATION

Academic Press Rapid Manuscript Reproduction

Proceedings of the 1981 ICN–UCLA Symposia
on Struture and DNA–Protein Interactions of Replication Origins
Held in Salt Lake City, Utah, on March 8–13, 1981

ICN–UCLA Symposia on Molecular and Cellular Biology
Volume XXII, 1981

THE INITIATION OF DNA REPLICATION

edited by

DAN S. RAY

Molecular Biology Institute
University of California, Los Angeles
Los Angeles, California

Series Editor
C. FRED FOX

Department of Microbiology and Molecular Biology
University of California, Los Angeles
Los Angeles, California

Managing Editor
FRANCES J. STUSSER
ICN–UCLA Symposia
University of California, Los Angeles
Los Angeles, California

ACADEMIC PRESS 1981
A Subsidiary of Harcourt Brace Jovanovich, Publishers
New York London Toronto Sydney San Francisco

COPYRIGHT © 1981, BY ACADEMIC PRESS, INC.
ALL RIGHTS RESERVED.
NO PART OF THIS PUBLICATION MAY BE REPRODUCED OR
TRANSMITTED IN ANY FORM OR BY ANY MEANS, ELECTRONIC
OR MECHANICAL, INCLUDING PHOTOCOPY, RECORDING, OR ANY
INFORMATION STORAGE AND RETRIEVAL SYSTEM, WITHOUT
PERMISSION IN WRITING FROM THE PUBLISHER.

ACADEMIC PRESS, INC.
111 Fifth Avenue, New York, New York 10003

United Kingdom Edition published by
ACADEMIC PRESS, INC. (LONDON) LTD.
24/28 Oval Road, London NW1 7DX

Library of Congress Cataloging in Publication Data
Main entry under title:

The Initiation of DNA replication.

(ICN-UCLA symposium on molecular and cellular
biology; v. 22)
  Papers from a meeting held in March 1981, at Salt
Lake City, Utah, sponsored by ICN Pharmaceuticals, Inc.
  Includes index.
  1. DNA replication--Congresses. I. Ray, Dan S.
II. ICN Pharmaceuticals, Inc. III. Series. [DNLM:
1. DNA replication--Congresses. WS 460 A243]
QH450.I56   574.87'3282   81-17541
ISBN 0-12-583580-9   AACR2

PRINTED IN THE UNITED STATES OF AMERICA

81 82 83 84   9 8 7 6 5 4 3 2 1

CONTENTS

Contributors   xi
Preface   xix

I. BACTERIAL REPLICATION ORIGINS

1. *Escherichia coli* Origin of Replication: Structural Organization of the Region Essential for Autonomous Replication and the Recognition Frame Model   1
Yukinori Hirota, Atsuhiro Oka, Kazunori Sugimoto, Kiyozo Asada, Hitoshi Sasaki, and Mituru Takanami

2. The DNA Replication Origin Region of the Enterobacteriaceae   13
Judith W. Zyskind, Nancy E. Harding, Yutaka Takeda, Joseph M. Cleary, and Douglas W. Smith

3. Appendix: The Consensus Sequence of the Bacterial Origin   26
Judith W. Zyskind, Douglas W. Smith, Yukinori Hirota, and Mituru Takanami

4. Mapping of Promoters in the Replication Origin Region of the *E. coli* Chromosome   29
Masayuki Morita, Kazunori Sugimoto, Atsuhiro Oka, Mituru Takanami, and Yukinori Hirota

5. Transcription and Translation Events in the *oriC* Region of the *E. coli* Chromosome   37
Flemming G. Hansen, Susanne Koefoed, Kaspar von Meyenburg, and Tove Atlung

6. Genes, Transcriptional Units, and Functional Sites in and around the *E. coli* Replication Origin 57
Heinz Lother, Hans-Jörg Buhk, Giovanna Morelli, Barbara Heimann, Trinad Chakraborty, and Walter Messer

7. Specific Binding of *Escherichia coli* Replicative Origin DNA to Membrane Preparations 79
W. Hendrickson, H. Yamaki, J. Murchie, M. King, D. Boyd, and M. Schaechter

II. PLASMID REPLICATION ORIGINS

8. Dissection of the Replication Region Controlling Incompatibility, Copy Number, and Initiation of DNA Synthesis in the Resistance Plasmids, R100, and R1 91
Thomas Ryder, Jonathan Rosen, Karen Armstrong, Dan Davison, Eiichi Ohtsubo, and Hisako Ohtsubo

9. Direct Repeats of Nucleotide Sequences Are Involved in Plasmid Replication and Incompatibility 113
David M. Stalker, Avigdor Shafferman, Aslihan Tolun, Roberto Kolter, Shengli Yang, and Donald R. Helinski

10. Incompatibility of IncFII R Plasmid NR1 125
Alan M. Easton, Padmini Sampathkumar, and Robert H. Rownd

11. Genetic Analysis of pMB1 Replication 143
Gianni Cesareni, Luisa Castagnoli, and Rosa M. Lacatena

12. Cloning of DNA Synthesis Initiation Determinants of ColE1 into Single-Stranded DNA Phages 157
Nobuo Nomura and Dan S. Ray

III. BACTERIOPHAGE REPLICATION ORIGINS

13. DNA Initiation Determinants of Bacteriophage M13 and of Chimeric Derivatives Carrying Foreign Replication Determinants 169
Dan S. Ray, Joseph M. Cleary, Jane C. Hines, Myoung Hee Kim, Michael Strathearn, Laurie S. Kaguni, and Margaret Roark

14. Essential Features of the Origin of Bacteriophage φX174 RF DNA Replication: A Recognition and a Key Sequence for φX Gene A Protein 195
P. D. Baas, F. Heidekamp, A. D. M. Van Mansfeld, H. S. Jansz, G. A. Van der Marel, G. H. Veeneman, and J. H. Van Boom

15. Viral DNA Sequences and Proteins Important in the φX174
DNA Synthesis 211
*Peter Weisbeek, Arie van der Ende, Harrie van der Avoort,
Renske Teertstra, Fons van Mansfeld, and Simon Langeveld*

16. Conservation and Divergence in Single-Stranded Phage DNA
Secondary Structure: Relations to Origins of DNA Replication 233
Thomas D. Edlind and Garret M. Ihler

17. Primary and Secondary Replication Signals in Bacteriophage λ
and IS5 Insertion Element Initiation Systems 245
Gerd Hobom, Manfred Kröger, Bodo Rak, and Monika Lusky

18. Interaction of Bacteriophage Lambda O Protein with the
Lambda Origin Sequence 263
Toshiki Tsurimoto and Kenichi Matsubara

19. On the Role of Recombination and Topoisomerase in Primary
and Secondary Initiation of T4 DNA Replication 277
*Gisela Mosig, Andreas Luder, Lee Rowen, Paul Macdonald, and
Susan Bock*

IV. INITIATION GENES IN BACTERIA

20. Analysis of Seven *dnaA* Suppressor Loci in *Escherichia coli* 297
Tove Atlung

21. Suppression of Amber Mutations in the *dnaA* Gene of
Escherichia coli K-12 by Secondary Mutations in *rpoB* 315
Nancy A. Schaus, Kathy O'Day, and Andrew Wright

22. Evidence That the *Escherichia coli DNAZ* Product, a
polymerization Protein, Interacts *in Vivo* with the *dnaA* Product, an
Initiation Protein 325
*James R. Walker, William G. Haldenwang, Joyce A. Ramsey,
and Aleksandra Blinkowa*

23. REC-Dependent DNA Replication in *E. coli*: Interaction
between REC-Dependent and Normal DNA Replication Genes 337
Karl G. Lark, Cynthia A. Lark, and Edward A. Meenen

24. An Alternative DNA Initiation Pathway in *E. coli* 361
*Tokio Kogoma, Ted Albert Torrey, Nelda L. Subia, and
Gavin G. Pickett*

25. Genetic Analysis of DNA Replication in *Salmonella
typhimurium* 375
Russell Maurer, Barbara C. Osmond, and David Botstein

V. ENZYMES AND PROTEINS IN INITIATION OF PHAGE AND BACTERIAL CHROMOSOMES

26. Initiation of Bacteriophage T7 DNA Replication 387
Stanley Tabor, Michael J. Engler, Carl W. Fuller, Robert L. Lechner, Steven W. Matson, Louis J. Romano, Haruo Saito, Fuyuhiko Tamanoi, and Charles C. Richardson

27. The Primosome in ϕX174 Replication 409
Peter M. J. Burgers, Robert L. Low, Joan Kobori, Robert Fuller, Jon Kaguni, Mark M. Stayton, Jim Flynn, LeRoy Bertsch, Karol Taylor, and Arthur Kornberg

28. Role of the β Subunit of the *Escherichia coli* DNA Polymerase III Holoenzyme in the Initiation of DNA Elongation 425
Kyung Oh Johanson and Charles S. McHenry

29. Priming of Phage ϕ29 Replication by Protein p3, Covalently Linked to the 5' Ends of the DNA 437
Margarita Salas Falgueras, Miquel A. Peñalva, Juan A. Garcia, José M. Hermoso, and José M. Sogo

VI. EUKARYOTIC REPLICATIONS ORIGINS

30. Functional Components of the *Saccharomyces cerevisiae* Chromosomes—Replication Origins and Centromeric Sequences 455
Clarence S. M. Chan, Gregory Maine, and Bik-Kwoon Tye

31. DNA Sequences That Allow the Replication and Segregation of Yeast Chromosomes 473
Dan T. Stinchcomb, Carl Mann, Eric Selker, and Ronald W. Davis

32. Sequencing and Subcloning Analysis of Autonomously Replicating Sequences from Yeast Chromosomal DNA 489
Gary Tschumper and John Carbon

33. Replication Origins Used *in Vivo* in Yeast 501
Carol S. Newlon, Rodney J. Devenish, Peter A. Suci, and C. J. Roffis

34. Replication Properties of Trp1-Rl-Circle: A High Copy Number Yeast Chromosomal DNA Plasmid 517
John F. Scott and Connie M. Brajkovich

35. *In Vitro* Replication of Yeast 2-μm Plasmid DNA 529
Akio Sugino, Hitoshi Kojo, Barry D. Greenberg, Patrick O. Brown, and Kwang C. Kim

CONTENTS

36. Replication in Monkey Cells of Plasmid DNA Containing the Minimal SV40 Origin 555
    R. Marc Learned, Richard M. Myers, and Robert Tjian

37. Locations of 5'-Ends of Nascent DNA at the Origin of Simian Virus 40 DNA Replication 567
    Ronald T. Hay and Melvin L. DePamphilis

38. Replication Directed by a Cloned Adenovirus Origin 581
    George D. Pearson, Kuan-Chih Chow, Jeffry L. Corden, and Jerry A. Harpst

39. Eukaryotic Origins: Studies of a Cloned Segment from *Xenopus laevis* and Comparisons with Human Blur Clones 597
    J. Herbert Taylor and Shinichi Watanabe

40. Regions of *Tetrahymena* rDNA Allowing Autonomous Replication of Plasmids in Yeast 607
    Gyorgy B. Kiss, Anthony A. Amin, and Ronald E. Pearlman

41. High-Frequency Transformation of Mouse L Cells 615
    Friedrich Grummt, Erika Dinkl, Ursula Wolf, and Werner Goebel

Index 625

CONTRIBUTORS

Numbers in parentheses indicate the chapter numbers.

Anthony A. Amin (40), *Department of Biology, York University, 4700 Keele Street, Downsview, Ontario M3J 1P3, Canada*

Karen Armstrong (8), *Department of Microbiology, Health Sciences Center, State University of New York, Stony Brook, New York 11794*

Kiyozo Asada (1), *Institute for Chemical Research, Kyoto University, Kyoto, Japan 611*

Tove Atlung (5, 20), *University Institute of Microbiology, University of Copenhagen, Oster Farimagsgade 2A, 1353 Copenhagen, Denmark*

P. D. Baas (14), *Institute of Molecular Biology, State University of Utrecht, Transitorium 3, 8 Padualaan, 3508 TB Utrecht, The Netherlands*

LeRoy Bertsch (27), *Department of Biochemistry, Stanford University School of Medicine, Stanford, California 94305*

Aleksandra Blinkowa (22), *Department of Microbiology, University of Texas, Austin, Texas 78712*

David Botstein (25), *Department of Biology, Massachusetts Institute of Technology, Room 56-743, Cambridge, Massachusetts 02139*

D. Boyd (7), *Department of Molecular Biology and Microbiology, Tufts University School of Medicine, 136 Harrison Avenue, Boston, Massachusetts 02111*

Connie M. Brajkovich (34), *Molecular Biology Institute, University of California, Los Angeles, California 90024*

Patrick O. Brown (35), *Department of Biochemistry, University of Chicago, 920 East 58th Street, Chicago, Illinois 60637*

Hans-Jörg Buhk (6), *Max-Planck Institut für Molekulare Genetik, Berlin 33 Dahlem, Ihnestr. 63-73, Federal Republic of Germany*

Peter M. J. Burgers (27), *Department of Biochemistry, Stanford University School of Medicine, Stanford, California 94305*
John Carbon (32), *Department of Biological Sciences, University of California, Santa Barbara, California 93106*
Luisa Castagnoli (11), *European Molecular Biology Laboratory, Postfach 10.2209, 6900 Heidelberg, Federal Republic of Germany*
Gianni Cesareni (11), *European Molecular Biology Laboratory, Postfach 10.2209, 6900 Heidelberg, West Germany*
Trinad Chakraborty (6), *Max-Planck-Institut für Molekulare Genetik, Berlin 33 Dahlem, Ihnestr. 63-73, Federal Republic of Germany*
Clarence S. M. Chan (30), *Section of Biochemistry, Molecular and Cell Biology, Wing Hall, Cornell University, Ithaca, New York 14853*
Kuan-Chih Chow (38), *Department of Biochemistry and Biophysics, Oregon State University, Corvallis, Oregon 97331*
Joseph M. Cleary (2, 13), *Department of Biology C-016, University of California at San Diego, La Jolla, California 92093*
Jeffry L. Corden (38), *Institut de Chimie Biologique, Faculté de Medecine, 11 rue Humann, 67085 Strasbourg Cedex, France*
Ronald W. Davis (31), *Department of Biochemistry, Stanford University School of Medicine, Stanford, California 94305*
Dan Davison (8), *Department of Microbiology, Health Sciences Center, State University of New York, Stony Brook, New York 11794*
Melvin L. DePamphilis (37), *Department of Biological Chemistry, Harvard Medical School, Boston, Massachusetts 02115*
Rodney J. Devenish (33), *Department of Zoology, University of Iowa, Iowa City, Iowa 52244*
Erika Dinkl (41), *Institut für Biochemie der Universität Würzburg, Röntenring 11, D-87 Würzburg, Federal Republic of Germany*
Alan M. Easton (10), *Laboratory of Molecular Biology, 1525 Linden Drive, University of Wisconsin, Madison, Wisconsin 53706*
Thomas D. Edlind (16), *Department of Medical Biochemistry, Texas A&M University, College of Medicine, College Station, Texas 77843*
Michael J. Engler (26), *Department of Biological Chemistry, Harvard Medical School, 25 Shattuck Street, Boston, Massachusetts 02115*
Margarita Salas Falgueras (29), *Centro de Biologia Molecular, Universidad Autónoma, Facultad de Ciencias, Canto Blanco, Madrid-34, Spain*
Jim Flynn (27), *Department of Biochemistry, Stanford University School of Medicine, Stanford, California 94305*
Carl W. Fuller (26), *Department of Biological Chemistry, Harvard Medical School, 25 Shattuck Street, Boston, Massachusetts 02115*
Robert Fuller (27), *Department of Biological Chemistry, Stanford University School of Medicine, Stanford, California 94305*
Juan A. Garcia (29), *Centro de Biologia Molecular, Universidad Autónoma, Facultad de Ciencias, Canto Blanco, Madrid-34, Spain*

Werner Goebel (41), *Institut für Mikrobiologie und Genetik der Universität Würzburg, Röntgenring 11, D-8700 Würzburg, Federal Republic of Germany*

Barry D. Greenberg (35), *Laboratory of Molecular Genetics, NIEHS, P. O. Box 12233, Research Triangle Park, North Carolina 27709*

Friedrich Grummt (41), *Institut für Biochemie der Universität Würzburg, Röntgenring 11, D-87 Würzburg, Federal Republic of Germany*

Willaim G. Haldenwang (22), *Department of Microbiology, University of Kansas, Lawrence, Kansas*

Flemming G. Hansen (5), *Department of Microbiology, Building 221, The Technical University of Denmark, 2800 Lyngby-Copenhagen, Denmark*

Nancy E. Harding (2), *Department of Biology C-016, University of California San Diego, La Jolla, California 92093*

Jerry A. Harpst (38), *Department of Biochemistry and Biophysics, Oregon State University, Corvallis, Oregon 97331*

Ronald T. Hay (37), *Department of Biological Chemistry, Harvard Medical School, Boston, Massachusetts 02115*

F. Heidekamp (14), *Institute of Molecular Biology, State University of Utrecht, Transitorium 3, 8 Padualaan, 3508 TB Utrecht, The Netherlands*

Barbara Heimann (6), *Max-Planck-Institut für Molekulare Genetik, Berlin 33 Dahlem, Ihnestr. 63-73, Federal Republic of Germany*

Donald R. Helinski (9), *Biology Department B-022, University of California San Diego, La Jolla, California 92093*

W. Hendrickson (7), *Department of Molecular Biology and Microbiology, Tufts University School of Medicine, 136 Harrison Avenue, Boston, Massachusetts 02111*

José M. Hermoso (29), *Centro de Biologia Molecular, Universidad Autónoma, Facultad de Ciencias, Canto Blanco, Madrid-34, Spain*

Jane C. Hines (13), *Molecular Biology Institute, University of California, Los Angeles, California 90024*

Yukinori Hirota (1, 3, 4), *National Institute of Genetics, Mishima, Shizuoka 411 Japan*

Gerd Hobom (17), *Universität Freiburg, Institut für Biologie III, Schanzlestrasse 1, 7800 Freiburg, Federal Republic of Germany*

Garret M. Ihler (16), *Department of Medical Biochemistry, College of Medicine, Texas A&M University, College Station, Texas 77834*

H. S. Jansz (14), *Laboratory for Physiological Chemistry, State University of Utrecht, 24a Vondellaan, 3521 GG Utrecht, The Netherlands*

Kyung Oh Johanson (28), *Department of Biochemistry and Molecular Biology, University of Texas Medical School, P.O. Box 20708, Houston, Texas 77025*

Laurie S. Kaguni (13), *Department of Pathology, Stanford University Medical Center, Stanford, California 94305*

Jon Kaguni (27), *Department of Biochemistry, Stanford University School of Medicine, Stanford, California 94305*

Kwang C. Kim (35), *Laboratory of Molecular Genetics, NIEHS, P.O. Box 12233, Research Triangle Park, North Carolina 27709*

Myoung Hee Kim (13), *Molecular Biology Institute, University of California, Los Angeles, California, 90024*

M. King (7), *Department of Molecular Biology and Microbiology, Tufts University School of Medicine, 136 Harrison Avenue, Boston, Massachusetts 02111*

Gyorgy B. Kiss (40), *Institute of Genetics, Biological Research Center, Szeged, Hungary*

Joan Kobori (27), *Department of Biochemistry, Stanford University School of Medicine, Stanford, California 94305*

Susanne Koefoed (5), *Department of Microbiology, Building 211, The Technical University of Denmark, 2800 Lyngby-Copenhagen, Denmark*

Tokio Kogoma (24), *Department of Biology, University of New Mexico, Albuquerque, New Mexico 87131*

Hitoshi Kojo (35), *Research Laboratories, Fujisawa Pharmaceutical Company Ltd., Yodogawa-Ku, Osaka 532, Japan*

Roberto Kolter (9), *Department of Biological Sciences, Stanford University, Stanford, California 94305*

Arthur Kornberg (27), *Department of Biochemistry, Stanford University School of Medicine, Stanford, California 94305*

Manfred Kröger (17), *Universität Freiburg, Institut für Biologie III, Schanzlestrasse 1, 7800 Freiburg, Federal Republic of Germany*

Rosa M. Lacatena (11), *European Molecular Biology Laboratory, Postfach 10.2209, 6900 Heidelberg, Federal Republic of Germany*

Simon Langeveld (15), *Institute of Molecular Biology, State University of Utrecht, 8 Padualaan, 3584 CH Utrecht, The Netherlands*

Cynthia A. Lark (23), *Department of Biology, University of Utah, 401 Life Science Building, Salt Lake City, Utah 84112*

Karl G. Lark (23), *Department of Biology, University of Utah, 401 Life Science Building, Salt Lake City, Utah 84112*

R. Marc Learned (36), *Department of Biochemistry, University of California, Berkeley, California 94720*

Robert L. Lechner (26), *Department of Biological Chemistry, Harvard Medical School, 25 Shattuck Street, Boston, Massachusetts 02115*

Heinz Lother (6), *Max-Planck-Institut für Molekulare Genetik, Berlin 33 Dahlem, Ihnestr. 63-73, Federal Republic of Germany*

Robert L. Low (27), *Department of Biochemistry, Stanford University School of Medicine, Stanford, California 94305*

Andreas Luder (19), *Department of Molecular Biology, Vanderbilt University, Nashville, Tennessee 37235*

Monika Lusky (17), *Virus Laboratory, Wendell M. Stanley Hall, University of California, Berkeley, California 94720*

Paul Macdonald (19), *Department of Molecular Biology, Vanderbilt University, Nashville, Tennessee 37235*

Gregory Maine (30), *Section of Biochemistry, Molecular and Cell Biology, Wing Hall, Cornell University, Ithaca, New York 14853*

Carl Mann (31), *Department of Biochemistry, Stanford University School of Medicine, Stanford, California 94305*

Steven W. Matson (26), *Department of Biological Chemistry, Harvard Medical School, 25 Shattuck Street, Boston, Massachusetts 02115*

Kenichi Matsubara (18), *Laboratory of Molecular Genetics, University of Osaka Medical School, Kita-Ku, Osaka, 530 Japan*

Russell Maurer (25), *Department of Biology, Massachusetts Institute of Technology, Room 56-743, Cambridge, Massachusetts 02139*

Charles S. McHenry (28), *Department of Biochemistry and Molecular Biology, University of Texas Medical School, P.O. Box 10708, Houston, Texas 77025*

Edward A. Meenen (23), *Department of Biology, University of Utah, 401 Life Science Building, Salt Lake City, Utah 84112*

Walter Messer (6), *Max-Planck-Institut für Molekulare Genetik, Berlin 33 Dahlem, Ihnestr. 63-73, Federal Republic of Germany*

Giovanna Morelli (6), *Max-Planck-Institut für Molekulare Genetik, Berlin 33 Dahlem, Ihnestr. 63-73, Federal Republic of Germany*

Masayuki Morita (4), *Institute for Chemical Research, Kyoto University, Kyoto, Japan 611*

Gisela Mosig (19), *Department of Molecular Biology, Vanderbilt University, Nashville, Tennessee 37235*

J. Murchie (7), *Department of Molecular Biology and Microbiology, Tufts University School of Medicine, 136 Harrison Avenue, Boston, Massachusetts 02111*

Richard M. Myers (36), *Department of Biochemistry, University of California, Berkeley, California 94720*

Carol S. Newlon (33), *Department of Zoology, University of Iowa, Iowa City, Iowa 52244*

Nobuo Nomura (12), *Molecular Biology Institute, University of California, Los Angeles, California 90024*

Kathy O'Day (21), *Department of Molecular Biology and Microbiology, Tufts University School of Medicine, 136 Harrison Avenue, Boston, Massachusetts 02111*

Eiichi Ohtsubo (8), *Department of Microbiology, Health Sciences Center, State University of New York, Stony Brook, New York 11794*

Hisako Ohtsubo (8), *Department of Microbiology, Health Sciences Center, State University of New York, Stony Brook, New York 11794*

Atsuhiro Oka (1, 4), *Institute for Chemical Research, Kyoto University, Kyoto, Japan 611*

Barbara C. Osmond (25), *Department of Biology, Massachusetts Institute of Technology, Room 56-743, Cambridge, Massachusetts 02139*

Ronald E. Pearlman (40), *Department of Biology, York University, 4700 Keele Street, Downsview, Ontario M3J 1P3, Canada*

George D. Pearson (38), *Department of Biochemistry and Biophysics, Oregon State University, Corvallis, Oregon 97331*

Miguel A. Peñalva (29), *Centro de Biologia Molecular, Universidad Autonoma, Facultad de Ciencias, Canto Blanco, Madrid-34, Spain*

Gavin G. Pickett (24), *Biology Department, University of New Mexico, Albuquerque, New Mexico 87131*

Bodo Rak (17), *Universität Freiburg, Institut für Biologie III, Schanzlestrasse 1, 7800 Freiburg, Federal Republic of Germany*

Joyce A. Ramsey (22), *Department of Microbiology, University of Texas, Austin, Texas 78712*

Dan S. Ray (12, 13), *Molecular Biology Institute, University of California, Los Angeles, California 90024*

Charles C. Richardson (26), *Department of Biological Chemistry, Harvard Medical School, 25 Shattuck Street, Boston, Massachusetts 02115*

Margaret Roark (13), *Molecular Biology Institute, University of California, Los Angeles, California 90024*

C. J. Roffis (33), *Grinnell College, Grinnell, Iowa 50112*

Louis J. Romano (26), *Department of Biological Chemistry, Harvard Medical School, 25 Shattuck Street, Boston, Massachusetts 02115*

Jonathan Rosen (8), *Department of Pathology, Stanford University School of Medicine, Stanford, California 94305*

Lee Rowen (19), *Department of Molecular Biology, Vanderbilt University, Nashville, Tennessee 37235*

Robert H. Rownd (10), *Laboratory of Molecular Biology, University of Wisconsin, 1525 Linden Drive, Madison, Wisconsin 53706*

Thomas Ryder (8), *Department of Microbiology, Health Sciences Center, State University of New York, Stony Brook, New York 11794*

Haruo Saito (26), *Department of Biological Chemistry, Harvard Medical School, 25 Shattuck Street, Boston, Massachusetts 02115*

Padmini Sampathkumar (10), *Laboratory of Molecular Biology, University of Wisconsin, 1525 Linden Drive, Madison, Wisconsin 53706*

Hitoshi Sasaki (1), *Institute for Chemical Research, Kyoto University, Kyoto Japan 611*

M. Schaechter (7), *Department of Molecular Biology and Microbiology, Tufts University School of Medicine, 136 Harrison Avenue, Boston, Massachusetts 02111*

Nancy A. Schaus (21), *Department of Molecular Biophysics and Biochemistry, Yale University, 333 Cedar Street, New Haven, Connecticut 06510*

John F. Scott (34), *Molecular Biology Institute, University of California, Los Angeles, California 90024*

Eric Selker (31), *Goethe Institute, Wasserburgerstrasse 54, 8018 Grafing, Federal Republic of Germany*

Avigdor Shafferman (9), *Biology Department B-022, University of California at San Diego, La Jolla, California 92093*

Douglas W. Smith (2, 3), *Department of Biology C-016, University of California at San Diego, La Jolla, California 92093*

José M. Sogo (29), *Centro de Biologia Molecular, Universidad Autónoma, Facultad de Ciencias, Canto Blanco, Madrid-34, Spain*

David M. Stalker (9), *Biology Department B-022, University of California at San Diego, La Jolla, California 92093*

Mark M. Stayton (27), *Department of Biochemistry, Stanford University School of Medicine, Stanford, California 94305*

Dan T. Stinchcomb (31), *Department of Biochemistry, Stanford University School of Medicine, Stanford, California 94305*

Michael Strathearn (13), *Molecular Biology Institute, University of California, Los Angeles, California 90024*

Nelda L. Subia (24), *Biology Department, University of New Mexico, Albuquerque, New Mexico 87131*

Peter A. Suci (33), *Department of Zoology, University of Iowa, Iowa City, Iowa 52244*

Kazunori Sugimoto (1, 4), *Institute for Chemical Research, Kyoto University, Kyoto, Japan 611*

Akio Sugino (35), *Laboratory of Molecular Genetics, NIEHS, P.O. Box 12233, Research Triangle Park, North Carolina 27709*

Stanley Tabor (26), *Department of Biological Chemistry, Harvard Medical School, 25 Shattuck Street, Boston, Massachusetts 02115*

Mituru Takanami (1, 3, 4), *Institute for Chemical Research, Kyoto University, Kyoto, Japan, 611*

Yutaka Takeda (2), *Department of Biology C-016, University of California at San Diego, La Jolla, California 92093,*

Fuyuhiko Tamanoi (26), *Department of Biological Chemistry, Harvard Medical School, 25 Shattuck Street, Boston, Massachusetts 02115*

J. Herbert Taylor (39), *Institute of Molecular Biophysics, Florida State University, Tallahassee, Florida 32306*

Karol Taylor (27), *Department of Biochemistry, University of Gdansk, 24 Kladki, 80-822 Gdansk, Poland*

Renske Teertstra (15), *Department of Molecular Cell Biology, State University of Utrecht, 8 Padualaan, 3584 CH Utrecht, The Netherlands*

Robert Tjian (36), *Department of Biochemistry, University of California, Berkeley, California 94720*

Aslihan Tolun (9), *Biology Department B-022, University of California at San Diego, La Jolla, California 92093*

Ted Albert Torrey (24), *Biology Department, University of New Mexico, Albuquerque, New Mexico 87131*

Gary Tschumper (32), *Department of Biological Sciences, University of California, Santa Barbara, California 93106*

Toshiki Tsurimoto (18), *Laboratory of Molecular Genetics, University of Osaka Medical School, Kita-Ku, Osaka, 530 Japan*

Bik-Kwoon Tye (30), *Section of Biochemistry, Molecular and Cell Biology, Wing Hall, Cornell University, Ithaca, New York 14853*

J. H. Van Boom (14), *Department of Organic Chemistry, State University of Leiden, Wassenaarseweg 76, Leiden, The Netherlands*

Harrie Van der Avoort (15), *Department of Molecular Cell Biology, State University of Utrecht, 8 Padualaan, 3584 CH Utrecht, The Netherlands*

Arie Van der Ende (15), *Institute of Molecular Cell Biology, State University of Utrecht, 8 Padualaan, 3584 Ch Utrecht, The Netherlands*

G. A. Van der Marel (14), *Department of Organic Chemistry, State University of Leiden, Wassenaarseweg 76, Leiden, The Netherlands*

A. D. M. Van Mansfeld (14), *Institute of Molecular Biology, State University of Utrecht, Transitorium 3, 8 Padualaan, 3508 TB Utrecht, The Netherlands*

Fons Van Mansfeld (15), *Institute of Molecular Cell Biology, State University of Utrecht, 8 Padualaan, 3584 CH Utrecht, The Netherlands*

G. J. Veeneman (14), *Department of Organic Chemistry, State University of Leiden, Wassenaarseweg 76, Leiden, The Netherlands*

Kaspar von Meyenburg (5), *Department of Microbiology, Building 211, The Technical University of Denmark, 2800 Lyngby-Copenhagen, Denmark*

James R. Walker (22), *Department of Microbiology, University of Texas, Austin, Texas 78712*

Shinichi Watanabe (39), *McArdle Institute, University of Wisconsin, Madison, Wisconsin 53706*

Peter Weisbeek (15), *Department of Molecular Cell Biology, State University of Utrecht, 8 Padualaan, 3584 CH Utrecht, The Netherlands*

Ursula Wolf (41), *Institut für Biochemie der Universität Würzburg, Röntegenring 11, D-87 Würzburg, Federal Republic of Germany*

Andrew Wright (21), *Department of Molecular Biology and Microbiology, Tufts University School of Medicine, 136 Harrison Avenue, Boston, Massachusetts 02111*

H. Yamaki (7), *Department of Molecular Biology and Microbiology, Tufts University School of Medicine, 136 Harrison Avenue, Boston, Massachusetts 02111*

Shengli Yang (9), *Biology Department B-022, University of California at San Diego, La Jolla, California 92093*

Judith W. Zyskind (2, 3), *Department of Biology C-016, University of California, at San Diego, La Jolla, California 92093*

PREFACE

One of the most precisely regulated events in living cells is the initiation of DNA replication. Initiation occurs at one or more sites called replication origins. In general, chromosomes of prokaryotic cells and their phages and plasmids have a single unique origin, while eukaryotic cells have multiple origins per chromosome. Recognition of the nucleotide sequence of an origin by specific proteins is likely to be crucial in regulating initiation.

Through the use of recombinant DNA techniques, various replication origins and replication determinants have been cloned and are now available in large amounts for both genetic and biochemical analysis. Tremendous progress has also been made in the purification of replication proteins. In the case of *E. coli*, the availability of large quantities of small phage chromosomes has facilitated the purification of the components of three distinct initiation systems. The rapid progress in these areas has now set the stage for detailed genetic and biochemical analysis of these critically important regulatory mechanisms.

Because considerable progress has been made with a variety of systems ranging from phage and plasmids to eukaryotic viruses and chromosomal segments, it seemed appropriate to bring together workers from diverse areas concerned with regulation of initiation. It was our hope that we could begin to address some of the unanswered questions in this area: Are there specific protein recognition sites within an origin? How many proteins interact at an origin and do they interact in a specific temporal sequence? Can origins be subdivided into distinct functional domains? Are there similar mechanisms of initiation in diverse biological systems? What are the specific biochemical steps in DNA chain initiation and how are they catalyzed?

This volume includes articles from plenary session speakers and poster presentations. The papers are arranged in groups according to their organization in the meeting. The meeting was held in March 1981 at Salt Lake City, Utah, and was one of the 1981 ICN–UCLA Symposia on Molecular and Cellular Biology. The series, sponsored by ICN Pharmaceuticals, Inc., is organized through the Molecular Biology Institute of the University of California, Los Angeles.

I greatly appreciate the assistance of various colleagues in chairing sessions and contributing to the intellectual vigor and excitement of the meeting. I am indebted to the ICN-UCLA Symposia staff for their handling of the logistics and organizational aspects of this meeting and the production of this volume.

Financial support for this meeting was provided in part by Monsanto and NIH contract 263-MD-035893, jointly sponsored by the National Cancer Institute, National Institute of General Medical Sciences, National Institute of Allergy and Infectious Diseases, and Fogarty International Center.

Dan S. Ray

ESCHERICHIA COLI ORIGIN OF REPLICATION: Structural organization of the region essential for autonomous replication and the recognition frame model

Yukinori Hirota

National Institute of Genetics,
Mishima, JAPAN 411

Atsuhiro Oka
Kazunori Sugimoto
Kiyozo Asada
Hitoshi Sasaki
Mituru Takanami

Institute for Chemical Research, Kyoto University, Kyoto, JAPAN 611

ABSTRACT The replication origin region of the *E. coli* chromosome has been cloned(1-6), and the precise location of the region carrying autonomous replicating function(defined *ori*) has been determined(7). The left boundary of *ori* is between positions 23(A) and 35(T) and the right boundary is between positions 267(C) and 268(A). The maximum size of the *ori* segment is therefore 245 base pairs(bp) long.

A series of mutations having short sequences inserted or deleted were introduced in the vicinity of *Bgl*II, *Bam*HI, *Ava*II, and *Hind*III sites within the 245 bp sequence. Base substitution mutations C to T(or G to A) were also introduced by *in vitro* mutagenesis which used bisulfite, and the correlation between the phenotypes and nucleotide sequences of mutant were analyzed. We could identify two categories of regions: one is the region in which even a single base substitution destroys Ori function and the other is the region in which insertion or deletion of short sequences, but not base-substitutions, destroys Ori function.

On the basis of these observations, we propose the following model for the structural organization of *ori*(recognition frame model): the recognition sequences that determine the sites where the components involved in the replication initiation recognize and the spacer sequences(or distance sequences) that maintain the recognition sequences at precise distances are connected with each other in a chimeric fashion(8).

According to this model, the striking effect caused by the shift of recognition sequences within *ori* can be explained. The *ori* stretch should provide precise topological information for the formation of an initiation complex by multiple components which are involved in the initiation of replication.

MATERIALS AND METHODS

E. coli K12 strains and plasmids: *E. coli* strains used were C600(F⁻ *thr thi lacY tonA supE*), GM31(F⁻ *dcm thr leu his thi ara lacY galK galT xyl mtl strA tonA tsx supE*)(9) and W3110*polA* (F⁻ *thyA polA*)(10). The *ori*-pBR322 recombinants used for deletion and insertion experiments at restriction sites are pTSO182(7), pTSO196(7), pTSO236(7) and pKA22(constructed by K. Asada). These plasmids, respectively, contain a single susceptible site for *Hind*III, *Bam*HI, *Bgl*II and *Ava*II only in the *ori* region. The *ori*-pBR322 recombinants used for base substitution experiments were pTSO293 and pKA22. pTSO293 contains a single susceptible site for each of *Bam*HI, *Hind*III, *Acc*I, *Eco*RI and *Pst*I. The former two sites are located within *ori* and the latter three sites at the outside of *ori*. pKA22 contains a single *Xho*I site at the outside of *ori* in addition to the *Ava*II site within *ori*.

In vitro mutagenesis: The procedures for introducing deletions and insertions at restriction sites have been described(7).

Base substitution-type mutations were obtained by applying the principle of the bisulfite method described by Sortle and Nathans(11). Bisulfite-catalyzed deamination of C to U specifically occurs in single-stranded polynucleotides(12, 13), so that circular heteroduplex molecules containing single-stranded gaps in appropriate sites were constructed and mutagenized. For example, mutagenesis in the region between *Bam*HI and *Hind*III sites within *ori* was performed by the following procedure. pTSO293 was digested with *Bam*HI and *Hind*III, and the longer fragment was isolated by gel electrophoresis. The fragment was mixed with pTSO293 linearized by digestion with *Eco*RI and after heat-denaturation, the solution was annealed to form circular heteroduplexes. The heteroduplexes are expected to contain a gap between the *Bam*HI and *Hind*III sites on one strand and nick at the *Eco*RI site on the other strand. For mutagenesis in the vicinity of restriction sites, pTSO293 or pKA22 was linearized by digestion with an enzyme(*Ava*II for pKA22), and several nucleotides were removed from the ends by exonucleolysis with *Micrococcal luteus* DNA polymerase I (Miles Lab, Inc.). The digested molecules were mixed with the plasmids linearized by other enzymes(e.g. *Xho*I for pKA22), and circular heteroduplex molecules containing various sizes of single-stranded gaps in the vicinity of restriction sites were obtained by annealing. The heteroduplexes thus prepared were treated at 37°C for 0.5 to 4 hrs in 3 M sodium sulfite-sodium bisulfite (pH 6), and then isolated as described by Sortle and Nathans(11). The products were repaired by T4 DNA polymerase and transformation was carried out. Ap^r transformants from pTSO293 derivatives and Km^r transformants from pKA22 derivatives were subjected to clone analysis, and Ori function of each clone was assayed as described(7). By sequence analysis of the mutagenized plasmids, base substitutions (C to T or G to A) were only detected in the expected regions.

Other procedures: Methods for bacterial transformation, preparation of plasmid DNA, clone analysis, restriction analysis and sequence determination have previously been described (7, 14).

INTRODUCTION

In bacteria, the ability of the cell to maintain its genetic content through successive generations is regulated through initiation of DNA replication. To initiate DNA replication, the initiation proteins recognize, bind, and act on specific sites at the DNA replication origin. What characterizes these sites? What aspects of the origin structure do the initiator proteins see? Recent studies on cloning and sequencing of the DNA replication origins (*ori*) in enteric bacteira(1-7) brought a major breakthough to approach these problems. In the cloned *ori* segment of *E. coli*, the precise location and size of the *ori* segment has been determined: the minimal size of *ori*-DNA(7), the 245 bp is required for the determination of the initiation of DNA replication. All the defective *ori* mutations were found to lie within the *ori* regions(8). Based on these results, we proposed "recognition frame model" of DNA replication origin(8). In this report, we shall present further evidence in support of the model.

Definition of *E. coli* origin of DNA replication (*ori*): *E. coli* origin of DNA replication(*ori*) is defined as a DNA fragment derived from the *E. coli* chromosome which is able to replicate autonomously in the *E. coli* cell. When non-replicating DNA was joined with *ori*, the non-replicating DNA replicates under the control of the *ori*. The abbreviation for the phenotype of *ori* which replicates autonomously in an *E. coli* cell is Ori$^+$, and that of the defective phenotype is Ori$^-$. A chimeric plasmid, pBR322(15) carrying *E. coli ori* autonomously replicates on *E. coli* F$^-$ *polA*$^-$ cell, but pBR322 is not(5,7,15). Thus, the *ori* of this chimeric plasmid which is able to replicate autonomously on F$^-$ *polA*$^-$ cell is defined as *ori*$^+$ and its phenotype as Ori$^+$, and the defective replication of the mutant is defined as *ori*$^-$ and its phenotype as Ori$^-$.

RESULTS AND DISCUSSION

We have approached this problem by extensive isolation of mutants of the *E. coli* DNA replication origin (*ori*) using *in vitro* gene manipulation technology, followed by determination of the nucleotide sequences of each mutant having either active (Ori$^+$) or defective (Ori$^-$) function. Thus, we detected the sequence alteration(s) which results either the loss or the maintainance of the Ori-function.

<u>Effects of insertion and deletion of short sequences within *ori*</u>. We constructed 3 derivatives from the chimeric plasmids carrying wild type *ori*-segments, but the *Bam*HI, *Hind*III or *Bgl*II sites, were eliminated from the vector DNA: pTSO196 which carried a single *Bam*HI site corresponding to *Bam*HI-3 of the *ori*-segment, pTSO182 which carried a single *Hind*III site corresponding to *Hind*III-1 of the *ori* segment, and pTSO236 which carried a single *Bgl*II site corresponding to *Bgl*II-2 of the *ori* segment(7)(Fig.1). Using these plasmids a series of mutations having short sequences inserted (Ω) or deleted (Δ) in the vicinity of restriction sites of the *ori*-segment, either *Bgl*II-2, *Bam*HI-3, *Ava*II-1 or *Hind*III-1 within the 245 base pairs were constructed. Linear DNA molecules were generated by *Bam*HI digestion of pTSO196, *Hind*III digestion of pTSO182 and *Bgl*II digestion of pTSO236, respectively. The linear molecules thus generated were subjected to either repair synthesis or nuclease S1 digestion. Then, the products were joined by T4 DNA ligase and transformed into C600 cells. By these procedures, it is easy to identify the mutant plasmids. Nucleotide sequences of the mutants thus isolated were examined. The function of *ori* of the mutants was determined by the capacity to replicate autonomously on the F$^-$ *polA1*$^-$ mutant (Ori$^+$) or not (Ori$^-$) (5, 7).

1 *ESCHERICHIA COLI* ORIGIN OF REPLICATION

FIG.1 Base substitution type mutations within the ori region. Capital and case letters represent conservative(consensus) and non-conservative residues deduced by comparison of the sequences of five bacterial origins(Zyskind et al., in this volume). (+) and (*) indicate mutations which show no significant effect and a slight effect on Ori function, respectively. (-) indicates mutations that result in defective Ori. pMU254, pMU315 and pMU1105 carry a single base change, respectively. pMU243 contains an additional two substitutions of C at positions 158 and 196, but the former overlaps with an Ori^+ substitution and the latter falls in the non-conservative region of Zyskind et al. Various Ori^- mutants carrying multi-base changes have been isolated, but these are not indicated.

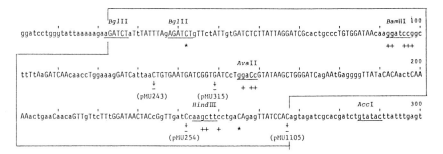

TABLE 1 Mutations induced in the vicinity of the four restriction sites within the ori region. Identical residues are indicated by dots and deleted residues by triangles, respectively. The position of insertion and inserted sequences are above the wild type sequences.

BglII Site (Positions 38-43)

Insertion	(Ori^+)	GATC
Wild	(Ori^+)	--AAGATCTATTTATTTAG<u>AGATC</u>TGTT--
Substitution	(Ori^*)	....................T....
Deletion	(Ori^-)	...............ΔΔΔΔΔΔΔ...
Deletion	(Ori^-)	..ΔΔΔΔΔΔΔΔΔΔΔΔΔΔΔ........

BamHI Site (Positions 92-97)

Insertion	(Ori^-)	GATC
Wild	(Ori^+)	--AACAA<u>GGATC</u>CGGCTTTTAAGATC--
Substitution	(Ori^+_+)	......A...................
Substitution	(Ori^+_+)	......A...................
Substitution	(Ori^+_+)	.........T................
Substitution	(Ori^+_+)	...........T..............
Substitution	(Ori^+)	...............A..........
Deletion	(Ori^-)	......ΔΔΔΔ................
Deletion	(Ori^-)	.....ΔΔΔΔΔΔΔΔΔΔΔΔΔΔΔΔ.....

AvaII Site (Positions 155-159)

Insertion	(Ori^-)	AC
Insertion	(Ori^-)	GAC
Wild	(Ori^+)	--GGTGATCCT<u>GGACC</u>GTATAAGCT--
Substitution	(Ori^+_+)	............A.............
Substitution	(Ori^+_+)	............T.............
Substitution	(Ori^-)	............TT............
Deletion	(Ori^-)	..........ΔΔΔ.............
Deletion	(Ori^-)	..........ΔΔΔΔ............
Deletion	(Ori^-)	..........ΔΔΔΔ............
Deletion	(Ori^-)	..........ΔΔΔΔΔ...........
Deletion	(Ori^-)	..........ΔΔΔΔΔΔ..........
Deletion	(Ori^-)	........ΔΔΔΔΔΔΔΔ..........
Deletion	(Ori^-)	........ΔΔΔΔΔΔΔΔ..........

HindIII Site (Positions 244-249)

Insertion	(Ori^-)	AGCT
Wild	(Ori^+)	--GATCC<u>AAGCTT</u>CCTGACAGAGTTA--
Substitution	(Ori^+_+)	........A.................
Substitution	(Ori^+_+)	...........T..............
Substitution	(Ori^-)	........T..T..............
Substitution	(Ori^+_+)	...T......................
Substitution	(Ori^-)	...T....T.................
Deletion	(Ori^-)	......ΔΔΔΔ................
Deletion	(Ori^-)	.......ΔΔΔΔΔΔΔ............
Deletion	(Ori^+_+)	.....ΔΔΔΔΔΔΔΔ.............
Deletion	(Ori^+)	.....ΔΔΔΔΔΔΔΔΔΔΔΔ.........

The correlation between the phenotypes of the mutants and the nucleotide sequences are summarized in Table 1. The following mutants carrying an insertion (Ω) or deletion (Δ) mutation exhibited the Ori⁻ phenotype: Δ16 base pairs (bp) at the BglII-2 site; Ω4bp, Δ4bp or Δ15bp at the BamHI-3 site; Ω2bp, Ω3bp, Δ3bp, Δ4bp, Δ5bp, Δ6bp, Δ7bp or Δ8bp at the AvaII-1 site; and Ω4bp, Δ5bp and Δ8bp at the HindIII-1 site.

All plasmids prepared from repaired molecules were found to have a 4bp insertion at the respective sites and, in the case of nuclease S1 digestion, mutants carrying deletions larger then 4bp at the ligated sites were found. Presumably, this due to excess activity of nuclease S1.

However, two exceptional mutants which remained Ori⁺ were found: an Ori⁺ mutant with an inserted GATC sequence at the BglII-2 site between positions 42(C) and 43(T), and the other Ori⁺ mutant with 12bp deleted at the HindIII-1 site between 244(A) and 257(G). A simple interpretation of these results is that the nucleotide sequences essential for the Ori⁺ phenotype were conserved in these Ori⁺ mutants at the fixed positions of the ori-segment.

As shown in Fig. 2, an Ori⁺ derivative containing a four base pair insertion mutation at the BglII-2 site created the same sequence at positions 24(A) to 32(A) as the wild type sequence AT·T·T·TA. The other Ori⁺ derivative of pTSO193 containing the 12 base pair deletion mutation created the same sequence, CA········CAC, at positions 255(C) to 267(C). No such homology is recognized in the other Ori⁻ derivatives.

Furthermore, when the nucleotide sequences of the mutants with insertions or deletions at BglII-2, BamHI-3, AvaII-2 or HindIII-1 sites are aligned for comparison with the wild type ori sequence, striking differences among the bases at the positions, as the result of insertions or the deletions, are demonstrated (7, 8)(Fig.2). It is likely that the presence of primary nucleotide sequences such as, AT·T·T·TA and CA········CAC at fixed positions is indispensable for the ori segment to function. These observations led us to hypothesize the presence of two different classes of nucleotide segments having separate functions, as follows. Class A, *recognition sequences*: sequences that determine the recognition site for the initiation proteins. The nucleotide sequence has to be specific for normal functioning. For example, the presence of the primary sequences AT·T·T·TA and CA········CAC at the fixed positions in the ori-segment is indispensable for the Ori⁺ function. We assign the role of these sequences as "recognition" sequences, and a base substitution mutation in the sequences causes defective ori-function. Class B, *distance or spacer sequences*: sequences that determine the precise distance between the recognition sequences. The region around

FIG.2 The differences among the bases (·) at the positions as the result of insertions or deletions shown in Table 1. The sequences of derivations with insertions or deletions at HindIII site 1 (a), the sequence of a derivative with an insertion at BglII site 2 (b), the sequence of a derivative with an insertion or deletion at the left end (C) or the right end (d) of AvaII, are aligned for comparison with the original ori sequence. A dot at a residue indicates the altered base from E. coli ori⁺, and a letter at a residue indicates the base conserved original ori residue at the position. HindIII site 1 (a) and BglII site 2 (b) have a line above. The two arrows indicate the left (c) and right (d) ends of the maximum size of ori determined. The sequence of Salmonella typhimurium ori (12) is also presented for comparison.

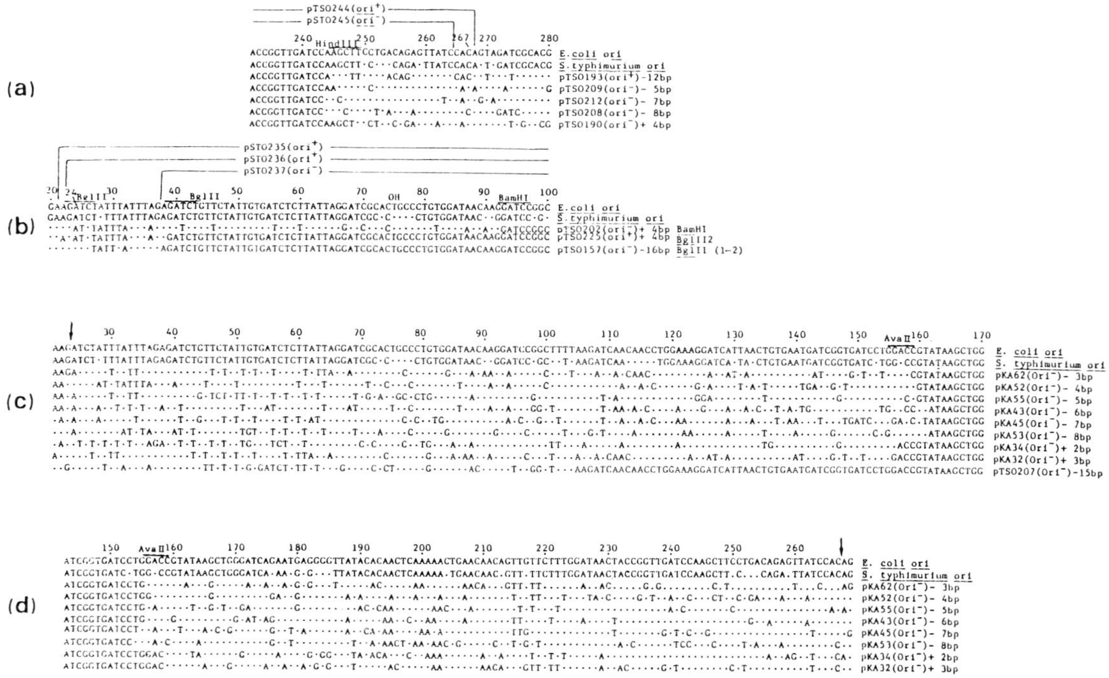

the HindIII-1 site plays a role in the maintenance of the appropriate distance beteen two sequences at both side of HindIII-1 site. Same argument could be applied to the BglII-2 AvaII-1 or BamHI-3 site. Thus, we assign the role of these sequences as a part of "distance" or "spacer" sequences, and a base substitution mutation of the sequences does not affect ori-function. However, deletion or insertion of nucleotides at the sequences may cause the alteration of an appropriate distance and destroy ori function.

Effect of base substitution at the restriction sites of BglII-2, BamHI-3, AvaII-1 or HindIII-1: Another explanation of the Ori-defective phenotype of the mutants having insertion or deletion mutations at the AvaII-1 or the BamHI-3 site is as follows. These sites are not spacer sequences but recognition sequences which are indispensable for the Ori$^+$ function. A substitution mutation of the site does change the sequence but the distance or number of nucleotides of the site remains unchanged. This possibility could be examined by testing phenotype of the mutants of ori having substitution mutations at the restriction sites. Sodium sulfite mutagenesis was done to isolate base substitution mutations (11). The sodium sulfite-mutagenesis is known to act on single stranded DNA yielding C to T (also G to A) (12, 13). Plasmids containing a single stranded region at the restriction sites of BglII-2, BamHI-3, AvaII-1 or HindIII-1 were constructed, and sulfite mutagenesis done. Plasmid DNAs were isolated from the transformant colonies randomly, and the DNAs which lost the susceptibility to those restriction enzymes were isolated.

As shown in Table 1, 12 separate mutants thus isolated carried substitution mutation(s) at the restriction sites. All the 12 mutants thus isolated had the Ori$^+$ phenotype. From the result, we conclude that the restriction sites, BglII-2, BamHI-3, AvaII-1 or HindIII-1 could be modified without loss of Ori function. It is concluded therefore that these sites serve as spacer sequences.

Effect of base substitution targeted on the 245 base pairs:
Based on the model, base substitution mutations that occur in the recognition sequence, are expected to produce defective recognition sequences. Base substitution mutations occurring in the spacer sequences, should not produce the Ori$^-$ phenotype. In other wards, the roles of the sequences within the defined ori-segment could be allocated by isolation of base substitution mutants to find if the mutants produce Ori$^+$ phenotype or Ori$^-$ phenotype. Thus, we subjected the 245 bp, ori segment to sodium sulfite mutagenesis.

The results obtained thus far are summarized in Fig. 1.

Apri ori, the following possibilities could be considered: 1) the phenotype of a single base substitution mutation producing either Ori$^+$ or Ori$^-$ phenotype can be expressed independently. Thus the combination of two substitution mutations of ori$^-$ and ori$^+$ or ori$^-$ and ori$^-$ should produce the Ori$^-$ phenotype. The combination of two mutations producing the Ori$^+$ phenotype produces Ori$^+$ phenotype. Alternatively, 2) no prediction of phenotype could be made. By inspection of the results, we found former possibility is the case. For example, a mutant, pMU202, having Ori$^-$ phenotype, carried two substitution mutations: one mutation was at 242(C), substituted to T, and the other mutation was 247(C), substituted to T. The latter mutation of 247T is already known to produce Ori$^+$ phenotype (Table 1). Thus, the double mutant carrying the 242(T) mutation having Ori$^-$ phenotype and the 247(T) mutation having Ori$^+$ phenotype produced a Ori$^+$ phenotype. We conclude that possibility (1) is the case. The following single substitution mutations producing Ori$^-$ phenotype have been found: 134(C) substituted to T, 149(G) substituted to A, 242(C) substituted to T, and 267(C) substituted to T. We conclude that the bases in these positions are indispensable, and are part of recognition sequences.

Diversity of DNA replication origins of the replicons in E. coli and other bacterial species. Many nucleotide sequences of the ori-region of replicons in E. coli are known, and the origin sequence of temperate coli-phages, T-series coli-phages, single stranded coli-phages, coli-plasmids, and that of the E. coli cells demonstrated differences (16-20). The nucleotide sequence of the lamboid phage origins, for example, are within the coding region of a gene essential for replication (19), but the E. coli origin appears to be in a non-coding area. On the other hand, the DNA replication origins of different species of enteric bacteria, such as E. coli, Salmonella typhimurium, Enterobacter aerogenes, Klebsiella pneumoniae and Erwinia carotebora (21, 22) having a striking nucleotide sequence homology. This homology strongly suggests the presence of a common mechanism for control of DNA initiation and cell division of these bacteria. It should be noted that all the origin regions of these bacteria were cloned on E. coli, so that these ori-segments must carry all the information to regulate DNA replication and cell division in E. coli cells. Thus the replication origins of these bacterial species could be considered as multisite-Ori$^+$ mutants of E. coli (21).

By collaboration with Ziskind and Smith, sequences of five bacterial origins (22) and of E. coli Ori$^+$ and Ori$^-$ mutants (7, 8) including substitution, insertion, and deletion mutations were compared. The consensus sequence was established (23). It sould be stressed that all the bases indispensable in positions 134(C), 149(G), 142(C) and 267(C) (Fig. 1) are conserved among the bacterial origin sequences. Furthermore, the conserved and non-conserved sequences are clustered (22). An Ori^+ mutant having a 4 base-insertion at the BglII-2 site carried the sequence of AT·T·T·TA from 24(A) to 32A, and another Ori^+ mutant having a 12 base deletion at the $Hind$III-1 site carried the common sequences CA........CAC from 255(C) to 267(C) (7, 8). No contradictory results were obtained so far. These results suggest, but do not prove, that the conserved sequences correspond to the recognition sequences. It is likely that the majority of the nucleotide sequences conserved in these bacteria are indispensable.

Concluding remark, the recognition frame: Studies of the structural and functional analysis of the DNA replication origin of bacteria have produced a model of the structural organization of DNA replication origin, the recognition frame model. The role of the structure of the origin DNA is explained to form "initiation complex" which is constructed from or composed of from multiple initiation proteins organized in an appropriate topological arrangement. The original DNA consists of two sequences which functions are different from each other: the recognition sequences and the spacer sequences. The recognition sequences are recognized and bound specifically by the respective initiation proteins and the distance sequences connected the recognition sequences tandemly to constitute the replication origin containing 245 bp. The order of the arrangement of recognition sequences and spacer sequences between the recognition sequences must be determined precisely. This structure, named the *recognition frame*, could determine the topological arrangement of individual initiation proteins to form initiation complex. In the complex, the initiation reaction could be processed. Thus, we suggest the role of novel nucleotide sequences which determine the distance for the determination of the topological structure of initiation complex.

Sequences which belong to this category are already known in the promoter sequences: promoter sequences contain at least two specific sequences, one is located at the -35th position upstream and the other is located at the -10th position upstream of the initiation site of transcription. Evidence has been presented that the arrangement of these two sequences in a precise distance is essential for promoter function (24, 25).

The maximum size of *ori* is 245bp and it is obviously far more complex than the promoter sequence. Several gene products essential for DNA initiation, *i.e.* gene products of dnaA, dnaB, dnaC, dnaI and dnaP are known to be involved in DNA initiation in *E. coli* (26). Therefore, several recognition sequences which specify these initiation proteins must be found on the *ori* region, and several distance sequences are required to constitute the recognition frame of the *ori*-region.

This work was supported by research grants from the Ministry of Education, Science and Culture of the Japanese Government.

REFERENCES

1. Yasuda, S., and Hirota, Y. *Proc. Natl. Acad. Sci. U.S.A. 74,* 5458-5462 (1977).
2. von Meyenberg, K., Hansen, F.G., Nielsen, L.D., and Riise, E., *Molec. gen. Genet., 160,* 287-295 (1978).
3. Messer, W., Meijer, M., Bergmans, H.F.N., Hansen, F.G. von Meyenberg, K., Beck, E., and Schaller, H., *Cold Spring Harbor Symp. Quant. Biol., 43,* 139-146 (1979).
4. Hirota, Y., Yasuda, S., Yamada, M., Nishimura, A., Sugimoto, H., Sugisaki, A., Oka, A., and Takanami, M., *Cold Spring Harbor Symp. Quant. Biol., 43,* 29-38 (1979).
5. Sugimoto, K., Oka, A., Sugisaki, H., Takanami, M., Nishimura, A., Yasuda, S., and Hirota, Y., *Proc. Nat. Acad. Sci. U.S.A., 76,* 575-579 (1979).
6. Meijer, M., Beck, F., Hansen, F.G., Bergmans, H.E.N., Messer, W., von Meyenberg, K., and Schaller, H., *Proc. Nat. Acad. Sci. U.S.A., 76,* 580-584 (1979).
7. Oka, A., Sugimoto, K., Takanami, M., and Hirota, Y., *Molec. gen. Genet., 178,* 9-20 (1980).
8. Hirota, Y., Yamada, M., Nishimura, A., Sugimoto, A., and Takanami, M., *Progress Nucleic Acid Res. Molec. Biol., 26,* 33-48 (1981).
9. Marinus, M.G., *Molec. gen. Genet., 127,* 47-55 (1973).
10. DeLucia, P., and Cairns, J., *Nature, 224,* 1164-1166 (1969).
11. Shortle, D., and Nathans, D., *Proc. Natl. Acad. Sci. U.S.A., 75,* 2170-2174 (1978).
12. Shapiro, R., Braverman, B., Louis, J.B., and Servicis, R.E., *J. Biol., Chem., 248,* 4060-4064 (1973).
13. Sono, M., Wataya, Y., and Hayatsu, H., *J. Am. Chem. Soc., 95,* 4745-4749. (1973).
14. Maxam, A., and Gilbert, W., *Proc. Natl. Acad. Sci. U.S.A., 74,* 560-564 (1977).
15. Boliver, F., Rodriguez, R.I., Greene, P.J., Betlach, M.C., Heyneker, H.L., and Boyer, H.W., *Gene, 2,* 95-113 (1977).
16. Fiddes, J.C., Barrell, B.G., and Godson, G.N., *Proc. Nat.*

Acad. Sci., U.S.A., 75, 1081-1085 (1978).
17. Sims, J., and Dressler, D., *Proc. Nat. Acad. Sci., 75,* 3094-3098 (1978).
18. Denniston-Thompson, K., Morre, D.D., Krugar, K.E., Furth, M.E., and Blattner, *Science, 198,* 1051-1055 (1977).
19. Hobom, G., Grosschedl, R., Lsuky, M., Scherer, G., Schwarz, E., and Kossel, H., *Cold Spring Harbor Symp. Quant. Biol., 43,* 165-174 (1979).
20. Tomizawa, J., Ohmori, H., Bird, R.E., *Proc. Nat. Acad. Sci. U.S.A., 74,* 1865-1869 (1977).
21. Ziskind, J.W., and Smith, D.W., *Proc. Nat. Acad. Sci. U.S.A., 77,* 2460-2464 (1980).
22. Ziskind, J.W., Harding, N.E., Takeda, Y., Cleary, J.M., and Smith, D.W., *in this volume* (1981).
23. Ziskind, J.W., Smith, D., Hirota, Y., and Takanami, M., *in Appendix of this volume* (1981).
24. Gilbert, W., *In RNA polymerase (ed.* R. Losick and M. Chamberlin, P.P. 193-205 (1976). Cold Spring Harbor Laboratory.
25. Siebenlist, U., Simpson, R.B., and Gilbert, W., *Cell, 20,* 269-281 (1980).
26. Kornberg, A., *in DNA Replication* (1979), W.H. Freeman and Company.

THE DNA REPLICATION ORIGIN REGION OF THE ENTEROBACTERIACEAE

Judith W. Zyskind
Nancy E. Harding
Yutaka Takeda
Joseph M. Cleary
Douglas W. Smith

Department of Biology, C-016
University of California, San Diego
La Jolla, California

ABSTRACT

The *Sal*I fragments containing the chromosomal origin of replication (*oriC*) of *Enterobacter aerogenes*, *Klebsiella pneumoniae*, and *Erwinia carotovora* have been isolated using a high copy number cloning vehicle in an *Escherichia coli polA* mutant. The size of the *Sal*I fragments containing *oriC* ranged from 7.9 kilobase pairs (kb) for *E. carotovora* to 17.5 kb for *E. aerogenes*. The plasmids derived from *E. carotovora* were the only *oriC* plasmids that could transform an *E. coli polA*+ strain. However, smaller deletion derivative plasmids still containing the *oriC* region of *E. aerogenes* and *K. pneumoniae* could transform an *E. coli polA*+ strain. This suggests that there is a chromosomal region near but separate from *oriC* which is lethal when many copies are present in a cell. Genetic and physical maps of these *oriC* regions have been constructed. The nucleotide sequence of the origin regions of *K. pneumoniae*, *E. aerogenes*, and *E. carotovora* were determined and compared to those of *E. coli* and *Salmonella typhimurium*. The five origins were almost congruent in that very few single base insertions or deletions were needed to align them. The high frequency of GATC sites found in the *E. coli* and *S. typhimurium* origins is also found in the three new origins and may be responsible for sequence conservation. The regions of nonhomology between the five *oriC* sequences appear in clusters, with significantly long regions of total homology. The fact that these new replication origins function in *E. coli* argues that the mechanisms and gene products involved in the initiation of DNA replication also have been conserved extensively in the gram-negative bacterial species studied here. Possible promoter sites based on *oriC* sequence comparisons and known promoter information (7) are discussed, and in vivo experiments searching for *oriC* promoters are described.

INTRODUCTION

Conserved stretches of DNA which code for control regions such as promoters and operators contain recognition and binding sites for proteins. A comparison of nucleotide sequences of available prokaryotic promoters reveals that two regions centered at 35 and 10 (Pribnow Box, 1) base pairs (bp) before the start site of transcription are highly conserved (1 - 4). Several chemical and photochemical probes have identified specific points of contact between RNA polymerase and the -10 and -35 regions (e.g. 4-7). The assumption that binding sites for DNA initiation proteins are conserved and exhibit similar patterns to that of promoters can be examined by comparing the nucleotide sequences of different bacterial DNA replication origins which still function in *Escherichia coli*. Protein binding sites would be expected to appear as clusters of conserved nucleotides flanked by nonconserved regions. Here we compare the nucleotide sequences of origins from five different bacteria belonging to the family Enterobacteriaceae. This family has been tentatively divided into five tribes (8) and three of the tribes are represented in this comparison. *E. coli* and *Salmonella typhimurium* are members of the tribe Eschericheae, *Enterobacter aerogenes* and *Klebsiella pneumoniae* are members of the tribe Klebsielleae, and *Erwinia carotovora* is a member of the tribe Erwineae. The pattern of homology seen here in these five origin sequences is a clustering of nucleotide differences, interspersed with conserved regions up to 19 bp long.

RESULTS

Construction and Properties of Plasmids Containing the *Enterobacter aerogenes* Origin.

Homology between the *E. aerogenes* origin and that of *S. typhimurium* was demonstrated by Southern blot hybridization (9) of restriction enzyme digests of *E. aerogenes* strain SD1 chromosomal DNA to a probe containing an 8.6 kb *S. typhimurium* DNA fragment carrying *oriC* and the *asnA* gene. The probe hybridized to a single *E. aerogenes* 17.5 kb *Sal*I fragment and three *Pst*I fragments of 3.1, 2.4, and 1.2 kb (10).

The cloning vehicle we employed, pMK2004 (11), carries three antibiotic resistance genes, *amp*, *kan*, and *tet*. Insertion into the *Sal*I site inactivates the *tet* gene. This plasmid is present at about 40 copies per *E. coli* chromosome, and this copy number can be increased by growth in the presence of chloramphenicol (11). The ColE1 origin in this plasmid requires DNA polymerase I, the *polA* gene product, for replication and, therefore, pMK2004 does not replicate in *polA*

2 DNA REPLICATION ORIGIN REGION OF THE ENTEROBACTERIACEAE

mutants.

A partial SalI digest of *E. aerogenes* strain SD1 chromosomal DNA was ligated to a SalI digest of pMK2004 and used to transform an *E. coli* polA1 r_k^- strain selecting for resistance to kanamycin (Km) and ampicillin (Ap). The colonies were screened for sensitivity to the mutagenic agent methylmethane sulfonate (MMS). Cells carrying a *polA1* mutation are incapable of normal DNA repair and as a result are more sensitive to MMS than are *polA+* cells (12). The eleven Km^r Ap^r colonies which were isolated fall into two groups. The MMS^r group contained eight plasmids which did not hybridize by Southern blotting analysis (9) to an *S. typhimurium* origin probe. The three BamHI fragments (417 bp total) from the *S. typhimurium* origin region (13) labeled with [γ-^{32}P]ATP by phage T4 polynucleotide kinase is referred to here and elsewhere throughout the text as the *S. typhimurium* origin probe. These putative *polA+* plasmids shared four common SalI fragments of 3.05, 1.75, 1.35, and 0.45 kb. Only three colonies were sensitive to MMS and they contained plasmids which hybridized to the *S. typhimurium* origin probe. The three plasmids contained a common SalI insert of 17.5 kb (see Fig. 1), the same size as the SalI fragment in *E. aerogenes* chromosomal DNA which hybridized to an *S. typhimurium* probe carrying *oriC* and the *asnA* gene (see above). Unfortunately, these plasmids would not transform *E. coli* polA+ strains. Genetic experiments with these plasmids could not be performed directly because the *E. coli* mutants available all carried the *polA+* gene. However, when pMK2004 was eliminated from one of these plasmids by religating a SalI digest of the plasmid, we could show complementation of *E. coli* asnA and uncB mutants with the resulting self-ligated 17.5 kb SalI fragment (10). These two genes flank the origin of *S. typhimurium* (13, 18) and *E. coli* (Fig. 1; 19, 20). Derivative *oriC+* plasmids were constructed containing either a 2.4 kb PstI fragment (pNH305) or BamHI fragments 0.1, 0.2, and 1.95 kb (pNH326) inserted into pMK2004. These plasmids, pNH305 and pNH326, transform both *E. coli* polA+ and polA- strains but carry no known *E. aerogenes* genes.

Several restriction sites have been conserved between the *E. aerogenes* origin, and those of *E. coli* and *S. typhimurium* (Fig. 2). The most striking observation is that three *E. aerogenes* BamHI fragments are identical in size to the three BamHI fragments D (220 bp), E (106 bp), and F (91 bp) encompassing the origin of *S. typhimurium*. The nucleotide sequence of the *E. aerogenes* origin has been determined (21) and is shown in Fig. 3. The extent of conservation of the *E. aerogenes* origin with the other five origins is shown in Table 1.

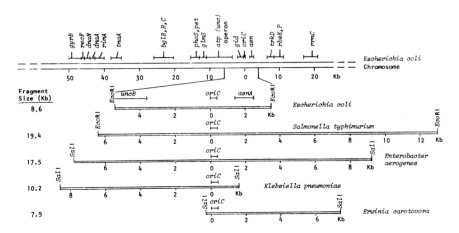

FIGURE 1. Cloned Bacterial Origin Regions. The genetic map locations are from von Meyenburg and Hansen (14). The *E. coli* EcoRI fragment containing *oriC* has been cloned by several groups (15-17) and the *S. typhimurium* EcoRI fragment by Zyskind, et al (18).

Construction and Properties of Plasmids Containing the *Klebsiella pneumoniae* Origin. The *K. pneumoniae* origin was cloned using the same protocol described above except that a complete rather than a partial SalI digest of *K. pneumoniae* strain M5al chromosomal DNA was used. The five Km^r Ap^r colonies obtained were all MMS^s. The plasmids which were isolated from 2 of the 5 colonies hybridized to the *S. typhimurium* origin probe and contained a 10.2 kb SalI fragment in common (10, see Fig. 1). These plasmids did not transform *E. coli* strain AI214 *polA*$^+$. However, when a 4.8 kb PstI fragment from the 10.2 kb SalI fragment was subcloned into pMK2004, the resulting plasmid did transform both *E. coli polA*$^+$ and *polA*$^-$ strains. The nucleotide sequence of *K. pneumoniae* origin was determined (21, Fig. 3) and the number of nucleotides in common between the *K. pneumoniae* origin and the other origin sequences is shown in Table 1. The *K. pneumoniae* origin is unique in that the origin is contained within a single BamHI fragment of 311 bp, making the *K. pneumoniae* origin easy to manipulate (10).

Construction and Properties of Plasmids Containing the *Erwinia carotovora* Origin. The isolation of plasmids containing the *E. carotovora* origin (22) followed the same protocol as that described above except that a complete SalI digest of *E. carotovora* strain EC153 was used. Of the nine Km^r Ap^r

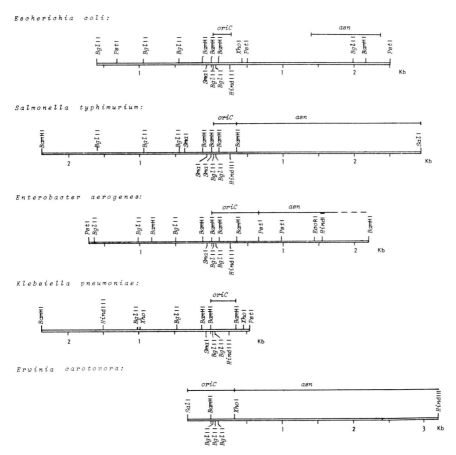

FIGURE 2. Restriction Sites in the Origin Region of Five Different Bacteria. See Refs. 16 and 20, and Refs. 13 and 18, for the *E. coli* and *S. typhimurium* data, respectively. The alignment of the origins is based on the nucleotide sequences.

colonies analyzed, three were MMS^r, and contained plasmids which did not hybridize to the *S. typhimurium* origin probe. These plasmids presumably carry the *E. carotovora* polA$^+$ gene. The other six Km^r Ap^r colonies were MMS^s. They contained plasmids which hybridized to the *S. typhimurium* origin probe and had a 7.9 kb *Sal*I insert in common (see Fig. 1 for a comparison with other origins). These plasmids are able to transform *E. coli* polA$^+$ cells, unlike the *oriC*-containing plasmids isolated from *E. aerogenes* and *K. pneumoniae*. Transformation of these plasmids into appropriate *E. coli* mutants showed that the plasmids with the *E. carotovora* 7.9 kb *Sal*I fragment complement *asnA* but not *uncB*. Deletion plasmids were

```
BamHI  -100                                                        SmaI                            SmaI
GGATCCTGATAAAACATGGTAAT-TGCCTCGCATAACGCGGTGTGAAAATGGATTGAAGCCCGGGCGGTGGATTCTACTCAACTTTAGCCGATGGAGAAA-GCCCGG
.....G..........................-.....................A......................C.................GT.G.C.T.......-.A..T.
....................G..C..........................T.................G...................................GCT.TT.C........G.GG...
....................GG.A..................A....T.....A....................C..............................TT.C.....-..G...
...GG.AC.GCTC.TC.CATTAT.C.GAATG.....GTG.CGAACG.TCGTT...TGCC.TA...T..G.................GAG.C.AC..CC..GT..GTT

 BamHI    10           20      BglII 30        BglII         50           60         70           80         90 BamHI 100
GGATCCGGGCTATTAAAAAGAAGATCTTTTTATTTAGAGATCTGTTCTATTGTGATCTCTTATTAGGATCGCGCCAGGCTGTGGATAACCCGGATCCTGT
.....T..G..........................A................................................A.TGCC...........AA.......G.C
.....T.AG.......................-...............................................ACT.TCTA........GT......ACG
.....T.AG.............................T.......................................GCTTGTCT........GT.A....GCG
.....TC.T..A.T...GA.........A.....C..T-....A............................TCGTGTTG......GTGATTATTCA.

101                                                       AvaII
AATAAAGATCAATGCGTTGGAAAGGATCACTAGCTGTGAATGATCGGTGATCGTGGTCCGATAAGCTGGGATCAAAACGGGTACTTATACACAACTCAA
TT.T..........CAACC.............................C...A..............G...T.A.GGG.............
.T.T..........ACGC..AAG........A.T..............C..T.................G..T.AAGGG......G....G....
GT.T.G......CCGT..AAG........G..TT..................................T.AAGGG..........G.A...
.T..........GAGAA...CGTT....CT..C............................C.A.C.A.........TT..T.T.GGT..........GGA...

201                                        HindIII
AAAGTGAACAACGGTTATTCTTTGGATAACTACCGGTTGATCCAAGCTTTCCACCAGATTTATCCACAA-TGGATCGCACGATCTTTACACTTATTTGAGT
...C..........A...G..........................C.TGA....G.........G.-.A.............G..T..........
....CAT..TC..............................................TG.G...GG......GAA.A.G.TGCG..........AAAC.CA....
....TTCAGG.........A.................................TA.G..T.G........G-AA...AA.T.T..........A..C.CA....C
...AC.C..TCG....G..A..G........G.C.TTA.ACCAGAA.TA.G..T.G.........T-.CA.CT..CGA..A....AC....C......

301         BamHI                                       370
AAATTAATCCAGGATCCGAGCCAAATCTCCGCTGGATCTTCCGGAATCTCATGTTCAAGGATGTTGATCT - Salmonella typhimurium
......C...C.......C......TTCT...T..C..............G..G..A.....A.........  - Escherichia coli
....CTT..T........C.C......C....GG.C..........A......G......G....CA..A..T.  - Enterobacter aerogenes
..G..TT...........................G..C...........T.........G....C..A..T.  - Klebsiella pneumoniae
......C...C......T......TTCT..A..C.........A....CT.G...C.G.GT.....C.....  - Erwinia carotovora
```

FIGURE 3. Nucleotide Sequence of the *oriC* Region in *Salmonella typhimurium* (13), *Escherichia coli* (16, 20), *Enterobacter aerogenes*, *Klebsiella pneumoniae*, and *Erwinia carotovora*. .: nucleotide is identical to that found in *S. typhimurium*. -: a deletion is imposed to permit maximal sequence homology between origins. Representative restriction sites in *S. typhimurium* are presented. GATC sequences are underlined. The numbering of nucleotide position is that first used for *E. coli* (16, 20). The minimal origin of *E. coli* (23) is enclosed within the box. The upper left end is the 5' end.

constructed to permit ordering of *oriC* and the *asnA* gene. Restriction analysis of these deletion plasmids indicate that the *E. carotovora* origin region does not contain many of the restriction sites conserved between the origin regions of *E. coli*, *S. typhimurium*, *E. aerogenes*, and *K. pneumoniae* (Fig. 2). Sequence analysis revealed that the *E. carotovora* origin is the most divergent among these five members of the family Enterobacteriaceae (22, Fig. 3 and Table 1).

DISCUSSION

One puzzling feature of these cloned bacterial replication origins is that the plasmids containing either the 17.5 kb *Sal*I fragment isolated from *E. aerogenes* (Fig. 1) or the 10.2 kb *Sal*I fragment from *K. pneumoniae* inserted into pMK2004 do not detectably transform *E. coli* $polA^+$ strains, whereas those containing the 7.9 kb *Sal*I fragment from

TABLE 1
DEGREE OF HOMOLOGY BETWEEN THE BACTERIAL ORIGIN SEQUENCES

	S. typh.	E. coli	E. aero.	K. pneu.	E. caro.	
S. typh.		86.9	80.8	83.8	59.4	O
E. coli	85.4		78.6	80.0	60.7	U T
E. aero.	81.7	81.7		85.2	53.7	S
K. pneu.	79.3	80.9	86.0		53.7	I D
E. caro.	67.9	70.7	70.3	72.0		E

I N S I D E

The degree or extent of homology is expressed in percentage. Inside refers to nucleotides within the box (245 bp) of Fig. 3 (the minimal origin), outside refers to nucleotides outside the box (231 bp) in Fig. 3. *S. typh.*, *E. aero.*, *K. pneu.*, and *E. caro.* are *Salmonella typhimurium*, *Enterobacter aerogenes*, *Klebsiella pneumoniae*, and *Erwinia carotovora*, respectively.

E. carotovora transform *E. coli* $polA^+$ cells at a normal frequency. This property can be explained if the *E. aerogenes* and *K. pneumoniae* SalI fragments carry a region of DNA which is lethal when present at high copy number in a cell and if this region is absent in the *E. carotovora* plasmids. K. Yamaguchi (personal communication) has isolated a HaeII fragment near *uncB* in *E. coli* which is lethal to a cell when present at a high copy number. From the genetic and physical maps of these plasmids, this region is carried on the *E. aerogenes* and *K. pneumoniae* plasmids but is not found in the 7.9 kb SalI fragment from *E. carotovora* supporting the above hypothesis. A prediction that follows is that the EcoRI fragment containing either the *E. coli* or *S. typhimurium* origins (see Fig. 1) cannot be cloned in a high copy number plasmid such as pMK2004 in an *E. coli* $polA^+$ strain.

Very few single base insertions or deletions are needed in order to completely align the nucleotide sequences of the five bacterial origins shown in Fig. 3. Within the box two deletions are included in the *E. carotovora* sequence and one deletion in each of the *E. aerogenes* and *K. pneumoniae* sequences. The box drawn around the sequence from +23 to +267 bp is the minimal origin as defined in *E. coli* (23). The sequences within the box are relatively A,T rich; the %(A+T) for *S. typhimurium* is 58.7, for *E. coli* is 59.2, for *E. aerogenes* is 59.8, for *K. pneumoniae* is 60.2, and for *E. carotovora* is 63.0.

All five origins contain an abundance of GATC sites. Only one such sequence every 256 bp is expected at random, and yet, within the boxed region of 245 bp there are 11 GATC sites in *S. typhimurium*, *E. coli*, *E. aerogenes*, and *K. pneumoniae* and 9 in *E. carotovora*. This sequence, GATC, is the site of methylation by the *E. coli dam* methylase (24) and appears to be involved in mismatch repair (25,26). We have previously proposed (13) that this repair system preferentially preserves the nucleotide sequence of the bacterial origin.

The percent of the number of base pairs in common between any two species (Table 1) has been computed from the sequences presented in Fig. 3. This comparison has been made both inside and outside the box drawn around the minimal origin. Members of the same tribe show the greatest homology both inside and outside the box. The second level of homology is between members of the two tribes Escherichaeae and Klebsielleae. *E. carotovora* appears to be equally distant from the other four species. The sequence conservation in the *E. carotovora* origin is greater within the box than outside the box.

The majority of nucleotide differences appear in clusters, interspersed with regions of total homology as long as 19 bp. The pattern of nucleotide differences in these five origins is not at all like the pattern observed in genes coding for proteins where most changes occur in the third position of a codon (e.g. 27-29). The pattern of homology observed in Fig. 3 closely resembles the distribution of homologous regions found by Oppenheim, et al. (6) in the promoter-operator region of the *trp* operon shown in Fig. 4. The two regions important in promoter function are conserved between the four *trp* promoters: the 6 bp sequence around -35 and the larger region centered at -10. The *trp* operator overlaps the *trp* promoter in the -10 region and thus places additional constraints on this sequence.

One of the most striking observations concerning the conserved regions within the origins is that there are four 9 bp repeats, two in opposite orientation to the other two. The distribution, location, and sequences of these repeats are shown in Fig. 5. These repeats do not appear to be related to repeats found in the *E. coli* plasmids R6K, RK2, and F or phage λ origins (30) and are probably either binding sites for a protein or involved in secondary structure.

The region from base 20 to base 71 is highly conserved, with four GATC sequences regularly spaced between very A,T-rich regions. Extensive intrastrand doublehelical structure is possible between these four GATC regions, since GATC is a

2 DNA REPLICATION ORIGIN REGION OF THE ENTEROBACTERIACEAE

```
             ┌─────────── E. coli Promoter ───────────┐
                              ┌──── E. coli Operator ────┐
                                                            trp mRNA →
                 -40        -30        -20        -10        +1
                  |          |          |          |          |
S. typhimurium- AAATAGGTGTTGACATTATTCCATCGAACTAGTTAACTAGTACGAAAGTTCACA
      E. coli-  ..TG..C........A.TAATC.....................C........G
   K. aerogenes- ...C.A.G......T.........T..................C........
   S. marcescens- ...AGA.G......T..GCCTTCG....C.............AC........G
```

FIGURE 4. Comparison of the Nucleotide Sequences of the *trp* Operator and Promoter Regions of 4 Bacteria belonging to the Family Enterobacteriaceae (6). *K. aerogenes* is *Klebsiella aerogenes* and *S. marcescens* is *Serratia marcescens*. Nucleotides identical to those of the *S. typhimurium* sequence are indicated with a dot.

nucleotide palindrome.

The consensus sequence of the bacterial origin presented as an appendix following this paper (36) is derived from the sequences of naturally occurring bacterial origins presented here and elsewhere (13,16,20) and from mutations in the *E. coli* origin (23, 31). The extensive analysis of $oriC^+$ and $oriC^-$ mutations in the *E. coli* origin undertaken by Oka, et al. (23) and Hirota, et al. (31) include substitutions, insertions, and deletions. One of the most interesting aspects of this consensus sequence is that deletions and insertions in the *Bam*HI (at base 95), *Ava*II, and *Hin*dIII sites are $oriC^-$, even though these three sites are in nonconserved regions. Nucleotide differences occur at every position in these sites, and yet insertions of as few as 2 bases in, for example, the *Ava*II site inactivate origin function. The requirement for spacial arrangement at these three locations suggests that they could be between binding sites of one or more proteins. In the 54 promoters whose sequences are known, the separation between the -10 and -35 regions varies at most by 3 bp (4). When this distance is further reduced by the deletion of 1 bp, as in the case of the *tyrT* tRNA promoter (32), a down promoter mutation is produced, suggesting that there is a required distance between the two regions where RNA polymerase is known to bind. Considerable indirect evidence for transcriptional activity in the bacterial origin has been obtained (33, for review see ref. 34). Also, the *Hin*d III site within the origin sequence is somewhat protected by RNA polymerase (B. Munson, personal communication). We examined the regions around the *Bam*HI (at base 95), *Ava*II, and *Hin*dIII sites for possible promoters, looking for sequence homology with the Pribnow box and the -35 region and requiring a spacing of 16 to 19 bp. All three restriction sites were centered in conserved promoter-like sequences (Fig. 6). The T at position

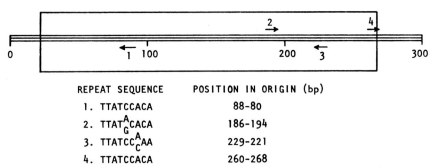

REPEAT SEQUENCE	POSITION IN ORIGIN (bp)
1. TTATCCACA	88-80
2. TTAT^A_GCACA	186-194
3. TTATCC^A_CAA	229-221
4. TTATCCACA	260-268

FIGURE 5. Location and Sequences of Conserved 9 bp Repeats in the Bacterial Origins. The 5' end of each sequence is on the left and the arrows are 5' to 3'.

-8 is a T in 54 different promoter sequences (4) and appears to be important for high transcriptional activity (see up and down promoter mutants summarized in ref. 4), although weak promoter activity remains when this T is mutated (32). Thus, if promoter function is in fact related to these sequences they would likely perform as down promoter mutants because of the differences at position -8 (Fig. 6). The putative promoters #1 and #2 (Fig. 6) are at opposite ends of the origin and include most of the repeats #1 and #4 (Fig. 5). The direction of transcription from these two regions would be in opposite directions towards the outside of the origin starting around bases 74 and 273. At least one of these promoters (#2) might be used for the in vitro transcription products described by Messer et al. (33).

We have examined in vivo promoter activity in the *S. typhimurium* origin using the promoter cloning vehicle pMC489 (35). Promoter activity within *Bam*HI fragments can be detected by insertion into the single *Bam*HI site of this plasmid and assaying for β-galactosidase activity. No in vivo promoter activity within the origin could be detected although transcription from the putative promoter #1 (Fig. 6) would not be assayed because it contains a *Bam*HI site. The approach we are taking is to look for possible up promoter mutants using these plasmids. The *Bam*HI fragment from bases -106 to +6 just outside of the *S. typhimurium* origin (Fig. 3) contains a promoter in one orientation and a terminator in the other orientation, with the direction of transcription from the promoter being away from the origin (J. Zyskind, unpublished data). This fragment would prevent transcription from entering the origin.

The consensus sequence in the following appendix (36) represents the primary structure required for origin function (initiation of rounds of DNA replication) and when viewed in this manner there are four caveats which must be kept in mind.

	Position in Origin	-35		-10
		T T G A C A	. . . . 16 to 19 bases . .	T A T A A T
1)	112 to 82	T T G A T C	. . (BamHI) . .19 bases . .	T A T C C A
2)	237 to 266	T T g a t C	. . (HindIII) .18 bases . .	T A T C C A
3)	136 to 166	G T G A A T	. . (AvaII) . .19 bases . .	T A T A A G

FIGURE 6. Possible Promoters in the Bacterial Origins. Largest letters at the top designate highly conserved positions ($\geq 46\%$) at the -10 and -35 regions in the 54 promoters compared by Siebenlist et al. (4). In the origin sequences, capital letters are positions conserved in all 5 origins and lower case letters are positions conserved in 4 of the 5 origins. The 5' end of each sequence is on the left and the 5' to 3' sequence on the strand opposite to that shown in Fig. 3 is given for possible promoter #1.

First, naturally occurring origins may contain compensating changes where, for example, two simultaneous nucleotide changes result in a functional origin. This would be most obvious in a region where secondary structure is important, as has been observed in the phage G4 origin (N. Godson, personal communication). Second, some $oriC^+$ mutants may not function as origins in the chromosome. The effect of these mutations on origin function was measured by the ability or inability of the mutated region to enable ColE1 derived plasmids to transform and replicate in *E. coli polA-* strains (23). Even though plasmid replication in this assay is dependent upon the *dnaA* gene product (37) and other replication proteins (33), this assay may not include all initiation and segregation events required for complete chromosomal origin function. For example, the adenine at position 268 is not an adenine in the new sequences brought contiguous by the $oriC^+$ deletion plasmids pTS0193 and pTS0236 (23). The adenine may be required for origin function in the chromosome, however, as it is conserved in all five naturally occurring origins. Further, this adenine is found in the repeat #4 (Fig. 5). This repeat is conserved in all five origins (Fig. 3) and is a precise 9 bp inverted repeat of #1 (Fig. 5) which is also conserved in all five sequences. Repeat #4 also contains the Pribnow Box of the putative promoter #2 (Fig. 6). The $oriC^+$ deletion in pTS0193 changes 6 of the bases in this 9 bp repeat, yielding an origin which is functional in the plasmid but which might not be completely functional in the bacterial chromosome. Third, since this same assay was used for isolating chromosomal origins from other bacterial species, nonconserved regions may not necessarily reflect non-required sequences in *E. coli*, in that these origins may not be able to replace the *E. coli* origin in the chromosome. Fourth, all possible base changes which still retain origin function have most probably not yet been found. Sequences of other bacterial origins should strengthen the consensus sequence (36) in this regard. We have isolated the *Vibrio (Beneckea) harveyi* origin of replication and are currently determining its nucleotide sequence. This species is distantly related to the enterics, and the nucleotide sequence of its origin should provide important new information about required nucleotides in the consensus sequence.

ACKNOWLEDGMENTS

We thank Barry Chelm and David Stalker for reading the manuscript and suggesting improvements. This work was supported by a grant from the National Institutes of Health (GM21978).

REFERENCES

1. Pribnow, D. (1975). *Proc. Natl. Acad. Sci. USA* 72, 784.
2. Schaller, H., Gray, C. and Herrmann, K. (1975). *Proc. Natl. Acad. Sci. USA* 72, 737.
3. Rosenberg, M., and Court, D. (1979). *Ann. Rev. Genet.* 13, 319.
4. Siebenlist, U., Simpson, R.B., and Gilbert, W. (1980). *Cell* 20, 269.
5. Johnsrud, L. (1978). *Proc. Natl. Acad. Sci. USA* 75, 5314.
6. Oppenheim, D.S., Bennett, G.M., and Yanofsky, C. (1980). *J. Mol. Biol.* 144, 133.
7. Siebenlist, U., and Gilbert, W. (1980). *Proc. Natl. Acad. Sci. USA* 77, 122.
8. "Bergey's Manual of Determinative Bacteriology" (1974). (R.E. Buchanan, ed.), p. 291. Waverly Press Inc., Baltimore.
9. Southern, E.M. (1975). *J. Mol. Biol.* 98, 503.
10. Harding, N.E., Cleary, J.M., Smith, D.W., and Zyskind, J.W. (1981). Manuscript in preparation.
11. Kahn, M., Kolter, R., Thomas, C., Figurski, D., Meyer, R., Remault, E. and Helinski, D.R. (1979). *Methods Enzymol.* 68, 268.
12. Gross, J., and Gross, M. (1969). *Nature (London)* 224, 1166.
13. Zyskind, J.W., and Smith, D.W. (1980). *Proc. Natl. Acad. Sci. USA* 77, 2460.
14. Meyenburg, K. von, and Hansen, F.G. (1980)."Mechanistic Studies of DNA Replication and Genetic Recombination", (B. Alberts, ed.), p. 137, Academic Press, NY.
15. Yasuda, S., and Hirota, Y. (1977). *Proc. Natl. Acad. Sci. USA* 74, 5458.

16. Meijer, M., Beck, E., Hansen, F.G., Bergmans, H.E.N., Messer, W., Meyenburg, K. von, and Schaller, H. (1979). *Proc. Natl. Acad. Sci. USA 76*, 580.
17. Leonard, A.C., Weinberger, M., Munson, B.R., and Helmstetter, C.E. (1980). "Mechanistic Studies of DNA Replication and Genetic Recombination", (B. Alberts, ed.), p. 171, Academic Press, NY.
18. Zyskind, J.W., Deen, L.T., and Smith, D.W. (1979). *Proc. Natl. Acad. Sci. USA 76*, 3097.
19. Meyenburg, K. von, Hansen, F.G., Nielsen, L.D., and Riise, E. (1978). *Molec. gen. Genet. 160*, 287.
20. Sugimoto, K., Oka, A., Sugisaki, H., Takanami, M., Nishimura, A., Yasuda, S., and Hirota, Y. (1979). *Proc. Natl. Acad. Sci. USA 76*, 575.
21. Cleary, J.M., Harding, N.E., Smith, D.W., and Zyskind, J.W. (1981). Manuscript in preparation.
22. Takeda, Y., Harding, N.E., Smith, D.W., and Zyskind, J.W. (1981). Manuscript in preparation.
23. Oka, A., Sugimoto, K., Takanami, M., and Hirota, Y. (1980). *Molec. gen. Genet. 178*, 9.
24. Geier, G., and Modrich, P. (1979). *J. Biol. Chem. 254*, 1408.
25. Wagner, R., Jr., and Meselson, M. (1976). *Proc. Natl. Acad. Sci. USA 73*, 4135.
26. Glickman, B., Van den Elsen, P., and Radman, M. (1978). *Mol. gen. Genet. 163*, 307.
27. Nichols, B.P., and Yanofsky, C. (1979). *Proc. Natl. Acad. Sci. USA 76*, 5244.
28. Crawford, I.P., Nichols, B.P., and Yanofsky, C. (1980). *J. Mol. Biol. 142*, 489.
29. Nichols, B.P., Miozzari, G.F., Cleemput, M. van, Bennett, G.N., and Yanofsky, C. (1980). *J. Mol. Biol. 142*, 503.
30. Stalker, D., Shafferman, A., Tolun, A., Kolter, R., Yang, S., and Helinski, D. (1981). This volume.
31. Hirota, Y., Oka, A., Sugimoto, K., Morita, M., and Takanami, M. (1981). This volume.
32. Berman, M.L., and Landy, A. (1979). *Proc. Natl. Acad. Sci. USA 76*, 4303.
33. Lother, H., Bühk, H., Morelli, G., Heimann, B., Chakraborty, T., and Messer, W. (1981). This volume.
34. Lark, K.G. (1979). "Biological Regulation and Development" (R. Goldberger, ed.), Vol. 1, p. 201, Plenum Press, NY.
35. Casadaban, M.J., and Cohen, S.N. (1980). *J. Mol. Biol. 138*, 179.
36. Zyskind, J.W., Smith, D.W., Hirota, Y., and Takanami, M. (1981). This volume.
37. Zyskind, J.W., Deen, L.T., Harding, N.E., Pritchard, R.H., and Smith, D.W. (1980). "Mechanistic Studies of DNA Replication and Genetic Recombination", (B. Alberts, ed.) p. 181, Academic Press, NY.

APPENDIX
THE CONSENSUS SEQUENCE OF THE BACTERIAL ORIGIN

Judith W. Zyskind and Douglas W. Smith

Department of Biology, C-016, University of California
at San Diego, La Jolla, California 92093

Yukinori Hirota

National Institute of Genetics, Mishima, Shizuoka 411, Japan

Mituru Takanami

Institute for Chemical Research
Kyoto University, Uji, Kyoto 611, Japan

The consensus sequence presented in Fig. 1 was derived from the sequences of five naturally occurring bacterial origins as well as *ori* mutants in *Escherichia coli*. The five naturally occurring origin sequences are *E. coli* (1,2), *Salmonella typhimurium* (3), *Enterobacter aerogenes* (4), *Klebsiella pneumoniae* (4), and *Erwinia carotovora* (4). The *E. coli ori*$^+$ and *ori*$^-$ mutants include substitution (5), insertion (6), and deletion (6,7) mutations. The consensus sequence is comprised of those nucleotides found most often in these naturally occurring and mutant bacterial origins, origins which function as replication origins in plasmids cloned in *E. coli*. The consensus sequence is discussed in the two preceding papers (4,5).

FIGURE 1. The Consensus Sequence of the Bacterial Origin. Naturally occurring bacterial origins are shown above the consensus sequence, and mutated *E. coli ori* sequence changes are shown below the consensus sequence. In the consensus sequence: larger capital letters: nucleotide is found in all functional bacterial origins presented; smaller capital letters: nucleotide is found in four of five, or five of six, functional bacterial origins presented; lower case letters: nucelotide is present in three of five, or four of six, functional bacterial origins presented; n: each nucleotide is present in at least one functional origin, or three of the four possible nucleotides are found, but each in no more than two functional bacterial origins presented. In the individual origin sequences: -: a deletion relative to the consensus sequence is present; .: nucleotide is the same as that of the consensus sequence. For the insertion and deletion derivative sequences: ΔΔΔ: region of *E. Coli* deletion

```
                    BglII                                BglII                                                        BamHI  100
Escherichia coli:    ..........A.........................................CAC..CCC  CAAG.....GGC
Salmonella typhimurium: ..............-......................................CGCCAGCC  CC.G.....TGT
Enterobacter aerogenes: ..............-......................................A.TCTCTA  ...G.....ACG
Klebsiella pneumoniae: ..............A....C..T-.............................G.T..TCT  ..A..GCG
Erwinia carotovora:   ..............A....C T-.............-A.................T.G..TTG  ..GATTATTCAT
                                                                50
CONSENSUS SEQUENCE: AAGATCTTtTATTAgAGATCtGTTcTATTgTGATCTCTTATTAGGATCGncntgnnnTGTGGATAAgtcgATccnnn
(Ori+ Insertion) pTSO225: TCT T..GAG TC                                                    AA  TTA
           Deletions:    ΔΔΔΔΔΔΔΔΔΔΔΔΔΔΔΔΔΔΔΔΔ : -7, Ori-                            -4, Ori-:  ΔΔΔΔΔΔΔΔ
                                               : -16, Ori-                         -15, Ori-:
         Insertions:                     GATC : +4, Ori+                          +4, Ori-:  GATC

T...      CAACC ..A...       .T..A           ..A..                    .G...                    A...
.A..      TGCG. ..A...       ...G             G ...                   .A..C  G.TAC             A...
..G       AC.C. AAG...       ..A.T           .T....                   .G..   .A...      G      ..A
..G       CC.T. AAG...       G..TG           .T....                   .A..   .T....            ..A
..A.      GA.AA ..CGTT       CT..C           .A.C.A                   TT     T....T            .GA
                                                                     AvaII
101                                                         150
aTTtAAGATCAAnngntTggnAagGATCacTAnCTGTGAATGATCGGTGATCCTGgtcgGTATAAGGCTGGATCAnAATgAggGgTTATACACAgctCAA
                                Ori-:T            Ori-:A           A TT
ΔΔΔΔΔΔΔ                                                        ΔΔΔΔΔΔΔ  : -3, Ori-
                                                                        : -4, Ori-
                                                                        : -6, Ori-
                                                                        : -7, Ori-
                                                                        : -8, Ori-
                                                            either AC or GAC : +2 or +3, both Ori-

.CT.A  ...A                     ........CCTGA....                      .G...                    A...
.GT.A.                          ........CC.C...T..                                               :Escherichia coli (1,2)
.G.AT. TC...                    ........TG....G....                                              :Salmonella typhimurium (3)
TT.AGG      .A..                ........TA....T.....                                             :Enterobacter aerogenes (4)
.A..C. TCG.   .A G              G C  TTA ACCAGA.TA..T..                                          :Klebsiella pneumoniae (4)
                                                                                                 :Erwinia carotovora (4)
                     HindIII 250
201
AAAncgnACaacGGTTaTTcTTTGGATAACTACcGGTTgATCCAagcTTtnagCAgagTATCCACA                         :CONSENSUS SEQUENCE
                        Ori-:T    AT T    T       Ori-:T
                                  GAG..ATCCA  GTAGATCG         G           :Ori+ Substitutions (5)
                        -8, Ori-:   -12, Ori+: ΔΔΔΔΔΔΔΔΔΔΔΔ  -2, Ori-: ΔΔ  :pTSO193 (Ori+ Deletion) (6)
                        -5, Ori-:    -7, Ori-:                             :Deletions (6,7)
                                              AGCT : +4, Ori-              :Insertions (6)
```

derivatives; ___ : specific *E. coli* deletion derivatives;
^: point of insertion for insertion derivatives. GATC sites
are underlined in the consensus sequence, and representative
restriction sites are noted. The sequence within the box is
the minimal bacterial origin based on deletion derivative
studies of the *E. coli* and *S. typhimurium* origins (3,6). The
numbering of nucleotide position is that first used for
E. coli (1,2), and the upper left end is the 5' end.

REFERENCES

1. Sugimoto, K., Oka, A., Sugisaki, H., Takanami, M., Nishimura, A., Yasuda, S., and Hirota, Y. (1979). *Proc. Natl. Acad. Sci. USA 76*, 575.
2. Meijer, M., Beck, E., Hansen, F. G., Bergmans, H. E. N., Messer, W., Meyenburg, K. von, and Schaller, H. (1979). *Proc. Natl. Acad. Sci. USA 76*, 580.
3. Zyskind, J. W., and Smith, D. W. (1980). *Proc. Natl. Acad. Sci. USA 77*, 2460.
4. Zyskind, J. W., Harding, N. E., Takeda, Y., Cleary, J. M., and Smith, D. W. (1981). This volume.
5. Hirota, Y., Oka, A., Sugimoto, K., Morita, M., and Takanami, M. (1981). This volume.
6. Oka, A., Sugimoto, K., Takanami, M., and Hirota, Y. (1980) *Molec. gen. Genet. 178*, 9.
7. Hirota, Y., Yamada, M., Nishimura, A., Oka, A., Sugimoto, K., Morita, M., and Takanami, M. (1981). *Prog. Nuc. Acid Res. Mol. Biol. 26*, 33.

MAPPING OF PROMOTERS IN THE REPLICATION ORIGIN REGION OF THE E. coli CHROMOSOME

Masayuki Morita
Kazunori Sugimoto
Atsuhiro Oka
Mituru Takanami

Institute for Chemical Research
Kyoto University
Uji, Kyoto

Yukinori Hirota

National Institute of Genetics
Mishima

ABSTRACT Promoters functioning *in vivo* in the region encompassing *ori* of the *E. coli* chromosome have been identified by using a promoter-cloning vecter pGA46 (An and Friesen: J. Bacteriol., *140*, 400-407, 1979)(*ori*: defined as the minimal region carrying autonomously replicating function and spanning from positions 23 to 267 in our nucleotide coordinate(Oka et al.: Molec. Gen. Genet., *178*, 9-20, 1980).

pGA46 contains a short DNA segment carrying three restriction sites, *Hind*III, *Bgl*II and *Pst*I, in place of the natural promoter for *tet* gene, so that various restriction fragments generated from the *ori* area were inserted in these sites directly or by using linkers, and insertion activation of the *tet* gene was analysed.

Active promotion was detected at five different sites, in which four were mapped outside, and one in the vicinity of the right boundary, of *ori*. Transcription promoted from these sites proceeds in the outward direction without entering *ori*, except for that from the one mapped in the promoter region of *proX* gene (positions 734 to 294, Hirota et al.: Progress Nucleic Acid Res. & Molec. Biol., *26*, 33-48, 1981). Transcription from this site goes across the *proX* gene region and enters the *ori* stretch.

INTRODUCTION

The replication origin region of the *E. coli* chromosome has been cloned(1-4), and the minimal stretch carrying the information for autonomous replication (defined *ori*) has been determined(5). *ori* is contained within 245 base-pairs spanning from positions 23 to 267 in our nucleotide coordinate(6). Analysis of the correlation between the *ori* sequence and Ori function has further demonstrated that the *ori* stretch is composed of several important sequences(recognition sequences) appropriately separated by flanking sequences(5,7,8). The next step is obviously the eludication of the functional role of the defined sequences, but so far little is known about the basic reactions involved in the initiation of replication. The replication of plasmids and bacteriophages has been shown to be initiated with RNA primers(9). It has also been suggested that RNA polymerase reaction is involved in the initiation of the *E.coli* chromosome, though the evidence is rather indirect (10,11). It seems in any event to be important to know the promoter sites in the vicinity of the *ori* stretch. The data should also serve information for the gene structures in the neighbouring region of *ori*.

In this paper, an attempt was made to map promoters functioning *in vivo* by using a promoter-cloning vector, pGA46, constructed by An and Friesen(12). In this vector, the natural promoter for its *tet* gene has been removed and replaced by a small fragment carrying three restriction sites, *Hind*III, *Bgl*II, and *Pst*I(in this order toward *tet* gene). The cells harboring this plasmid is thus sensitive to tetracycline(Tc) unless insertion activation of the *tet* gene occurs by cloning a promoter-carrying fragment. This plasmid is maintained in cells at a low copy number, so that the level of Tc resistance is able to assay in the wide range of Tc concentrations(12).

MATERIALS AND METHODS

Bacterial strains and plasmids: *E.coli* K12 cells used were C600 (F⁻*thr leu thi lac tonA supE*) and its plasmid-carrying derivatives. Plasmid pGA46 was donated by Dr.G.An. pBR322-recombinants as the sources of restriction fragments from the replication origin region were as follows: pTSO169(*Hae*III-1 to *Hae*III-2 in Fig.1 A)(5), pTSO116(*Hae*III-2 to *Hae*III-3 in Fig.1 A)(6), pTSO103 (*Bam*HI-3 to *Bam*HI-4 in Fig.1 A)(6), pMU2(Ori⁺ plasmid carrying *Hae*III-1 to *Hae*III-2 but with modified *Hind*III site at position 246(AAGCTT to AAACTT)(8), and pMY129(*Pst*I-1 to *Pst*I-2 in Fig.1 A)(donated by Dr. M.Yamada).

4 REPLICATION ORIGIN REGION OF THE E. COLI CHROMOSOME

Preparation of plasmids and restriction fragments : The methods used for transformation and preparation of plasmid DNA and their restriction fragments were essentially identical to those described previously(6).

Cloning of restriction fragments with pGA46 : The ends of restriction fragments were converted into blunt ends by T4 DNA polymerase and joined with a HindIII linker(CCAAGCTTGG, Collaborative Res. Inc.), and cloning with pGA46 was carried out. The detailed conditions will be described elsewhere. Fragments generated by combinations of HindIII, BglII and PstI were directly cloned in the promoter-cloning sites of pGA46. The ligation products were used to transform C600 cells in the presence of 20 µg/ml chloramphenicol(Cm) and either 5 µg/ml Tc or 20 µg/ml Tc. Plasmids were isolated from transformants and restriction analysis was performed to identify the cloned fragment and its polarity.

RESULTS AND DISCUSSION

Restriction fragments primarily assayed are shown in Fig.1 B. These fragments were inserted into the promoter-cloning sites of pGA46 by using a linker or directly, and transformation was carried out. The result of a typical experiment is shown in Table 1, in which the fragment numbers correspond to those given in Fig.1 B. Since the cloning procedure which uses the linker includes many steps, the cloning efficiency was not high, but the yield of Tc^r transformants depended on the species of fragments used. Restriction analysis of plasmids extracted from the Tc^r transformants with fragments 1, 2, 3 and 6 indeed indicated that plasmids contained expected fragments in proper orientation. In contrast, fragments 4 and 5 yielded little Tc^r transformants. Although a small number of colonies were obtained at 5 µg/ml Tc, analysis of plasmids indicated that unidentified fragments have been inserted instead of the subjected fragments. The result suggests that fragments 4 and 5 do not carry promotion activity. Fragments 2 and 6 yielded Tc^r transformants even at 20 µg/ml Tc, implying to contain a strong class promoter.

Another type of experiment carried out was analysis of deletion derivatives. The three restriction sites, HindIII, BglII and PstI, in the promoter-cloning sites of pGA46 are also located in the *ori* area, so that we could introduce deletions in the cloned fragments by using these sites. This made it possible to map promotion sites more precisely.

In the above analyses, however, the sequences at the junction sites of cloned fragments have been altered from those of the original sequences, so that data should be evaluated based on the consistent results with various overlapping fragments.

FIG.1 Promoter mapping in the replication origin region. Restriction sites in the ori-asn area are shown in (A). The ori stretch and proX and asn gene regions are indicated in boxes below the map. Examined fragments are shown in (B) by numbering. Active fragments identified for each direction are shown in (C). Broken lines represent fragments in which promotion activity was not detected. The regions carrying respective promoters are indicated by filled lines in (D).

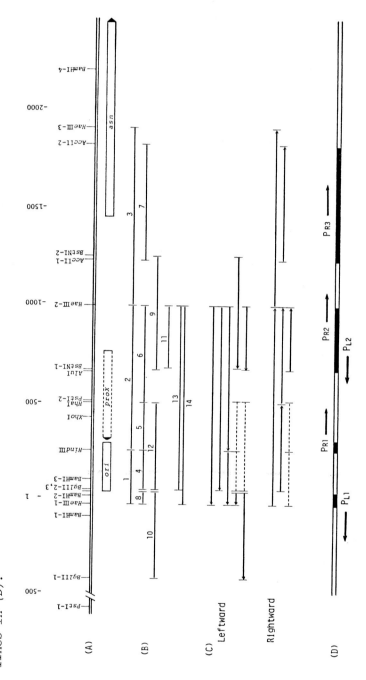

4 REPLICATION ORIGIN REGION OF THE E. COLI CHROMOSOME

TABLE 1 *Transformation by ligation products of pGA46-HindIII digests and HindIII linker-joined restriction fragments. About 2-5 pmol of fragments, of which the ends had been converted to blunt ends, were ligated with 30 times excesses of the HindIII linker. After HindIII digestion, the second ligation was carried out with 0.5 pmol of HindIII-digested pGA46, and the products (one sixth of the total) were used for transformation. Colonies were scored on selective agar plates containing Cm alone (20µg /ml) or Cm(20µg/ml) plus Tc(5µg or 20µg/ml).*

Fragment No.	Number of Transformants		
	Cm alone	Cm + Tc (5µg/ml)	Cm + Tc (20µg/ml)
1	~ 2 x 10^4	780	< 3
2	~ 2 x 10^4	759	154
3	~ 2 x 10^4	288	51
4	1 x 10^4	21	< 3
5	0.9 x 10^4	25	< 3
6	1.3 x 10^4	332	139

Though analysis is still in progress, active promotion was so far identified at five different regions: two were in the right to left direction and three in the left to right direction. Promoters in these regions were tentatively named PL1, PL2, PR1, PR2 and PR3, respectively (PL and PR denote leftward and rightward promoters)(Fig.1). PL1 was localized just outside of the left boundary of ori (HaeIII site at position -43 to BglII site at position 22) and PL2 in the region between BstNI site at position 668 and HaeIII site at position 973, respectively. The plasmid constructed by insertion of the region between BglII site at position 38 and PstI site at position 489 into the BglII-PstI sites of pGA46 did not confer any Tc resistance on the host cell. This implies, together with other data, that the ori area does not contain any activity promoting leftward transcription.

The location of PL2 is in the promoter region for the hypothetical proX gene(7). The transcription from pL2 seems to enter the ori stretch reading through this gene, as the leftward transcription was identified with both fragments 2 and 13. The transcription however seems to be weakened or to be terminated during passage through the ori stretch. In Fig.2, the yield of Tc^r transformants at different Tc concentrations has been compared with three plasmids respectively carrying fragments 2, 13 and 14 in the same orientation. Insertion of fragment 13 (BglII site at position 38 to HaeIII site at position 973) does not yield any colonies at 10 µg/ml Tc, in contrast to that of fragment 2 (HindIII site at position 244 to HaeIII site at position 973) which forms Tc^r colonies even at 20 µg/ml Tc. The colony

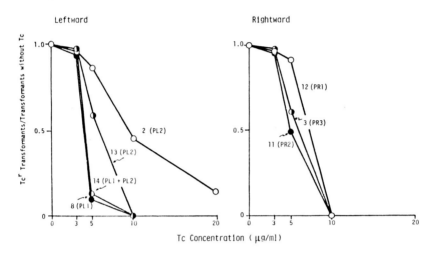

FIG.2 Formation of Tc^r transformants by pGA46 carrying different promoter fragments. Plasmids carrying indicated fragments in the proper orientations were introduced into C600 cells, and Tc^r transformants were scored on agar plates containing different concentrations of Tc. Fragment numbers correspond to those given in FIG.1.

forming ability of fragment 14 which carries both PL1 and PL2 (HaeIII site at position -43 to HaeIII site at position 973) was as weak as that carrying PL1 alone(fragment 8, HaeIII site at position -43 to BglII site at position 22). This implies that the transcription from PL2 is terminated during the passage of ori. However, the formation of Tc^r colonies is assumed to be also a function of the copy number of plasmids in cells, so that the above interpretation should be varified by in vitro transcription experiments.

As mentioned previously, no promotion activity was detected with fragment 4 (BglII site at position 38 to HindIII site at position 244) and fragment 5 (HindIII site at position 244 to HhaI site at position 476). Nevertheless, we could identify rightward transcription with fragment 12 (BglII site at position 38 to HhaI site at position 476) which was prepared from an Ori$^+$ base-substitution mutant in HindIII site at position 244(AAGCTT to AAACTT). The result is interpreted that, unless promotion activity is created by this base-substitution, a promoter(PR1) is located in the vicinity of the HindIII site and the cleavage at this site destroys promoter function. In this region however, the Pribnow box-like sequence seems to occur at only outside of ori(TATACT at position 286). More precise mapping and characterization of this promoter is required.

PR2 and PR3 were identified in the regions between AluI site at position 658 and HaeIII site at position 973 and between

AccII site at position 1218 and AccII site at position 1794, respectively. As *asn* gene has been assigned from positions 1434 to 2423(M.Nakamura et al.: manuscript in preparation), it is likely that at least one of these promoters contributes the transcription of *asn* gene.

To summarize the above results, five promotion sites were identified by using an *in vivo* promoter assay system, and four were mapped outside of *ori* and one in the vicinity of the right boundary of *ori*. Transcription promoted at four sites proceeded in the outward direction without entering *ori*, and only that promoted at the one mapped in the promoter region of *proX* gene entered the *ori* stretch transcribing through this gene region.

In order to eludicate the role of RNA polymerase reaction in the initiation of replication, of course, more detailed analysis on the potential promoters and reactions involved in RNA primer synthesis are required. A series of experiments along this line is in progress, and the result will be reported elsewhere.

This work was supported by research grants from the Ministry of Education, Science and Culture of the Japanese Government.

REFERENCES

1. Yasuda,S., and Hirota,Y., *Proc. Natl. Acad. Sci. U.S.A.*, 74, 5458-5462 (1977).
2. Messer,W., Bergmans,H.E.N., Meijer,M., and Womarck,J.E., *Molec. Gen. Genet.*, 162, 269-275 (1978).
3. von Meyenburg,K., Hansen,F.G., Nielsen,L.D., and Riise,E., *Molec. Gen. Genet.*, 160, 287-295 (1978).
4. Miki,T., Hiraga,S., Nagata,T., and Yura,T., *Proc. Natl. Acad. Sci. U.S.A.*, 75, 5099-5103 (1978).
5. Oka,A., Sugimoto,K., Takanami,M., and Hirota,Y., *Molec. Gen. Genet.*, 178, 9-20 (1980).
6. Sugimoto,K., Oka,A., Sugisaki,H., Takanami,M., Nishimura,A., Yasuda,S., and Hirota,Y., *Proc. Natl. Acad. Sci. U.S.A.*, 76, 575-579 (1979).
7. Hirota,Y., Yamada,M., Nishimura,A., Oka,A., Sugimoto,A., and Takanami,M., *Progress Nucleic Acid Res. Molec. Biol.*, 26, 33-48 (1981).
8. Hirota,Y., Oka,A., Sugimoto,K., and Takanami,M., *this volume*.
9. Tomizawa,J., and Selzer,G., *Ann. Rev. Biochem.*, 48, 999-1034 (1979).
10. Lark,K.G., *J. Molec. Biol.*, 64, 47-60 (1972).
11. von Meyenburg,K., Hansen,F.G., Riise,E., Bergmans,H.E.N., Meijer,M., and Messer,W., *Cold Spring Harbor Symp. Quant. Biol.*, XLIII, 121-128 (1978).
12. An,G., and Friesen,J.D., *J. Bacteriol.*, 140, 400-407 (1979).

TRANSCRIPTION AND TRANSLATION EVENTS IN THE oriC REGION OF THE E. COLI CHROMOSOME

Flemming G. Hansen
Susanne Koefoed
Kaspar von Meyenburg

Department of Microbiology
The Technical University of Denmark
Lyngby-Copenhagen, Denmark

Tove Atlung

University Institute of Microbiology
University of Copenhagen
Copenhagen K., Denmark

ABSTRACT DNA segments from the region of the chromosomal origin of replication, oriC, were cloned in plasmid pBR322. Polypeptides encoded on and expressed from these plasmids were analysed resulting in a precise location of genes in the vicinity of oriC. The DNA segments were inserted into the HindIII site of pBR322 thereby destroying the normal tet gene promotor. Therefore in vivo functioning promotors on the cloned chromosomal DNA could be identified by the restoration of expression of the tet gene resulting in a characteristic level of tetracycline resistance of the host strain. A very active promotor p1 was found between the genes for a 15.5 kD protein and the 37 kD asparagine synthetase approximately 730 bp clockwise of oriC. It yields leftwards transcription across the HindIII site into oriC, where it is effectively terminated. A promotor p3 resulting in leftwards transcription was found just counterclockwise of oriC. It is probably the gid gene promotor which might also contribute messenger for synthesis of a 28 kD polypeptide. None of the promotors appeared to be essential for the function of oriC.

INTRODUCTION

The origin of replication, oriC, of the E. coli chromosome has been thoroughly studied in the last few years (1, 2, 3, 4). The oriC has been sequenced (5, 6), and the minimal stretch of chromosomal DNA sufficient for initiation of DNA replication has been narrowed down to 256 basepairs (7). Minichromosomes encompassing these 256 bp spliced to a DNA fragment carrying the bla gene have properties very similar to those of the chromosomal origin (8).

Initiation of replication on the minichromosome as well as on the chromosome requires an RNA polymerase mediated transcription, as well as de novo protein synthesis. The minichromosomal oriC is dependent on functional dnaA and dnaC products (8, von Meyenburg et al., this volume).

In order to define the transcriptional event required for initiation of replication we have constructed a series of plasmids which allowed a critical test of promotion of transcription in the oriC region. A similar method as described by An and Friesen (9) was used. DNA fragments from the oriC region were spliced into the HindIII site of the cloning vector pBR322 (10). This destroys the normal tet gene promotor in pBR322 by separating its RNA polymerase binding site (-35 sequence) from the Pribnow box. The presence of active promotors on the cloned fragments was thus revealed by the expression of the tet gene yielding tetracycline resistance, assuring that no artificial (hybrid) promoter were formed by the insertion of the new DNA fragments. Improved mapping of the promotors was obtained by constructing deletions in vitro in the cloned DNA segments.

Information was also obtained on the relative strength of the various promotors through the analysis of the degree of the resulting tetracycline resistance of the cells harboring these plasmids.

RESULTS

A family of plasmids containing DNA from the oriC region of the E. coli chromosome cloned in pBR322.

In order to study genes and gene products in the oriC region in more detail, we have constructed plasmids consisting of pBR322 and various DNA segments from the oriC region (Fig. 1). DNA segments from the specialized transducing phage

5 TRANSCRIPTION AND TRANSLATION EVENTS IN THE *ORIC* REGION

λasn105 or the minichromosome pCM960 (8) were inserted in the HindIII site of pBR322 resulting in the plasmids pFH167, pFH321, pFH248, pFH270, pFH268, pFH271, pFH265 and pFH264 or into the HindIII and EcoRI sites resulting in pFH247. In these cases the tet promotor of pBR322 was destroyed (9) while the structural gene for tetracycline resistance tet remained intact. Plasmid pFH350 was constructed by cloning the 3.7 kb BamHI fragment left of oriC into the BamHI site in the tet gene of pBR322 thereby destroying this gene.

Strains carrying these plasmids (Fig. 1) were tested on solid rich medium with various concentrations of tetracycline. They exhibited widely different degrees of tetracycline resistance but in all cases it was lower than for the strain carrying pBR322 (Table 2). The tetracycline resistance of these strains (except the one carrying pFH350) indicated that the tet gene of pBR322 was transcribed from promotors present on the inserted fragments since the normal tet promotor on pBR322 was destroyed. The degree of tetracycline resistance was taken to be indicative of the strength of the promotors on the chromosomal DNA fragments. The 4.4 kb HindIII fragment in pFH167 - located on the chromosome between 3.7 and 8.2 kb-L (Fig. 1) and carrying part of the atp operon (ATP synthase; ref. 4) - showed leftward transcription; also the 1.5 kb HindIII fragment located between 1.95 and 3.5 kb-L - showed leftward transcription (pFH321) but very little if any rightward transcription (pFH248) out of the cloned fragment and into the tet gene. The same was the case for the 2.2 kb HindIII fragment from 0.244 kb-R to 1.95 kb-L on the chromosome (pFH270,pFH268). In agreement with this result, the DNA segment cloned in pFH271 showed leftward transcription. The DNA fragments present in pFH265 and pFH247 - representing 2.3 and 3.3 kb of DNA to the right of the 0.244 kb-R HindIII site - again showed leftward transcription but apparently none in the rightward direction (Fig. 1, Table 1).

Promotors in the 3.3 kb segment to the right of the HindIII site at 0.244 kb-R in oriC.

In order to localize more precisely the promotors in the HindIII fragment cloned in pFH265 deletions were created in the plasmids pFH265, pFH271 and pFH247 (Fig. 1,2) using restriction enzymes BglII, PvuI, ClaI and XhoI-SalI. The deletions in the chromosomal segments of these plasmids are shown in Fig. 2. The degree of tetracycline resistance of cells of strain CM987 carrying the respective plasmids was determined as descibed above (see Table 1) and is indicated in Fig. 2.

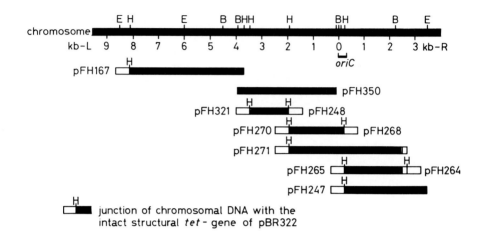

FIGURE 1. Plasmids carrying DNA from the oriC region of the E. coli chromosome. The upper part shows the resriction enzyme cleavage site map of the chromosomal DNA flanking oriC (from ref. 1, 4) as reference to which the chromosomal DNA fragments cloned in pBR322 should be compared to. Standard methods were used for plasmid and phage preparation, preparation of DNA, its cleavage with restriction enzymes and ligation (11) and transformation (12). Extent of chromosomal DNA segments cloned in pBR322 (10) is shown by black bars below; their junction with the tet gene when cloned into the HindIII site of pBR322 is indicated by the adjacent open bars. For example pFH321 and pFH248 carry the same 1.5 kb HindIII fragment, but in opposite orientation. Another HindIII fragment from λasn105 is present in pFH167 and a HindIII-EcoRI and a BamHI from the same source in pFH247 and pFH350, respectively. The entire minichromosome pCM960 (ref. 8) opened at the HindIII site at 1.95 kb-L is present in pFH271 and either of its two HindIII fragments were cloned in pFH270, pFH268, pFH264 and pFH265. For explanation of the hatched portion of DNA in these plasmids see legend to Fig. 2.

5 TRANSCRIPTION AND TRANSLATION EVENTS IN THE *ORIC* REGION

TABLE 1

TETRACYCLINE RESISTANCE OF STRAIN CM987
CONTAINING PLASMIDS CONSTRUCTED BY SPLICING HindIII DNA FRAGMENTS
INTO THE HindIII SITE OF pBR322.(a)

CM987 carrying plasmid	Size of colonies on plates containing the indicated concentration of tetracycline in µg/ml. (b)											"Level of tet resistance" (c)
	0	1	2	4	7	10	15	20	30	40	50	
pFH167	1	1	1	0.3								4
pFH350	1	<0.1										0
pFH321	1	1	1	1	0.3							7
pFH248	1	0.3	<0.1									1
pFH270	1	1	1	0.3	<0.1							4
pFH268	1	0.3	<0.1									1
pFH271	0.7	0.7	0.7	0.7	0.7	0.7	0.7	0.7	0.7	(d)	(d)	30
pFH265	1	1	1	1	1	1	1	1	(d)			20
pFH264	1	0.3	<0.1									1
pFH247	1	1	1	1	1	1	1	1	(d)			20
pBR322	1	1	1	1	1	1	1	1	1	1	1	>50

(a) The plasmids are present in strain CM987: (K-12, F-, asnA31, asnB32, relA1, spoT1, thi-1).

(b) Cells containing the plasmids (pregrown in rich medium, NY) were spotted at different dilutions on tetracycline plates. The diameter of single colonies is normalised to the colony diameter of CM987 (pBR322) cells at 0 µg/ml tetracycline. Values lower than 1 indicate decreased growth rate.

(c) Tet resistance in µg/ml. Corresponding values are given in Fig. 2 and Fig. 3.

(d) Colonies of the same size as on lower tetracycline concentrations appeared on these plates, but we observed a gradual decrease in number towards higher tetracycline concentrations.

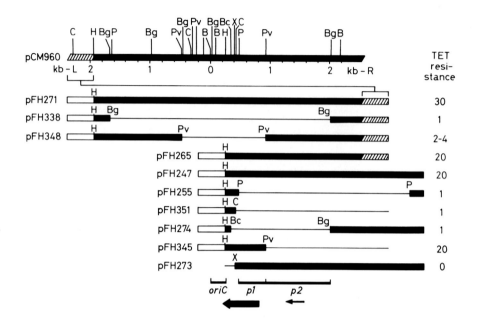

FIGURE 2. Deletion derivatives of plasmid pFH271, pFH265 and pFH247. The restriction map of the minichromosome pCM960 is shown at the top for reference. It consists entirely of chromosomal DNA (ref. 8). pFH271 contains the entire pCM960; however, due to the cloning with HindIII the normal context of the chromosomal DNA in pCM960 is disrupted at the HindIII site at 1.95 kb-L. Therefore, a small segment of chromosomal DNA adjacent to this HindIII site now appears displaced to the right of position 2.6 kb-R in pFH271 and similar derivatives. The restriction sites used to construct the different deletions are indicated: C: ClaI; H: HindIII Bg: BglII; P: PstI; Pv: PvuI; B: BamHI; Bc: BclI; X: XhoI. Characteristic levels of tetracycline resistance of strain CM987 carrying the different plasmids is indicated to the right of the respective plasmid's DNA map (for details see Table 1). p1 and p2 indicate position of promotors with the arrows showing the direction of transcription.

5 TRANSCRIPTION AND TRANSLATION EVENTS IN THE *ORIC* REGION   43

Cells with pFH255, pFH274 or pFH351 were nearly as sensitive to tetracycline as pFH273 which is a XhoI-SalI deletion derivative of pFH247 in which the structural tet-gene is destroyed. This shows that no active leftward promotor is present within the segment from 0.244 kb-R to 0.488 kb-R. (The precise figures given here and in the following sections come from the sequence analysis of the oriC region (Bukh and Messer, unpublished results). Cells with pFH345 showed the same high tetracycline resistance as with pFH247 or pFH265 (Fig. 2). The leftward transcription giving this high tetracycline resistance must therefore start from a promotor p1 to the left of the PvuI site at 934 kb-R used to construct pFH345 and to the right of the PstI site at 0.488 kb-R used to construct pFH255 (Fig. 2). This transcription runs - on the chromosome - into a DNA segment which has been shown to be an essential part of the origin of replication oriC (7).

A second promotor p2 giving leftward transcription - though at a lower intensity than p1 - can be assigned to a segment between the PvuI site at 0.934 kb-R and the BglII site at 2.006 kb-R since (a) the presence of pFH338 resulted in tetracycline resistance though at a low level while pFH338 did not (Fig. 2) and (b) there is no transcription starting between 0.487 (PvuI) and 1.95 kb-L (HindIII) (see next section; Fig. 3). A promotor was also detected on this DNA segment clockwise of oriC which results in rightward transcription. It is most likely the promotor of the asnA gene (Fig.6; ref.4), which, however, shall not be dealt further with in this paper.

Promotors on the 2.2 kb segment from 0.244 kb-R to 1.95 kb-L.

In order to locate the start of transcription giving rise to the tet gene expression in pFH270 (Table 1, Fig. 1) deletions were created both in pFH270 and pFH271 using restriction enzymes BglII, ClaI and BamHI. The structure of the deletion derivatives and the level of tetracycline resistance of cells harboring the different plasmids is shown in Fig. 3. The deletions in pFH337, pFH338, pFH339 and pFH352 all eliminate the expression of the tet gene that is the leftwards transcription which in pFH270 resulted in resistance to 4 ug/ml tetracycline. The deletion in pFH355 did not eliminate this transcription (Fig. 3). Thus, a promotor designated p3 could be located within the coordinates 0.022 kb-R and 0.102 kb-L since this is the only fragment absent from all the tetracycline sensitive deletion derivatives of pFH270 and pFH271 (Fig. 3). These results also showed that there is no active leftward promotor on pFH339 and pFH337 lying between 1.95 (HindIII) and 0.104 kb-L (BamHI) and between 0.022 kb-R (BglII) and 0.244 kb-R (HindIII).

Above we showed that there is a promotor pl giving effective leftwards transcription located between 0.934 and 0.488 kb-R (Fig. 2). This DNA segment is present on the plasmids pFH339 and pFH352 (Fig. 3). The absence of expression of the tet gene on these two plasmids indicates that the leftward transcription starting at pl does not - or only to a minute extent (pFH352) - pass through oriC; thus, this transcription is efficiently terminated within a 152 bp segment between 0.244 (HindIII) and 0.092 kb-R (BamHI) inside oriC. We designate this terminator region tl (Fig. 3).

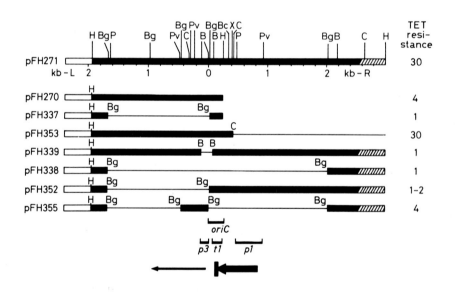

FIGURE 3. Deletion derivatives of plasmid pFH270 and pFH271. The restriction map of the chromosomal DNA in pFH271 is shown for reference. The restriction sites used to construct the different deletion derivatives are indicated as well as the tetracycline resistance characteristics. (For details see legend to Fig. 2). pl, p3: promotor locations; tl: terminator; direction and width of arrows indicates direction and intensity of transcription, respectively.

5 TRANSCRIPTION AND TRANSLATION EVENTS IN THE *ORIC* REGION

Copy number of pBR322 plasmids containing segments of the oriC region.

With these conclusions in mind one would expect the plasmids pFH353 and pFH271 to result in the same degree of tetracycline resistance as pFH270 and pFH355. The fivefold higher tetracycline resistance of the strains with plasmid pFH271 and pFH353 (Fig. 3) was therefore surprising. A major difference between these two sets of plasmids is the presence of an intact chromosomal origin of replication oriC on the former two plasmids while the latter two (like all the others) are missing a more or less extended segment of the 256 bp oriC site (7). In fact, of all the plasmids analysed here only pFH271 and pFH353 yielded upon transformation into W3110polA (E. coli K-12 F , thyA, polA1) ampicillin resistant colonies. This is a sensitive test for the presence of a functional replication origin, in this case oriC, on the pBR322 plasmid (7) since replication from the pBR322 origin depends on a functional DNA polymerase I (10). The correlation between the high level of tetracycline resistance and the oriC+ function of plasmid pFH271 and pFH353 suggested that these plasmids might be present in high copy numbers.

It has been shown that beta-lactamase activity correlated well with the copy number of R-plasmids and derivatives thereof (13). We therefore determined beta-lactamase activity in strain CM987 carrying the pBR322 plasmids containing the various segments from the oriC region (Table 2). Strains harboring pFH271 and pFH353 exhibited a 4-5 fold higher beta-lactamase level than those with the comparable oriC- plasmids (pFH339, pFH352 and others) and a 2-3 fold higher activity than with pBR322 (Table 2). Since pFH271 and pFH339 (Fig. 3) are identical except that the latter plasmid is lacking the two small BamHI fragments in and to the left of oriC (from 0.092 kb-R to 0.104 kb-L) we must conclude that pFH271 has a 4-5 fold higher copy number than pFH339. This must also be the case for pFH353. Thus, the unexpected high level of tetracycline resistance of cells carrying these plasmids can now readily be explained by a gene dosage effect due to the increased copy number of oriC+ plasmids.

This analysis shows that the chromosomal origin of replication is active even when integrated into the high copy number plasmid pBR322.

TABLE 2
beta-LACTAMASE ACTIVITY IN STRAIN CM987
CARRYING DIFFERENT PLASMIDS (a)

plasmid	beta-lactamase % of pBR322	Tet (b) resistance ug/ml
pFH271	200	30
pFH270	40	4
pFH337	50	1
pFH353	300	30
pFH339	50	1
pFH338	60	1
pFH352	50	1 - 2
pFH355	40	4

(a) beta-lactamase activity per unit cell mass was determined as described (13) except that the cells were toluenized which improved the reproducibility of the assay.
(b) Level of tetracycline resistance of cells carrying the respective plasmid (data from Table 1 and Fig. 2 and 3).

Localization of genes in the oriC region.

Through the analysis of specialized transducing phages λasn carrying segments of chromosomal DNA we had localized a number of genes in the oriC region (1, 4). In order to correlate the promotors defined in the present analysis with genes and gene products a more detailed study was made. We determined the gene products coded for by the chromosomal DNA on either side of oriC. To this end we compared the polypeptides synthesized from the chromosomal DNA of the specialized transducing phage λasn132 (1, 4; Fig.4) with the polypeptides expressed from the chromosomal DNA present on the different plasmids constructed here (Fig. 1, 2, 3). λasn132 carries the chromosomal DNA segment between 3.6 kb-R and 9.5 kb-L (4; Fig. 6). Six polypeptides representing six of the eight subunits of the membrane bound ATP synthase are expressed from this DNA (4) and are here used as molecular weight markers (Fig. 4). The respective genes of the atp operon (unc; 4) are located to the left of 4.0 kb-L (Fig. 6). The ATP subunits a, b, c, α and δ are expressed from plasmid pFH167 which contains 4.5 kb of chromosomal DNA left of 4.0 kb-L (Fig. 4).

5 TRANSCRIPTION AND TRANSLATION EVENTS IN THE *ORIC* REGION

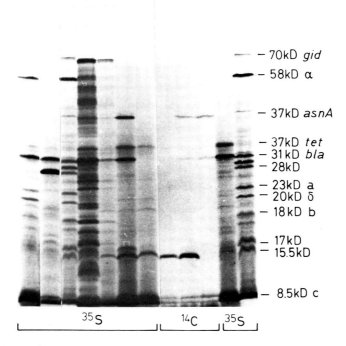

FIGURE 4. Polypeptides expressed from different pBR322 plasmids carrying chromosomal DNA segments from the oriC region (autoradiogram). The indicated plasmids were transferred into the maxicell strain CSR603 (E.coli K-12 recA1, uvrA6, phr-1; ref. 14). After irradiation with UV light of cultures of the resulting strains and incubation to allow preferential degradation of chromosomal DNA (14) proteins expressed from the plasmids were radioactively labeled with 35S-methionine or 14C-amino acid mixture. Samples were prepared and proteins separated by SDS polyacrylamide gel electrophoresis (15 % acrylamide, 0.1 % bisacrylamide) as described earlier (15). Samples containing radioactively labeled proteins expressed from the specialized transducing phage λasn132 (see legend to Fig. 5) carrying chromosomal DNA between 3.8 kb-R and 10 kb-L were included as standard. Identification of protein bands and/or molecular weight is indicated.

A 28 kD polypeptide was expressed from pFH248 besides the
tet, the bla and other (small) gene products specific for
pBR322 (Fig. 4). This 28 kD protein which is also expressed
from λasn132 (Fig. 4) had earlier been designated 25 kD (4);
it is coded for in the segment between 1.95 and 3.5 kb-L (Fig.
1) in accordance with the earlier allocation (4). A more
precise location is given below.

A 69 kD polypeptide is expressed from pFH353, pFH271 and
pFH270 (Fig. 4); it appears to be a fragment of the 70 kD
polypeptide expressed from λasn132 (Fig. 4) and λasn20
(Fig. 5). Since these plasmids are identical with respect to
the junction of chromosomal DNA with pBR322 on the left hand
side we conclude that a very small segment of the 70 kD gene
(gid) is to the left of the HindIII site at 1.95 kb-L
(Fig. 3). The major part of the gene lies to the right of this
site in accordance with the analysis of Tn10 insertion
mutations in this region (4) which are described in detail
below. The numerous bands in the pFH353 and pFH271 samples
indicate degradation of the 69 kD fragment polypeptide.

Three polypeptides with molecular weights of 15.5, 17 and
37 kD were expressed from pFH265 (Fig. 4) and pFH247 (not
shown). They must be encoded by the segment between 0.244 and
2.6 kb-R (Fig. 1). The 37 kD polypeptide had earlier been
assigned to the asnA gene (asparagine synthetase; 4; see
discussion). The 15.5 and 17 kD were weakly labeled when
synthesized in the presence of 35S-methionine; however in the
presence of 14C-amino acids the 15.5 kD protein labeling
increased manyfold (Fig. 4) indicating that this protein only
contains very few methionine residues and that it is actually
expressed rather strongly compared to the other proteins
synthesized from for example pFH265 (Fig. 4). The 15.5 kD
protein was also expressed from pFH345 while, neither the 17
kD nor the 37 kD were (Fig. 4). Therefore, the 15.5 kD protein
must be located between 0.244 (HindIII) and 0.934 kb-R
(PvuI) (Fig. 2). From the nucleotide sequence from this
region (6; Bukh and Messer, unpuplished results) a 15.5 kD
protein containing only one (N-terminal) methionine residue
can be read in the leftward direction, starting at 0.733
kb-R. In accordance with this allocation we found that pFH273
(Fig. 2) did not express the 15.5 kD protein (Fig. 4).

The 17 kD protein was expressed from pFH265 and pFH273
but neither from pFH345 nor from pFH348 (Fig. 4; confer Fig. 2)
Its gene must therefore lie across the PvuI site at 934
kb-R.

Insertion mutagenesis of the oriC region using the
tetracycline transposon Tn10 has been described by von
Meyenburg and Hansen (4). λasn20 which carries the segment
of chromosomal DNA between 6.5 kb-R and 4.4 kb-L (1, 4; see
Fig. 1) was used as target. Expression of proteins from

different λasn20::Tn10 was analysed (Fig. 5). Insertion of Tn10 at 2.08 kb-L abolishes the synthesis of the 28 kD polypeptide (insertion no. 120); insertion of Tn10 at 2.8 kb-L (insertion no. 402, not shown) did not, nor did an insertion at 1.98 kb-L (insertion no.420). The latter, however, eliminates synthesis of the 70 kD protein (gid) giving rise to a new shorter polypeptide of approximately 69 kD. Insertions located to the right of the HindIII site at (1.95 kb-L) also eliminated synthesis of the 70 kD protein and resulted in shorter and shorter polypeptides the further toward oriC they were located (Fig. 5). The 28 kD protein

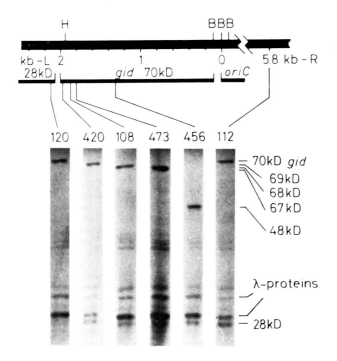

FIGURE 5. Polypeptides expressed from specialized transducing phages λasn20 (1, 4) carrying transposon Tn10 in the chromosomal DNA of the phage (autoradiogram). Insertion mutants of λasn20 (λasn20::Tn10) were isolated as described in ref. 4. Position of the Tn10 insertions (indicated by 3-digit figures) was determined by restriction analysis of the respective phage´s DNA. Polypeptides expressed from the phages were labeled with ^{35}S-methionine after infection of UV irradiated cells of strain 159 (E.coli K-12 uvrA157, galK2, rpsL200, λPaPa+; see ref. 15). Samples were prepared and proteins separated as described earlier (15).

was expressed from these phages with Tn10 in gid (insertion nos. 420, 108, 473, 456), at a decreased level as compared with the expression of the 28 kD protein from λasn20::Tn10 phages where the insertion into chromosomal DNA is outside this region (f.ex. insertion no. 112). Insertion of Tn10 at 1.7 kb-R and 2.3 kb-R resulted in Asn-phenotype and eliminated expression of the 37 kD polypeptide (not shown).

These results lead us to conclude that

(a) the gene for the 28 kD protein is located between 2.08 and 2.8 kb-L,

(b) the gene for the 70 kD polypeptide that is the earlier defined gid gene (4) is located between 0.1 and 2.0 kb-L and is transcribed leftwards, the C-terminal end just reaching across the HindIII site at 1.95 kb-L (in accordance with the results obtained from the analysis of proteins expressed from the different plasmids in the maxicell system (see above; Fig. 4),

(c) the 28 kD protein gene may be cotranscribed with the gid gene since Tn10 insertions in gid reduce 28 kD protein expression (Fig. 5).

DISCUSSION

The information obtained from the study of cells containing intact and defective origins of replication, oriC´s, cloned into pBR322 is summarized in Fig. 6.

Genes in the oriC Region

Genes encoding five polypeptides with molecular weights of 15.5, 17, 28, 37 and 70 kD, respectively, could be precisely located in the vicinity of oriC; these gene products are expressed both from chromosomal DNA fragments cloned into the pBR322 plasmid and from specialized transducing phages λasn. The asnA gene, encoding the 37 kD protein - the asparagine synthetase -, and two genes encoding the 15.5 and 17 kD protein, respectively, are located to the right of oriC (i. e. clockwise of oriC), while the genes for the 70 kD (gid) and the 28 kD protein are located to the left (Fig. 6).

The gid gene is transcribed leftward ; its N-terminal end must be coded for by a nucleotide sequence at 0.1 ± 0.03 kb-L as inferred from the size of the peptide fragments

5 TRANSCRIPTION AND TRANSLATION EVENTS IN THE *ORIC* REGION

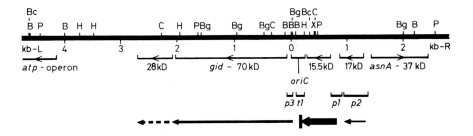

FIGURE 6. Location of genes and transcription signals in the vicinity of oriC. The position of the coding sequences for proteins expressed from the region of the E. coli chromosome are shown. The genes are indicated by the genetic symbol and/or by the size of the gene products. Segments to which promotors (p) or terminators (t) were allocated are indicated by brackets; direction and intensity of transcription from the different promotors is shown by the arrows.

synthesized by the gid::Tn10 mutations (Fig. 5). The 28 kD protein gene is adjacent to the gid gene and most likely cotranscribed with it.

The genes for the 15.5 and 17 kD proteins could be located between 0.244 and 0.934 kb-R and 0.5 and 1.35 kb-R, respectively. Within each of the two segments of chromosomal DNA one coding sequence was found for polypeptides consisting of 147 and 152 amino acid residues, respectively (6), both read in the leftwards direction, the former starting at 0.733 kb-R, the latter at 1.281 kb-R (Fig. 6).

The asnA gene has earlier - on the basis of the analysis of λasn phages (1) - been allocated to the right of position 1.1 kb-R (4); together with the data from the analysis of the Tn10 insertion mutations this shows that its start (or end) is located between 1.1 and 1.7 kb-R. The precise location of the start of its coding sequence can be inferred to be at 1.432 kb-R on the basis of the nucleotide sequence of minichromosomes (6). From the nucleotide sequence of the minichromosome pCM959 (Bukh and Messer, unpublished results) also the precise start of the coding sequence of the gid gene (70 kD) can be located, namely at 0.088 kb-L.

With the exception of the asnA gene the function of of these genes is unknown; neither of them is essential for growth or initiation of replication at oriC as indicated by the analysis of oriC-deletion strains (4).

Transcription signals in the oriC region

Three promotors p1, p2 and p3 have been identified (Fig. 6). All of them give leftwards transcription.

p1 is probably the promotor responsible for transcription of the 15,5 protein gene. It appears to be the most active one. The transcription starting at p1 enters oriC across the HindIII site at 0.244 kb-R and is efficiently terminated within oriC between 0.244 kb-R to 0.092 kb-R. This terminator region is denoted t1 (Fig. 6). Transcription starting at p1 may also be terminated at a second terminator t1* lying between the end of the 15.5 kD gene and the HindIII site at 0.244 kb-R. This indication comes from the comparison of the level of synthesis of the 15.5 kD and the tet protein from f.ex. plasmid pFH345 in which tet is cotranscribed with the 15.5 kD gene (Fig.2). The 15.5 kD protein is apparently synthesized at a considerably higher level than the tet protein (Fig. 5).

p2 may well be the promotor for the weakly expressed 17 kD protein gene.

p3 is probably the promotor of the gid gene. Its position matches well with the location of the N-terminal end of the gid gene (see above). If our conclusion of cotranscription of the 28 kD protein gene with the gid is correct, thus, p3 also contributes mRNA for synthesis of the 28 kD protein.

The location of promotors p1, p2 and p3 coincides with the allocation of promotors B, C2 and E by Morelli and coworkers (16) obtained by means of electron microscopic analysis of RNA polymerase binding to DNA from the oriC region (see also Messer et al., this volume). Their RNA polymerase binding site A near the HindIII site at 0.244 kb-R which on the basis of in vitro transcription experiments has been subdivided into a leftward promotor at 0.180 kb-R and a rightwards one at 0.310 kb-R (Messer et al., this volume) - was not detected in our in vivo search for active promotors. The left portion of promotor A coincides with our location of the terminator region t1 (Fig. 6). Leftward transcription out of this region at a level 4 times lower than the one from p3 would, however, not have been scored positive in our determinations of degrees of tetracycline resistance. With respect to the right portion of promotor A at 0.31 kb-R our results do not allow us to draw any conclusion.

5 TRANSCRIPTION AND TRANSLATION EVENTS IN THE *ORIC* REGION

Is There a Relationship between the Functioning of oriC and the Transription Events in the oriC Region?

Of the promotors defined in this study only p1 bears any direct relationship to oriC. Transcription starting at p1 enters oriC and is efficiently terminated within the oriC segment (Fig. 6). This transcription from p1 into oriC appears, however, not to be required for efficient oriC functioning. Plasmid pFH353 which does not carry p1 (or p2) exhibits the oriC+ phenotype in the polA1 strain as well as plasmid pFH271 (Fig. 3); it also results like pFH271 in high levels of beta-lactamase (Table 2) indicating increased copy number of the oriC-pBR322 chimera. Alternative leftward transcription entering oriC in pFH353 from pBR322 DNA and replacing the transcription from p1 can be excluded since pFH351 in which the 2.2 kb HindIII fragment (1.95 kb-L to 0.244 kb-R) has been deleted shows no tetracycline resistance.

It thus appears that none of the promotors identified in this study in the vicinity of oriC is essential for the functioning of oriC. This is in agreement with the finding that chromosomal DNA clockwise of the XhoI site at 0.417 kb-R is not essential for the functioning of oriC for example in mini-chromosomes pOC5, pOC24 and pOC34 (3, 8). However, in these cases no evidence had so far been presented which would have ruled out the possibility of a replacement transcription coming from the adjacent DNA fragment carrying an ampicillin resistance gene.

We can now conclude that high transcription intensity starting outside oriC is not required for initiation of replication at oriC. Since overwhelming evidence points to an absolute requirement for a transcriptional event (RNA polymerase dependent) in the initiation of chromosome replication we arrive at concluding that this "initiation" transcription is either a low intensity transcription originating outside or inside oriC or a high level transcription originating inside oriC which is also terminated inside oriC. Our results show that there is no high level transcription passing leftward through the BglII site at 0.022 kb-R (pFH337, Fig. 3) and the BamHI site at 0.092 kb-R (pFH339, Fig. 3) or rightward through the HindIII site at 0.244 kb-R (pFH268, Fig. 1, Table 1) which might originate within oriC.

The increased copy number of pBR322 oriC chimeras indicates that oriC is as efficiently used as the pBR322 origin in spite of the high copy number of the vector. We conclude that these two origins on these chimeras are fully compatible and independent of each other. Resulting in an additive effect on the copy number. We also conclude that E. coli has the capability to initiate replication at oriC many times in a cell cycle.

ACKNOWLEDGMENTS

The expert technical assistance of Birthe Jørgensen and Lise Sørensen is gratefully acknowledged. We also thank the staff at the lab for advice and suggestions. This work was supported by a grant from the Danish Natural Science Research Council.

REFERENCES

1. von Mey nburg, K., Hansen, F. G., Nielsen, L. D., and Riise, E. (1978). Molec. Gen. Genet. 160, 287.
2. Yasuda, S., and Hirota, Y. (1977). Proc. Nat. Acad. Sci. 74, 5458.
3. Messer, W., Bergmans, H. E. N., Meijer, M., Womack, J. E., Hansen, F. G., and von Meyenburg, K. (1978). Molec. Gen. Genet. 162, 269.
4. von Meyenburg, K., and Hansen, F. G. (1980). In Mechanistic Studies of DNA replication and Genetic Recombination. ICN-UCLA Symp. Mol. Cell. Biol. 19, 137.
5. Meijer, M., Beck, E., Hansen, F. G., Bergmans, H. E. N., Messer, W., von Meyenburg, K., and Schaller, H. (1979). Proc. Nat. Acad. Sci. 76, 580-584.
6. Sugimoto, K., Oka, A., Sugisaki, H., Takanami, M., Nishimura, A., Yasuda, Y., and Hirota, Y. (1979). Proc. Nat. Acad. Sci. 76, 757.
7. Oka, A., Sugimoto, K., Takanami, M., and Hirota, Y. (1980). Molec. Gen. Genet. 178, 9.
8. von Meyenburg, K., Hansen, F. G., Riise, E., Bergmans, H. E. N., Meijer, M., and Messer, W. (1979). Cold Spring Harbor Symp. Quant. Biol. 43, 121.
9. An, G., and Friesen, J. D. (1979). J. Bacteriol. 140, 400.

10. Bolivar, F., Rodriques, R. L., Greene, P. J., Beclach, M. C., Heynecker, H. L., Boyer, H. W., Cross, J. H., and Falkow, S. (1977). Gene 2, 95.
11. Timmis, K. N., Cohen, S. N., and Cabello, F. C. (1978). Prog. Mol. Subcell. Biol. 6, 1.
12. Mandel, J., and Higa, A. (1970). J. Mol. Biol. 73, 453.
13. Uhlin, B. E., and Nordstrom, K. (1977). Plasmid 1, 1.
14. Sancar, A., Hack, A. M., and Rupp, D. (1979). J. Bacteriol. 137, 692.
15. Hansen, F. G., and von Meyenburg, K. (1979). Mol. gen. genet. 175, 135.
16. Morelli, G., Bukh, H.-J., Fisseau, C., Lother, H., Yoshinaga, K., and Messer, W. (1981). Molec. Gen. Genet. submitted.

GENES, TRANSCRIPTIONAL UNITS AND FUNCTIONAL SITES
IN AND AROUND THE E. COLI REPLICATION ORIGIN

Heinz Lother, Hans-Jörg Buhk, Giovanna Morelli,
Barbara Heimann, Trinad Chakraborty and
Walter Messer

Max-Planck-Institut für molekulare Genetik
Berlin, West-Germany

ABSTRACT

The determination of the nucleotide sequence of the minichromosome pCM959 gave the precise location and size of the coding regions for 4 proteins which are located close to the E. coli replication origin, oriC. Transcriptional units in and close to oriC were analyzed by electron microscopy and by determining the length and the partial sequence of transcripts obtained in vitro. There is a promoter for each of the proteins. Two promoters are located in the center of oriC. They presumably promote the synthesis of RNA primers. Minichromosome replication is dependent on the products of the genes dnaA, dnaC, dnaI, and dnaB252.

INTRODUCTION

Chromosome replication in E. coli proceeds bidirectionally from a fixed point, the origin of replication, oriC, which is located at 83.5 min on the revised E. coli genetic map between gene asnA and the unc gene cluster (1). The analysis of the replication origin and of the initiation process was greatly facilitated by obtaining smaller replicons which contain oriC. F' plasmids containing the origin were isolated (2, 3, 4), oriC was obtained in specialized transducing phages λasn (5, 6, 7) and in single-stranded phages (8), and minichromosomes were constructed in vivo and in vitro which contain oriC as their only replication origin (6, 9 - 14). The replication origin was mapped precisely and its nucleotide sequence was determined (13, 15 - 17).

Mapping of genes and restriction sites in the chromosome segment surrounding the origin of replication (49.7 kb counterclockwise to 20.7 kb clockwise of oriC) has recently been reviewed (18). In this paper we describe the primary structure of the immediate vicinity of oriC (0.68 kb counterclockwise to 3.33 kb clockwise of oriC), transcriptional units within this segment, and functional aspects of origin function.

RESULTS AND DISCUSSION

I. NUCLEOTIDE SEQUENCE OF THE MINICHROMOSOME pCM959

The minichromosome pCM959 was obtained by in vivo recombination from a specialized transducing phage λasn. It contains exclusively E. coli chromosomal DNA, oriC as its replication origin, and asnA as a selectable marker (13, 14). The nucleotide sequence of oriC, and of a DNA segment of 2.2 kb clockwise of oriC had been determined (13, 15).

For a determination of the complete nucleotide sequence, we have first mapped restriction endonuclease recognition sites by analyzing restriction

6 GENES, TRANSCRIPTIONAL UNITS, AND FUNCTIONAL SITES

enzyme generated fragments on agarose or polyacrylamide gels, or by the limit digest technique using 5'-end labelled DNA (19). Suitable DNA fragments were selected and subjected to the chemical modification sequencing technique of Maxam and Gilbert (20, 21). Both strands of the plasmid were sequenced completely in order to eliminate any ambiguity. The complete nucleotide sequence of 4012 bp will be published elsewhere (Buhk and Messer, in preparation). A map of restriction sites in pCM959 is shown in Fig. 1.

In order to determine the point at which ring closure had occurred during the generation of pCM959 we determined the nucleotide sequence of the BglII fragment located at 0.50 to 0.96 kb counterclockwise of oriC (17). The position at which this sequence diverges from the pCM959 sequence is the point at which recombination had occurred. It is located between base pairs 3335 and 3336 of the pCM959 sequence.

The sequence has been searched with the aid of a computer. One of the aspects was the determination of regions coding for proteins. There are 4 proteins coded by pCM959 with molecular weights of 15813, 16880 and 56026, respectively, and the asnA product with a molecular weight of 36662. They will subsequently be referred to as 16 kD, 17 kD, 56 kD proteins, and asnA. Each of the coding regions is preceded by a promoter (see below) and a ribosome binding site (22, 23). The subunit molecular weight of the asnA product has been determined in SDS-polyacrylamide gels to be 39000 (18) which is in close agreement with the molecular weight obtained from the sequence. The 56 kD protein is a fusion product of a 21 kD and a 35 kD partial peptide since its coding region traverses the point of circularization of pCM959, without change of the reading frame. The 21 kD partial peptide may be the beginning of a 70 kD protein designated gid, which was mapped in this region (18). When λ phages, into which different minichromosomes were inserted, were used to infect minicells carrying the λ repressor overproducing plasmid pGY101 (24), proteins were synthesized, which corresponded in their molecular weights to the asnA product, the 16 kD and the 17 kD proteins, and the 70 kD protein. The positions of the coding regions for the different proteins are shown in Fig. 1.

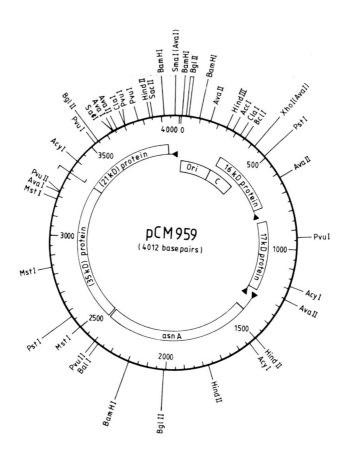

FIGURE 1. *Genetic and physical map of pCM959.*

At the end of each coding region inverted repeats are found which might serve as terminators of transcription (25). In the case of the 17 kD protein this invertedly repeated region is followed by 6 T residues. For two proteins, the asnA protein and the 16 kD protein, an invertedly repeated sequence is also found between the promoter and the initiation codon, which might serve a regulatory role (25).

6 GENES, TRANSCRIPTIONAL UNITS, AND FUNCTIONAL SITES 61

Whether these structures exert the proposed role *in vivo* was not established.

II. TRANSCRIPTIONAL UNITS IN AND AROUND oriC

The analysis of transcriptional units around the replication origin is important because of several reasons: i) We wanted to define the promoters from which mRNAs are transcribed for the proteins in the immediate vicinity of oriC in order to analyze a possible operon structure. ii) The knowledge of transcriptional units around oriC is important for an evaluation of mutations (mainly insertions and deletions) which might affect origin functions not only by eliminating a protein but also by affecting a transcriptional step which might modulate initiation. iii) The initiation of DNA replication in E. coli requires the function of RNA polymerase (26, 27), in addition to the products of several initiation genes. Therefore, also minichromosome replication requires functional RNA polymerase (14). There are several possibilities how RNA polymerase could act in the initiation of replication. a) The enzyme might synthesize a primer, as has been shown for colEI (28, 29, 30) and for complementary strand synthesis of M13 and fd (31, 32). b) RNA polymerase could be responsible for transcriptional activation of the origin as has been suggested for λ (33). c) An RNA transcript might serve a regulatory role or a structural role other than primer function in the initiation process.

Because of these reasons we have analyzed the transcriptional units in pCM959 using several complementary techniques.

A. *Mapping of RNA Polymerase Binding Sites by Electron Microscopy*

First a survey for the positions of promoters was done by mapping RNA polymerase binding sites with electron microscopic techniques. Conditions for RNA polymerase binding and preparation of the complexes were as described for phage ϕ29 (34). pCM959 DNA was cleaved at the unique SmaI or XhoI recognition sites either before or after incubation with RNA polymerase. Purified E. coli RNA polymerase was bound to pCM959 DNA at a molar ratio of 7 for covalently closed circular DNA or at a ratio of 14 for linear DNA in 30 mM triethanolamine-HCl buffer, pH 7.9, 50 mM KCl, 8 mM Mg-acetate. DNA-protein complexes were fixed with 0.1 % glutaraldehyde, free enzyme was removed by gel filtration, and the complexes were processed for electron microscopy.

Full length molecules containing 3 or more polymerase molecules bound were selected from the electron micrographs. The contour lengths of the molecules and the positions of RNA polymerase were measured, and the molecules were then oriented visually. The result of an experiment in which supercoiled pCM959 was incubated with RNA polymerase and subsequently cleaved with SmaI is shown in Fig. 2. The cumulative results of several experiments with different fragments are presented in Table I.

Table I. Positions of RNA polymerase in the analyzed molecules[a]

	A	B	C	D	E
position (bp)[b]	267	791	1344	2786	3991
±standard deviation	61	55	70	62	68
%polymerase molecules	11.5	20.0	24.7	14.0	13.7

[a]Total number of molecules analyzed: 454. Total number of polymerase molecules: 1549
[b]Position in the 4012 bp sequence, see Fig. 1.

Five RNA polymerase binding sites are found in pCM959 DNA by electron microscopic mapping. We designate them A - E following the pCM959 map. These positions in the pCM959 sequence are A: 267 ± 61 bp, B: 791 ± 55 bp, C: 1344 ± 70 bp, D: 2786 ± 62 bp, E: 3991 ± 68 bp (Table I).

6 GENES, TRANSCRIPTIONAL UNITS, AND FUNCTIONAL SITES

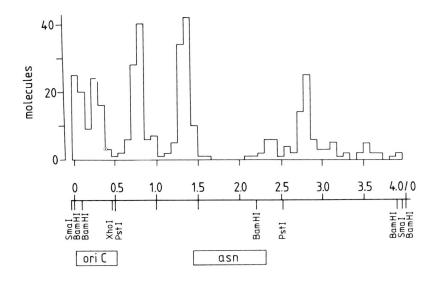

FIGURE 2. Positions of RNA polymerase molecules in pCM959. Supercoiled pCM959 DNA was incubated with RNA polymerase and, after fixation, cleaved with SmaI. The positions of RNA polymerase molecules are shown as measured from electron micrographs.

B. Promoters as Obtained from the Nucleotide Sequence

The nucleotide sequence of pCM959 (see section I) was searched with the help of a computer for regions which are similar to the model sequence for promoters (reviewed in 25, 35). An additional criterion was that promoters from which transcription of mRNA starts should be followed by a sequence which is complementary to the 3'-end of 16 S rRNA (22, 23), and by a coding sequence for protein. Using these criteria we identified the following sequences as potential promoters; they are aligned at the Pribnow box, the direction of transcription can be read from the nucleotide positions, and the letters refer to the binding sites determined by electron microscopy (see section IIA):

Model sequence:

```
          5'TGTTGACAATTT---------------TATAAT--------CAT
              7 9 3                    7 5 6
                ↓                        ↓
      B:  5'CAGGTAGATCCCAACGCGTTCACAGCGTACAATACGCCACTCTTA
              1 3 0 6                          1 3 4 3
                ↓                                ↓
      C₁: 5'ATGATTCATTCTATTTTAGCCTTCTTTTTTAATGAATCAAAAGTG
              1 3 4 5                          1 3 0 4
                ↓                                ↓
      C₂: 5'CACTTTTGATTCATTAAAAAAGAAGGCTAAAATAGAATGAATCAT
              3 9 9 0                          3 9 4 9
                ↓                                ↓
      E:  5'AAGTTGAGTAGAATCCACGGCCCGGGCTTCAATCCATTTTCATAC
              4 0 1 0                          3 9 7 2
                ↓                                ↓
      or: 5'GGTCTTTCTCAAGCCGACAAAGTTGAGTAGAATCCACGGCCCGGG
```

For binding sites B, C and E there was an excellent agreement between the positions found through search of the sequence and the positions at which RNA polymerase was mapped electron microscopically. In the region of binding site C two promoters are found by sequence analysis, pointing in opposite directions. We can thus assign promoter B to the 16 kD protein, promoter C_2 to the 17 kD protein, promoter C_1 to the asnA gene and promoter E to the 56 kD fusion protein. These promoters are indicated in Fig. 1 by triangles pointing in the direction of transcription. All proteins coded for by pCM959 have their own promoter.

Around binding site D several overlapping Pribnow box sequences can be found, however, no good homology exists at the corresponding -35 regions:

```
                              2 8 0 8      2 7 9 3
                                ↓            ↓
    D: 5'ATTGCTCTCGCAGAAAACCGGCGCTGCTATATTATGCTATTTTCCACCGAGAT
```

Binding site D is also located within the coding region of the 35 kD partial peptide. Therefore we conclude that this site does not represent a promoter which is functional in vivo.

C. In vitro Transcription from pCM959 Promoters

RNA polymerase binding site A suggests the existence of a promoter within the replication origin. For a precise analysis of transcription starting within oriC, and for an unambiguous definition of the promoters whose transcripts code for proteins, we have analyzed transcripts obtained in vitro from various restriction fragments of pCM959.

Restriction fragments obtained from pCM959 were separated by polyacrylamide gel electrophoresis, eluted from the gel and purified as described (21). These fragments were used as templates in in vitro transcription with purified E. coli RNA polymerase (36 - 38).

The sizes of the transcripts were analyzed by electrophoresis in polyacrylamide gels containing 7 M urea. $5'-^{32}PO_4$ end labelled restriction fragments were used as size standards.

Fig. 3 shows the result of such an analysis for oriC and the two promoters next to oriC, and also demonstrates the principle of the technique. As an example with the HaeIII fragment (positions 3970 to 635) as template transcripts with lengths of 225 and 340 nucleotides, respectively, are obtained. When this fragment is cleaved with HhaI and used as template, the 225 nucleotide transcript is still present, but instead of the 340 nucleotide transcript a shorter one of 175 nucleotides is found. Using different fragments as templates, the positions and the directions of transcription of the different promoters can be unambiguously determined.

In this analysis, the positions of the promoters for the two proteins (16 kD and 21 kD partial proteins) have been obtained more precisely. They corroborate the results obtained by electron microscopy, and the direction of transcription agrees with the one inferred from the pCM959 nucleotide sequence (see Table II for a summary of the results obtained with the different techniques).

Interestingly, two promoters are found within oriC, located approximately in the center of the origin sequence required for bidirectional replication (see

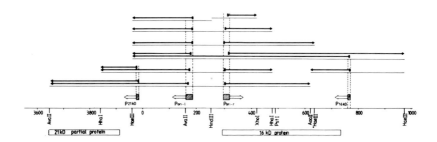

FIGURE 3. *In vitro transcripts obtained from the oriC segment. RNA transcripts (thick lines) were obtained in vitro with various restriction fragments (thin lines) as templates. Their length was determined, and they were arranged by aligning their ends to the ends of the templates. The HaeIII site at position 685 (*HaeIII) is not present in pCM959. This fragment was obtained from a similar plasmid.*

section III), and promoting transcription outward from the center of oriC. We designate them P ori-l and P ori-r.

A similar analysis was performed with restriction fragments which contain the promoter of the 17 kD protein and the asnA promoter. The promoter of the 17 kD protein was located at position 1318 ± 10 bp with counterclockwise transcription, the asnA promoter at position 1350 ± 10 bp promoting clockwise transcription. Both results are in excellent agreement with the electron microscopic mapping and the assignments obtained from the nucleotide sequence (see also Table II).

D. *Nucleotide Sequence of Transcripts Obtained in vitro*

In order to define the start of the transcripts, expecially those from P ori-l and P ori-r, precisely at the nucleotide level, we determined the nucleo-

6 GENES, TRANSCRIPTIONAL UNITS, AND FUNCTIONAL SITES

tide sequence of their 5'-terminal portion. RNA transcripts were obtained with restriction fragments as templates as described in section IIc. They were separated according to size by urea-polyacrylamide gel electrophoresis, and extracted from the gel (21). Their 5'-ends were labelled with $^{32}PO_4$ (21), and they were subjected to the enzymatic RNA sequencing procedure (39 - 41).

The nucleotide sequences of the transcripts were compared to and aligned with the nucleotide sequence of oriC and pCM959, respectively. The 5'-end of the P ori-r transcript is nucleotide 313, the 5'-end of the P ori-1 transcript is nucleotide 167 (and 168). See sections III and IV for a discussion of the implications of these results.

TABLE II. Positions of promoters in pCM959

Promoters	Positions[a] and direction of transcription obtained by			
	Electron Microscopy	Sequence analysis[b]	in vitro transcripts	
			Length[c]	Sequence[c]
P ori-1	267±61	—	162±11 ccw[d]	167 ccw
P ori-r		—	309±17 cw	313 cw
P 16 kD	791±55	763 ccw	755± 7 ccw	
P 17 kD	1344±70	1326 ccw	1318±10 ccw	
P asnA		1336 cw	1350±10 cw	
P (21 kD)	3991±68	3978 ccw	3994± 6 ccw	

[a] Positions in the 4012 bp nucleotide sequence, see Fig. 1.
[b] The position of the invariant T in the Pribnow box is given.
[c] Position of the 5'-end of the transcript.
[d] Counterclockwise (ccw) or clockwise (cw) direction of transcription

In Table II the positions of the promoters in pCM959 and the directions of transcription are summarized, as obtained with the different techniques. Within their limits of resolution the different methods give identical results. With the results reported here, not only the nucleotide sequence, but also the transcriptional units in and around oriC are known. The promoters defined here for the proteins around oriC are in excellent agreement with results obtained from the cloning of promoter containing fragments and their analysis in vivo (Hansen et al., this volume; Zyskind et al., this volume).

III. STRUCTURE OF THE REPLICATION ORIGIN

A. Definition of oriC

The replication origin of E. coli, oriC, has originally been defined as a segment of 422 base pairs (position 0 - 422 on pCM959) as the then smallest fragment which was able to promote minichromosome replication (13, 15, 17). The isolation and analysis of minichromosomes from Salmonella typhimurium suggested that the minimal size of oriC might be smaller (296 base pairs, Fig. 4, 42, 43). Subsequently the minimal DNA segment required for minichromosome establishment and replication was determined to extend from the BglII site at position 22, or the BglII site at position 38 to position 266 or 267, i.e. 232 or 245 base pairs (Fig. 4, 44). It has been shown that minichromosomes in which DNA clockwise of the XhoI site at position 417 was deleted replicated predominantly unidirectionally counterclockwise, and that for efficient bidirectional replication sequences to the right of, but close to, the XhoI site were required (45, 46). A DNA binding protein isolated from the membrane of E. coli (47) binds specifically to two sites in oriC, one of them located in the segment between the XhoI site (position 417) and the PstI site at position 488 (Fig. 4, 48). Probably it is the loss of this binding site which is responsible for unidirectional replication. We, therefore, interprete that oriC, the origin required for bidirectional replication, extends from the BglII site at position 22 or 38 to the PstI site at position 488.

6 GENES, TRANSCRIPTIONAL UNITS, AND FUNCTIONAL SITES   69

B. Functional Sites in oriC

So far we have localized two types of functional sites in oriC. The membrane-derived DNA binding protein B' recognizes specifically single stranded DNA from oriC. One recognition sequence is located around position 100 on the strand reading 3'-5' in the orientation of the pCM959 map (which is also the orientation of the E. coli genetic map, 1), the second recognition site is between the XhoI site (position 417) and the PstI site (position 488) on the DNA strand of opposite polarity (Fig. 4, 48).

The two origin promoters represent the second type of functional sites whose location is known within oriC. Both types of sites seem to indicate a remarkable degree of symmetry within oriC (Fig. 4). The center of this symmetry coincides approximately with the boundary of the minimal size oriC segment (Fig. 4). However, we do not want to suggest that all sites required for initiation are repeated in both halves of oriC.

Invertedly repeated sequences may be an indication for recognition sites of as yet unknown functions. Two sets of invertedly repeated sequences are found within oriC (Fig. 4). The center of symmetry of the first set coincides with the symmetry center defined by the promoters and protein B' recognition sites. The second set of inverted repeats is, however, almost completely contained in the left half of oriC (Fig. 4). In fact, so far only the isolated left half of oriC was shown to allow the replication of minichromosomes (16, 17, 44), whereas no minichromosomes were isolated in which the left part of oriC was deleted.

IV. ASPECTS OF ORIGIN FUNCTION

A. Role of RNA Polymerase

Within oriC sequences have been localized which show some homology to the complementary strand origin of phage G4 (13, 15). This led to the idea that the RNA primer for DNA replication might be synthesized by primase, the dnaG product (16, 17).

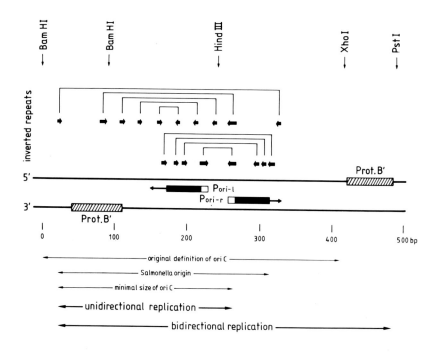

FIGURE 4. Structure of oriC.
Origin promoters are indicated by filled bars.
The transcripts start at the bases of the
arrows. Possible regulatory sites upstream
of the promoters are shown as open bars,
Corresponding invertedly repeated sequences
are connected by brackets.

6 GENES, TRANSCRIPTIONAL UNITS, AND FUNCTIONAL SITES

The function of RNA polymerase, in this case, would be transcriptional activation, presumably from a promoter outside oriC, as has been suggested for λ (33, 49).

However, transcriptional activation of oriC from an outside promoter is unlikely to be essential for initiation, although a modulating effect cannot be ruled out. The only promoter which could serve this function is the promoter of the 16 kD protein (see Fig. 1). Numerous derivatives of minichromosomes have been constructed in which the DNA segment containing P 16 kD was replaced by other sequences (9, 10, 12, 16, 17, 44).

The promoters within oriC are ideally located in order to promote the transcription of an RNA which is used as a primer. Transitions from RNA to DNA, i.e. inception points, were reported to occur in the left part of oriC at positions 71 and 108, respectively (50). If these are connected with the transcript from P ori-1, RNA primers of 103 and 66 nucleotides in length would result. Apparently, additional evidence is required to prove this concept and to locate an inceptor in the right half of oriC.

B. *Possible Control at Origin Promoters*

When the nucleotide sequences of P ori-1 and P ori-r are compared an extended sequence homology is observed around position -50 (Fig. 5). This sequence homology is also evident when E. coli origin promoters are compared to the corresponding segments in the Salmonella typhimurium origin (Fig. 5, 43). This region of homology is at the approximate position which is important for positive control in the promoters of the lac and gal operons (25). An attractive hypothesis is, therefore, that origin promoters are subject to positive control. A candidate for a protein exerting such a control is the dnaA product, for which a positive control function has been suggested (51 - 53) and which interacts with RNA polymerase (54).

```
                    -50                    -35                   -10       +1
                     ↓                      ↓                     ↓        ↓
1:  AAGCTTCCTGACAG AGTTATCCACAGTAGATCGCACGATCTGTATACTTATTTGAGTAAA  TTAAACCACGATCCCAGCCATTCTTCTGCCGGATCTTCCGGAAT
                   •  • •  ••••••••• ••  • •    • ••  ••  • • ••• •••        •   •••••••••  • •  • ••••• •••
2:  AAGCTTGGATCAACCGGTAGTTATCCAAAGAACAACTGTTGTTCAGTTTT TGAGTTGTGTATAACCCCTCATTCTGATCCCAGCTTATACGGTCCAGGATCA CCGATCA
                   •  • ••  ••••••• •        ••• • ••••    •   ••           •••••  •••     •   • •••• •••
3:  AAGCTTTCCACCAG ATTTATCCACAATGGATCGCACGATCTTTTATACTTATTTGAGTAAA  TTAATCCAGGATCCGAGCCAAATCTCCGCTGGATCTTCCGGAAT
                   •  ••• •  ••••••••• •   ••• • •  • •• ••• ••• •••        •••••  •••  •  •   •  ••••  •••
4:  AAGCTTGGATCAACCGGTAGTTATCCAAAGAATAACCGTTGTTCACTTTT TGAGTTGTGTATAAGTACCCGTTTTGATCCCAGCTTATACGGACCACGATCA CCGATCA
```

FIGURE 5. Comparison of origin promoters of E. coli and S. typhimurium.
1: E. coli P ori-r, 2: E. coli P ori-l,
3: S. typhimurium P ori-r,
4: S. typhimurium P ori-l.
Numbers refer to nucleotide positions relative to the start of the transcript in P ori-r. In this scale transcription at P ori-l starts at position +8.

Sequence homologies are also observed close to the start of the transcripts. Within each promoter these regions of homology show dyad symmetry (Fig. 5). It is possible that these regions represent operator-like structures (55), and that, therefore, origin promoters are also subject to negative control. Negative control of replication has been suggested before (56, 57).

C. Effects of Initiation Mutants on Minichromosome Replication

It has been shown that the replication of the minichromosome pOC2 depends not only on functional RNA polymerase and continuing protein synthesis but also on two of the gene products (dnaA and dnaC), which are required for the initiation of replication of the chromosome (14). Here we extend this investigation by analyzing the replication of minichromosomes in additional mutants. We have also chosen different minichromosomes containing different amounts of chromosomal DNA in order to exclude that smaller minichromosomes might be less stringent in their requirements.

The minichromosomes pOC2, pOC12, pOC24, and pOC34 were used in this experiment. They contain the chromosomal sequences given in brackets (O is, as in pCM959, the left end of oriC), in addition to am ampicillin resistance gene for selection (18): pOC2 (-6 kb to +3.5 kb), pOC12 (-6 kb to 4.1 kb, -1.5 kb to +0.5 kb, +3.4 kb to +3.5 kb), pOC24 (-2.3 kb to + 0.4 kb), and pOC34 (-3.7 kb to -3.3 kb and 0 to 0.4 kb). Strains containing the mutants dnaA204, dnaA5, dnaC201, dnaI208, or dnaB252, in addition to the recA mutation, were transformed with these plasmids. They were grown at 30° in minimal-glucose medium, and labelled with a 20 min pulse of ^{3}H-thymidine at 30° and at different times after a shift to 42°. The rate of chromosomal DNA synthesis was determined as acid precipitable radioactivity. ^{3}H-thymidine incorporated into plasmid DNA was measured with the DNA-DNA hybridization technique (58). The results are summarized in Table III.

In all mutants minichromosome replication stopped upon a shift to 42°. There seems to be a tendency that smaller minichromosomes (pOC24, pOC34) stop replication more slowly at the nonpermissive temperature, which is at present not explained. All four minichromosomes showed substantial replication in the dnaB252 mutant during the time period 10 - 30 min at 42°, but replication eventually stopped at 42°. Despite these more subtle differences it is evident that irrespective of their size, all minichromosomes depended on the gene products of all mutants tested for their initiation, i.e. the dnaA, dnaC, dnaI, and dnaB252 gene products.

ACKNOWLEDGMENTS

The expert technical assistance of Laura Blumenthal and Mike Hearne is gratefully acknowledged. This work was in part supported by grant No. Me 659/1 of the Deutsche Forschungsgemeinschaft

Table III

mutant	plasmid	relative rate of synthesis[a]			
		10–30 min at 42°		40–60 min at 42°	
		chromosome	plasmid	chromosome	plasmid
dnaA204	pOC2	0.60	0.08	0.01	< 0.01
	pOC12		0.06		0.03
	pOC24		0.22		0.04
	pOC34		0.24		< 0.01
dnaA5	pOC2	0.60	0.05	0.02	< 0.01
	pOC12		0.23		0.03
	pOC24		0.22		0.04
	pOC34		0.10		< 0.01
dnaC2	pOC2	0.42	0.08	0.01	0.03
	pOC12		0.05		0.02
	pOC24		0.20		0.01
	pOC34		0.17		< 0.01
dnaI208	pOC2		–		0.10
	pOC12	0.99	0.11	0.06	0.05
	pOC24		0.25		0.15
	pOC34		0.21		0.03
dnaB252	pOC2	0.63	0.46	0.04	0.08
	pOC12		0.40		0.10
	pOC24		0.24		0.09
	pOC34		0.34		0.03

[a] Rate of synthesis at 30° = 1

REFERENCES

1. Bachman, B. J., and Low, K. B., *Microbiol. Rev. 44*, 1 (1980).
2. Masters, M., *Mol. Gen. Genet. 143*, 105 (1975).
3. von Meyenburg, K., Hansen, F. G., Nielsen, L.D., and Jørgensen, P., *Mol. Gen. Genet. 158*, 101 (1977).
4. Hiraga, S., *Proc. Nat. Acad. Sci. USA 73*, 198 (1976).
5. von Meyenburg, K., Hansen, F. G., Nielsen,L.D., and Riise, E., *Mol. Gen. Genet. 160*, 287 (1978).
6. Miki, T., Hiraga, S., Nagata, T., and Yura, T., *Proc. Nat. Acad. Sci. USA 75*, 5099 (1978).
7. Soll, L., *Mol. Gen. Genet. 178*, 381 (1980).
8. Kaguni, J., La Verne, L. S., and Ray, D. S., *Proc. Nat. Acad. Sci. USA 76*, 6250 (1979).
9. Yasuda, S., and Hirota, Y., *Proc. Nat. Acad. Sci. USA 74*, 5458 (1977).
10. Messer, W., Bergmans, H.E.N., Meijer, M., Womack, J. E., Hansen, F. G., and von Meyenburg, K., *Mol. Gen. Genet. 162*, 269 (1978).
11. Diaz, R., and Pritchard, R. H., *Nature 275*, 561 (1978).
12. Leonard, A. C., Weinberger, M., Munson, B. R., and Helmstetter, C. E., *ICN-UCLA Symp. Molec. Cell. Biol. 19*, 171 (1980).
13. Meijer, M., Beck, E., Hansen, F. G., Bergmans, H.E.N., Messer, W., von Meyenburg, K., and Schaller, H., *Proc. Natl. Acad. Sci. USA 76*, 580 (1979).
14. von Meyenburg, K., Hansen, F. G., Riise, E., Bergmans, H.E.N., Meijer, M., and Messer, W., *Cold Spring Harbor Symp. Quant. Biol. 43*, 121, (1979).
15. Sugimoto, K., Oka, A., Sugisaki, H., Takanami, M., Nishimura, A., Yasuda, Y., and Hirota, Y., *Proc. Natl. Acad. Sci. USA 76*, 575 (1979).
16. Hirota, Y., Yasuda, S., Yamada, M., Nishimura, A., Sugimoto, K., Sugisaki, H., Oka, A., and Takanami, M., *Cold Spring Harbor Symp. Quant. Biol. 43*, 129 (1979).
17. Messer, W., Meijer, M., Bergmans, H.E.N., Hansen, F. G., von Meyenburg, K., Beck, E., and Schaller, H., *Cold Spring Harbor Symp. Quant. Biol. 43*, 139 (1979).

18. von Meyenburg, K., and Hansen, F. G., *ICN-UCLA Symp. Molec. Cell. Biol. 19*, 137 (1980).
19. Smith, H. O., and Birnstiel, M. L., *Nucl. Acids Res. 3*, 2387 (1976).
20. Maxam, A. M., and Gilbert, W., *Proc. Natl. Acad. Sci. USA 74*, 560 (1977).
21. Maxam, A. M., and Gilbert, W., *Meth. Enzymol. 65*, 499 (1980).
22. Shine, J., and Dalgarno, L., *Proc. Natl. Acad. Sci. USA 71*, 1342 (1974).
23. Steitz, J. A., *in* "Ribosomes, Structure, Function and Genetics" (G. Chambliss et al., ed.), University Park Press (1980).
24. Reeve, J. N., *Nature 276*, 728 (1978).
25. Rosenberg, M., and Court, D., *Ann. Rev. Genet. 13*, 319 (1979).
26. Lark, K. G., *J. Mol. Biol. 64*, 47 (1972).
27. Messer, W., *J. Bacteriol. 112*, 7 (1972).
28. Staudenbauer, W. L., *Mol. Gen. Genet. 145*, 273 (1976).
29. Itoh, T., and Tomizawa, J., *Cold Spring Harbor Symp. Quant. Biol. 43*, 409 (1978).
30. Itoh, T., and Tomizawa, *Proc. Natl. Acad. Sci. USA 77*, 2450 (1980).
31. Tabak, H. F., Griffith, J., Geider, K., Schaller, H., and Kornberg, A., *J. Biol. Chem. 249*, 3049 (1974).
32. Geider, K., Beck, E., and Schaller, H., *Proc. Natl. Acad. Sci. USA 75*, 645 (1978).
33. Dove, W. F., Inokuchi, H., and Stevens, F., *in* "The Bacteriophage lambda (A. D. Hershey, ed.), p. 747. Cold Spring Harbor Laboratory, Cold Spring Harbor, New York.
34. Sogo, J. M., Inciarte, M. R., Corral, J., Vinuela, E., and Salas, M., *J. Mol. Biol. 127*, 411 (1979).
35. Siebenlist, U., Simpson, R. B., and Gilbert, W., *Cell 20*, 269 (1980).
36. Otsuka, A., and Abelson, J., *Nature 276*, 689 (1978).
37. Meyer, B. J., Kleid, D. G., and Ptashne, M., *Proc. Natl. Acad. Sci. USA 72*, 4785 (1975).
38. Maniatis, T., Jeffrey, A., and Van de Sande, H., *Biochemistry 14*, 3787 (1975).
39. Simoncsits, A., Brownlee, G. G., Brown, R. S., Rubin, J. R., and Guilley, H. G., *Nature 269*, 833 (1977).

40. Donis-Keller, H., Maxam, A. M., and Gilbert, W., *Nucl. Acids Res. 4*, 2527 (1977).
41. Krupp, G., and Gross, H. J., *Nucl. Acids Res. 6*, 3481 (1979).
42. Zyskind, J. W., Deen, L. T., and Smith, D. W., *Proc. Natl. Adad. Sci. USA 76*, 3097 (1979).
43. Zyskind, J. W., and Smith, D. W., *Proc. Natl., Acad. Sci., USA 77*, 2460 (1980).
44. Oka, A., Sugimoto, K., Takanami, M., and Hirota, Y., *Mol. Gen. Genet. 178*, 9 (1980).
45. Meijer, M., and Messer, W., *J. Bacteriol. 143*, 1049 (1980).
46. Messer, W., Heimann, B., Meijer, M., and Hall, S., *ICN-UCLA Symp. Molec. Cell Biol. 19*, 161 (1980).
47. Jacq, A., and Kohiyama, M., *Eur. J. Biochem. 105*, 25 (1980).
48. Jacq, A., Lother, H., Messer, W., and Kohiyama, M., *ICN-UCLA Symp. Molec. Cell. Biol. 19*, 189 (1980).
49. Hobom, G., Grosschedl, R., Lusky, M., Scherer, G., Schwarz, E., and Kössel, H., *Cold Spring Harbor Symp. Quant. Biol. 43*, 165 (1979).
50. Okazaki, T., Hirose, S., Fujiyama, A., and Kohara, Y., *ICN-UCLA Symp. Molec. Biol. 19*, 429 (1980).
51. Zahn, G., and Messer, W., *Mol. Gen. Genet. 168*, 197 (1979).
52. Hansen, F. G., and Rasmussen, K. V., *Mol. Gen. Genet. 155*, 219 (1977).
53. Lycett, G. W., Orr, E., and Pritchard, R. H., *Mol. Gen. Genet. 178*, 329 (1980).
54. Bagdasarian, M. M., Izakowska, M., and Bagdasarian, M., *J. Bacteriol. 130*, 577 (1977).
55. Miller, J. H., and Reznikoff, W. S., "The Operon", Cold Spring Harbor Laboratory, Cold Spring Harbor, New York.
56. Pritchard, R. H., Barth, P. T., and Collins, J., *Symp. Soc. Gen. Microbiol. 19*, 263 (1969).
57. Tippe-Schindler, R., Zahn, G., and Messer, W., *Mol. Gen. Genet. 168*, 185 (1979).
58. Frey, J., Chandler, M., and Caro, L., *Mol. Gen. Genet. 174*, 117 (1979).

SPECIFIC BINDING OF ESCHERICHIA COLI
REPLICATIVE ORIGIN DNA TO MEMBRANE PREPARATIONS[1]

W. Hendrickson
H. Yamaki
J. Murchie
M. King
D. Boyd
M. Schaechter

Department of Molecular Biology and Microbiology
Tufts University School of Medicine
Boston, Massachusetts 02111

I. ABSTRACT

We have studied the binding of the replicative origin of Escherichia coli to membrane preparations. Origin DNA was fractionated by sucrose gradient centrifugation of a French pressure cell lysate. Several aspects of the binding kinetics were investigated, which have led to the development of a filter binding assay.

Nonspecific binding has been eliminated by competing calf thymus DNA. A 460 base pair region containing the origin of replication is required for binding of DNA restriction fragments to outer membrane preparations. This specific interaction is greatly enhanced by two proteins which remain bound to origin DNA after CsCl gradient purification.

II. INTRODUCTION

The attachment of the bacterial chromosome to the E. coli membrane may explain chromosome segregation into progeny cells and, more remotely, regulation of initiation of DNA synthesis. However, in the eighteen years since the

[1]This work was supported by grant No. AI09465 - UHPHS.

formulation of the replicon model, these are little more than alluring possibilities. Many reports have been published on the isolation of DNA-membrane complexes. Paradoxically, while these experiments have been repeated many times, they do not provide compelling evidence for membrane-DNA association _in vivo_. In our view, the data accumulated suggests only that bacterial DNA is _likely_ to be membrane associated. The involvement of the membrane in regulation is still an open question (1). In part this is due to the low degree of specificity reported for this association. Frequently the _replicative origin_ is reported to be attached to the membrane, but again, the degree of enrichment is disappointingly low (2).

Most of this work has dealt with the isolation of DNA-membrane fractions from disrupted cells. Recently, we have undertaken a different approach, that of reconstructing complexes from isolated components. While such an _in vitro_ reconstruction is even further removed from the intracellular condition, it allows us to purify each component individually and to eliminate many nonspecific interactions. We have now been able to develop methods that result in a degree of specificity not previously attained. Armed with detailed knowledge of specific DNA-membrane interactions _in vitro_ we should eventually be able to attack the problem of _in vivo_ binding in a much more deliberate fashion.

We have found that using a simple technique of cell disruption (French pressure cell) and fractionation on sucrose gradients in the presence of EDTA, we can recover a complex of origin containing DNA and certain proteins (3). This material, which we call "origin complex," is associated specifically at the _Eco_ Rl restriction fragment that contains the replicative origin. We have shown that this complex will bind to outer membrane preparations in the presence of magnesium. The complex can be further purified by centrifugation through cesium chloride gradients because the relevant proteins have the remarkable property of remaining bound to the DNA through this treatment (4). We report here further studies on the kinetics and specificity of binding of this "CsCl-complex" to membranes.

III. DEFINITIONS

The cellular fractions used in these studies have been operationally defined as follows:

7 SPECIFIC BINDING OF E. COLI REPLICATIVE ORIGIN DNA

Origin complex: a particle with unique sedimentation properties in sucrose gradients (300-400S, heavy density). These properties allow us to achieve a thirty fold enrichment of origin DNA. The complex consists of a 40 kilobase fragment of origin DNA and several proteins.

Rapidly sedimenting complex (RSC): a complex with a higher sedimentation coefficient than origin complex which cosediments with the outer membrane. It is about five fold enriched for origin DNA. The majority of DNA in this fraction is not from the origin, but this region is selectively labeled using a temperature sensitive initiation mutant.

CsCl-DNA: a preparation of either origin complex or rapidly sedimenting complex purified on CsCl gradients. The DNA containing fraction has a density similar to E. coli DNA and is highly enriched for two proteins of 75,000 and 55,000 daltons.

Outer membrane: the high density fraction of the membrane which contains lipopolysaccharide and cell wall. It has been further treated by CsCl in order to remove DNA.

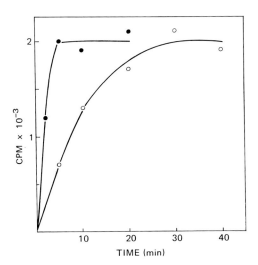

Figure 1. Time course of filter binding. "CsCl-DNA" purified from origin complex was incubated with 1 µl (0-0), or 5 µl (●-●) outer membrane in 50 µl of buffer containing 20 mM

Tris pH 7.2, 100 mM KCl, 1 mM EDTA, 5 mM MgCl2, 1 μg bovine serum albumin and 1 mM β-mercaptoethanol. After incubation at room temperature, samples were slowly passed through 8 mm nitrocellulose filters (Schleicher and Schuell), washed three times, dried and counted. Counts due to nonspecific sticking of DNA in the absence of outer membrane (3% of input) have been subtracted. Counts have been corrected for quenching.

IV. RESULTS

Kinetics of origin DNA-membrane binding. We have studied the kinetics of binding of "CsCl-DNA" to outer membrane preparations. Figure 1 shows the time course of the reaction. Maximum binding occurs within five minutes at high membrane to DNA ratios, while at lower ratios maximum binding requires up to thirty minutes. The binding reaction is not greatly influenced by temperature, and takes place at approximately the same rate at 0°, 22°, and 42°.

The extent of binding is dependent on the amount of both reactants added to the system. Increasing the concentration of either reactant while keeping the other constant results in simple saturation curves (Fig. 2A and B). At lower levels of membrane, the plateau value of DNA bound is proportional to the amount of membrane added. At higher levels of membrane (greater than 4 μg protein) the capacity of the system is apparently exceeded and little increase in DNA binding is seen. One goal of this work is to further fractionate the membrane in the hope of identifying the "receptor" for DNA attachment. Within the limits described above the assay will be useful in such studies.

Further analysis of the binding curves indicates that more than one species of each reactant is involved. Complex binding behavior can be illustrated by transforming the data to construct Scatchard plots. The data does not fit a straight line as expected for a simple bimolecular reaction (Fig. 3).

7 SPECIFIC BINDING OF E. COLI REPLICATIVE ORIGIN DNA

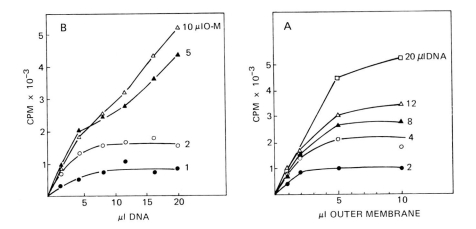

Figure 2. Filter binding Assay.
Various amounts of CsCl-origin complex DNA and outer membrane were incubated for 30 minutes as described in Figure 1. One µl of DNA contains approximately one nanogram and 500 c.p.m. One µl of outer membrane contains 0.8 µg protein as determined by Biorad assay. A) Binding of DNA as a function of outer membrane concentration; B) Binding as a function of DNA concentration.

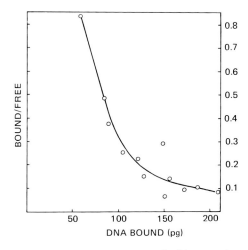

Figure 3. Scatchard Plot of Binding Data.
"CsCl-DNA" from a scaled up preparation and 5 µl

outer membrane were incubated and filtered as in Figure 2. To convert CPM to absolute amount of DNA, a portion of the sample was purified by phenol extraction and the DNA concentration determined by ultraviolet adsorptions. The number of high affinity binding sites per cell equivalent was then estimated from: (1) the intercept of the tangent to the upper part of the curve; (2) the number of origin molecules bound at this point calculated from the molecular weight and the purity of the origin fragment in the sample; and (3) the cell equivalents of outer membrane in the reaction calculated from the number of cells used and the percent recovery of outer membrane.

The membrane of E. coli has been shown to contain a number of DNA binding proteins (5, 6). Many of the proteins bind non-specifically to DNA. Our assay is likely to include binding by some of these proteins. Upwardly curved Scatchard plots can result from multiple components with different binding constants or from large ligand interactions at random sites on DNA. To distinguish among these possibilities, the binding reaction will have to be carried out with fully purified components. Nonetheless, the data allow a rough estimate of the number of high affinity DNA binding sites per cell equivalent of membrane. Their number can be estimated from the intercept of the tangent to the upper portion of the curve. In order to express the data in terms of actual number of reactants, we scaled up our preparation to determine the concentration of DNA and membrane. Using this material we find, as reported previously (4), that approximately one membrane site per cell has high affinity for origin DNA.

Some properties of the binding reaction. "CsCl-DNA" isolated from either origin complex or from rapidly sedimenting complex both behave similarly in our assays. Binding is enhanced five fold by magnesium (Table 1). Maximum binding is obtained with 3 mM to 10 mM $MgCl_2$. Several other divalent cations also stimulate binding to a similar extent.

TABLE 1. Effect of Divalent Cations and Polyamines on the Binding of "CsCl-DNA" to Outer Membrane Preparation

Cations (10 mM)	Percent bound
none	18
$MgCl_2$	100
$MnCl_2$	85
$CaCl_2$	84
$ZnCl_2$	29
$CuSO_4$	87
$CoCl_2$	106
$NiCl_2$	93
spermidine	54
putrescine	16

The ionic nature of the interaction is further suggested by a complete inhibition in binding in the presence of 0.4 M KCl. Heparin reverses the binding since membrane-DNA complexes are dissociated completely by the addition of 5 μg/ml heparin to the reaction mixture for five minutes prior to filtration.

Specificity of binding. If DNA binding takes place at both origin specific and nonspecific sites on the membrane, the nonspecific sites may be competed by added irrelevant DNA, such as calf thymus DNA. A competition experiment is shown in Figure 4.

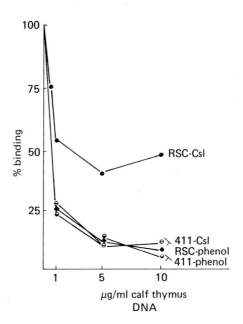

Figure 4. Competition by Calf Thymus DNA. Samples were incubated in the standard reaction mixture plus the indicated amount of calf thymus DNA. The values for binding without competitor are set at 100%. DNA preparations used in the assay are: "CsCl-DNA" purified from RSC, (●-●); Deproteinized "CsCl-RSC DNA," (◆-◆); CsCl purified λ411 DNA, (0-0); Pronase, phenol and CsCl purified λ411 DNA, (☻-☻).

Origin DNA and membrane were incubated in the presence of increasing amounts of calf thymus DNA. Approximately 50% of the DNA binds even at high concentration of competitor. This plateau value is also seen with different ratios of origin DNA and membrane. The specificity of this binding was studied further by restriction enzyme analysis.

Localization of binding sites on the DNA. After treatment of "CsCl-DNA" with restriction enzymes we find specific binding of the fragment that contains the origin of replication. We have previously reported this finding for Eco Rl endonuclease treated DNA (4), and have carried out further studies to localize the relevant region more closely. The Eco Rl fragment containing the origin is quite large (9.5 kb). Both Sma 1 and Xho 1 endonucleases make single cuts in this fragment at sites which are 460 base pairs apart (Fig. 5).

7 SPECIFIC BINDING OF E. COLI REPLICATIVE ORIGIN DNA

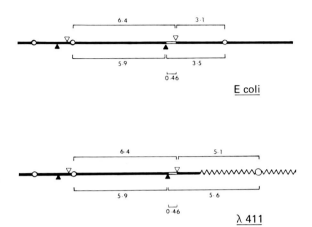

Figure 5. Restriction Map of the Origin Region.
Restriction endonuclease cleavage sites within
a 14 kilobase region spanning the origin are
shown. Chromosomal DNA (upper) and λ411 (lower)
were digested with Eco Rl (-O-), and either
Xho 1, (▽) or Sma 1 (▲). The size in kilobases
of fragments produced by the double digestions
is indicated. The origin region (▬▬),
and λDNA sequences, (⋀⋀⋀⋀) are indicated.

This 460 base pair region has been shown to contain the functional origin by its ability to permit replication of plasmids that contain no other origin (7,8). "CsCl-DNA" was digested with a combination of Eco Rl and one or the other of these enzymes. The samples were then incubated with outer membrane and calf thymus DNA, filtered, eluted from the filter and separated on agarose gels. As a control, "CsCl-DNA" was digested and run directly on the gel. We found that fragments containing the origin region bind to membranes to a much greater extent than those containing flanking stretches (Fig. 6). The enrichment of the origin containing fragment is enhanced with increasing amounts of calf thymus competitor. A maximum enrichment of greater than ten fold is obtained in the presence of 10 µg/ml calf thymus DNA. Although we have not yet studied the behavior of the 460 base pair fragment alone, we can conclude that it is essential for the specific binding of DNA to the membrane.

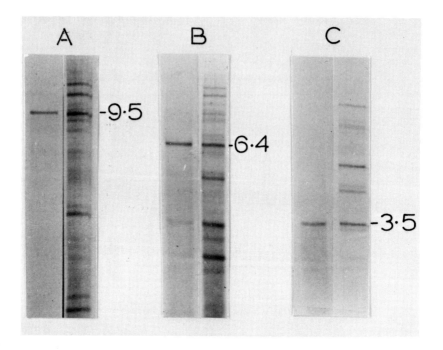

*Figure 6. Filter Binding of Restriction Fragments.
"CsCl-RSC DNA" was digested with restriction endonucleases. Samples were incubated with 5 µl outer membrane and 10 µg/ml calf thymus DNA. After filtration the DNA was eluted with 1% sarkosyl and electrophoresed on 0.7% agarose gels. As a control, one fifth of the DNA was applied to the gel directly without filtration (right slot of each pair). The size in kilobases of the origin containing fragments is indicated.
(A) Eco Rl; (B) Eco Rl-Xho l double digest;
(C) Eco Rl-Sma l double digest.*

Proteins of "CsCl-DNA" and their role in binding to membranes. "CsCl-DNA" preparations contain two peptide bands as shown by SDS polyacrylamide gel electrophoresis (R. Balakrishnan and T. Kusano, unpublished data). Their apparent molecular weights are 55,000 and 75,000. We are currently trying to assign them to particular genes, but have not been successful in early attempts.

We have studied the possible role of these proteins by comparing the binding behavior of "CsCl-DNA" before and after pronase, phenol extraction. Protein-free origin DNA has also been obtained by isolating DNA from a specialized transducing λ phage (λ411, J. Felton and A. Wright) that carries an extensive region of the E. coli origin.

Binding of "CsCl-DNA" and its protein free counterparts is shown in Figure 4. DNA preparations which contain the protein (upper curve) show a high level of specific binding to the membrane. Large concentrations of calf thymus DNA only suppress binding approximately 50%. This may reflect the heterogeneity of the DNA preparation. Deproteinized origin DNA is competed by calf thymus DNA to a much greater extent, about 95%. DNA isolated from λ411 is competed to the same extent as deproteinized "CsCl-DNA." The proteins do not appear to be associated with λ411 virion DNA since we cannot detect them after labeling with high specific activity ^{35}S-methionine. In addition, DNA isolation from gently lysed virions behaves the same way in binding assays after treatment with pronase and phenol.

Surprisingly, the residual, non-competed binding of both "CsCl-DNA" and deproteinized DNA takes place at the origin. This is seen by carrying out the binding reactions after restriction enzyme treatment of both protein-containing and protein-free DNA. At high concentrations of calf thymus DNA, the origin fragment of λ411 is enriched (data not shown). Thus, the presence of proteins (or one protein, since either alone could be sufficient) on the DNA does not affect the _specificity_ of its binding at the origin, merely the amount of binding observed.

V. DISCUSSION

We have developed assays for the specific binding of replicative origin DNA to membrane preparations. Membrane binding is carried out _in vitro_ from isolated DNA and membrane fractions. The number of membrane "receptors" for high affinity binding has been estimated at about one per cell. Of course, we do not know how many similar sites may exist but not be available for binding in our preparations.

When nonspecific binding is eliminated by competing calf thymus DNA, binding to membranes requires a 460 base pair region which contains the origin of replication. This specific interaction is greatly enhanced by proteins which remain bound to the DNA after CsCl treatment. One possible explanation for the effect of the CsCl proteins is that they increase the affinity of the origin for membrane receptors.

Without the protein, these receptors may bind origin DNA with low affinity. Another possibility is that the membrane receptors bind the proteins of CsCl-DNA directly. Some of these proteins may be present in the membrane preparation and may account for the specific low level binding of deproteinized origin DNA. Furthermore, there could be different origin binding complexes which differ in affinity and abundance.

A membrane protein which specifically binds single strand DNA at two sites at or near the origin has been purified by Jacq et al. (6). Its relationship to the proteins of our system is presently unclear since our preparations contain double stranded DNA and do not bind at one of the two sites of Jacq, et al. We hope to answer these questions by using the binding assay to purify the relevant membrane components.

REFERENCES

1. Leibowitz, P. J., and Schaechter, M. (1975) Int. Rev. Cytol. 41, 1.
2. Nicolaidis, A., Holland, I. (1978) J. Bact. 135, 178.
3. Nagai, K., Hendrickson, W., Balakrishnan, R., Yamaki, H., Boyd, D. and Schaechter, M. (1980) Proc. Natl. Acad. Sci. U.S.A. 77, 262.
4. Yamaki, H., Balakrishnan, R., Hendrickson, W. and Schaechter, M. (1980). In "Mechanistic Studies of DNA Replication and Genetic Recombination" (B. Alberts, ed.), p. 199, Academic Press, New York.
5. Jacq, A. and Kohiyama, M. (1980). Eur. J. Biochem. 105, 25.
6. Jacq, A., Lother, H., Messer, W. and Kohiyama, M. (1980). In "Mechanistic Studies of DNA Replication and Genetic Recombination" (B. Alberts, Ed.,), p. 189.
7. Oka, A., Sugimoto, K., Takanomi, M. and Hirota, Y. (1980). Mol. Gen. Genet. 178, 9.
8. Meijer, M., Beck, E., Hansen, F., Bergmans, H., Messer, W., von Mayenburg, K. and Schaller, H. (1979) Proc. Natl. Acad. Sci. U.S.A. 76, 580.

DISSECTION OF THE REPLICATION REGION CONTROLLING
INCOMPATIBILITY, COPY NUMBER, AND INITIATION OF DNA SYNTHESIS
IN THE RESISTANCE PLASMIDS, R100 AND R1[1]

Thomas Ryder, Jonathan Rosen[2], Karen Armstrong, Dan Davison,
Eiichi Ohtsubo, and Hisako Ohtsubo

Department of Microbiology, State University of New York at
Stony Brook, Stony Brook, New York 11794

ABSTRACT

We have constructed a genetic map of the replication and incompatibility regions of R100 and R1 based on a comparison of the nucleotide sequence of small multi-copy derivatives of each, and by analyzing several gene products produced *in vivo* or *in vitro* from the replication region. These gene products include both RNA transcripts and proteins. We define genes and functional sites involved in replication control, including promoters, terminators, and sequences which may be involved in origin function. The distribution of base differences between the R100 and R1 derivatives may provide clues to the function of the gene products and sites affected by these differences. We have tried to construct models which incorporate these functions to predict how the proteins, transcripts, and functional sites might interact to control replication. We have begun to test specific aspects of these models by constructing plasmids with mutations at specific locations within the replication region.

---

[1]This research was supported by United States Public Health Service Grants to E.O. (GM22007) and H.O. (GM26779), by an NSF National Needs Postdoctoral Fellowship to K.A. (SPI-793986), and by partial support to T.R. under an NIH training grant (CA-09176).
[2]Present address: Dept. of Pathology, Stanford University School of Medicine, Stanford, CA 94305

INTRODUCTION

R100 and R1 are large, self-transmissible plasmids coding for multiple drug resistance (1). They are mutually incompatible in *Escherichia coli* K12, belonging to the FII incompatibility group, and they are structurally similar, indicating that they share a common ancestor (2,3). R100 and R1 are normally maintained at a low copy number. Mutants of each plasmid have been isolated with increased copy number, suggesting that replication is regulated under negative control expressed by the plasmid itself (4-6). By analyzing several copy number mutants of R1 with various phenotypes, Nordström et al. (reviewed in ref. 7) conclude that at least one protein is involved in copy number control, that there are important target sites required for replication, and that copy number control and incompatibility are closely related phenomena. A temperature sensitive mutant of a closely related plasmid, Rms201, is defective in replication at the restrictive temperature, suggesting that some plasmid encoded function may be required for replication (8).

A region or function of R100 which was necessary for integrative suppression of an *E. coli* ts-*dnaA* phenotype was roughly mapped by Yoshikawa and called *repA* (12). Comparing heteroduplexes between a number of deletion derivatives of R1 and R100, Ohtsubo et al. (9) determined that a 2.5 kilobase (kb) region, including the origin of replication (10,11) was sufficient to express autonomous replication. Another observation from these studies was that a small region of non-homology (∼250 base pairs) between R1 and R100 existed within this 2.5 kb region. Cloning analysis by several labs has shown that most of this 2.5 kb region is involved in replication control and that it contains sequences which express incompatibility and copy number control as well as sequences required for replication (13-16).

We have been working on defining the genes and sites within the replication region which are involved in regulating plasmid DNA replication. By comparing the nucleotide sequence from the replication regions of two small copy number mutants of R100 and R1, pSM1 and pTR1 respectively, we have been able to construct a map of coding frames and sites which are likely to be involved in replication control (17-19). We have been able to identify proteins produced *in vivo* on the basis of the size and amino acid composition of the coding frames predicted from the DNA sequence. We have also identified three RNA transcripts produced from the replication regions of both pSM1 and pTR1 (20). One transcript in particular, RNAI, may have interesting implications for replication control.

8 DISSECTION OF THE REPLICATION REGION

MATERIALS AND METHODS

Techniques and strategies for obtaining the DNA sequence of pSM1 and pTR1 have been described (17-19). *In vitro* transcription and RNA sequencing are described in ref. 20. Proteins obtained from purified mini-cells (47) were analyzed by SDS-polyacrylamide gel electrophoresis (48). Plasmids discussed in this paper and relevant references are shown in Fig. 1.

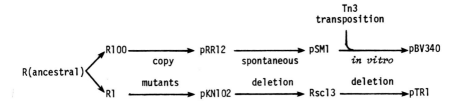

Figure 1. A pedigree of the plasmids described in this paper. Original references are: pKN102 (4), pRR12 (5), Rsc13 (44), pSM1 (45), and pTR1 (19). The construction of pBV340 will be described elsewhere (H.O.,B.V.,& E.O., in prep.).

RESULTS

Nucleotide Sequence Containing the Region Necessary for Replication. Rsc13 and pSM1 were both produced by successive deletions which eliminated most of their parental plasmid sequences (9,21). Rsc13 is 7.8 kb in length and contains a complete copy of Tn3, which is 5 kb. Most of the rest of Rsc13 consists of the 2.5 kb replication region. pSM1 (5.6 kb) does not contain Tn3 so Rsc13 and pSM1 are homologous only in their replication regions. Within this 2.5 kb sequence is the 250 base region of non-homology first identified by heteroduplex analysis. Rsc13 and pSM1 share one *Bgl*II site and three *Pst*I sites in the replication region. When Rsc13 was digested with *Bgl*II and *Bam*HI and ligation performed at low DNA concentration, the 3.8 kb ampicillin resistance plasmid pTR1 was isolated.

Fig. 2 shows that pTR1 contains only part of the region defined as the replication region of pSM1 and Rsc13. pTR1 does however retain sufficient information to express controlled, autonomous replication as it is inherited stably in bacterial cells and expresses incompatibility against a pTR1 derivative with an additional selective marker (19). pSM1 also expresses incompatibility against itself since it cannot be maintained in the same cell with a pSM1 derivative which expresses tetracycline resistance.

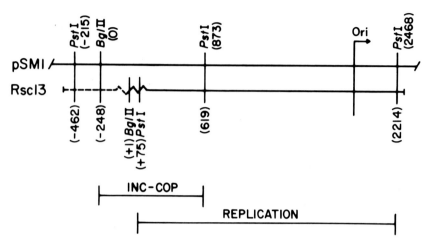

Figure 2. Skeleton of the replication region of pSM1 and Rsc13, showing the origin and direction of replication (10), and an approximate assignment of the minimal replication region and a region expressing incompatibility and copy number control (inc-cop) as deduced from cloning analysis (13,14). The inc-cop and replication regions overlap, and part of the inc-cop region is non-homologous (sawtooth line) between R1 and R100 (18,19). The numbers above the pSM1 line are the nucleotide coordinates of Rosen et al. (18). Positive numbers below the Rsc13 line show the base coordinates of pTR1 (Fig. 3). Sequences present in Rsc13 but not pTR1 are shown by a dashed line. These sequences are shown in Fig. 7.

Figure 3. Nucleotide sequence of one strand of the replication region of pTR1 with a few bases of Tn3 sequences. In circular pTR1, base 1 is joined to base 3754. Bases which are different in pSM1 are shown directly below the pTR1 sequence. Bases 1-125 of pTR1 are part of the non-homologous region shown in Fig. 2. Nucleotide 126 of pTR1 is aligned with nucleotide 380 of pSM1 (18). The amino acid sequence specified by the major coding frames is shown and the amino acids which are different in pSM1 are shown above the pTR1 amino acid sequence. The two coding frames which we feel are most likely to produce functional proteins are enclosed in boxes. The box in the lower part of the figure shows one standard deviation about the position where the origin was mapped (10,17). The locations of the RNA transcripts produced from the replication region are shown. Facing arrows beneath the sequence show dyad symmetries.

8 DISSECTION OF THE REPLICATION REGION

The region containing the origin of replication has been mapped (see Fig. 2). Both plasmids replicate unidirectionally from this origin (10,11).

We have previously described the nucleotide sequence of the region necessary for replication of pSM1 (17,18). We have also recently obtained the entire nucleotide sequence of pTR1 and have compared it to the corresponding region of pSM1 (19). These results are shown in Fig. 3. The first nucleotide of pTR1 sequence in Fig. 3 corresponds to the site where the *Bgl*II and *Bam*HI cohesive ends where ligated together in the

construction of pTR1. The first 2372 nucleotides of pTR1 in
Fig. 3 correspond to the replication region. The rest of
pTR1 consists of 1382 base pairs of Tn3 sequences. Therefore,
pTR1 is 3754 base pairs in length. Fig. 3 also compares the
pTR1 sequence with base differences in the pSM1 replication
region. The first 125 bases of pTR1 are part of the nonhomologous region shown in Fig. 2. Nucleotides 126-2372 of pTR1
are essentially homologous to 380-2626 of pSM1 with the only
differences being the base substitutions shown in Fig. 3.

Transcripts Produced from the Replication Region. Using an
in vitro transcription system, we have so far identified three
RNA transcripts produced from both pSM1 and pTR1 (20). We

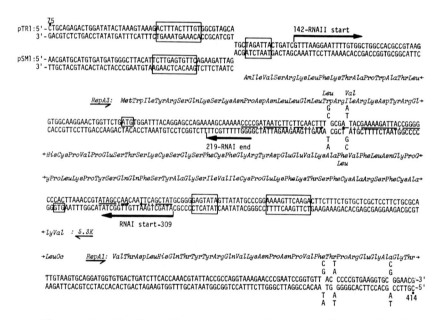

*Figure 4. Nucleotide sequence of pSM1 and pTR1 in the
region where the minimal replication region and the inc-cop
region overlap (Fig 2). The numbers shown correspond to
nucleotide positions in Fig. 3. Sequence differences in pTR1
and pSM1 are shown above and below, respectively, the base
pairs which are common to both. The large and small boxes
preceding the RNA starts show the -35 and Pribnow sequences,
which are features of procaryotic promoters (22). The 5' end
of both transcripts and the 3' end of RNAI are shown. The
amino acid sequences of two overlapping coding frames, RepA3
and 5.3K, are shown. The direction of the amino terminal →
carboxy terminal polarity of each coding frame is shown by arrows. The beginning of the RepA1 coding frame is also shown.*

8 DISSECTION OF THE REPLICATION REGION

have mapped these by obtaining as much of the RNA sequence as possible and comparing it to the DNA sequence. The location of each is shown in Fig. 3.

RNAII is a long transcript which begins at pTR1 position 142 and extends to the right (Fig. 3). Homology to the -35 region and Pribnow sequence, which are consensus features of promoter sites (22), can be found preceeding the start of RNAII. It is interesting that the -35 sequence of RNAII from pSM1 and pTR1 lies in the non-homologous region (Fig. 4).

RNAI, a 91 nucleotide transcript, begins at base position 309 and terminates at 219. RNAII and RNAI are transcribed from the same region of DNA sequence, but in opposite directions (Fig. 4). The sequence of RNAI, including two possible secondary structures, is shown in Fig. 5. RNAI has several very interesting features:

1. RNAI is transcribed from a region known from cloning to express incompatibility and copy number control. There are two base differences between the RNAI sequence from pSM1 and pTR1 and these occur at the top of the large secondary structure.

2. There are several statistically significant homologies between RNAI sequences and DNA sequences in the origin region.

3. The sequences at the bottom of the large stem structure possess several features known to be involved in transcription termination in procaryotes (22). These include a base paired G-C rich stem followed by a sequence rich in U.

4. When the transcription reactions were done in the presence of glycerol, which should destabilize the base pairing of the secondary structure, a new transcript, RNAIx, was observed in addition to RNAI. (RNAI:RNAIx > 10:1). RNAIx has the same 5' end as RNAI but is 200-300 nucleotides long.

When the large origin-containing *Pst*I fragment of pSM1 or pTR1 is used as a substrate for transcription, a transcript which begins at base 2062 and usually terminates at 2212 is observed. We call this transcript RNAIII. The sequence at which it terminates, TAATCAATAT, is very similar to the sequence at the ρ-dependent termination sites of λt_{R1}, CAATCAA (23), and tRNA^{Tyr}, CAATCAATAT (24). The DNA sequence from which RNAIII is transcribed includes a sequence which is homologous to 13 bases of RNAI (20).

Coding Frames in the Replication Region. Many open reading frames in the replication regions of pTR1 and pSM1 could produce proteins larger than 5,000 daltons. Of these, most can be ruled out because many base substitutions would cause amino acid changes or because nonsense codons appear in one frame but not the other when the two plasmids are compared. Furthermore, most of the coding frames are not preceded by the

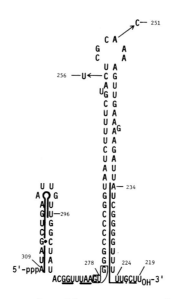

Figure 5. The nucleotide sequence of RNAI from pSM1. The base differences in RNAI from pTR1 are shown by arrows. Numbers refer to Fig. 3. The 5' end (underlined) is complementary to the DNA sequence between 2113-2126. The 3' end (underlined) is complementary to the origin region as described in the text. The initiator codon for the 5.3K protein is boxed. The adjacent sequence GGUUUAA might act as a ribosome binding site as described in the text.

features usually found at ribosome binding sites. These include complementarity to the 3' end of the 16S rRNA, the Shine-Delgarno sequence (25), and a UAA or a UGA sequence in the vicinity of the initiator codon (26). Two protein coding frames which are consistent with most criteria applicable to true coding frames are enclosed in boxes in Fig. 3.

The RepA1 coding frame could produce a protein of 33,000 daltons. In this coding frame there are 49 base pair differences between the pSM1 and pTR1 sequences. However, these changes result in only eight amino acid differences in RepA1 from these two plasmids since 39 of the differences are single base changes which produce synonomous codons. RepA1 is preceeded by a four base complementarity to 16S rRNA, AGGA. It can be shown that RepA1 is actually synthesized and involved in replication control of pSM1, *in vivo*, by the following results: A.) pSM1 was ligated to a ColE1 derivative and deletions were made within the RepA1 coding frame by digesting with *Sal*I or with *Sma*I. pSM1 has two *Sal*I sites in RepA1 at locations corresponding to 573 and 597 in Fig. 3. pSM1 also has two *Sma*I sites in RepA1, at 748 and 1036. (As shown,

8 DISSECTION OF THE REPLICATION REGION

pTR1 loses one *Sal*I and one *Sma*I site due to single base changes.) The pSM1-ColE1 plasmids with these *Sal*I or *Sma*I fragments deleted are defective in replication in an *E. coli polA* host. B.) RepA1 spans a *Pst*I site between two fragments which must be rejoined in the parental orientation to permit autonomous replication in cloning experiments (13-15).
C.) Proteins labeled with radioactive amino acids were isolated from purified mini-cells containing pSM1 or various pSM1 fragments cloned into a ColE1 vector. These proteins were analyzed by SDS-polyacrylamide gel electrophoresis and a band corresponding in size to RepA1 has been detected (43).

A coding frame for a 6,700 dalton protein, RepA3, is found in a region known from cloning to express incompatibility and copy number control (14-16). This coding frame is preceeded by two possible Shine Delgarno sequences, AAGGA or GGU. Because of the location of the coding frame, RepA3 could be a repressor involved in regulating incompatibility and copy number. Two base differences between pTR1 and pSM1 are found in this region and these would result in amino acid changes in RepA3. These amino acid changes might correspond, in part, to the increased copy number of pKN102 and pRR12 (Fig. 1). Codon frequency analysis of RepA3 shows a bias in favor of codons which are rarely used in the chromosomal genes of enteric bacteria (27). Furthermore, both the pSM1 and the pTR1 base change results in a new rare codon in the RepA3 sequence. The possible significance of these observations will be considered later. The DNA sequence from both pSM1 and pTR1 predicts that RepA3 does not contain histidine. A band corresponding to a 7,000 dalton polypeptide has recently been identified which does not appear on SDS-polyacrylamide gels when the mini-cell preparations are labeled with [^{3}H]-His. We are presently trying to determine if this band corresponds to RepA3 by determining the amino acid sequence of the 7,000 dalton polypeptide.

As mentioned previously, RNAI can be extended to give RNAIx. We noticed that a 5,300 dalton protein (5.3K) could be translated from RNAIx, between nucleotides 280 and 131 in Fig. 4. This coding frame also contains the two single base changes in this region; however, one base change does not result in an amino acid difference. Based on its location in the inc-cop region, 5.3K, if it exists, is another candidate for the repressor function. However, as described below, one interesting possibility is that initiation for translation of 5.3K is a significant aspect of replication control whether the protein is functional or not. The sequence preceeding the GUG initiator codon does contain a UAA codon and there is a three base homology to 16S rRNA, GGU. These two sequences are within a sequence GGUUAA. Sequences of the form PuPuUUUPuPu are found preceding many phage coding frames (46) and this may signify some regulatory characteristic.

An open reading frame for a 14,000 dalton protein, RepA4, is found in the origin region of both pSM1 and pTR1 (Fig. 3 and refs. 17-19). It is difficult to predict if this protein is made, as the coding frame is not preceded by a Shine Delgarno sequence or by a nonsense codon. 14% of the amino acids would be proline and RepA4 from pTR1 would have four His residues, while RepA4 from pSM1 would only have one His. If RepA4 is made, it may be possible to identify it based on these features of the amino acid composition. In any event, RepA4 is not required for replication because, as described below, part of the RepA4 coding frame can be deleted.

Origin. The origin of replication of pSM1 was determined by electron microscopic analysis of replicating plasmid molecules. The box in the lower portion of Fig. 3 includes nucleotides within one standard deviation of the position where it was mapped. Two standard deviations include nucleotides between 1519-2202 (10,17). Rosen et al. determined that within this region there are several dyad symmetries which could possibly form secondary structure similar to those shown in Fig. 6 (17). The base differences between pTR1 and pSM1 would not affect the intrastrand base pairing of any of these secondary structures. A deletion in R1*drd-19* reported by Oertel et al. (28) to prevent R1 replication corresponds to base positions 1730-1944 of pTR1. This deletion would remove most of the large stem structure in Fig. 6. These facts suggest that one or more of the secondary structures may normally play some role in determining origin function. However, we have recently constructed a series of plasmids in which the replication region of pSM1 has been trimmed from the ends by Tn3 insertion. One such pSM1 derivative, pBV340, is structurally equivalent to a deletion in the origin region of pTR1 between bases 1779 and 2372. As shown in Fig. 6, this would eliminate almost half of the large stem structure. Therefore, sequences between 1730-1779 may be very important for replication. In addition, about half of RepA4 and the entire region coding for RNAIII are deleted.

Computer analysis shows very little significant homology between the origins of these plasmids and the origins of other replicons which have been described. However, several small patches of homology with other origins can be found which may be interesting. For example, the sequence GAAAAAC, which is found in the inceptor region of ColE1 DNA synthesis in the *in vitro* system of Itoh and Tomizawa (29), can be found between 1575-1581 and between 1686-1692 of the sequence in Fig. 3. Furthermore, this sequence is homologous to the 3'

8 DISSECTION OF THE REPLICATION REGION

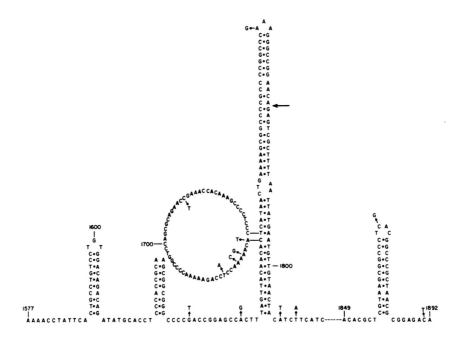

Figure 6. Possible secondary structure in the origin region of pTR1. These stems correspond to the dyad symmetries in the origin region in Fig. 3. Base differences in pSM1 are shown. The arrow shows the point of Tn3 insertion in pBV340.

end of RNAI. A particularly good match is found between RNAI and the sequence near 1686-1692:

3'HO-UUCGUUUUUGGGGCUA-5'
5'-AG AAAAACCCCGGT-3'

The first small hairpin in Fig. 6 is similar in structure to small hairpins found near the dnaG initiation site in several phages (30). Though the sequences of the hairpins are not homologous, they do have these features in common: the sequence CTGCC appears in the stem and the sequence TGT appears at the top of the hairpin. Small stems of similar structure can be found at the origins of other replicons and may be involved in dnaG action (31,32). We do not yet know what constitutes the minimal origin region so these observations, though interesting, remain speculative until the origin region is better defined.

Nucleotide Sequence and a Coding Frame Outside the Minimal Replication Region. A possible coding frame for an 11,000 dalton protein, RepA2, has also been found in the inc-cop region of pSM1 (Fig. 7 and ref. 18). Since the RepA2 coding frame spans the 250 base region of non-homology between R100 and R1, we were interested in comparing the sequences of pSM1 and Rsc13 in this region. As shown in Fig. 7, the transition

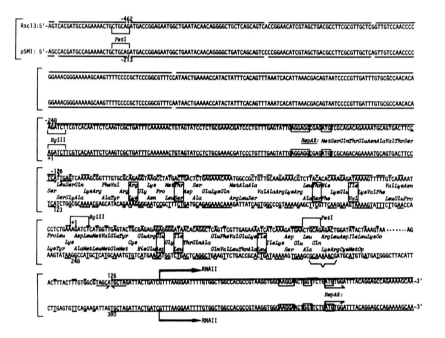

Figure 7. Nucleotide sequence of Rsc13 and pSM1 in the vicinity of the non-homologous region. Rsc13 sequences with negative numbers are deleted from pTR1 (Fig. 2). Rsc13 is overlined and pSM1 is underlined where these two sequences are identical. Gaps in these lines show base differences. Small boxes in the DNA sequence show the initiator codons and possible Shine Delgarno sequences of RepA2 and RepA3. Facing arrows beneath the DNA sequence show dyad symmetries at the junctions between homologous and non-homologous sequences. The amino acid sequence of RepA2 from Rsc13 and from pSM1 is shown between the DNA sequences. Boxes show conservative amino acid substitutions. Note that the sequence of pSM1 shown by the bracket: }, is GCAAAAACG, instead of GCAAAACG as previously reported (18). The additional base, A, changes the reading frame of RepA2 makes RepA2 shorter than previously reported. The additional A becomes pSM1 position 332. All pSM1 positions ≥332 in the original coordinates increase by 1. However, references to pSM1 positions in this paper use the original coordinates.

8 DISSECTION OF THE REPLICATION REGION

between homologous and non-homologous sequences is abrupt at both ends of the non-homologous region (NHR). The flanking sequences are nearly identical between pSM1 and Rsc13, but within the NHR, the two sequences are only 42% homologous, not including the seven base pair gap shown in Fig. 7. The DNA sequence of Rsc13 predicts a coding frame in this region for an 11,000 dalton protein. Interestingly, the ribosome binding site and the first eleven amino acids of RepA2 from pSM1 and from Rsc13 are in the homologous region. Within the part of the NHR which codes for RepA2 from Rsc13, the two DNA sequences are 45% homologous. Where RepA2 from Rsc13 and pSM1 can be aligned, they share identical amino acids at 38% of the codons within the NHR. Therefore, both RepA2 proteins may be subject to similar selective pressures since in the absence of selection, this degree of DNA sequence divergence should be distributed randomly and the amino acid homology would be less than 18%. Furthermore, many of the amino acid differences between these two proteins are conservative substitutions and in several places, amino acids are displaced by only one codon. These facts suggest the interesting possibility that these two proteins, though widely diverged, are synthesized *in vivo* from both plasmids and may have similar functions. The DNA sequence predicts that RepA2 does not contain tryptophan. SDS-gel electrophoresis shows that a band corresponding to a protein of this size does label with [^{35}S]-Met but not with [^{3}H]-Trp. The results of Molin et al. (33) suggest that an 11,000 dalton protein is also produced from the inc-cop region of R1 and the DNA sequence of R1 predicts a coding frame in this region (34). The nature of the sequence divergence in this region will be considered in the Discussion.

DISCUSSION

We have attempted to construct testable models for replication control based upon the results described above, and which are consistent with the genetic analysis outlined in the introduction. A summary of these results is shown in Fig. 8.

Interactions in the Origin Region. Since the gene products which we presume to be involved in replication control, including RepA1 and the products of the inc-cop region, are encoded outside the origin region itself, we are interested in the manner in which these gene products influence events at the origin. We think that RepA1 is probably involved in initiating DNA synthesis. Deletions in RepA1 cause pSM1 replication to be defective. Of the gene products in the

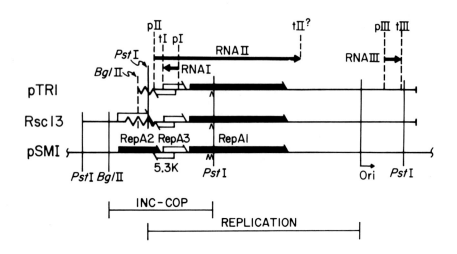

Figure 8. Some of the genes and sites which might be involved in replication control. Coding frames described in the text (RepA's, 5.3K) are shown by heavy arrows. Arrows corresponding to proteins which have been verified in vivo are filled in. The location of RNA transcripts is shown. Promoters and terminators for each are designated pI, tI, etc. The arrow at the origin (Ori) shows the direction of replication. The sawtooth line shows the region of non-homology between R1 (Rsc13, pTR1) and R100 (pSM1). SalI sites are shown by carets (Λ).

replication region, RepA1 is the best candidate for a required function because it is so highly conserved between pSM1 and pTR1.

One simple possibility is that RepA1 binds to the origin and promotes the assembly of an initiation complex. One observation that addresses this possibility is that of Wang and Iyer (35) that many R plasmids, including R100 and R1, may encode a function which interacts with or perhaps substitutes for dnaB. RepA1 may act to stabilize a dnaB-origin complex.

Another possible role for RepA1 might be to prevent termination of RNAII transcription. We don't presently know the normal termination point for RNAII but the DNA sequence between 1380-1430 resembles a transcription termination signal as the RNA could form a secondary structure with a G-C rich stem followed by a string of U's. Transcription of RNAII into the origin region might provide a primer or stabilize some conformation of the DNA which corresponds to

8 DISSECTION OF THE REPLICATION REGION

origin competence. The evidence of Diaz et al. (36) that rifampicin reduces but does not abolish replication of an R1 derivative *in vitro* is consistent with the possibility of a transcriptional activation event which increases the efficiency of initiation of DNA replication (37). In one report, a *Sal*I fragment from pRR12 which would contain part of the replication region, including the origin but none of the inc-cop region (Fig. 8) was cloned into pBR313 (15). This cloned origin could not be rescued in a *E. coli polA* host by complementation from intact pRR12 sequences. RNAII is one good candidate for a function which can act in cis on the origin region, but whose expression depends on sequences outside the origin-containing *Sal*I fragment. It is alternatively possible that RepA1 itself could be predominantly cis-acting because there is some evidence that RepA1 is unstable *in vivo* (18,19). Instability could result in cis-dominance if the rate of RepA1 binding is limited by diffusion.

Other possible interactions in the origin region involve hybridization of RNAI to complementary sequences in the origin region. The possibility that the base changes at the top of the large secondary structure of RNAI relax repression suggests the following scheme:

    1. In the repressed state, RNAI is prevented from hybridizing to the origin because it is bound by some repressor protein, perhaps a plasmid-encoded protein. The base changes in RNAI reduce the strength of this binding in copy mutants.

    2. At some frequency, RNAI dissociates from the repressor and is free to hybridize to a complementary sequence in the DNA, for example, at one of the GAAAAAC sequences mentioned above. This R loop formation would be promoted and stabilized by some protein(s), possibly RepA1.

    3. Some replication complex takes advantage of the strand separation afforded by the R loop formation to promote initiation. For example, one of the GAAAAAC sequences is adjacent to the small hairpin which resembles those near dnaG initiation sites (1576-1611, Fig. 3).

Interactions in the Inc-Cop Region. The models described above assumed that the correct gene products were available when they were required. Based on the structure of the map in Fig. 8, the availability of these gene products is likely to be controlled entirely within the inc-cop region, which is consistent with the fact that the replication control of these plasmids is predominantly negative (4-6) and that negative control functions have been mapped to this region (14-16). Previously, we proposed that the base differences between pSM1 and pTR1 at base positions 251 and 256 (Fig. 3)

corresponded at least in part, to the original pRR12 and
pKN102 copy mutations (18). Recent sequencing of R1 (34)
supports this prediction and suggests that the base change
G→T (251) corresponds to the pRR12 mutation and the change
G→A (256) corresponds to the pKN102 mutation. As discussed
below, it is possible that each of these changes simultaneously affects several aspects of replication control.

One possibility for a regulatory mechanism might be an
autorepressor system such as that proposed by Sompayrac and
Maaløe (38), in which a repressor regulates the rate of
transcription of a polycistronic message which encodes both
the repressor itself and some function which is required for
replication. This type of regulation has been described for
λdv (39). A similar situation would exist for R1 and R100
if RepA3 exists and binds near pII (Fig. 8), preventing
transcription of RNAII, which could code for RepA1 and RepA3.
An example of this scheme might be:

1. RepA3 is normally bound to a site near pII.
2. RepA3 dissociates from pII at a certain frequency.
The rate at which the repressor and pII reassociate is reduced by inhibitor dilution (38,40) as the cell grows.
3. Unbound pII promotes the transcription of RNAII.
4. RepA1 and RepA3 are translated from RNAII but the
effective concentration of RepA1 rises more rapidly than the
effective concentration of RepA3. In this way a sufficient
amount of RepA1 can accumulate before transcription is again
repressed. Two observations from the DNA sequence address
this possibility:

    a. The region preceeding the RepA1 coding frame
    possesses a four base Shine Delgarno sequence, AGGA.
    RepA3 is preceeded by a five base Shine Delgarno se-
    quence AAGGA, but this is 10 bases away from the initia-
    tor codon and may not function efficiently. A shorter
    homology to 16S rRNA, GGU, is found closer to the
    initiator codon. Thus translation may be initiated more
    efficiently for RepA1 than for RepA3.

    b. The codon usage in RepA3 is shifted in favor of
    codons which are rarely used in *E. coli*. Thus, the rate
    of translation of RepA3 may be retarded relative to the
    rate of translation of RepA1, whose codon usage is
    normal. The fact that the amino acid changes in RepA3
    of both pSM1 and pTR1 are also changes to unusual co-
    dons may mean that RepA3 from these plasmids is pro-
    duced at a lower rate than from wild-type in addition to
    being structurally defective.

Thus, the production of repressor might initially increase
more slowly than the production of activator. This type of
regulation, combined with rapid degradation of RepA1 could
result in brief peaks of RepA1 activity which correspond to

8 DISSECTION OF THE REPLICATION REGION

replication-competent states. This aspect of the proposed model differs slightly from that of Sompayrac and Maaløe which predicts a constant concentration of activator across the cell cycle, with initiation-competence corresponding to accumulation of a critical mass of activator.

The possibility that the target site for repressor action is near pII is uncertain in view of the fact that the pTR1 and pSM1 sequences are identical between the end of the non-homologous region (NHR) and the beginning of RepA3. There is genetic evidence that both the pRR12 (pSM1) and the pKN102 (pTR1) copy mutations have cis-dominant character, implicating a target site mutation (7,15). However, when the pSM1 and pTR1 sequences in this region are compared to the sequence reported for R1 (34), there is one difference. The sequence of pTR1 and pSM1 between 160-165 (Figs. 3,4) is CTGGCC whereas the R1 sequence in the corresponding region is CTGCC, deleting one G. Both pTR1 and pSM1 are cut by *Hae*III at this site (18,19) so the presence of both G's in these plasmids is confirmed. Therefore, it is possible that pSM1 and pTR1 have identical target site mutations, the insertion of an extra base. However, it is also possible that the R1 stock used by Stougaard et al. for sequencing has a single base deletion which is phenotypically silent.

Since RNAI also contains the two base changes corresponding to the copy number changes, and as both changes occur at the top of a stable secondary structure, RNAI is a strong candidate for a regulatory element as well. The base changes in RNAI may correspond to the target site mutations. One model in which RNAI acts as a target site for a protein repressor was described above.

Another model we have considered for control exerted by RNAI takes advantage of the relationship between RNA secondary structure and transcription termination seen in attenuator systems (41). RNAI does have features found at ribosome binding sites, though ribosome binding may be inefficient. As described above, translation of nascent RNAI would result in an extended transcript, RNAIx. Based on the length of RNAIx observed *in vitro*, its transcription would extend into the pII region. As shown schematically in Fig. 9, this might influence the expression of RNAII. This scheme suggests a way in which transcription of RNAII could be regulated whether or not pII is a site for repressor binding. For example, some protein may bind to the large stem of nascent RNAI, helping to stabilize it. This binding may compete with ribosome binding, with the equilibrium state normally in favor of bound repressor.

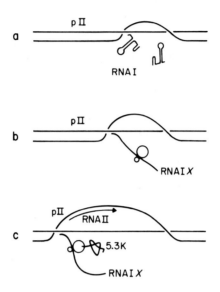

Figure 9. One possible involvement of RNAI in replication control. In this model: a.) RNAI usually terminates at tI, producing the transcript shown in Fig. 5. b.) If a ribosome succeeds in binding to nascent RNAI, transcription extends beyond the normal termination point, producing RNAIx. c.) Transcription of RNAIx through pII results in transcription of RNAII, which may promote initiation by various mechanisms as described in the text.

Non-Homologous Region (NHR). The conservation of the amino acid sequence of RepA2 from pSM1 and Rsc13 relative to the DNA sequence conservation in the NHR suggests that these two proteins may have similar functions, but it is difficult to predict what that function might be. One proposal for the function of this protein from R1 involves copy number control (33).

The sequence of Rsc13 and pSM1 in the NHR is an interesting evolutionary study. Calculations based on those described by Kimura (42) show that the sequence from the NHR of pSM1 and Rsc13 coding for RepA2 has diverged over a period of time at least ten times longer than that over which the RepA1's from pSM1 and pTR1 have diverged. It seems likely that the RepA1 sequences represent the extent to which R100 and R1 have diverged. The transition between homologous and non-homologous sequences is abrupt and occurs within short

8 DISSECTION OF THE REPLICATION REGION

inverted repeats at both ends of the NHR (Fig. 7 and ref. 19). These facts suggest the intriguing possibility that R1 and R100 differ in the NHR by a substitution of DNA which is evolutionarily related to the DNA which was replaced.

NOTE:

Larger copies of the DNA sequence figures (Figs. 3,4, & 7) are available upon request to the authors.

ACKNOWLEDGMENTS

We thank Kate Nyman for critical reading of the manuscript. We also thank Sandra Donaldson and Linda Hollmann for typing the manuscript, Mary Anne Huntington for the illustrations, and Jeff Demian for the photographs.

REFERENCES

1. Meynell, E., Meynell, G.G., and Datta, N. *Bacteriol. Rev.* 32, 55 (1968).
2. Datta, N., In *"Microbiology* 1974" (D. Slessinger, ed.) p. 9. American Society for Microbiology, Washington, D.C. (1974).
3. Sharp, P.A., Cohen, S.N., and Davidson, N., *J. Mol. Biol.* 75, 235 (1973).
4. Nordström, K., Ingram, L.C., and Lundbäck, A., *J. Bacteriol.* 110, 562 (1972).
5. Morris, C.F., Hashimoto, H., Mickel, S., and Rownd, R., *J. Bacteriol.* 118, 855 (1974).
6. Uhlin, B.E., and Nordström, K., *J. Bacteriol.* 124, 641 (1975).
7. Gustafsson, P., Dressig, H., Molin, S., Nordström, K., and Uhlin, B.E., *Cold Spring Harbor Symp. Quant. Biol.* 43, 419 (1978).
8. Ike, Y., Hashimoto, H., Motohashi, K., Fujisawa, N., and Mitsuhashi, S., *J. Bacteriol.* 141, 577 (1980).
9. Ohtsubo, E., Rosenbloom, M., Schrempf, H., Goebel, W., and Rosen, J., *Mol. Gen. Genet.* 159, 131 (1978).
10. Ohtsubo, E., Feingold, J., Ohtsubo, H., Mickel, S., and Bauer, W., *Plasmid* 1, 8 (1977).
11. Silver, L., Chandler, M., delaTour, E.B., and Caro, L., *J. Bacteriol.* 131, 929 (1977).
12. Yoshikawa, M., *J. Bacteriol.* 118, 1123 (1974).
13. Kollek, R., Oertel, W., and Goebel, W., *Mol. Gen. Genet.* 162, 51 (1978).

14. Taylor, D.P., and Cohen, S.N., *J. Bacteriol.* <u>137</u>, 92 (1979).
15. Miki, T., Easton, A.M., and Rownd, R.H., *J. Bacteriol.* <u>141</u>, 87 (1980).
16. Molin, S., and Nordström, K., *J. Bacteriol.* <u>141</u>, 111 (1980).
17. Rosen, J., Ohtsubo, H., and Ohtsubo, E., *Mol. Gen. Genet.* <u>171</u>, 287 (1979).
18. Rosen, J., Ryder, T., Inokuchi, H., Ohtsubo, H., and Ohtsubo, E., *Mol. Gen. Genet.* <u>179</u>, 527 (1980).
19. Ryder, T., Rosen, J., Ohtsubo, E., and Ohtsubo, H., Submitted to *J. Bacteriol.*
20. Rosen, J., Ryder, T., Ohtsubo, H., and Ohtsubo, E., *Nature*, in press (1981).
21. Mickel, S., Ohtsubo, E., and Bauer, W., *Gene* <u>2</u>, 193 (1977).
22. Rosenberg, M., and Court, D., *Ann. Rev. Genet.* <u>13</u>, 319 (1979).
23. Rosenberg, M., Court, D., Shimatake, H., Brady, C., and Wulf, D.L., *Nature* <u>272</u>, 414 (1978).
24. Küpper, H., Sekiya, T., Rosenberg, M., Egan, J., and Landy, A., *Nature* <u>272</u>, 423 (1978).
25. Shine, J., and Delgarno, K., *Proc. Natl. Acad. Sci. U.S.A.* <u>71</u>, 1342 (1974).
26. Atkins, J.F., *Nucleic Acids Res.* <u>7</u>, 1035 (1979).
27. Nichols, B.P., Miozzari, G.F., van Cleemput, M., Bennett, G., and Yanofsky, C. *J. Mol. Biol.* <u>142</u>, 503 (1980).
28. Oertel, W., Kollek, R., Beck, E., and Goebel, W., *Mol. Gen. Genet.* <u>171</u>, 277 (1979).
29. Itoh, T. and Tomizawa, J., *Proc. Natl. Acad. Sci. U.S.A.* <u>77</u>, 2450 (1980).
30. Sims, J., Capon, D., and Dressler, D., *J. Biol. Chem.* <u>254</u>, 12615 (1979).
31. Sugimoto, K., Oka, A., Sugisaki, H., Takanami, M., Nishimura, A., Yasuda, Y., and Hirota, Y., *Proc. Natl. Acad. Sci. U.S.A.* <u>76</u>, 575 (1979).
32. Grosschedl, R., and Hobom, G., *Nature* <u>277</u>, 621 (1979).
33. Molin, S., Stougaard, P., Light, J., Nordström, M., and Nordström, K., *Mol. Gen. Genet.* <u>181</u>, 123 (1981).
34. Stougaard, P., Molin, S., Nordström, K., and Hansen, F.G., *Mol. Gen. Genet.* <u>181</u>, 116 (1981).
35. Wang, P.Y., and Iyer, V.N., *Plasmid* <u>1</u>, 19 (1977).
36. Diaz, R., Nordström, K., and Staudenbauer, W.L. *Nature* <u>289</u>, 326 (1981).
37. Dove, W.F., Inokuchi, H., and Stevens, W.F. *In* "The Bacteriophage Lambda" (A.D. Hershey, ed.) p. 147 Cold Spring Harbor Laboratory, Cold Spring Harbor, New York (1971).

38. Sompayrac, L., and Maaløe, O., *Nature New Biol.* 241, 133 (1973).
39. Matsubara, K., *J. Mol. Biol.* 102, 427 (1976).
40. Pritchard, R.H., Barth, P.T. and Collins, J., *Symp. Soc. Gen. Microbiol.* 19, 263 (1969).
41. Lee, F., and Yanofsky, C., *Proc. Natl. Acad. Sci. U.S.A.* 74, 4365 (1977).
42. Kimura, M., *Proc. Natl. Acad. Sci. U.S.A.* 78, 454 (1981).
43. Armstrong, K., Rosen, J., Ryder, T., Ohtsubo, E., and Ohtsubo, H., *In* "Molecular Biology, Pathogenicity and Ecology of Bacterial Plasmids" (S.B. Levy, ed.) Plenum Publishing Inc. (1981).
44. Goebel, W., and Bonewald, R., *J. Bacteriol.* 123, 658 (1975).
45. Mickel, S. and Bauer, W., *J. Bacteriol.* 127, 644 (1976).
46. Grunberg-Manago, M., and Gros, F., *Prog. Nucleic Acid Res. Mol. Biol.* 20, 209 (1977).
47. Meagher, R.B., Tait, R.C., Betlach, M., and Boyer, H.W., *Cell* 10, 521 (1977).
48. Tegtmeyer, P., Schwartz, M., Collins, J.K., and Rundell, K., *J. Virol.* 16, 168 (1975).

DIRECT REPEATS OF NUCLEOTIDE SEQUENCES ARE INVOLVED IN PLASMID REPLICATION AND INCOMPATIBILITY

David M. Stalker, Avigdor Shafferman, Aslihan Tolun, Roberto Kolter, Shengli Yang and Donald R. Helinski

Department of Biology B-022, University of California, San Diego, La Jolla, CA 92093

ABSTRACT

We have been concerned with the structural features of the replication/incompatibility region of three plasmid elements R6K (38 kb), RK2 (56 kb) and F (94.5 kb). Plasmid R6K exhibits three origins of replication (designated α, β and γ) both *in vivo* and in an *in vitro* replication system. An R6K specified protein, π protein, is required for the initiation of R6K DNA replication. The complete nucleotide sequence has been obtained for the R6K γ-replicon consisting of the γ-origin and the π gene. The R6K γ-origin has been localized to a 260 bp region containing seven 22 bp direct repeats. Direct repeats of nucleotide sequences also have been identified in the replication origin region of plasmid RK2 and in the *inc*C region of mini-F, a derivative of the F plasmid. Eight 17 bp direct repeats are found in the RK2 origin region while five 22 bp direct repeats reside in the mini-F *inc*C region. The direct repeats determine incompatibility for these two plasmid systems. In addition to their requirement for an active R6K γ-origin, the 22 bp direct repeats of R6K also play a role in R6K specified incompatibility. Thus, in the case of three different plasmids, a striking feature of the essential replication region is the presence of direct repeats. In addition, these repeats play a role in plasmid incompatibility for all three systems.

INTRODUCTION

Plasmids have been found to determine a wide variety of phenotypic traits of both gram-positive and gram-negative bacteria. These covalently-closed circular DNA elements vary considerably in size and copy number. Unlike many bacterial viruses, the replication of plasmid DNA is regulated and this process assures their stable maintenance in the extrachromosomal state with a copy number characteristic of the particular plasmid. The hereditary stability of plasmids in a growing population of cells suggests the operation of a segregation apparatus distributing at least one copy of a plasmid molecule to each daughter cell during cell division. Competition of similar plasmids for a replication and/or putative segregation apparatus may account for the phenomenon of plasmid incompatibility.

Detailed analysis of plasmid DNA segments essential for replication and stable maintenance in a bacterial host can be obtained by the application of recombinant DNA techniques, rapid nucleotide sequence analysis and the development of *in vitro* systems for plasmid replication. These different approaches have been used in the analysis of the essential region for plasmid R6K DNA replication. In addition, segments of the replication regions of plasmids R6K, F and RK2 that are involved in plasmid incompatibility have been isolated and characterized. A striking feature in the essential region of replication of these three plasmid elements is the presence of direct repeats of nucleotide sequences. The major structural features of the essential region for plasmid R6K replication and the important role of the direct repeats in both plasmid replication and incompatibility will be considered in this article.

RESULTS

Plasmid R6K. Plasmid R6K is a naturally occurring 38 kilobase genetic element that specifies resistance to the antibiotics ampicillin and streptomycin. This plasmid exists at 10-15 copies per chromosome in *E. coli* (1). Electron microscopic analysis of replicative intermediates of R6K revealed three origins of replication (designated α, β, γ) and a fixed assymetric terminus of replication (Fig. 1). In *E. coli*, the α and β origins exhibit a sequential, bidirectional mode of replication toward the assymetric terminus of replication (2, 3). The frequency of *in vivo* usage of these three origins of replication differs from the frequency observed in an *in vitro* system developed for the replication of plasmid R6K. *In vivo*, the α and β origins are predominantly used while all three origins are utilized with approximately equal frequency *in vitro* (4, 5).

The use of the restriction endonuclease Hind III allowed us to delete regions of plasmid R6K not essential for replication (6-8). The replication region of R6K is contained in a DNA segment approximately 4 kb in length and includes the adjacent Hind III fragments 2, 4, 9 and 15 (Fig. 1). A number of autonomously replicating low molecular weight derivatives of R6K have been obtained which have in

9 DIRECT REPEATS OF NUCLEOTIDE SEQUENCES

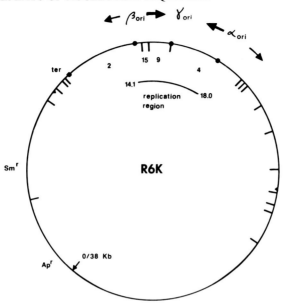

Figure 1. Physical and genetic map of plasmid R6K. Hash marks on the plasmid refer to the position of Hind III sites and 2, 15, 9 and 4 refer to the Hind III DNA fragments which span the origin of R6K DNA replication. *ter* refers to the assymetric terminus of replication. The streptomycin and ampicillin resistance genes are indicated by Sm^r and Ap^r respectively. The single BamHI restriction site is located by an arrow (↓).

common Hind III fragments 9 and 15 and a portion of Hind III fragment 4 (7). This minimal replicon, shown in Fig. 2, consists of two separable components: a DNA segment containing the γ-origin of replication and a structural gene *pir*, that specifies a product acting *in trans* to support the replication of the γ-origin (9). Insertion of DNA fragments between Hind III fragments 4 and 9 inactivates the γ-origin of replication. Attempts to derive minimal replicons for the α and β origins have been unsuccessful. Development of an *in vitro* system for the replication of plasmid R6K and its derivatives identified a 35,000 dalton polypeptide as the trans-acting product of the *pir* gene. This protein, designated π, is required for initiation of R6K DNA replication both *in vitro* and *in vivo* (10, 11).

Nucleotide sequence techniques were used to determine the entire π gene γ-origin replicon consisting of 1583 bp (12, 13). The minimal γ-origin component of this replicon, defined by insertion of the Tn5 transposon, was found to

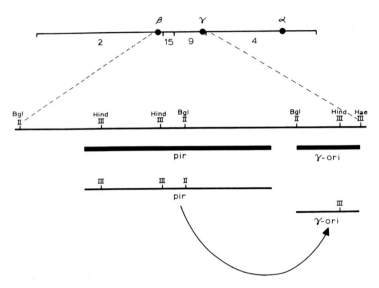

Figure 2. Unique features of the R6K replication region. Relevant restriction endonuclease cleavage sites are shown. *pir* refers to the structural gene region that specifies the π protein. The region containing the functional γ-origin is indicated.

consist of a 260 bp region extending from a short distance to the left of the Hind III 4/9 junction to the Bgl II site in Hind III fragment 9 (Fig. 3) (14). This region contains seven tandem 22 bp direct repeats and a 100 bp region rich in AT. If three or more of these direct repeats are removed, functional γ-origin activity is lost (14). An eighth 22 bp direct repeat is located near the putative promotor for the *pir* gene (13).

Analysis of the 1583 bp sequence for the presence of translation initiation and termination codons revealed a single open reading frame that begins in Hind III fragment 9, runs through Hind III fragment 15 and terminates in the adjacent Hind III fragment 2 (13). A polypeptide of 35,000 molecular weight was deduced from the putative amino acid sequence and this is in agreement with the size estimate of the π protein produced in *E. coli* minicells.

Role of the π Protein in R6K Replication. Both *in vivo* and *in vitro* evidence have been obtained for the essential role of the π protein in the initiation of plasmid R6K DNA replication (7, 9, 11). A role for the 22 bp direct repeats in γ-origin activity has been established by deletion analysis in

9 DIRECT REPEATS OF NUCLEOTIDE SEQUENCES   117

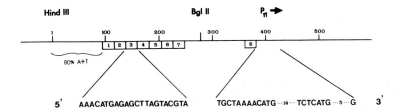

Figure 3. Schematic representation of the nucleotide sequence of the γ-origin region. The eight 22 bp direct repeats are indicated by the boxes. Numbers above the line refer to base pair positions. Pπ is the location of the promotor for the π structural gene. The sequence of one of the direct repeats and the π promotor region are displayed in detail.

the γ-origin region (14). If π protein functions as a regulatory protein, then control of the initiation frequency of the γ-origin could involve a relatively simple circuit (Fig. 4) consisting of the interaction of π with the seven tandem direct repeats within the γ-origin. Autoregulated expression of the π structural gene would be mediated by interaction of π protein with the eighth direct repeat near the putative π promotor.

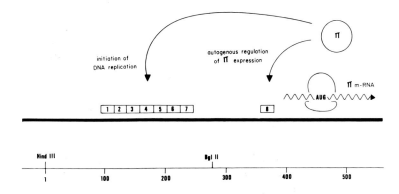

Figure 4. Model for the role of the π protein in the initiation of DNA replication of plasmid R6K. The 22 bp direct repeats are indicated by the numbered boxes.

If the expression of π is autoregulated, then a change in the number of *pir* genes per cell should not affect linearly the concentration of π protein in that cell. *E. coli* strains containing one copy of the *pir* gene per cell were obtained by lysogenization with a λ-*pir* hybrid phage (9, 15). 20 to 40 copies per cell of the *pir* gene were obtained by transformation of *E. coli* with a Col El-*pir* hybrid plasmid. The amount of π per cell was monitored by the *in vitro* R6K replication assay using extracts prepared from λ-*pir* lysogens and Col El-*pir* transformants. Similar amounts of π were recovered from both cell types (15). These results indicate that π concentration in the cell remains relatively constant despite a 20-fold increase in the *pir* gene dosage. The activity of the *pir* gene promotor was monitored by fusing a DNA sequence containing the putative π promotor to the lac Z gene (15, 16). The amount of β-galactosidase expression in cells carrying this *pir lac* fusion plasmid reflects the level of π promotor activity. If π protein is provided *in trans* in a cell containing the *pir lac* fusion plasmid, a significant reduction in the level of β-galactosidase expression is observed (15). These two experimental conditions indicate that the π protein interacts with its own promotor region thereby regulating its own expression.

A fundamental role assigned to π in the working model (Fig. 4) is its ability to regulate positively the frequency of initiation of R6K DNA replication. This regulatory role of π was tested by placing π expression under different promotors and assaying for the effects of different cellular levels of π on the copy number of pRK526 (9), an R6K derivative containing the γ-origin. A DNA fragment containing the entire *pir* structural gene, but deleted for the *pir* promotor region and the first two N-terminal amino acids of the π protein, was isolated. This fragment was fused in the correct translational reading frame to a tryptophan promotor fragment containing the seven N-terminal codons of trpE (15). This fragment also provides a promotor, a ribosomal binding site and a translational start signal. Cells carrying this fusion plasmid were transformed with the R6K γ-origin plasmid pRK526. By varying conditions for tryptophan expression, the copy number of pRK526 could be determined for varying cellular concentrations of π protein. The cellular concentration of π protein was assayed both by measuring the synthesis of π protein in minicells and employing the *in vitro* R6K replication assay. Results show that varying the concentration of π in the cell over a 70-95 fold range has no effect on the copy number of the R6K γ-origin plasmid (15). This observation argues against a positive regulatory role for π protein in determining the frequency of initiation events.

<u>Requirements for α and β Origin Activity.</u> The α and β origins of replication for plasmid R6K are the predominant origins

utilized *in vivo* (2-4). Segments of DNA containing either of these two origins cannot be isolated even when the π protein is supplied *in trans*. To test the hypotheses that the π protein requirement is limited to γ-origin function, site-specific insertions were generated *in vitro* by insertion of a small DNA fragment within the *pir* structural gene. The data from these experiments are summarized in Figure 5. In the construction of plasmid pAS808, a 58 bp Hind III fragment was inserted in the junction of Hind III fragments 9 and 15. The insertion of this small DNA fragment results in an inactive π protein. Plasmid pAS808 could be transformed into a *polA* strain of *E. coli* only when a functional π protein was supplied *in trans*. This experiment demonstrates the requirement for a functional π protein for replication from the α and β origins as well as the γ-origin.

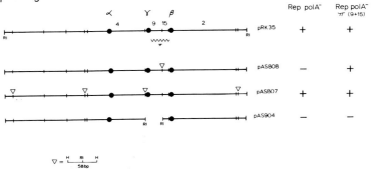

Figure 5. Ability of plasmids containing mutant R6K replication regions to be maintained in *E. coli* DNA polymerase I deficient strains. Positions of the α, β and γ origins of replication are indicated with respect to the Hind III fragments. Insertion sites for a 58 bp Hind III fragment are indicated (∇). EcoRI restriction sites are designated RI. Rep refers to the replication of the plasmids in the presence or absence of functional π protein.

To test whether the γ-origin region is required *in cis* for α and β origin activity, plasmid pAS904, a deletion mutant of the R6K replicon missing Hind III fragment 9, was constructed (Fig. 5). Deletion of this DNA fragment from the R6K replication region removes the seven 22 bp direct repeats essential for functional γ-origin activity and the majority of the π protein N-terminal region. Plasmid pAS904, which contains both the α and β origins of replication, but lacking the γ-origin and π N-terminal regions, cannot be maintained in *E. coli* even when functional π protein and γ-origin sequences are supplied *in trans*. Thus, replication of R6K from the α and β origins requires a functional π protein and a *cis* interaction which is supplied by the γ-origin region. It is conceivable that the proposed interaction of π protein with the direct repeats in the γ-origin region activates the α and β origins either via a transcriptional activation event or by promoting synthesis of RNA transcripts that are subsequently processed into functional initiation primers specific for the α and β origins.

Plasmid Incompatibility is Specified by Direct Repeats of Nucleotide Sequences. The γ-origin region of R6K containing the seven 22 bp direct repeats is also involved in the expression of R6K incompatibility. Various DNA segments containing these direct repeat units have been cloned into compatible plasmid replicons such as ColE1 and pACYC184. Plasmids carrying these R6K direct repeat units were found to eliminate either the parent R6K plasmid molecule or smaller autonomously self-replicating derivatives of R6K.

Direct repeats have also been identified in an incompatibility region of plasmid mini-F, a low molecular weight derivative of the F plasmid (17). The mini-F *incC* region of approximately 600 bp (45.8 - 46.4 kb on the F plasmid map) has been cloned onto a ColE1 replicon and the resulting plasmid shown to express incompatibility with mini-F derivatives (M. Kahn, unpublished observations). Mutations obtained in this region also result in a higher mini-F plasmid copy number and a loss of incompatibility properties (18). The *incC* region was subjected to nucleotide sequence analysis to identify the incompatibility determinant expressed by this DNA segment (19). As summarized in Fig. 6, the prominent features of this 543 bp region include the presence of five 22 bp direct repeats and open reading frames for two small putative

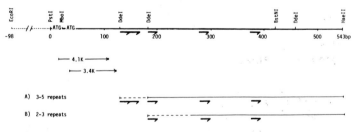

Figure 6. Relevant features of the mini-F *incC* region as determined from the nucleotide sequence. Restriction endonuclease sites are indicated. The 22 bp mini-F direct repeats are indicated by arrows (→). ATG refers to the start positions of the two small putative polypeptides. The region shown by a dotted line is a segment of ColE1 DNA. Solid lines in A and B indicate regions of DNA which are unambiguously present while dashed lines show regions that may be present.

9 DIRECT REPEATS OF NUCLEOTIDE SEQUENCES

polypeptides. Deletion derivatives (type A in Fig. 6) of the $incC$ region lacking the translational start signals for these two putative polypeptides exhibited no decrease in the expression of incompatibility by the $incC$ region. However, deletions retaining only two or three of the 22 bp direct repeats (type B) exhibited markedly decreased incompatibility properties.

To test directly whether the direct repeats express incompatibility, a 58 bp Dde I fragment containing two of the 22 bp repeats was cloned into various sites in pACYC184. These hybrid plasmids expressed incompatibility not only against the mini-F plasmid but also against F'lac. When two copies of this fragment were inserted in tandem, expression of incompatibility was considerably stronger (19).

Nucleotide sequence analysis of the origin of replication of the broad host range plasmid RK2 also revealed the presence of direct repeats of nucleotide sequences (20). A 617 bp sequence of the RK2 origin region contains eight 17 bp direct repeats. The sequence of these direct repeats is shown in Figure 7. Five of the repeat units reside in a 393 bp region functioning as a minimal RK2 origin of replication while the remaining three are contained in a 140 bp segment adjacent to this functional RK2 origin and contribute to the stability of RK2 origin plasmids (21). When DNA segments containing these RK2 direct repeat units were tested against RK2 replicons, it was found that all eight of the 17 bp direct repeats appear to be involved in RK2 specified plasmid incompatibility (20, 21).

DISCUSSION

Direct repeats of nucleotide sequences have been shown to play a vital part in the replication and incompatibility properties of plasmids R6K, mini-F and the broad host range plasmid RK2. Families of direct repeats from each of these plasmid systems are displayed in Fig. 7. Also included in this figure are the four 18 bp direct repeats located within the bacteriophage lambda origin of replication (22). Extensive regions of nucleotide sequence containing direct repeats are not, however, ubiquitous to all prokaryotic replicons. The origin ($oriC$) regions of the chromosome of many of the Enterobacteriacae and the replication region of plasmid R1 lack such nucleotide sequences (23-25).

It is of interest that the hexanucleotide TGAGGG is present in the direct repeats of these four prokaryotic systems. This sequence is also found once in the $oriC$ region of *E. coli.* (23). Six of the seven 22 bp direct repeats located in the R6K γ-origin region, however, contain the hexanucleotide sequence TGAGAG. The eighth repeat, located near the promotor for the *pir* gene (Fig. 3) contains a TGAGTG hexanucleotide sequence. Conservation of the TGAGPuG hexanucleotide core within the repeats may identify a common host component utilized by these four prokaryotic replicons for their stable maintenance in *E. coli*. At present the biochemical nature of the role of these direct repeats in replication or origin activity and incompatibility is unknown. Clearly the repeats can serve as binding sites for plasmid specific or host proteins involved in the replication and/or plasmid

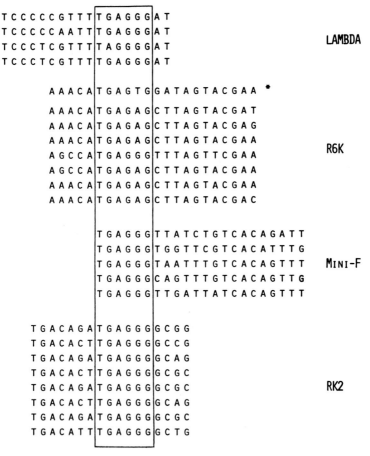

Figure 7. Families of direct repeat units from 4 different prokaryotic replicons are displayed. Sequences are aligned by a box to show the presence of a 6 bp homologous region. The eighth direct repeat of plasmid R6K that is near the π promotor region is indicated (*).

partitioning process. Indirect evidence has been obtained for the binding of the essential π protein to the R6K direct repeats. Additionally, the direct repeats may facilitate association of the plasmid with a replication and/or plasmid partitioning membrane site. Finally, it is possible that RNA transcripts of the regions containing direct repeat sequences account for their role in replication and/or incompatibility.

ACKNOWLEDGMENT

This work was supported by the National Institutes of Health and the National Science Foundation. Avigdor Shafferman is a recipient of J.F. Fogarty International Research Fellowship; David M. Stalker is a recipient of an N.I.H. Postdoctoral Fellowship; Aslıhan Tolun is supported by a Damon Runyon-Walter Winchell Cancer Fund (DR6-294-F).

REFERENCES

1. Kontomichalou, P., Mitani, M. and Clowes, R.C. (1970). J. Bacteriol. 104, 33-44.

2. Lovett, M.L., Sparks, R.B. and Helinski, D.R. (1975). Proc. Natl. Acad. Sci. USA 72, 2905-2909.

3. Crosa, J.H., Luttrop, L.K., Heffron, F. and Falkow, S. (1975). Mol. Gen. Genet. 140, 39-50.

4. Crosa, J.H. (1980). J. Biol. Chem. 255, 11075-11077.

5. Inuzuka, N., Inuzuka, M. and Helinski, D.R. (1980). J. Biol. Chem. 255, 11071-11074.

6. Crosa, J.H., Luttrop, L.K. and Falkow, S. (1978). J. Mol. Biol. 124, 443-468.

7. Kolter, R. and Helinski, D.R. (1978). Plasmid 1, 571-580.

8. Kolter, R. and Helinski, D.R. (1978). J. Mol. Biol. 124, 425-444.

9. Kolter, R., Inuzuka, M. and Helinski, D.R. (1978). Cell 15, 1199-1208.

10. Inuzuka, M. and Helinski, D.R. (1978). Biochemistry 17, 2567-2573.

11. Inuzuka, M. and Helinski, D.R. (1978). Proc. Natl. Acad. Sci. USA 75, 5381-5385.

12. Stalker, D.M., Kolter, R. and Helinski, D.R. (1979). Proc. Natl. Acad. Sci. USA 76, 1150-1154.

13. Stalker, D.M., Kolter, R. and Helinski, D.R. (submitted for publication).

14. Kolter, R. and Helinski, D.R. (submitted for publication).

15. Shafferman, A., Kolter, R., Stalker, D.M. and Helinski, D.R. (submitted for publication).

16. Casadaban, M. and Cohen, S.N. (1980). J. Mol. Biol. 138, 179-207.

17. Lovett, M.A. and Helinski, D.R. (1975). J. Bacteriol. 127, 982-987.

18. Manis, J.J. and Kline, B.C. (1978). Plasmid 1, 492-507.

19. Tolun, A. and Helinski, D.R. (1981). Cell 24, (in press).

20. Stalker, D.M., Thomas, C.M. and Helinski, D.R. (1981). Mol. Gen. Genet. 181, 8-12.

21. Thomas, C.M., Stalker, D.M. and Helinski, D.R. (1981). Mol. Gen. Genet. 181, 107.

22. Moore, D.D., Denniston-Thompson, K., Kruger, K.E., Furth, M.E., Williams, B.G., Daniels, D.L. and Blattner, F.R. (1978). CSHSQB 43, 155-164.

23. Zyskind, J.W., Harding, N.E., Takeda, Y., Cleary, J.M. and Smith, D.W. (this volume).

24. Oertel, W., Kolleck, R., Beck, E. and Goebel, W. (1979). Mol. Gen. Genet. 171, 277-285.

25. Rosen, J., Ohtsubo, H. and Ohtsubo, E. (1979). Mol. Gen. Genet. 171, 287-293.

INCOMPATIBILITY OF IncFII R PLASMID NR1

Alan M. Easton, Padmini Sampathkumar and Robert H. Rownd

Laboratory of Molecular Biology and Department of Biochemistry
University of Wisconsin, Madison, Wisconsin

ABSTRACT The region of the IncFII R plasmid NR1 which is sufficient for autonomous replication consists of two adjacent PstI fragments of sizes 1.1 and 1.6 kilobases (kb). The origin of DNA replication has been mapped within the PstI 1.6 kb fragment (1,2,3). Previously it has been shown that the functions of incompatibility, copy number control and is a target site for incompatibility reside on the PstI 1.1 kb fragment (4,5). We now describe experiments in which the lacZ gene has been cloned adjacent to the incompatibility region such that beta-galactosidase is produced as a result of transcription which originates in the incompatibility region and proceeds toward the origin of replication. Higher levels of beta-galactosidase were produced if the incompatibility region used was derived from the copy mutants pRR12 or pRR21 rather than from NR1. The amount of this transcription was found to decrease when extra copies of the incompatibility region were introduced into cell. The control of this transcription may, therefore, be a feature of the incompatibility mechanism of NR1.

We have identified five RNA transcripts produced in vitro from the 2.7 kb replicator region. Their coding regions and directions of transcription have been determined. One of the RNA species, about 1150 bases long, has been shown to originate within the PstI 1.1 kb fragment and to extend into the PstI 1.6 kb fragment. This is probably the species whose transcription is controlled by the incompatibility product. We speculate that this transcript has a role in the control of plasmid replication and that by regulating its initiation or extension, the incompatibility product can bring about plasmid exclusion.

INTRODUCTION

The majority of low copy bacterial plasmids are stably maintained in a population of host cells. Plasmid inheritance must be controlled by some mechanism which ensures that each daughter cell receives at least one copy of the plasmid and that the plasmid is replicated during each cell cycle. The molecular mechanisms by which these processes occur are not known. The proposal of Jacob et al (6) involves a positively acting substance or site for which plasmids

compete. Most available data, however, are consistent with a
negative control model. Pritchard and his colleagues (7,8)
have proposed that the number of new initiations of DNA
replication could be controlled by a replicon specified repressor which interacts with a site on the DNA molecule.
This model can account for the control of plasmid replication
and for the inability of plasmids which share the same replication machinery to coexist in the same cell line. Mutations
in either the repressor or in its target site could give rise
to less well regulated control and thus to copy number mutants.
Several of these copy mutants have been isolated and many,
but not all, have been found to have altered incompatibility
properties (9).

The R plasmid NR1 (also called R100 and R222) is a 90
kilobase (kb) self-transmissable drug resistance plasmid which
belongs in the incompatibility group IncFII. The locations
on the plasmid of resistance genes and the cleavage sites for
several restriction endonucleases have been mapped (10,11).
In this communication we present evidence for a model in which
the incompatibility product decreases the synthesis of a
transcript which may be important to plasmid replication.

RESULTS AND DISCUSSION

Identification of DNA Restriction Fragments Which Encode Functions of Autonomous Replication and Incompatibility of Plasmid NR1.

Restriction fragments of NR1 capable of mediating autonomous replication were identified in sequential in vitro deletion experiments some of which have been previously presented (11). Initially, small plasmids were obtained which
used EcoRI fragment B as the replicator (Fig. 1). From these,
nonessential PstI fragments were deleted leaving replicators
consisting of two adjacent PstI fragment of sizes 1.1 kb and
1.6 kb. The PstI 1.1 kb fragment contains the incompatibility
product (see below) while the PstI 1.6 kb fragment contains
the origin of replication (1,2). Nonessential Sau3A fragments
have been deleted from PstI 1.1 kb fragment reducing the
replicator region to less than 2.1 kb in size. Since this
smallest NR1 derivative still expresses incompatibility and
is excluded from the cells by NR1 (data not shown) the incompatibility functions must be within **the 500 base pair** segment
of the PstI 1.1 fragment adjacent to the PstI 1.6 fragment.

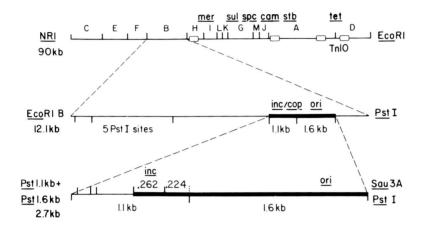

FIGURE 1. Subcloning of the Inc FII R Plasmid NR1. Restriction Fragments of NR1 capable of mediating autonomous replication were identified in sequential subcloning experiments in which replicator fragments were ligated to nonreplicating drug resistance fragments. The incompatibility (inc) and copy control functions (cop) have been mapped to the PstI 1.1 kb fragment (Table 1). The origin of replication (ori) is within the PstI 1.6 kb fragment (1,2,3). Non-essential Sau3A fragments have been deleted from the PstI 1.1 fragment leaving a replicator of about 2100 base pairs. This smallest derivative retains the abilities to exclude and be excluded indicating that these incompatibility functions lie within the 500 base pairs of PstI 1.1 kb fragment which remain. In Figure 1, the thicker lines represent fragments essential for replication.

TABLE 1. Properties of Mutant and Hybrid Replicators Derived from NR1

Resident plasmid[a]	NR1	pRR933	pRR955	pRR966	pRR942	pRR12	pRR1919
Source of PstI 1.1	NR1	NR1	NR1	pRR12	pRR12	pRR12	pRR21
Source of PstI 1.6	NR1	NR1	pRR12	NR1	pRR12	pRR12	pRR21
Copy number[b]	1.0	1.0	0.8	7.7	7.0	2.8	7.6[c]
Percent of transformants harboring resident plasmids after transformation by[d]							
pBR322 + PstI 1.1 (NR1)	0	0	0	100	100	100	0
pBR322 + PstI 1.1 (pRR12)	100	46	42	0	0	0	90
pBR322 + PstI 1.1 (pRR21)[e]	0	0	NT	NT	100	NT	0
pBR322	100	90	71	100	100	100	100

[a] Except for NR1 and pRR12, the 90 kb parent plasmids, all resident plasmids in this experiment were PstI generated plasmids with the structure shown in Figure 2.

[b] Copy numbers were determined by assays of the activity of chloramphenicol acetyl transferase in crude sonicates of cells harboring the plasmids.

[c] The copy number of pRR1919 was determined by measuring ratios of covalently closed circular DNA to chromosomal DNA in ethidium bromide-cesium chloride gradients. By this method there are $7.6 \pm .6$ copies of pRR1919 per cell, while there are 24 copies of pRR942 per cell, decidedly higher than determined by the other method.

[d] Incompatibility tests were performed as follows: E. coli KP435 (recA) cells harboring the resident plasmid were made competent to receive plasmid DNA. After uptake of DNA, the cells were cultured in drug-free broth and appropriate dilutions were then spread on nutrient agar plates containing a single drug to which resistance was conferred only by the donor plasmid. Ten individual transformants were streaked onto drug-free Penassay plates. From each of these streaks ten single colonies were picked and examined for drug resistance by replica plating. Low copy number resident plasmids which lack a region involved in plasmid stability (stb) (Figure 1) are lost from cells spontaneously in the course of the experiment. NR1 derivatives lacking stb are more sensitive to the incompatibility of pRR12 than is the parent NR1.

[e] This plasmid was generously provided by Dr. D.P. Taylor (14).

Mapping the Incompatibility Gene and a Target Site for its Action.

The plasmids pRR12 and pRR21 are copy mutants (cop^-) of NR1 (12,13). We have generated small plasmid derivatives from pRR12 in the same manner as described above for NR1. In addition, the individual PstI 1.1 kb and 1.6 kb fragments of both NR1 and pRR12 have been inserted into the PstI site on the cloning vehicle pBR322 (11). Similar plasmid constructs with fragments from pRR21 were provided by Dr. D.P. Taylor (14).

Incompatibility tests were performed by introducing these pBR322 derivatives into cells which harbored replicators from NR1, pRR12 or pRR21. As reported before (4,11) only the pBR322 carrying the PstI 1.1 kb fragment caused exclusion of the resident plasmid. This mapped the incompatibility function to this fragment. The pBR322 derivatives carrying the PstI 1.1 kb fragment of NR1 excluded NR1 and pRR21 from the cells but did not exclude pRR12. On the other hand, the pBR322 plasmid carrying the PstI 1.1 fragment of pRR12 excluded only the pRR12 residents and not the resident plasmids derived from NR1 or pRR21. The same specificity for exclusion was seen for NR1 and pRR12 even if they were held coresident in the same cell. Further, hybrid replicators were constructed which contained one PstI replicator fragment from NR1 and the other from PRR12. The incompatibility and copy number properties of these hybrid plasmids were the same as those of the parent of the PstI 1.1 kb fragment. This showed that a copy control function and a locus involved in sensitivity to exclusion are on the PstI 1.1 kb fragment (Table 1).

To explain these observations we assume a negative control model. In this scheme, the PstI 1.1 kb fragment produces an incompatibility repressor which acts on the molecule to cause plasmid exclusion. The target site for the repressor lies on the PstI 1.1 kb fragment. NR1 and pRR12 must differ in both components, the repressor and the target site. It is not known in which component the pRR21 mutation lies.

Effect of the Incompatibility Product on Transcription in the Replicator Region of NR1.

Molin and Nordstrom (15) have shown that the incompatibility product acts to shut off the DNA replication of the IncFII plasmids. We have defined a target site for this product which is distinct from the origin of replication itself. It was of interest to examine the possibility that the incompatibility substance acted on its target site to control transcription which might be involved in plasmid maintenance as in mechanisms described for control of λdv plasmid replication (16). The DNA sequence of the replicator region of pRR12 (determined by Rosen et al. (17)) showed that long open reading frames for proteins would be found only on transcripts which were read from the PstI 1.1 kb fragment toward the origin of replication. Such transcripts could provide products or cis activation necessary for initiation of DNA synthesis.

The incompatibility regions of NR1, pRR12 and pRR21 were inserted into the phage λRS205 next to a copy of the lacZ gene (Fig. 2). This phage, designed by Dr. K. Bertrand and Dr. W.S. Reznikoff, carries a copy of the beta-galactosidase gene along with its ribosome binding site but lacking its promoter. There are single recognition sites for EcoRI and SalI upstream from the ribosome binding site. Insertion of DNA at these sites can bring the expression of beta-galactosidase under the control of promoters on the inserted DNA. The plasmids used as a source for the incompatibility region were of the structure designated pRR933 shown in Fig. 2. These plasmids consisted of the two essential PstI fragments attached to a 2.1 kb PstI fragment (PstI cam) which encodes resistance to chloramphenicol. The enzyme SalI cleaves twice in the PstI 1.1 kb fragment at sites within 50 base pairs of the PstI site which divides the PstI 1.1 kb and PstI 1.6 kb fragments. The EcoRI site used is external to the replicator, in the gene for chloramphenicol acetyltransferase. Using these two restriction endonucleases, the entire incompatibility region could be inserted into the phage such that the target site for incompatibility was proximal to the lacZ gene and that transcription which in the plasmid would proceed toward the origin of replication would, in the phage, read through the lacZ gene.

The strain NK5031 produces no beta-galactosidase. Lysogens were constructed from this strain using the phages λRS205, λRS205 with the lactose promoter (lacP$^+$) inserted upstream of the lacZ gene, and the three types of inc-lac phages. The number of beta-galactosidase units produced from these lysogens are presented in Table 2. Lysogens with prophages with incompatibility region inserts produced more beta-galactosidase than the background levels produced in lysogens with the λRS205 prophage. The levels were higher if the incompatibility region inserts were derived from copy mutants rather than from NR1. Three or four independently constructed lysogens of each type were found to produce the same level of beta-galactosidase.

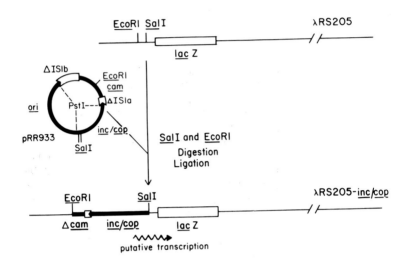

FIGURE 2. Construction of lambda phages carrying the incompatibility region adjacent to the lacZ gene. The region in which the incompatibility product was believed to act was cloned adjacent to a lacZ gene which lacks its own promoter. The locations of SalI and EcoRI sites were convenient for this purpose. One of the SalI sites in the NR1 replicator is 50 base pairs from the PstI site dividing the two essential PstI fragments. The EcoRI site is external to the replicator, in the gene for chloramphenicol acetyltransferase. The phage λ RS205 was constructed and provided by Dr. K. Bertrand and Dr. W. S. Reznikoff.

TABLE 2. β-Galactosidase Units Produced in Lysogens of
λ-inc/cop-lacZ phages

	Prophage in lysogen				
Plasmid in lysogen	λRS205	λRS205 inc-NR1	λRS205 inc-pRR12	λRS205 inc-pRR21	λRS205 lacP⁺
None	66	220±10	504±7	1577±117	1768
pBR322 + PstI 1.1 NR1	54	56±9	144±20	264±38	1433
pBR322 + PstI 1.1 pRR12	55	95±7	83±19	408±18	1830
pBR322 + PstI 1.1 pRR21	77	63±6	144±4	233±40	1679
pBR322	57	245±24	611	1616±145	

Beta-galactosidase units produced from lysogens made with λRS205 derivatives were measured in standard assays (18). The inc-lacZ fusions have the structure shown at the bottom of Figure 2. The strain used was NK5031, which carries a lacO,lacZ deletion and produces 0 beta-galactosidase units. The low levels produced from λRS205 may be due to read through transcription from phage or chromosomal genes. The introduction of pBR322 plasmids carrying a PstI 1.1 kb fragment resulted in reductions of the beta-galactosidase levels made in inc-lac lysogens. When the plasmids were cured from the cells the levels returned to those characteristic of the lysogen (data not shown).

The pBR322 plasmids carrying the individual PstI 1.1 kb fragments of NR1, pRR12 or pRR21 were introduced into the lysogens to test the effect of the extra copies of the incompatibility fragment on the levels of beta-galactosidase production. The introduction of these plasmids resulted in reduced levels of beta-galactosidase in each type of lysogen. When the plasmids were cured from the cells the amounts of beta-galactosidase produced returned to the level characteristic of the lysogen (data not shown). These same plasmids did not affect transcription which started from the lactose promoter or the basal level of transcription in λRS205. The plasmid pBR322 itself and derivatives carrying the PstI 1.6 kb fragments did not affect the level of beta-galactosidase production in the inc-lac lysogens.

These observations suggest that the incompatibility fragment encodes a repressor which acts specifically to reduce the transcription which is initiated in the incompatibility region. Moreover, for a given lysogen, the particular pBR322 derivatives which caused the greater decrease in the amount of beta-galactosidase were those which displayed the stronger capacity to exclude the plasmid from which the fragment in the prophage had been derived. Thus, the plasmid NR1 is excluded from cells to a greater extent by the pBR322 plasmid containing the PstI 1.1 kb fragment of NR1 than it is by the pBR322 derivative carrying the PstI 1.1 kb fragment of pRR12 (Table 1). From a prophage with the NR1 incompatibility region as a target the beta-galactosidase production was decreased more by the pBR322 derivative which causes the greater exclusion of NR1 (Table 2). Similarly, pRR12 is excluded more by its own incompatibility fragment when present on a high copy vector than it is by that of NR1 and the levels of transcription from the prophage insert derived from pRR12 was decreased more by the pBR322 plasmid which carries the incompatibility fragment of pRR12 (compare Table 1 and Table 2). There is then, a correlation between the specificity of plasmid exclusion and the depression of this particular transcription activity. This strengthens the premise that the incompatibility product is a transcription repressor and that the reduction of the transcriptional activity is related to plasmid exclusion.

As mentioned, above, there was a higher level of transcription through the lacZ gene in prophages carrying the incompatibility regions from copy mutants than there was in those with the incompatibility region of NR1. The amount of this transcription may therefore be related to copy number control. It should be noted that the incompatibility product would be made from the fragment in the prophage and would be

10 INCOMPATIBILITY OF IncFII R PLASMID NR1

acting on the resident target site. The transcription activity from the incompatibility region may reflect the balance between the amount and strength of the repressor and the sensitivity of the target site to its action. Differences in promoter strength between the mutants could also be a part of the cause for variations in the transcription levels.

Mapping of RNA Transcripts made in vitro from the Replicator Region of NR1.

In elucidating a model for incompatibility involving transcriptional control, it is important to understand the transcriptional organization of the replicator region. Toward this end, we have examined RNA transcripts made in vitro by RNA polymerase on the DNA template of plasmid pRR933 (structure shown in Fig. 2) in which the PstI 1.1 kb and 1.6 kb fragments have been ligated to a PstI fragment containing the chloramphenicol acetyltransferase gene. In addition, some smaller derivatives of this plasmid in which nonessential Sau3A fragments have been deleted were also used as templates in these studies.

Using pRR933 as template DNA, six RNA transcripts (1150, 720, 480, 380, 95 and 85 bases long) have been identified in vitro (Fig. 3). In order to determine their coding regions, the individual RNA transcripts have been isolated from gels and hybridized to various restriction fragments within the replicator region. The 720 base long transcript was observed to hybridize only to the PstI cam fragment and is probably the transcript for the chloramphenicol acetyltransferase gene. This transcript is smaller when produced from templates cleaved with EcoRI or PvuII and the sizes of the truncated transcripts correspond to those expected from the location of these restriction sites with respect to the promoter for the cam gene (Fig. 3) (19). The remaining five transcripts show no homology to the PstI cam fragment and hybridize to various restriction fragments within the PstI 1.1 kb and 1.6 kb fragments. These results are summarized in Fig. 4.

It can be seen from Fig. 4 that both RNA transcripts C and E hybridize only to the PstI 1.1 kb fragment. Of these, transcript C (380 bases) hybridizes to both the 331 base pair (bp) and 262 bp Sau3A fragments (Fig. 4) and is absent in transcripts made from templates which are deleted in the 331 bp Sau3A fragment (data not shown). This transcript must, therefore, extend rightwards from within the 331 bp Sau3A fragment to the 262 bp Sau3A fragment (Fig. 5). It is not

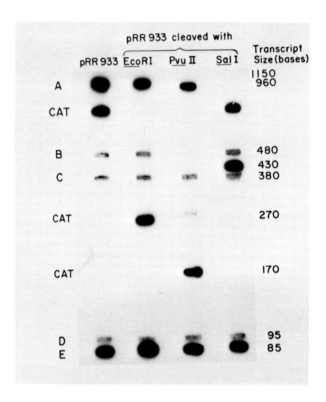

FIGURE 3. In vitro RNA transcripts from the replicator region of NR1. 5 µg of purified linear (plasmid fragment) or supercoiled (whole plasmid) DNA template was transcribed in vitro according to Weisblum et al. (20). The RNA was labeled with α^{32}P-ATP (24 Ci/mmole), purified over a G-50 Sephadex column, denatured using glyoxal and separated on 2.5% acrylamide-0.5% agarose composite gels according to McMaster and Carmichael (21).

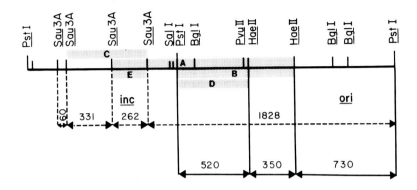

FIGURE 4. Hybridization of RNA transcripts to PstI 1.1 and PstI 1.6 kb fragments. α^{32}P-labeled RNA transcripts separated by gel electrophoresis were cut out of the gel, extracted with 0.2 M NaCl, ethanol precipitated, and hybridized to either linear (plasmid fragment) or supercoiled (whole plasmid) DNA fixed to nitrocellulose filters. Hybridizations were performed according to Wahl et al. (22).

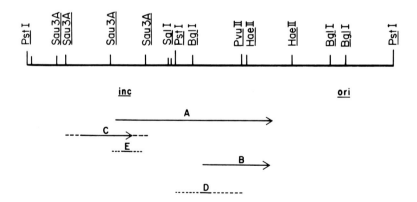

FIGURE 5. Map of the in vitro RNA transcripts.

essential for autonomous replication or incompatibility since the 331 bp Sau3A fragment coding for this RNA can be deleted from plasmids without loss of these functions. Transcript E (85 bases) hybridizes only to the 262 bp Sau3A fragment which is known to contain the incompatibility function. Perhaps this RNA is involved in the expression of this function.

The largest of the RNA transcripts, A (1150 bases), hybridizes to both the PstI 1.1 kb and 1.6 kb fragments. It does not hybridize to the 331 bp Sau3A fragment to the left of the 262 bp Sau3A fragment (Fig. 4) and is not observed in transcripts made from plasmids deleted in the 331 bp Sau3A fragment (data not shown). These observations suggest that RNA A probably originates within the 331 bp Sau3A fragment close to the Sau3A site between the 331 and 262 bp fragments and extends toward the origin of replication. In order to verify this, plasmid DNA pRR933 cleaved with enzymes PvuII or SalI was used as template for the in vitro synthesis of the RNA transcripts. The truncated transcript produced from the template DNA cleaved with SalI was shorter than that produced from the template cleaved with PvuII confirming that the transcript extends from the PstI 1.1 kb fragment to the PstI 1.6 kb fragment. Based on these results, RNA A must originate about 430 bases to the left of the SalI sites and extend about 720 bases to the right (Fig. 5). Since this is the only in vitro transcript that extends across the SalI sites, it is probably the species whose transcription is controlled by the incompatibility product suggesting an important role for this transcript in the control of replication. One such role may be as a primer RNA in DNA synthesis. The 3' end of this transcript is about 500 bases away from the origin of replication determined by Ohtsubo et al. (1) for the plasmid pRR12. However, the origin of replication mapped by Synenki et al. (2) for another IncFII R plasmid, R6-5, is only 40 bases away from the 3' end of transcript A. In addition, preliminary data from this laboratory (3) place the origin of replication for NR1 in a similar location to that determined by Synenki et al. (2). Given this ambiguity in the location of the origin of replication, it is not known if this in vitro product actually serves as a primer RNA in vivo.

Transcripts B and D hybridize only to the PstI 1.6 kb fragment. Of these, the larger transcript B (480 bases) hybridizes to both the 520 bp and 350 bp HaeII fragments within the PstI 1.6 kb fragment. This transcript is smaller when synthesized from template DNA cleaved with PvuII (Fig. 3) and the 280 base long truncated transcript hybridizes only to the

520 bp HaeII fragment within the PstI 1.6 kb fragment. This transcript therefore originates 280 bases to the left of the PvuII site and extends 200 bases to the right, terminating close to the 3' end of transcript A. Since it is read in the same direction as transcript A and terminates close to the 3' end of transcript A, it could conceivably also serve as a primer for DNA replication. Transcript D (95 bases), also encoded within the PstI 1.6 kb fragment, has no known function.

Model for incompatibility mechanism of R plasmid NR1

Comparison of the incompatibility properties of NR1 and of pRR12 have provided insight into the mechanism of Inc FII incompatibility. Both the diffusible incompatibility product and its target site differ between NR1 and pRR12, as depicted in Fig. 6. The two products act preferentially on the plasmid DNA from which they are encoded. The target site is near or within the sequence from which the product is encoded, a region distinct from the origin of replication. In this model, the incompatibility substance reduces the production of a transcript which starts in the region which contains the incompatibility target site and is read toward the origin of replication. We speculate that this transcript plays an essential role in DNA replication. It could encode a required protein product, serve as a primer for DNA synthesis, or by some other means activate initiation of DNA replication. The incompatibility product could shut off replication by diminishing the amount of this transcript.

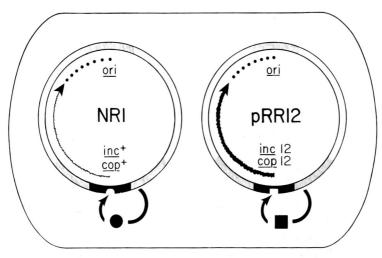

FIGURE 6. Model for incompatibility mechanism of R plasmid NR1.

ACKNOWLEDGMENTS

This work was supported in part by U.S. Public Health Service Research Grants GM14398 and GM26527 and U.S. Public Health Service Research Training Grants from the National Institute of General Medical Sciences.

REFERENCES

1. Ohtsubo, E. Feingold, J., Ohtsubo, H., Mickel, S. and Bauer, W. (1977). Plasmid 1, 8.
2. Synenki, R.M., Nordheim, A. and Timmis K.N. (1979). Mol. Gen. Genet. 168, 27.
3. Kasner, J.P. and Rownd, R.H., unpublished observations.
4. Rownd, R.H., Easton, A.M., Barton, C.R., Womble, D.D., McKell, J., Sampathkumar, P. and Luchow, V.A. (1980). In Mechanistic Studies of DNA replication and Genetic Recombination, B. Alberts and C.F. Fox, eds. ICN-UCLA Symposia on Molecular and Cellular Biology Vol. XIX p. 311.
5. Timmis, K.N., Andres, I. and Slocombe, P.M. (1978) Nature 273, 27.
6. Jacob, F., Brenner, S. and Cuzin, F. (1963). Cold Spring Harbor Symp. Quant. Biol. 28, 329.
7. Pritchard, R.H., Barth, P.T. and Collins, J. (1969). Symp. Soc. Gen. Microbiol. 19, 263.
8. Pritchard, R.H. (1978). In DNA Synthesis: Present and Future, I. Molineux and M. Kohiyama, eds., Plenum Publishing Corp. New York, p. 1.
9. Uhlin, B.E. and Nordstrom, K. (1975). J. Bacteriol. 124, 641.
10. Miki, T., Easton, A.M. and Rownd, R.H. (1978). Mol. Gen. Genet. 158, 217.
11. Miki, T., Easton, A.M. and Rownd, R.H. (1980). J. Bacteriol. 141, 87.
12. Morris, C.F., Hashimoto, H. Mickel, S. and Rownd, R.H. (1974). J. Bacteriol. 118, 855.
13. Taylor, D.P., Greenberg, J. and Rownd, R.H. (1977). J. Bacteriol. 132, 986.
14. Taylor, D.P., Cohen, S.N. (1979). J. Bacteriol. 137, 92.
15. Molin, S. and Nordstrom, K. (1979). J. Bacteriol. 141, 111.
16. Berg, D.E. and Kellenberger-Gujer, G. (1974). Virology 62, 234.
17. Rosen, J., Ryder, T., Inokuchi, H., Ohtsubo, H. and Ohtsubo, E. (1980). Mol. Gen. Genet. 179, 527.
18. Miller, J.H. (1972). Experiments in Molecular Genetics C.S.H.L. p. 353.

19. Alton, N.K. and Vapnek, D. (1979). Nature 282, 864.
20. Weisblum, B., Graham, M.Y., Gryczan, T. and Dubnau, D. (1979). J. Bacteriol. 137, 635.
21. McMaster, G.K. and Carmichael, G.G. (1977). Proc. Natl. Acad. Sci. 74, 4835.
22. Wahl, G.M., Stern, M. and Stark, G.R. (1979). Proc. Natl. Acad. Sci. 76, 3683.

GENETIC ANALYSIS OF pMB1 REPLICATION

Gianni Cesareni
Luisa Castagnoli
Rosa M. Lacatena

European Molecular Biology Laboratory
Heidelberg
Germany

I. ABSTRACT

We show how a plasmid-λ phage hybrid can be conveniently used to isolate plasmids defective in either replication or its control. The versatility of the system has allowed the isolation and mapping of 35 mutants. These mutations have been aligned with the DNA sequence by marker rescue with deletions whose end points in the replication region are known. The phenotype of our mutants defective in the control of copy number is not compatible with a simple repressor-operator model for the control of pMB1 replication.

II. INTRODUCTION

The plasmid pMB1 belongs to a group of non cojugative plasmids whose replication is controlled in a relaxed mode and strictly depends on DNA polymerase I. The best known plasmid of this group is colE1. ColE1 and pMB1 are incompatible, they have more than 95% sequence homology in the replication region and they do not show any major difference in their mode of replication. The initation of DNA synthesis *in vitro* is sensitive to the RNA polymerase inhibitor rifampicin suggesting that transcription is an essential requirement for "colE1 type" replication (1). *In vivo* (2) and *in vitro* (1) studies indicate that no plasmid encoded protein acts as a positive regulator

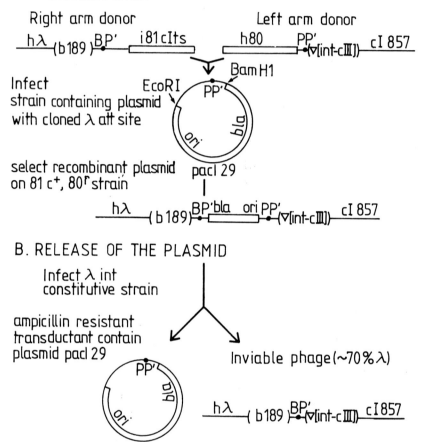

FIGURE 1. Int mediated insertion of the plasmid pacl29 into λ.

A) Scheme of the operations involved in the selective integration of the plasmid between the two shortened arms. Only recombinants containing the plasmid will plate on the selective strain resistant to h80 and carrying an i81 prophage. The remaining recombinant chromosomes are in fact too short (30% deletion) to be packaged.

B) Scheme of the release of the plasmid from chromosome in an integrase constitutive strain.

11 GENETIC ANALYSIS OF pMB1 REPLICATION

for plasmid replication. However the involvement of small polypeptide(s) in the control of copy number can not be ruled out.

The development of *in vitro* systems for colE1 replication (1, 3) has allowed the dissection of this process and the identification among other features of two transcripts whose implication in the initiation of DNA synthesis (4) or its control (5) has been suggested. Unfortunately the lack of point mutants defective in plasmid replication does not allow to identify which one of these features are essential for plasmid replication. It is clear however that the isolation of plasmid mutants unable to replicate and the correlation of their *in vivo* and *in vitro* phenotypes to the change in the DNA sequence is the most promising approach to the understanding of the mechanism of this process. Only one colE1 point mutant defective in replication has been isolated and sequenced so far (6).

We describe here a detailed genetic analysis of plasmid pac129, a small derivative of pMB1. This was made possible by the development of a new system (7) that allows the exploitation of λ genetic technology to study plasmid functions. The insertion *in vitro* into a plasmid of a fragment of λ DNA containing the λ*att* site allows the selective integration of the plasmid into a λ chromosome via *int* mediated recombination (Fig. 1). The insertion of the plasmid into the phage confers on the hybrid (phasmid) the new property of growing on a homoimmune lysogen (virulence). This phenotype, that was shown to be caused by plasmid replication proved to be very useful in the isolation and genetic mapping of the mutants. The details of the phasmid system will be described elsewhere (7). The advantage of this system with respect to similar ones already described resides in its versatility. The insertion of the plasmid into the λ chromosome is reversible and the plasmid can be readily excised from λ sequences by infection of a bacterial strain harbouring an *int* constitutive prophage (Fig. 1). Both integration and release procedures are selective and can be carried out on a large number of clones by replica-plating techniques.

Using this technology we have isolated and mapped 35 mutants defective either in plasmid replication (29) or in the control of copy number (6).

III. METHODS

General microbiological techniques are according to Miller (8). Details about the insertion and release of the plasmid into and from the hybrid phasmid will be described (7).

IV. RESULTS

A. *Phasmid Virulence*

The *int* mediated integration of the plasmid pac129 between the two short arms of the donor phages (Fig. 1) confers on the hybrid the property of growing on a homoimmune lysogen (virulence). This is in agreement with what was already described for other systems (9) where a replication origin has been inserted *in vitro* into a λ chromosome. We observed that phasmids carrying an i81 or i21 right arms can form plaques at high efficiency (0.5-1) on a lawn of homoimmune lysogen whereas phasmid with an iλ right arm do not. In the latter case virulence is visualized by plating on a double indicator lawn, obtained by mixing a permissive strain and an immune one. On the mixed indicator lawn, virulent phasmid produce clear plaques whereas non virulent phasmids give turbid plaques. The following observations suggest that the virulent phenotype is caused by the increase in the copy number of λ operators driven by plasmid replication.

(1) The resident prophage is eventually derepressed and can complement mutations in the N gene of the incoming phasmid.

(2) Virulence is abolished when a plasmid of the same compatibility group is present in the tester lysogen.

(3) Virulence does not depend on plasmid orientation.

(4) Virulence is conferred by a variety of pMB1 or colE1 plasmid derivatives whose only common sequence is a region of about 1 kb near the origin of replication.

B. *Isolation of vir Mutants*

The relationship between phasmid virulence and the process of plasmid replication provides a tool for the isolation of phasmid mutants defective in plasmid replication. Such phasmids should not be able to overcome repression by the infected lysogen and will give rise to turbid plaques on a double indicator lawn. A mutagenized lysate of phasmid Φ1 (Fig. 2) was plated on a mixed indicator lawn and turbid plaques were purified and tested again (Table 1). In order to exclude the possibility that mutations in the phasmid immunity region might interfere with the observed phenotypes, a non mutagenized right arm was introduced by crossing the mutants with phage 585 (Fig. 2). In a similar way substitution of the left arm did not affect the *vir* phenotype indicating that the mutations map within plasmid sequences.

TABLE I. *Frequency of vir and svir Mutants after Nitroso guanidine and Ultraviolet (UV) Light Mutagenesis*

Mutagen Phenotype	Nitrosoguanidine	UV
vir	$3 \cdot 10^{-3}$	$5 \cdot 10^{-4}$
svir	10^{-3}	ND

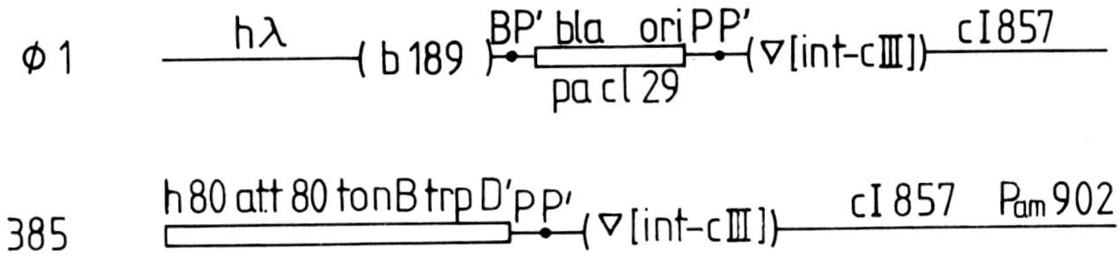

Φ1 —hλ—(b189)—BP' bla ori PP' / pa cl29—(∇[int-cIII])—cI857—

385 —h80 att 80 tonB trp D' PP'—(∇[int-cIII])—cI857—Pam902—

Figure 2. *Phasmid Strains*

C. Isolation of svir Mutants

The wild-type phasmid Φ1 makes turbid plaques on a double indicator lawn if the tester lysogen contains plasmid of the same compatibility group. When a mutagenized phasmid lysate is plated on the same mixed indicator clear plaques appear at a frequency of 10^{-3} (Table I). In an attempt to isolate plasmid mutants defective in the control of plasmid replication or in its incompatibility properties we have purified some of these mutants (*svir*). Out of ten independent mutations isolated four map in the immunity region of the phasmid. The remaining six mutations map within the plasmid sequences as shown by the persistence of their *svir* phenotype when non mutagenized left or right arms are introduced.

D. Copy Number of vir and svir Mutants

The major advantage of the system that we have described is the possibility of *in vivo* excision of the plasmid from the phage when the product of gene *int* is provided in trans (Fig. 1). In order to test whether *vir* and *svir* mutants affect pMB1 replication we have excised some of the plasmids by infecting a bacterial strain which carries an *int* constitutive prophage. Selection at 32°C on ampicillin plates (50 μg/ml) yields colonies containing a plasmid identical in size to the plasmid pac129 that was originally inserted in the phasmid Φ1. If *vir* mutants are defective in plasmid replication we would predict that they should not be able to excise a self replicating plasmid and confer ampicillin resistance to the infected host. When *vir* mutants are used in the release experiment in fact the efficiency of plasmid release drops by a factor of 10^{-1}-10^{-3} with respect to the wild type efficiency.

It has been shown that the degree of ampicillin resistance of a bacterial strain is proportional to the number of ß-lactamase (*bla*) genes present inside the cell (10). It follows that the measurement of the concentration of ampicillin at which the bacterial titer decreases gives a good estimate of the copy number of the plasmid which codes for the ß-lactamase enzyme. Such experiments carried out on bacteria harbouring different *vir* mutants (Table II) indicate that *vir* mutations decrease the copy number of the plasmid to values ranging from 1/10 to 1/30 of the wild type copy number. In the case of *vir* 040 we were able to show that the plasmid does not replicate and that the few ampicillin resistant clones obtained carry the plasmid integrated into the chromosome.

11 GENETIC ANALYSIS OF pMB1 REPLICATION

On the contrary, *svir* mutants have copy numbers 4-5 times higher than the wild type and are probably mutated in the mechanism controlling the copy number. As shown in Table II, *svir* mutants can not be complemented in trans by the related plasmid pMB9. Bacteria harbouring both plasmids are resistant to concentrations of ampicillin only slightly lower than bacteria harbouring *svir* mutants alone.

TABLE II. Copy Number of vir and svir Mutants

	Resistance Level[a]	
Plasmid	No Plasmid Coresident	Plasmid pMB9 Coresident
pacl 29 w.t.	1.5	1
vir 040	0.05	
vir 042	0.15	
vir 041	0.15	
vir 045	0.10	
svir 2	6.5	5.5
svir 5	7	6
svir 7	7	6
svir 8	7	6

[a]*Resistance level is defined as the concentration of ampicillin (mg/ml) at which bacterial titer drops by a factor of 10.*

E. Svir Mutants are Compatible with the Related Plasmid pMB9

If a repressor-operator interaction represents the basic mechanism underlying the control of copy number of the plasmid pMB1, non complementable copy number mutants are candidate for operator mutants. Such mutants should produce large amounts of repressor due to the high number of copies of repressor genes present in the cell. Under these conditions the replication of a wild type plasmid of the same group should be prevented with the result of strong incompatibility between the two and segregation of the wild type.

To test this prediction we constructed bacteria harbouring two plasmids by infection of an *int* constitutive lysogen containing the related plasmid pMB9 with an average of 3 phasmids per cell. The presence of the two plasmids in the infected bacteria was monitored by plating at different times on antibiotic plates. The result of the experiment in Fig. 3 shows that bacteria containing pMB9 are unstable when infected by a phasmid carrying the wild type pacl 29 and tend to lose the infecting plasmid. A different result is obtained when either *svir* or *svir*8 are introduced in the *int* constitutive strain. In this case both the incoming and the resident plasmids are stable.

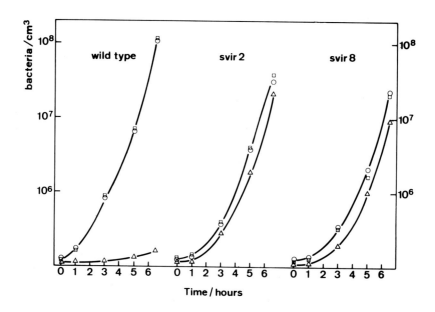

FIGURE 3. *Coreplication of pMB9 and pacl29 svir mutants. The integrase constitutive strain EQ84 containing the plasmid pMB9 (confers resistance to tetracycline) was infected at a multiplicity of 3 with the wild type phasmid ø1 or two different svir mutants. The infected culture was diluted in L broth and titrated, after different times of incubation at 35°C, on L plates (□) or on L plates containing either 50 µg/ml of ampicillin (Δ) or 10 µg/ml of tetracyclin (o).*

11 GENETIC ANALYSIS OF pMB1 REPLICATION

More conventional experiments in which the heterozygous bacteria were constructed by cotransformation gave similar results showing that pMB9-pac129 *svir* heterozygotes are stable at least for 100 generations (not shown).

These experiments suggest that *svir* mutants are not mutations in the target of a diffusible repressor.

F. Deletion Mapping

Although the insertion of the plasmid in the phage chromosome allows the use of λ technology (11,12) to select deletion mutants that are defective in the replication of the plasmid, we were unsuccessful in this approach. Out of fifty independent *vir⁻ bla⁺* deletions isolated by selecting phasmids resistant to chelating agents (11), none was able to split our point mutants in classes by marker rescue. We constructed therefore, by means of a series of *in vivo* and *in vitro* manipulations the set of deletions shown in Fig. 4A (Cesareni & Lacatena, manuscript in preparation).

The end points of these deletions in the replication origin are well defined and correspond to one of the *Hpa*II restriction sites originally used in the *in vitro* construction. A series of crosses of the type shown in Fig. 4B allowed us to construct a physical map of the replication origin of the plasmid pMB1 (Fig. 5).

IV. DISCUSSION

We have described an advantageous system which exploits the *vir* phenotype in order to isolate plasmid mutants altered in the mechanism of replication or its control. The selective and reversible integration of the plasmid into λ readily allows its insertion and excision from the phage itself by replica plating of a large number of clones. In principle this system can be extended to any high copy number replicon containing a λ *att* site inserted *in vitro*. However lambda type immunities other than the one we used here (λ) might prove to be more useful in the case of replicons with a copy number lower than that of pMB1.

The application of these techniques to a small derivative of plasmid pMB1 has allowed the isolation of 35 replication mutants. The mutations have been aligned with the DNA sequence

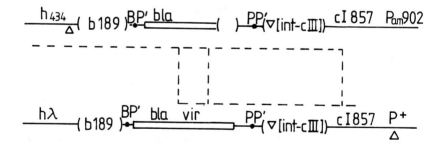

FIGURE 4. *Deletion mapping of replication mutants.*
A. Structure of the phasmids used in the deletion mapping. The numbers above the arrows refer to the distances in nucleotides of the end points of the deletions from the origin of DNA synthesis (13).
B. Scheme of the marker rescue cross: after selection for the external markers h_{434} P the vir phenotype was screened directly on the selective plates containing the double indicator described in the text.

11 GENETIC ANALYSIS OF pMB1 REPLICATION

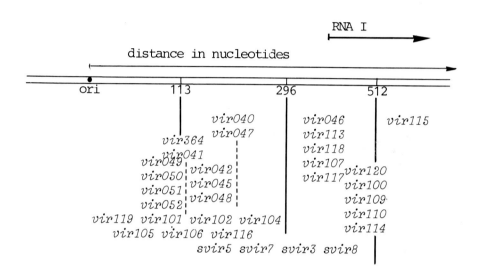

FIGURE 5. *Physical map of the replication region of the plasmid pMB1.*

This figure combines a series of mapping data obtained by four point crosses (7) and deletion mapping (see text). Classes of mutants divided by broken lines have been ordered by four point crosses. Unbroken lines divide classes of mutants obtained by deletion mapping. vir mutants that gave less than 10^{-4} vir⁺ recombinants (among hP recombinants, Fig. 4B) in a cross with a given deletion were considered to be overlapped by that deletion.

by crossing them with deletions constructed *in vitro*. Vir mutants (replication defective) are not clustered and have been found in all the intervals defined by the end points of the deletions (Fig. 5). This is in agreement with previous results (14) that suggested that most of the 600 nucleotide region upstream from the origin of replication is essential for plasmid replication. Furthermore this result proves that the screening for *vir* mutants is not selective for a small subclass of replication mutants. *svir* (high copy number) mutants are all situated on the left of nucleotide 512 (Fig. 5).

Transcomplementable high copy number mutants that map in a different region of colE1 or pMB1 replication origins have already been described (15,16). Our screening for *svir* mutants was devised in the attempt to select mutations affecting specifically the targets of these diffusible negative controlling elements. Indeed all five of the *svir* mutants tested were found to be trans dominant as expected for operator mutants. Contrary to the predictions based on a simple repressor operator model our *svir* mutants do not interact negatively with a co-resident wild type plasmid and they can replicate in the same cell for several generations. It would be difficult to explain these results on the basis of a repressor operator model even assuming that the repressor is self-regulating its expression.

We propose therefore that a classical interaction between a diffusible repressor and a cis acting element can not be the only mechanism regulating pMB1 replication and that a different and probably more complex mechanism must be invoked.

ACKNOWLEDGMENTS

We want to thank Noreen Murray and Lulla Melli for discussions and comments on the manuscript and Wendy Moses for carefully typing it. During this work M.R.L. was supported by a Blancheford-Ludovisi fellowship and L.C. by a Cenci-Bolognetti postdoctoral grant.

REFERENCES

1. Sakakibara, Y., and Tomizawa, J. (1974). Proc. Natl. Acad. Sci. USA 71, 802-806.
2. Donoghue, D.J., and Sharp, P.A. (1978). J. Bacteriol. 133, 1287-1294.
3. Staudenbauer, W.L. (1976) Molec. Gen. Genet. 145, 273-280.
4. Itoh, T., and Tomizawa, J. (1980). Proc. Natl. Acad. Sci. USA 77, 2450-2454.
5. Conrad, S.E., and Campbell, J.L. (1979) Cell 18, 61-71.
6. Naito, S., and Uchidea, H. (1980) Proc. Natl. Acad. Sci. USA 77, 6744-6748.
7. Brenner, S., Cesareni, G. and Karn, J. manuscript in prep.

8. Miller, J.H. (1972). *Experiments in Molecular Genetics*, New York, Cold Spring Harbor Laboratory.
9. Windass, J.D., and Brammar, W.J. (1979) *Molec. Gen. Genet. 172*, 329-337.
10. Uhlin, B.E., and Nordström, K. (1977) *Plasmid 1*, 1-7.
11. Parkinson, J.S., and Davis, R.W. (1971) *J. Mol. Biol. 56*, 425-428.
12. Enquist, L., and Sternberg, N. (1973) *Methods in Enzymology 68*, 281-298.
13. Sutcliffe, J.G. (1978) *Cold Spring Harbor Symp. on Quant. Biol. 43*, 77-90.
14. Backman, K., Betlach, M., Boyer, H.W., and Yanofsky, S. (1978) *Cold Spring Harbor Symp. on Quant. Biol. 43*, 63-76.
15. Twigg, A.J., and Sherrat, D. (1980) *Nature 283*, 216-218.
16. Shepard, H.M., Gelfand, D.H., and Polinski, B. (1979) *Cell 18*, 267-275.

CLONING OF DNA SYNTHESIS INITIATION
DETERMINANTS OF ColE1 PLASMID INTO SINGLE-STRANDED DNA PHAGES

Nobuo Nomura
Dan S. Ray

Department of Biology
and
Molecular Biology Institute
University of California
Los Angeles, California

I. ABSTRACT

*Eco*RI-cleaved ColE1 DNA and *Hae*II restriction fragments of ColE1 DNA were cloned into the intergenic region of filamentous DNA phages and were analyzed with regard to their ability to promote DNA strand initiation on a single-stranded DNA template. Two sites, which direct conversion of chimeric phage single-stranded DNA to the parental replicative form in the presence of rifampicin, have been found on the L-strand (leading strand) of the *Hae*II E fragment and on the H-strand (lagging strand) of the *Hae*II C fragment. These DNA synthesis initiation determinants have been named *rriA* (*r*ifampicin-*r*esistant-*i*nitiation) and *rriB*, respectively. The functions of both *rriA* and *rriB* are dependent on the *dnaB* gene product. It is postulated that *rriA* might be a determinant (origin) for initiation of lagging strand synthesis during unidirectional replication of ColE1 DNA.

II. INTRODUCTION

Plasmid ColE1 replication initiates at a unique origin and the replication fork moves unidirectionally (1,2). Tomizawa and coworkers have shown that the 6S L-fragment is synthesized first (3-5), and the 5' end of primer RNA and the transition point from primer RNA to DNA of the L-strand fragment have been determined (6-8). Although RNA polymerase participates in synthesis of the 6S L-fragment by making an RNA primer (3,4), RNA polymerase is not required for further replication (5). Because the *dnaG* gene is involved in replication of ColE1 DNA (9,10), RNA priming of H-strand (lagging strand) initiation may possibly be catalyzed by primase (*dnaG* gene product). Also, RNA primer(s) for leading strand replication after 6S L-fragment synthesis, if any, might be made by primase. It seems reasonable to suppose that during replication two strands of DNA are somewhat unwound and that each strand serves as a template for synthesis of the other strand. Actually, it has been observed that after 6S L-fragment synthesis, the early replicative intermediate contains a D-loop (displacement loop) structure (11). This suggests that initiation of H-strand synthesis might occur on a single-stranded DNA (SS DNA) template and that genetic information necessary for initiation of H-strand replication might be contained in the L-strand. A unique method to assay DNA strand initiation of a fragment in a single-stranded form has been developed (12). This method takes advantage of the fact that the first step of M13 phage replication, conversion of phage SS DNA to the parental replicative form (RF) DNA, requires RNA polymerase for priming (13). Therefore, RNA polymerase-independent RNA primer synthesis due to a cloned fragment can be detected by measuring the efficiency of conversion of the infecting phage DNA to a duplex form under conditions in which RNA polymerase activity is inhibited. The validity of this method has been demonstrated by the successful expression of the complementary strand origin of bacteriophage G4 inserted into the intergenic region of M13 (12). Thus, we have cloned the entire ColE1 genome and *Hae*II restriction fragments of ColE1 DNA into the intergenic region of single-stranded DNA phages and screened for DNA sequences which direct DNA synthesis initiation by a rifampicin-resistant mechanism. We present here results that indicate the existence of sites (named $rriA$ and $rriB$) on the L-strand of the *Hae*II E fragment and the H-strand of the *Hae*II C fragment which are capable of promoting DNA strand initiation. The possible biological roles of both $rriA$ and $rriB$ are discussed.

III. MATERIALS AND METHODS

A. *Bacterial and Phage Strains*

The bacterial strains used are derived from *Escherichia coli* K12. The strains used are PC3 (*dnaG3, leu-6, thyA47, str-153*) (14); RL114 (F⁺ derivative of PC3); RL115 (temperature-resistant revertant of RL114); BT1029 (F⁺, *dnaB, thy, arg, endA2, polA1, rpsL*) (15); RL116 (temperature-resistant revertant of BT1029); W1485 (F⁺, *supE42*) (16); and LA6 (F⁺, *pro, thi, lacY, gal, ara∆766, rpsL, supE*) (17). The fl derivative R199 phage (18) was a gift of K. Horiuchi, Rockefeller University, New York. M13Go*ri*1 has been described (12).

B. *Construction and Isolation of M13E1 and f1E1 Phages*

EcoRI cut ColE1 DNA was ligated with EcoRI cut R199 RF DNA which has a single EcoRI site in the intergenic region (18) (Figure 1). Also, fragments from a HaeII digest of ColE1 DNA were ligated with unit length linear M13 RF DNA which was produced by HaeII partial digestion (19) (Figure 1). Calcium chloride-treated LA6 cells (20) were then transfected with DNA from the ligation reaction and plated for plaque formation. Screening of chimeric phages and the restriction analysis of chimera molecules will be published elsewhere.

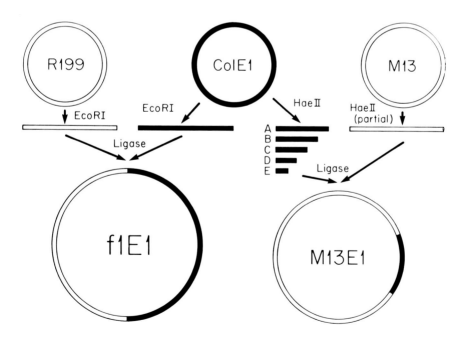

FIGURE 1. *Construction of f1E1 and M13E1 phages. ColE1 DNA was cut with EcoRI or HaeII and cloned into f1 derivative R199 (18) or M13.*

C. *Assay of DNA Strand Initiation Activity*

^{32}P- or ^{3}H-labeled phages are prepared as described previously (19). A culture of *E. coli* LA6 was grown in an M9 glucose minimal medium containing thymidine at 2 μg/ml and 0.5% casamino acids at 37°C (21). At an optical density of 0.5 at 595 nm, the culture was divided into aliquots (5 ml each) and 50 μl of rifampicin at 40 mg/ml in dimethyl sulfoxide was added. After 10 min at 37°C, cultures were infected with ^{32}P-labeled f1E1 or M13E1 chimeric phages and ^{3}H-labeled M13 phage. After an additional 20 min at 37°C, an equal volume of ice-cold NaCl/EDTA/Tris/NaCN buffer (21) was added and the mixture was kept in ice for 10 min. The bacteria were centrifuged and washed twice. Cell pellets were resuspended in 0.5 ml of NaCl/EDTA/Tris buffer (21) and incubated with 50 μl of lysozyme at 4 mg/ml for 10 min at 37°C. Then, 25 μl of 10% Sarkosyl NL97 was added and incubation was continued for 10 min at 37°C. The resulting clear and highly viscous lysate was layered on top of 5-20% sucrose gradients containing 1 M NaCl, 10 mM Tris·HCl (pH 8.0), and 1 mM EDTA. Sedimentation was for 4.5 h at 5°C and 40,000 rpm in a Beckman SW40 rotor. Fractions were collected from the top directly into scintillation vials for measurement of radioactivity (21). The efficiency of conversion of ^{32}P-labeled chimera SS DNA to parental RF DNA was compared with that of ^{3}H-labeled M13 phage DNA which was added as an internal marker (19). RL114, RL115, BT1029 and RL116 strains were used to test *dnaG* and *dnaB* gene requirement (19). These cells were grown at 32°C and the temperature was shifted to 42°C after addition of rifampicin. The temperature was kept at 42°C during phage infection. Phage infection, lysis and sedimentation analysis were carried out in the same fashion as described above.

IV. RESULTS

The entire genome of ColE1 was inserted into the intergenic region of the duplex form of f1 derivative phage R199 (18) in both orientations (Figure 1). f1E1-2 and f1E1-12 are two representative isolates of f1E1 phages in which the L-strand or the H-strand of the entire ColE1 genome has been ligated to the viral strand of R199, respectively. In order to test the capability of cloned ColE1 DNA to direct DNA strand initiation by a rifampicin-resistant mechanism, f1E1-2, f1E1-12 and M13 phages were labeled with ^{32}P or ^{3}H and used to infect *E. coli* cells in the presence of rifampicin. If DNA synthesis

12 CLONING OF DNA SYNTHESIS INITIATION DETERMINANTS

from the complementary strand origin of M13 or an origin within the cloned DNA occurs, ^{32}P or ^{3}H label contained in the infecting viral DNA will be converted to RF DNA. Although essentially no parental RF DNA synthesis of M13 was observed, fIE1-2 and fIE1-12 phage SS DNAs were converted to RF DNA in the presence of rifampicin (Table 1). Conversion of fIE1-2 and fIE1-12 phage SS DNAs to parental RF in the presence of rifampicin suggests that RNA primer synthesis occurs on each ColE1 strand by a rifampicin-resistant mechanism, probably involving primase (*dnaG* gene product).

In order to map the DNA sequences which promote DNA synthesis initiation, *Hae*II restriction fragments (A, B, C, D and E) of ColE1 have been cloned into the intergenic region of M13 phage in both orientations (Figure 1). As shown in Table 1, the DNA strand initiation activities were detected in the L-strand of the *Hae*II E fragment and the H-strand of the *Hae*II C fragment. These two sites, which direct conversion of chimeric phage single-stranded DNA to parental replicative form in the presence of rifampicin, have been named *rriA* (*r*ifampicin-*r*esistant-*i*nitiation) and *rriB*, respectively.

As both the *dnaG* and the *dnaB* gene products are required for *in vitro* replication of the early replicative intermediate of ColE1 (22,23), it was asked whether the *dnaB* gene is involved in *rriA* and *rriB* functions. The results are shown in Table 2. M13Gori1 phage DNA, which was used as a control, was not converted to parental RF DNA in RL114 (*dnaG*ts) but was

TABLE 1. Rifampicin-Resistant Initiation Activity

ColE1 DNA	*H-strand*	*L-strand*
Entire genome	+	+
HaeII A fragment	−	−
HaeII B	−	−
HaeII C	+	−
HaeII D	−	−
HaeII E	−	+

Conversion of ^{32}P-labeled chimera SS DNA to parental RF DNA was compared with that of ^{3}H-labeled M13 phage DNA which was added as an internal marker. (+) indicates rifampicin-resistant DNA strand initiation.

TABLE 2. *rriA* Function is *dnaB* and *dnaG* Dependent

	RL114	RL115	BT1029	RL116
M13Gori1	−	+	+	+
M13E1-7[a]	−	+	−	+
flE1-2	n.t.	n.t.	−	n.t.

In order to test that *rriA* is dependent on both the *dnaG* and the *dnaB* gene, RL114 ($dnaG^{ts}$), RL115 (temperature-resistant revertant of RL114), BT1029 ($dnaB^{ts}$) and RL116 (temperature-resistant revertant of BT1029) strains were infected at 42°C in the presence of rifampicin. (+) indicates synthesis of the complementary strand, and (−) indicates no conversion of phage SS DNA to parental RF DNA. n.t. = not tested.

[a] The L-strand of the *HaeII* E fragment has been joined with the viral strand of M13 (19).

converted to parental RF DNA in the BT1029 strain ($dnaB^{ts}$) at the restrictive temperature, consistent with the known initiation mechanism at the complementary strand origin of G4 (24). However, conversion of M13E1-7 to parental RF is blocked in both RL114 and BT1029 (Table 2) (19). This result indicates that the *dnaG* and *dnaB* gene products are required for expression of *rriA*. Also, *rriB* has been shown to require the *dnaB* gene product for its function (preliminary results).

V. DISCUSSION

*Hae*II fragments of ColE1 have been cloned into the 507 base pair intergenic region of bacteriophage M13. M13E1-30 and M13E1-55 phage SS DNA, in which the H-strand of the *Hae*II C fragment and the L-strand of the *Hae*II E fragment have been ligated to the viral strand of M13, were converted to parental RF DNA in the presence of rifampicin (Table 1). These results indicate the presence of determinants of DNA strand initiation on these fragments (Figure 2). Because the loci promote DNA synthesis initiation by a rifampicin-resistant mechanism, we propose to name the determinants *rriA* and *rriB* (*r*ifampicin-*r*esistant-*i*nitiation). *rriA* was called *rri-1* formerly (19). We have shown here that the functions of *rriA* and *rriB* are dependent on the *dnaB* gene product.

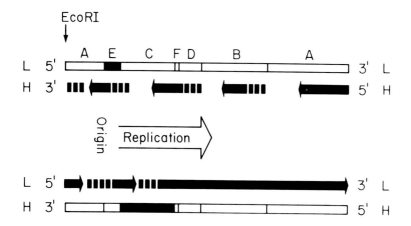

FIGURE 2. *A model of ColE1 replication. The ColE1 genome is aligned at the single EcoRI site. The L-strand and H-strand are placed at top and bottom (▭). The 5' → 3' polarity of the L-strand is from left to right. HaeII fragments (A,B,C,D,E and F)(31), "origin" of replication and the direction of the replication fork movement (1,2) are indicated. The L-strand of HaeII E fragment and the H-strand of the HaeII C fragment are blackened (■) to emphasize the presence of rriA and rriB. Possible DNA fragments as well as the 6S L-fragment are shown as arrows (→ and ←). (■■■) indicates RNA primers. The positions of H-strand initiations are arbitrary.*

Two mechanisms of *dnaG*-dependent primer synthesis have been identified by *in vitro* replication studies of small single-stranded DNA phage (25,26). In ØX174 complementary strand synthesis, primer RNA is polymerized at various positions, having little or no specificity. It has been postulated that a multiprotein complex, "primosome", (27), known to contain the *dnaB* and *dnaC* gene products, proteins i, n, n', n'' and primase is assembled at or near the n' recognition locus in the untranslated region between genes F and G of ØX174 (28). Primosome moves along the template strand in the 5' → 3' direction, initiating synthesis of RNA transcripts (27). This mechanism probably accounts for the multiple initiation sites of RNA primers. Alternatively, in G4 complementary strand synthesis, primer RNA is synthesized at a unique DNA sequence (origin) without participation of the *dnaB* protein (29). Because the function of *rriA* in ColE1 H-strand initiation is dependent on both the *dnaG* and *dnaB* gene products, the mechanism of DNA synthesis initiation due to *rriA* resembles the "ØX174 pathway" rather than the "G4 pathway". The finding that M13E1-7 SS DNA has effector activity for ATP hydrolysis catalyzed by the n' protein (Robert Low, personal communication) (30) supports this mechanism. Moreover, the complementary strand of M13E1-7 was synthesized by purified proteins which are used to synthesize the complementary strand of ØX174 phage. Therefore, we postulate that *rriA* is a protein n' recognition locus necessary for primosome construction and that primosome migrates on the L-strand in the 5' → 3' direction, laying down primer RNA for the discontinuous replication of the H-strand (Figure 2).

Although polIII holoenzyme has been shown to replace DNA polymerase I after 6S L-fragment synthesis (32), it is not

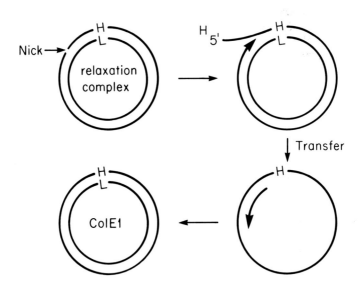

FIGURE 3. *A model of ColE1 transfer. The 3' end of the nick which is found in the relaxation complex is extended and the displaced H-strand is transferred to the recipient cells (34). After circularization, the L-strand is synthesized by the function of rriB and duplex ColE1 DNA is produced.*

clear whether the 6S L-fragment is extended directly or another RNA primer synthesis is required. Because the 3' end of 6S L-fragment is mapped within the *Hae*II C fragment, the existence of the *rriB* in that restriction fragment might mean that a new RNA primer for DNA polymerase III holoenzyme is synthesized. *rriB*, as well as *rriA*, has been shown to retain effector activity for ATP hydrolysis by the n' protein (Robert Low, personal communication) (30). Also, complementary strand DNA synthesis of M13E1-30 phage requires the same proteins as that of ØX174. Thus after 6S L-fragment synthesis, primosome might be assembled at or near *rriB* and participate in RNA primer synthesis for DNA polymerase III holoenzyme (Figure 2). Alternatively, *rriB* might not be involved in L-strand synthesis but rather may be relevant to "repliconation" (33). Repliconation is the process whereby transferred DNA in the recipient is prepared for vegetative replication (33). By analogy to F DNA transfer, it has been speculated that the single-strand of ColE1 DNA is mobilized in transfer (34) (Figure 3). The *nic* locus (33), to which the protein components of ColE1 relaxation complex are bound, has been mapped on the H-strand (35). It has been postulated that the *nic* locus is the initiation site of transfer DNA replication and that the displaced H-strand is transferred to the recipient cells (34). We propose that *rriB* might serve as a determinant (origin) for conversion of the transferred H-strand to ColE1 duplex DNA (Figure 3).

ACKNOWLEDGMENT

This work was supported by Research Grant AI 10752-08 from the National Institutes of Health.

REFERENCES

1. Inselberg, J., *Proc. Natl. Acad. Sci. U.S.A. 71*, 2256 (1974).
2. Lovett, M.A., Katz, L., and Helinski, D.R. *Nature (London) 251*, 337 (1974).
3. Sakakibara, Y., and Tomizawa, J., *Proc. Natl. Acad. Sci. U.S.A. 71*, 802 (1974).
4. Sakakibara, Y., and Tomizawa, J., *Proc. Natl. Acad. Sci. U.S.A. 71*, 1403 (1974).
5. Tomizawa, J., *Nature (London) 257*, 253 (1975).
6. Tomizawa, J., Ohmori, H., and Bird, R.E., *Proc. Natl. Acad. Sci. U.S.A. 74*, 1865 (1977).

7. Bird, R.E., and Tomizawa, J., *J. Mol. Biol.* 120, 137 (1978).
8. Itoh, T., and Tomizawa, J., *Proc. Natl. Acad. Sci. U.S.A.* 77, 2450 (1980).
9. Collins, J., Williams, P., and Helinski, D.R., *Mol. Gen. Genet.* 136, 273 (1975).
10. Staudenbauer, W.L., Scherzinger, E., and Lanka, E., *Mol. Gen. Genet.* 177, 113 (1979).
11. Tomizawa, J., Sakakibara, Y., and Kakefuda, T., *Proc. Natl. Acad. Sci. U.S.A.* 71, 2260 (1974).
12. Kaguni, J., and Ray, D.S., *J. Mol. Biol.* 135, 863 (1979).
13. Brutlag, D., Schekman, R., and Kornberg, A., *Proc. Natl. Acad. Sci. U.S.A.* 68, 2826 (1971).
14. Wechsler, J.A., and Gross, J.D., *Mol. Gen. Genet.* 113, 273 (1971).
15. Wechsler, J.A., Nüsslein, V., Otto, B., Klein, A., Bonhoeffer, F., Herrmann, R., Gloger, L., and Schaller, H., *J. Bacteriol.* 113, 1381 (1973).
16. Bachmann, B., *Bacteriol. Rev.* 36, 525 (1972).
17. Kaguni, J., LaVerne, L.S., and Ray, D.S., *Proc. Natl. Acad. Sci. U.S.A.* 76, 6250 (1979).
18. Boeke, J.D., Vovis, G.F., and Zinder, N.D., *Proc. Natl. Acad. Sci. U.S.A.* 76, 2699 (1979).
19. Nomura, N. and Ray, D.S., *Proc. Natl. Acad. Sci. U.S.A.* 77, 6566 (1980).
20. Cohen, S.N., Chang, A.C.Y., and Hsu, L., *Proc. Natl. Acad. Sci. U.S.A.* 69, 2110 (1972).
21. Ray, D.S., and Schekman, R.W., *Biochim. Biophys. Acta* 179, 398 (1969).
22. Staudenbauer, W.L., Lanka, E., and Schuster, H., *Mol. Gen. Genet.* 162, 243 (1978).
23. Tomizawa, J., in "DNA Synthesis: Present and Future" (I. Molineux and M. Kohiyama, eds.), p. 797. Plenum Press, New York, (1978).
24. Bouché, J.P., Zechel, K., and Kornberg, A., *J. Biol. Chem.* 250, 5995 (1975).
25. Meyer, R.R., Schlomai, J., Kobori, J., Bates, D.L., Rowen, L., McMacken, R., Ueda, K., and Kornberg, A., *Cold Spring Harbor Symp. Quant. Biol.* 43, 289 (1978).
26. Wickner, S., in "The Single-Stranded DNA Phages" (D.T. Denhardt, D. Dressler and D.S. Ray, eds.), p. 255, Cold Spring Harbor Laboratory, New York, (1978).
27. Arai, K., and Kornberg, A., *Proc. Natl. Acad. Sci. U.S.A.* 78, 69 (1981).
28. Shlomai, J., and Kornberg, A., *Proc. Natl. Acad. Sci. U.S.A.* 77, 799 (1980).
29. Bouché, J.P., Rowen, L., and Kornberg, A., *J. Biol. Chem.* 253, 765 (1978).

30. Zipursky, S.L., and Marians, K.J., *Proc. Natl. Acad. Sci. U.S.A. 77*, 6521 (1980).
31. Oka, A., and Takanami, M., *Nature (London) 264*, 193 (1976).
32. Staudenbauer, W.L., *in* "DNA Synthesis: Present and Future" (I. Molineux and M. Kohiyama, eds.), p. 827, Plenum Press, New York, (1978).
33. Clark, A.J., and Warren, G.J., *Ann. Rev. Genet. 13*, 99 (1979).
34. Warren, G.J., Twigg, A.J., and Sherratt, D.J., *Nature (London) 274*, 259 (1978).
35. Bastia, D., *J. Mol. Biol. 124*, 601 (1978).

DNA INITIATION DETERMINANTS OF BACTERIOPHAGE M13
AND OF CHIMERIC DERIVATIVES CARRYING
FOREIGN REPLICATION DETERMINANTS

Dan S. Ray
Joseph M. Cleary
Jane C. Hines
Myoung Hee Kim
Michael Strathearn
Laurie S. Kaguni
Margaret Roark

Molecular Biology Institute
and
Department of Biology
University of California
Los Angeles, California 90024

I. ABSTRACT

Specific DNA sequences located in a 508 base pair intergenic space in the genome of bacteriophage M13 specify the initiation of each of the two strands of the duplex replicative form (RF). Initiation of the viral strand of the RF involves nicking of the DNA at a specific site within the intergenic space. The complementary strand is initiated by RNA polymerase-dependent priming at a site 24 nucleotides away from the viral strand origin. This region of the viral DNA has the potential for forming secondary structures, and specific potential hairpins have been postulated to be involved in origin function.

Fragments of the M13 RF have been cloned into the plasmid pBR322 and found to be capable of directing M13-dependent replication of the chimeric plasmid in *polA* hosts. A 142 base pair *Hae*III fragment of M13 RF is sufficient to direct M13-dependent plasmid replication although transformation by plasmids containing this fragment is very inefficient.

Plasmids carrying an additional 40 base pairs of M13 DNA show a 10^4-fold increase in transformation frequency.

Deletions have been introduced into the origin region of M13 by *in vitro* techniques. Plaque-forming phages have been produced which contain deletions of both the DNA sequence specifying the RNA primer for the complementary strand and the hairpins postulated to constitute part of the complementary strand origin.

Replication determinants of bacteriophages G4 and ØX174, the plasmid ColE1, the *E. coli* replication origin and the *Salmonella typhimurium his* operon have been cloned into M13 and derivative cloning vectors and found to be capable of substituting for the M13 complementary strand origin in the presence of rifampicin, a specific inhibitor of M13 complementary strand initiation. This technique provides a means for identifying the determinants that specify a single DNA strand initiation either within a replication origin or along the length of the chromosomal DNA (i.e., determinants for the initiation of discontinuous synthesis).

II. INTRODUCTION

DNA chains can be initiated in *E. coli* by at least three distinct mechanisms involving only bacterial proteins (1). In each case initiation involves the synthesis of an RNA primer followed by elongation by DNA polymerase III holoenzyme. The purification of the enzymes involved in these different initiation mechanisms has taken advantage of the availability of the single stranded DNA phages as sources of homogeneous templates for *in vitro* DNA synthesis. The phages M13, G4, and ØX174 all have a circular single stranded DNA genome of about 5-6,000 nucleotides in length. These DNAs are all readily converted to their duplex replicative forms (RF) by extracts of *E. coli*. By studying the conversion of the single stranded (SS) DNA to the parental RF, it has been possible to define the requirements for initiation of a single strand without the complication of the synthesis of the other strand.

In the case of M13, initiation involves the rifampicin-sensitive RNA polymerase of *E. coli* and occurs at a unique site on circular single stranded M13 DNA coated with single strand binding (*ssb*) protein (2). This mechanism of initiation is clearly distinct from those used by G4 (3-5) and ØX174 (1,6,7) where RNA primers are synthesized by the action of the rifampicin-resistant *dnaG* primase. However, these latter two viral DNAs are distinguished by their different requirements for additional factors. G4 initiates at a unique site on the *ssb* protein-coated DNA by the formation of an RNA primer by the *dnaG* primase. Initiation on *ssb* protein-coated ØX174 DNA involves several proteins in addition to *dnaG* primase, including the *dnaB* and *dnaC* proteins and proteins i, n, n' and n''. In addition, initiation on the ØX174 genome occurs at multiple sites with little, if any, specificity.

How is it that these different single stranded DNA templates are able to appropriate such distinctly different enzymatic machinery for their initiation? How is it that initiation on ØX174 templates occurs at virtually random sites on the ØX genome yet this same mechanism is not available to M13 DNA when RNA polymerase is inhibited by rifampicin? The answers to these intriguing questions must surely lie in the interactions between specific determinants within each of these DNA molecules and the various proteins involved in each initiation system. Furthermore, since *E. coli* did not evolve for the convenience of these single stranded phages, the specific DNA-protein interactions that occur during initiation on phage single-stranded DNA templates likely reflect mechanisms utilized in the initiation and replication of the bacterial chromosome itself. In this paper we will attempt to more precisely define some of the sequence determinants in the initiation mechanisms used by single-stranded DNA phages. In addition, we hope to show the enormous value of M13-type phages for cloning and analyzing DNA initiation determinants from the bacterial chromosome and from other phages and plasmids.

III. RESULTS

A. *Expression of the M13 Origin Cloned into the Plasmid pBR322*

A 508 base pair intergenic region between genes II and IV in the M13 RF (8) contains the origins of both the viral and the complementary strands. The viral strand origin is defined by the site of action of the gene II protein (gIIp), a site-specific endonuclease that nicks the RF (9) to provide a 3' hydroxyl terminus for the initiation of viral strand synthesis by the rolling circle mechanism. This nicking site is contained within the loop of a potential small hairpin structure of the viral DNA (Figure 1). The complementary strand origin is defined by its specific RNA primer (10), also shown in Figure 1.

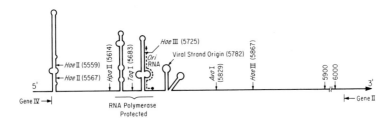

FIGURE 1. *Physical map of the intergenic region containing the M13 viral and complementary strand origins. The dashed line represents the RNA primer for the complementary strand. The viral strand origin (gene II protein nicking site) is indicated by an arrow. The nucleotide coordinates of various restriction sites are shown in parenthesis.*

It is important to distinguish here between the origin of a DNA strand and the DNA sequence that specifies the initiation of that strand. This distinction is akin to that made between the site of initiation of an mRNA and the promoter sequence that specifies that initiation.

Thus, while we know where the M13 viral and complementary strands initiate, we do not know the precise sequences that determine each initiation. One aim of the present work is the identification of the initiation determinants for each of the strands of M13 DNA. As one approach, we have cloned fragments of the M13 origin region into the plasmid cloning vector pBR322 (11). Chimeric plasmids capable of replication under the control of cloned M13 sequences can be directly selected under conditions non-permissive for the pBR322 replicon, i.e., in *polA* mutant hosts.

Restriction endonuclease *Hpa*II was chosen to digest M13 RF because one of the fragments produced, M13 *Hpa*II F, spans 75% of the intergenic region including major hairpins in the vicinity of the replication origins (12). The cohesive ends of the fragments in a total *Hpa*II digest were filled in by DNA polymerase I and *Bam*H1 linkers added by blunt-end ligation. These fragments were digested with *Bam*H1 and ligated into the *Bam*H1 site of pBR322. The ligated DNA was used to transform an M13-infected *E. coli polA* mutant strain to ampicillin resistance. The isolate pJMC110 (Figure 2) contains the 381 base-pair *Hpa*II F fragment inserted into the *Bam*H1 site of pBR322.

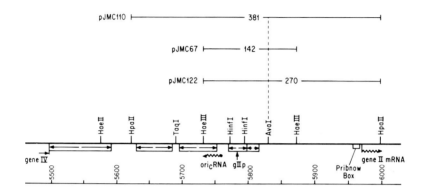

FIGURE 2. Physical map of the M13 intergenic region showing specific fragments cloned into the plasmid pBR322 (11). Regions of potential secondary structure are indicated by boxed arrows.

In order to establish that this chimeric plasmid contains functional M13 initiation determinants, purified plasmid DNA was used to transform $polA^+$, $polA$ or M13-infected $polA$ strains. The results of such an experiment are shown in Table 1.

TABLE 1.   Transformation Assay for <u>oriM13</u> Function in pJMC110

	DNA transformation frequency*		
Strain	M13 bla6 RF	pBR322	pJMC110
K37 $polA^+$	1.3×10^{-3}	5.0×10^{-3}	3.4×10^{-3}
H560 polA	1.1×10^{-3}	$<1 \times 10^{-7}$	$<1 \times 10^{-7}$
H560 polA, M13 infected	8.2×10^{-4}	$<1 \times 10^{-7}$	9.0×10^{-4}

*Transformation frequency is defined as Ap^R colony-forming units (CFU)/μg of DNA divided by total cells per incubation. $<1 \times 10^{-7}$ indicates that no transformants were observed.

These results show that pBR322 DNA alone is unable to transform *polA* strains to ampicillin resistance. However, RF DNA from M13*bla*6, an M13 Ap^R transducing phage (13) containing the β-lactamase gene from Tn*3*, efficiently transforms both *polA*$^+$ and *polA* strains. In addition, it is observed that prior infection by M13 phage has little effect on the cells' transformability. The plasmid pJMC110 also transforms *polA* cells with similar efficiency, but it is specifically dependent on prior infection by an M13 helper phage to provide M13 gene products. Samples from cultures prepared from individual drug-resistant colonies were examined directly by agarose gel electrophoresis. It was found (11) that pJMC110 is present at a substantial copy number comparable to that of the M13 RF which is normally present at approximately 100 copies per cell. These results indicate that the M13 *Hpa*II F fragment is sufficient for replication of the chimeric plasmid in a *polA* host.

In order to further localize the DNA sequences responsible for determining the M13-dependent replication of these chimeric plasmids, we have examined chimeric plasmids carrying sub-fragments of the *Hpa*II F fragment. Two such chimeras, pJMC67 and pJMC122 (Figure 2) are of particular interest. The plasmid pJMC67 carries the 142 base pair *Hae*III G fragment which contains both the viral strand origin and the sequence that codes for the complementary strand RNA primer. This plasmid was tested for a functional M13 replication determinant by the transformation assay described in Table 1 and was found to transform M13-infected *polA* cells 10^4-fold less efficiently than the plasmid carrying the 381 base pair *Hpa*II F fragment (Table 2). Furthermore, plasmid pJMC66, which

Table 2. Transformation assay for *ori* M13 functions

Plasmid DNA	M13 insert	DNA transformation frequency	
		K37	H560 (M13 infected)
pJMC110	*Hpa*II F	2.5×10^{-3}	9.7×10^{-4}
pJMC66	*Hae*III G	3.5×10^{-3}	2.0×10^{-7}
pJMC67	*Hae*III G	2.6×10^{-3}	2.5×10^{-7}
pJMC122	*Hae*III G - *Hpa*II F hybrid	2.8×10^{-3}	1.1×10^{-3}
pJMC221	*Hae*III G - *Hpa*II F hybrid	3.8×10^{-3}	9.7×10^{-4}
pBR322	None	4.5×10^{-3}	$<1 \times 10^{-7}$

Transformation frequencies are as described for Table 1.

carries the 142 base pair *Hae*III G fragment in the opposite
orientation in the *Bam*H1 site of the cloning vector, also
transforms M13-infected *polA* cells at this greatly reduced
efficiency. Surprisingly, the rare transformants obtained
with these DNAs contain the chimeric plasmids at a high copy
number similar to that of pJMC110 (11). Re-isolation of
plasmid DNA from these rare transformants and repeated transformation of M13-infected *polA* cells indicates that these rare
transformation events are not the result of a selection for a
mutant plasmid DNA. The re-isolated plasmid DNA transforms
cells with an efficiency similar to the original DNA. Likewise, we have shown that the rare transformation events do not
represent the successful transformation of revertants of the
polA mutation. The rare transformed cells were found to still
retain the *polA* mutation. On the basis of these results, we
conclude that the *Hae*III G fragment is sufficient for M13-
dependent replication of the chimeric plasmids <u>once replication is established</u>. However, additional sequences are
required for efficient establishment of replication.

The lack of efficient transformation by plasmids carrying
the *Hae*III G fragment suggested that additional sequences
within the M13 *Hpa*II F fragment, but outside of the M13 *Hae*III
G fragment, are important for efficient transformation by the
chimeric plasmids. The plasmid pJMC122 carries a 270 base
pair insert (Figure 2) which includes sequences to the right
of the central *Hae*III G fragment. This chimeric plasmid and
one (pJMC221) carrying the same insert in the opposite
orientation, were both found to transform M13-infected *polA*
cells at the same efficiency as the larger plasmid pJMC110
(see Table 2).

More precise localization of the DNA sequence required
for efficient transformation of M13-infected *polA* cells has
been obtained (14) by *in vitro* deletion of DNA sequences at
either end of the inserted fragment in pJMC122 as shown in
Figure 3. The M13 sequences retained in these plasmids are
shown in Figure 4. The plasmid pJMC122 Δ60-63 contains only
an additional 40 base pairs of DNA beyond the right-hand end
of the *Hae*III G fragment, yet it transforms M13-infected *polA*
cells as efficiently as the parental plasmid pJMC122. The
plasmid pJMC122 Δ90-11, deleted up to the *Hae*III site so that
it is equivalent to pJMC67, transforms with an efficiency
10^4-fold lower. These results indicate that sequence information within the 40 base pairs to the right of the *Hae*III G
fragment are strongly required for efficient M13-dependent
transformation of *polA* cells.

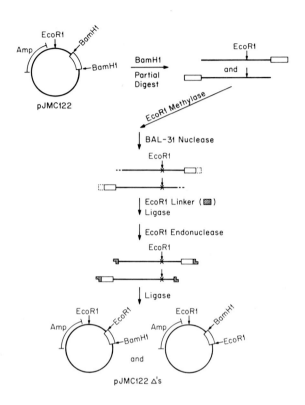

FIGURE 3. *In vitro deletion at either end of the M13 insert in the plasmid pJMC122 (14).*

Deletions at the left-hand end of the insert in pJMC122 suggest that these chimeric plasmids may be replicating by a hybrid mechanism in which one strand is initiated using the M13 viral strand initiation mechanism and the other strand is initiated by sequences within the plasmid cloning vector (14). Deletions pJMC122 Δ30-16 and pJMC122 Δ30-36 completely remove the DNA sequence coding for the primer for the complementary strand and extend to the base of the hairpin containing the viral strand origin. Results obtained with M13 chimeric phage containing fragments of ColE1 DNA are consistent with such a hybrid replication mechanism (Nomura and Ray, this volume). Specific sequences within ColE1 DNA were found to be capable of substituting for the M13 complementary strand origin. We therefore suggest that the M13-dependent replication of the chimeric plasmids described here utilizes only the viral strand origin of M13. Consequently, the essential M13 sequences defined in these experiments likely relate only to the

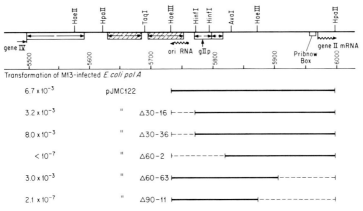

FIGURE 4. *Physical map and replication properties of deletion derivatives of the plasmid pJMC122 (14). Dashed lines indicate sequences that were deleted from the M13 insert in pJMC122. The ability of each plasmid to transform M13-infected polA cells is given in the left-hand column in units of ampicillin resistant CFU/μg of plasmid DNA divided by the total number of cells per incubation. $<1 \times 10^{-7}$ indicates that no transformants were detected.*

viral strand initiation mechanism. How the 40 base pair sequence to the right of the *Hae*III G fragment might relate to viral strand synthesis, if it does, is not yet understood.

B. *Deletions in the M13 Complementary Strand Origin*

Our second approach to the identification of specific sequences involved in the initiation of M13 DNA replication is the introduction of deletions into the origin region (15). Initially we have taken advantage of the existence of a single *Asu*I site in M13 RF located near the RNA-DNA junction between the RNA primer and the complementary DNA strand of the RF. Deletions were introduced as shown in Figure 5. M13 RF DNA was cleaved with *Asu*I and subjected to limited exonucleolytic degradation with BAL31. The treated molecules were then circularized by blunt-end ligation with T4 DNA ligase and used to transform CaCl$_2$-treated *E. coli* cells. The resulting plaques were picked and individual isolates selected for

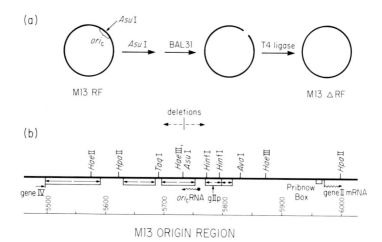

FIGURE 5. *In vitro deletion of sequences around the AsuI site in M13 RF DNA. (a) M13 RF was cleaved at the single AsuI site, partially digested with BAL31 and the termini joined by blunt-end ligation with T4 ligase (15). (b) Location of the AsuI site in the M13 origin region.*

further growth and DNA sequence analysis by the di-deoxynucleotide sequencing technique.

   Figure 6 shows the deletion maps of several individual isolates. Deletions range from one base pair up to 201 base pairs. In some cases, a 12 base pair linker containing an *Eco*RI site was inserted at the site of the deletion, and the extra nucleotides added to the M13 viral DNA as a result are shown in parentheses. Isolates ΔE101, ΔE5 and ΔE3 all contain this entire linker sequence while isolates ΔE48 and ΔE13 contain only an additional 10 nucleotides of the linker sequence. Isolate ΔE4 contains the entire 12 nucleotides of the linker sequence plus an additional 17 nucleotide sequence of unknown origin. It is also unclear how the single base deletion in Δ134 might have arisen.

   The extent of the deletions recovered is unexpected. In the largest deletion, ΔE4, the entire sequence specifying the RNA primer for the complementary strand and both of the large hairpins downstream from the start of the RNA primer have been deleted. These hairpins have previously been implicated in

13 DNA INITIATION DETERMINANTS OF BACTERIOPHAGE M13

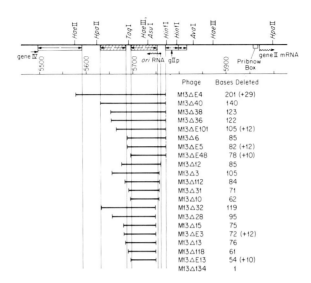

FIGURE 6. *Deletion map of the M13 origin region. Individual isolates produced as indicated in Figure 5 were sequenced by the Sanger dideoxy-substitution technique (29). Horizontal bars indicate the sequences deleted in each isolate. In some cases, a 12 base pair EcoRI linker was introduced at the site of the deletion and is so indicated by the letter E preceding the isolate number. The number of nucleotides deleted in each isolate is given in the right-hand column. Numbers in parenthesis indicate nucleotides added.*

complementary strand initiation (16). Both of these structures are protected from nuclease degradation by the specific binding of RNA polymerase to viral DNA coated with the *ssb* protein. In all of the deletions, including the largest, the phage retains the ability to replicate and to form a plaque. In general, the plaques produced by these phage are faint compared to the wild type, and the phage yield in liquid culture is reduced approximately 10-fold. Both *in vivo* and *in vitro* studies of parental RF synthesis indicate that these deletion mutants are defective in the initial events in the conversion of the mutant viral DNAs to the parental RF (15 and unpublished results).

It should be noted that the deletions shown in Figure 6 are not symmetric about the *Asu*I site. Eight of the deletions stop precisely at the base of the potential hairpin structure containing the sequence nicked by the gene II protein during viral strand initiation. Deletions that extended beyond this point were possibly lethal due to the loss of sequence information required for gene II nicking. These results suggest that the left boundary of the gene II protein nicking site must extend to the base of this hairpin.

C. *Construction of a Single Strand Initiation Vector*

Because of the greatly reduced ability of the mutant M13 phages to propagate, it was of interest to know whether or not a DNA sequence containing a foreign single-strand initiation determinant might be capable of rescuing the defect in these phages. This possibility would open the door to the direct selection of DNA sequences capable of promoting DNA chain initiation. It was with this possibility in mind that deletion phages were also constructed with *Eco*RI linkers inserted at the site of the deletion in the region of the normal M13 complementary strand origin, as shown in Figure 7. Mutants having a reduced capacity for *s*ingle *s*trand *i*nitiation are designated as *ssi*⁻ mutants.

As a test of the possible direct selection of chimeric phages carrying cloned sequences capable of promoting single strand initiation, we selected the G4 complementary strand

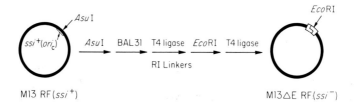

FIGURE 7. *Construction of vectors for cloning ssi⁺ determinants. Deletions are introduced into the M13 origin region as in Figure 6 except that EcoRI linkers are inserted at the deletion site to provide a receptor site for foreign DNA fragments having EcoRI ends. M13 RF is capable of initiating complementary strand synthesis and is therefore designated as ssi⁺. Deletion mutants having a reduced ability to initiate synthesis of the complementary strand are designated as ssi⁻.*

origin as a DNA expected to give a significant rescue of the defect contained in the M13 deletion phages. In previous experiments (17) we have cloned the G4 complementary strand origin into wild type M13 and found that the G4 origin can substitute for the M13 origin when initiation at the latter origin is inhibited by rifampicin. In addition, the G4 complementary strand origin sequence is widely separated from the G4 viral strand origin sequence (18) and is flanked (13) by a single *Eco*RI site and an *Sst*I site (Figure 8).

In order to obtain the G4 complementary strand origin on an *Eco*RI fragment suitable for insertion into an M13 deletion phage carrying a single *Eco*RI site, M13G*ori*1 RF (which carries this region of the G4 genome) was cleaved with *Sst*I, treated briefly with *BAL*31 and *Eco*RI linkers added by blunt-end ligation with T4 ligase. The resulting DNA was cleaved with *Eco*RI and ligated to *Eco*RI-cut RF DNA from several different M13ΔE phages. The ligated DNA was then used to transfect CaCl$_2$-treated *E. coli* and individual plaques were screened for G4 DNA inserts by plaque hybridization (13). A physical map of one such isolate, M13G*ori*101, is shown in Figure 8.

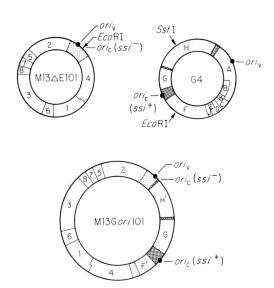

FIGURE 8. *Physical maps of M13ΔE101, G4, and the chimeric derivative M13Gori101. Viral strand origins are indicated by* $\overline{ori}_v$, *and complementary strand origins by* $\overline{ori}_c$.

In order to use the M13ΔE phages as selectable cloning vectors for DNA initiation determinants (promoters of single strand initiation), the RF DNA of the rescued phage should give a clearly distinguishable plaque type upon transformation of $CaCl_2$-treated cells. A comparison of plaques produced by the M13ΔE101 RF and that of M13G*ori*101 RF shows a striking difference (Figure 9). The plaques of M13G*ori*101 RF are considerably larger and clearer. In a mixture of plaques produced by M13ΔE101 and M13G*ori*101 RFs the M13G*ori*101 plaques are easily distinguishable. In addition to the effect on plaque morphology, the yield of the M13G*ori*101 phage in liquid culture was found to be restored to the level of the M13 wild type. These results suggest that the M13ΔE phages can be used for the direct selection of DNA sequences that promote single strand initiation.

D. *Cloning Rifampicin-Resistant Initiation Determinants*

Single strand initiation determinants that specify initiation by a rifampicin-resistant mechanism have been cloned into M13 previously (17,19,20) and found to confer drug resistance on the SS → RF reaction. In the case of wild type M13 this reaction is sensitive to rifampicin since the primer for the M13 complementary strand is synthesized by the rifampicin-sensitive RNA polymerase of *E. coli* (21). Chimeric M13 phages carrying rifampicin-resistant initiation (*rri*)

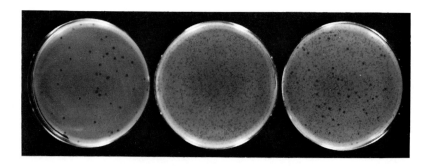

FIGURE 9. *Plaque morphologies of M13ΔE101 and M13Gori101 produced by transformation of $CaCl_2$-treated E. coli cells. Left, M13Gori101 RF; center, M13ΔE101; right, M13Gori101 and M13ΔE101 RFs.*

determinants have available to them an alternate pathway which allows parental RF formation even in the presence of rifampicin. In the case of the G4 complementary strand origin cloned into M13, initiation can be mediated by either the RNA polymerase-dependent mechanism at the normal M13 origin or by a $dnaG$ primase-dependent mechanism at the G4 origin (4,5). This ability to substitute a foreign replication determinant for the usual M13 complementary strand origin provides an assay for the expression of $\underline{r}$ifampicin-$\underline{r}$esistant $\underline{i}$nitiation (rri^+) determinants (Figure 10). (rri^+ determinants represent a sub-class of ssi^+ determinants.) Radioactively labeled chimeric phages are used to infect cells in either the presence or absence of rifampicin. Only when the inserted fragment carries an rri^+ determinant is there a significant conversion of the phage SS DNA to RF in the presence of rifampicin. Using this assay for rri^+ sequences, we have identified such determinants in DNAs from several sources (Table 3).

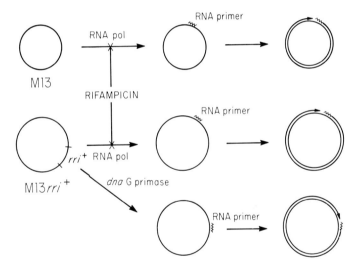

FIGURE 10. *Rifampicin-resistant pathway for RF formation by chimeric M13 derivatives carrying $\underline{ri}fampicin$ $\underline{r}esistant$ $\underline{i}nitiation$ ($\underline{rri}^+$) determinants.*

Table 3. *DNAs containing rri^+ sequences*

1.	G4
2.	ØX174
3.	ColE1
4.	E. coli oriC
5.	Salmonella his region

E. *Initiation Determinants in ColE1 and ØX174*

Two rri^+ sequences were identified in the genome of ColE1 and are described in detail in this volume (Nomura and Ray). These determinants specify initiation mechanisms similar to that of the ØX174 complementary strand in which initiation involves the n' ATPase and the *dnaB* protein (22). The finding of n' ATPase effector sites in the genome of ColE1 suggests that the ØX174-type initiation mechanism is also utilized by ColE1. This mechanism is especially attractive for the synthesis of the lagging strand during the unidirectional replication of the plasmid DNA.

Various fragments of ØX174 DNA have been cloned into M13 vectors in order to more precisely define the ØX174 DNA sequence that specifies the ØX174-type of initiation (19,23). Fragments spanning the intergenic space between genes F and G have been found to confer upon M13 the ØX174-type of initiation capability. Among the rri^+ M13-ØX174 chimeras which we have constructed, the one carrying the smallest ØX174 fragment contains a 180 base pair insert from the *Hin*fI site at position 2264 in the ØX174 RF to the *Alu*I site at position 2444 (Figure 11). This particular fragment of ØX174 RF was cloned into the single *Pst* site in the M13 vector M13*bla*61 (13). The protruding end at the *Hin*fI site on the fragment was filled in by polymerase I, and *Pst* linkers were added by blunt-end ligation with T4 ligase. Clones were obtained with this fragment inserted in each of the two possible orientations. The isolate M13ØX81 contains the fragment in the orientation in which the viral strand of the ØX174 DNA is ligated to the M13*bla*61 viral strand. The insert is contained in the opposite orientation in M13ØX80.

^{32}P-labeled M13ØX80 and M13ØX81 phages were used to infect *E. coli* in the presence and absence of rifampicin. As shown in Figure 12, the M13ØX81 phage is able to form the parental RF in the presence of rifampicin while M13ØX80 cannot. Expression of rifampicin-resistant initiation by M13ØX81 in which the viral strand of the cloned ØX174 fragment is present in the viral strand of the chimeric phage is consistent with its normal expression in ØX174 viral DNA. The lack of expression of the complementary strand of this fragment joined to the viral strand of M13*bla*61 indicates that the recognition of the initiation determinant depends on the specific DNA sequence of the viral strand in this region.

F. *Initiation Determinants in the E. coli Origin*

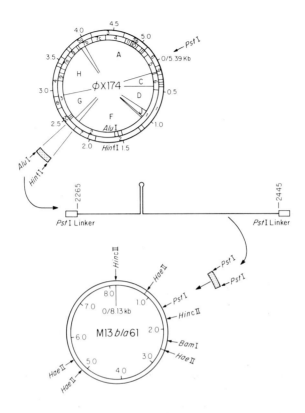

FIGURE 11. *Cloning a 180 base pair rri⁺ determinant from ØX174 RF (23). PstI linkers were joined to the 180 base pair AluI-HinfI restriction fragment purified by gel electrophoresis. The fragment was treated with PstI and inserted into the PstI site of M13bla61 RF (13). Isolates carrying ØX174 sequences were detected by plaque hybridization and further characterized by DNA sequence analysis.*

We have also identified initiation determinants within the replication origin (*oriC*) of *E. coli*. In order to obtain the *E. coli* origin on a relatively small restriction fragment, we have cloned *oriC* from the plasmid pJS5 (24). This plasmid consists entirely of *E. coli* DNA from the origin region and is derived from the plasmid pSY211 (25). The latter plasmid was reduced in size by *in vitro* deletion of 0.4 and 2.2 kb *Bam* fragments on one side of *oriC*. As a consequence of this

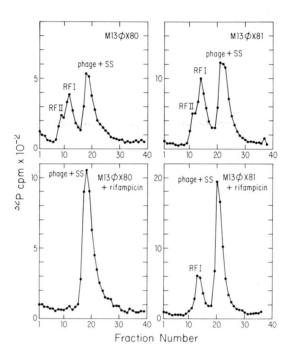

FIGURE 12. *Parental RF formation by ^{32}P-labeled M13ØX80 and M13ØX81 in the presence of rifampicin. Isolate M13ØX81 contains the viral strand of the insert shown in Figure 11 while M13ØX80 contains the complementary strand. Radioactive phages were used to infect E. coli in the presence or absence of rifampicin at 400 µg/ml. Cells were harvested, lysed and sedimented through sucrose gradients as described (17,23). Sedimentation is from left to right.*

deletion, a *Bcl*I restriction site has been brought close to *oriC*, and another *Bcl*I site already existed on the opposite side of *oriC*. Thus, pJS5 contains *oriC* on a single 500 base pair *Bcl*I restriction fragment. This *Bcl*I fragment was cloned into the single *Bam*HI site of the vector M13*bla*61, taking advantage of the ability of *Bcl*I-cut fragments to be joined by cohesive-end ligation to *Bam*HI termini (Figure 13). Two representative isolates, M13*oriC*52 and M13*oriC*74, contained the *Bcl*I insert in opposite orientations and were selected for further study.

As shown in Figure 14, both the M13*oriC*52 and M13*oriC*74 phages undergo parental RF formation at a significant but

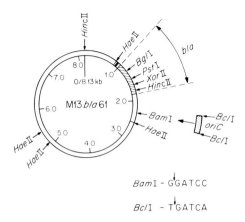

FIGURE 13. *Cloning a 500 base pair fragment from the E. coli replication origin (oriC) into the BamI site of M13bla61. The 500 base pair BclI fragment was ligated to BamI-cut M13bla61 RF and used to transform $CaCl_2$-treated E. coli cells. Individual isolates were obtained with the insert in either orientation.*

reduced level in the presence of rifampicin. Synthesis of parental RF in the presence of drug is considerably less than that observed for the rifampicin-resistant M13Gori1 phage, but at least twice that of the rifampicin-sensitive M13 phage. No difference is observed with regard to the orientation of the *oriC* fragment. This relatively weak rifampicin-resistant initiation of the M13*oriC* phages suggests that the single stranded *oriC* DNA presented to the cell in the chimeric DNA may not serve as an efficient template for initiation. The complexity of the *E. coli* initiation mechanism may possibly preclude a strong initiation on a template that is largely or entirely single stranded. However, like other M13*oriC* phages (24,26), the cloned *oriC* DNA is capable of replicating the double stranded RF DNA as shown by the ability of these phages to undergo double-strand replication in *E. coli rep* mutants. The significance of the weak rifampicin-resistant initiation on the single-stranded forms of *oriC* is still uncertain.

G. *Chromosomal Initiation Determinants*

If the ØX174-type of initiation mechanism is indeed a good model for the mechanism of discontinuous DNA synthesis of

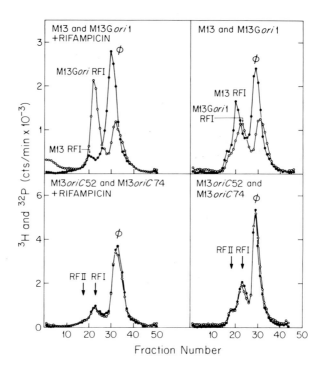

FIGURE 14. *Parental RF formation by ^{3}H- or ^{32}P-labeled M13, M13Gori1, M13oriC52 and M13oriC74 in the presence or absence of rifampicin as in Figure 12. In each case cells were infected with a pair of phages, one labeled with ^{32}P and the other with ^{3}H.*

the bacterial chromosome, then ØX174-type initiation determinants may exist at multiple sites on the chromosome. In order to avoid the possible confusion that could be introduced by sequences from the bacterial replication origin, we have examined cloned chromosomal sequences from regions far removed from the origin of replication.

The M13 Ho167 and M13 Ho168 phages are recombinant transducing phages carrying 5400 and 5600 base pairs of DNA from the *His* OGD region of *S. typhimurium* in opposite orientations (27). A third phage, M13 Ho176, contains the *his* DNA in the same orientation as M13 Ho168 but contains only 3300 base pairs of the *his* region. We have examined all three of these phages for their ability to initiation SS → RF replication in the presence of rifampicin.

Figures 15 and 16 show that all three phages strongly initiate SS → RF replication in the presence of rifampicin. In each case, parental RF is synthesized in the presence of rifampicin in an amount similar to that made in the absence of drug. These results provide strong support for the existence of single-strand initiation determinants at points on the bacterial chromosome outside of the origin region.

IV. DISCUSSION

The nucleotide sequence is known for many prokaryotic replication origins and in some cases the origins of the individual strands have been identified. In the case of bacteriophage M13 the viral strand is initiated at the gene II protein nicking site in the viral strand, and the complementary strand is initiated by the synthesis of an RNA primer

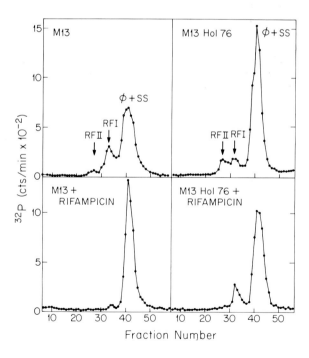

FIGURE 15. Parental RF formation by ^{32}P-labeled M13 and M13Hol76 in the presence or absence of rifampicin as in Figure 12.

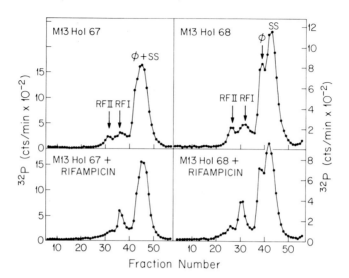

FIGURE 16. *Parental RF formation by ^{32}P-labeled M13Hol67 and M13Hol68 phages in the presence or absence of rifampicin as in Figure 12.*

starting at a site 24 nucleotides away from the viral strand origin. Yet, even in this well defined origin the precise DNA sequences that determine each initiation are still unknown.

In order to avoid confusion with the term "origin", we have introduced the term <u>single strand initiation determinant</u>, or ssi^+ determinant, to refer to the sequence that specifies the initiation of a single DNA strand. Such determinants are, in some cases, analogous to promoters of transcription, the DNA sequences that determine the initiation of specific messenger RNAs. However, they also include specific recognition sequences that determine initiation in other ways, such as the ØX174 and M13 nicking sites or the ØX174 primasome loading site (28).

We consider the ssi^+ determinant of a single DNA strand to carry the necessary information for the appropriation of specific enzymatic machinery for the initiation of that strand at one or more sites. In the simplest cases, as in M13 or G4 complementary strand initiation, a single initiation at a unique site is specified by the ssi^+ determinant. Where initiation is known to occur by a rifampicin-resistant mechanism, we have termed such determinants rri^+ determinants (20).

In a more complex case, as for ØX174, initiation may occur at multiple sites, possibly having little specificity. Thus, although ØX174 initiates at multiple sites, the initiation mechanism is a direct consequence of specific DNA-protein recognition at the ØX174 rri^+ determinant (28).

In the work presented here we have begun to define the initiation determinants for the M13 viral and complementary strands. The viral strand determinant may prove to be no more than a recognition sequence for the site specific nicking enzyme specified by the M13 gene II. Based on deletion analysis of the M13 origin region, the left-hand boundary of the M13 gene II protein nicking site must extend to the base of the hairpin structure containing this site. The right-hand boundary remains to be identified.

The finding of a 40 nucleotide sequence located 88 nucleotides to the right of the gene II protein nicking site and which strongly enhances the transformation by chimeric plasmids containing fragments of the M13 origin region was unexpected. This sequence also confers upon plasmids the ability to interfere with M13 growth. The precise role of this sequence remains to be established.

The determinant of M13 complementary strand synthesis is still unknown. The fact that M13 phages carrying deletions that span the entire DNA sequence that specifies the primer RNA for the complementary strand suggests that there may be secondary initiation determinants for the complementary strand or that the initiation determinant lies upstream and possibly overlaps the gene II protein nicking site. Further *in vitro* alterations of the origin region will hopefully resolve this question.

The greatly reduced ability of these deleted M13 phages to yield progeny phage and to produce a normal plaque has provided a direct means for selecting chimeric phages carrying ssi^+ determinants. The strong rescue of this defect by the cloned G4 complementary strand origin is highly promising for the future use of these new vectors.

Using the ability to screen for the rifampicin-resistant conversion of chimeric phage DNA to RF, we have detected rri^+ determinants in the DNAs of G4, ØX174 and ColEl. Weak rri^+ sequences are also present in the origin region of *E. coli*.

In an attempt to identify sequences that specify the initiation of Okazaki fragments, we have examined cloned chromosomal sequences located far from the bacterial replication origin. Such sequences offer the advantage of avoiding possible confusion with determinants from the replication origin. Specific sequences contained within the *his* operon region of *Salmonella typhimurium* give a very strong rri^+ response in *E. coli*. Further subcloning of these sequences is in progress in order to answer important questions such as: Are there families of sequences involved in the initiation of Okazaki fragments? Are such sequences distributed asymmetrically between the two strands in a given region? Are there distinct host gene requirements for initiation at different initiation determinants? The M13 deletion phages described here will hopefully provide a means for isolating chromosomal ssi^+ determinants and allow these questions to be addressed.

ACKNOWLEDGMENTS

This work was supported by a research grant (AI 10752) and by training grants (GM 07104 and GM 07185) from the National Institutes of Health.

REFERENCES

1. Schekman, R., Weiner, A., and Kornberg, A. (1974). *Science 184*, 987.
2. Tabak, H.F., Griffith, J., Geider, K., Schaller, H., and Kornberg, A. (1974). *J. Biol. Chem. 249*, 3049.
3. Bouché, J.-P., Zechel, K., and Kornberg, A. (1975). *J. Biol. Chem. 250*, 5995.
4. Bouché, J.-P., Rowen, L., and Kornberg, A. (1978). *J. Biol. Chem. 253*, 765.
5. Wickner, S. (1977). *Proc. Natl. Acad. Sci. U.S.A. 74*, 2815.
6. McMacken, R., and Kornberg, A. (1978). *J. Biol. Chem. 253*, 3313.
7. Wickner, S., and Hurwitz, J. (1974). *Proc. Natl. Acad. Sci. U.S.A. 71*, 4120.
8. Vovis, G.F., Horiuchi, K., and Zinder, N.D. (1975). *J. Virol. 16*, 674.
9. Meyer, T.F., Geider, K., Kurz, C., and Schaller, H. (1979). *Nature 278*, 365.
10. Geider, K., Beck, E., and Schaller, H. (1978). *Proc. Natl. Acad. Sci. U.S.A. 75*, 645.
11. Cleary, J.M., and Ray, D.S. (1980). *Proc. Natl. Acad. Sci. U.S.A. 77*, 4638.

12. Suggs, S.V., and Ray, D.S. (1979). *Cold Spring Harbor Symp. Quant. Biol. 43*, 379.
13. Hines, J.C., and Ray, D.S. (1980). *Gene 11*, 207.
14. Cleary, J.M., and Ray, D.S. (1981). Submitted for publication.
15. Kim, M.H., Hines, J.C., and Ray, D.S. (1981). Submitted for publication.
16. Gray, C.P., Sommer, R., Polke, C., Beck, E., and Schaller, H. (1978). *Proc. Natl. Acad. Sci. U.S.A. 75*, 50.
17. Kaguni, J., and Ray, D.S. (1979). *J. Mol. Biol. 135*, 863.
18. Ray, D.S., and Dueber, J. (1977). *J. Mol. Biol. 113*, 651.
19. Kaguni, J., LaVerne, L.S., Strathearn, M., and Ray, D.S., *in* "DNA - Recombination, Interactions and Repair" (S. Zadrazil and J. Sponar, eds.), p. 35. Pergamon Press, New York (1980).
20. Nomura, N., and Ray, D.S. (1980). *Proc. Natl. Acad. Sci. U.S.A. 77*, 6566.
21. Geider, K., and Kornberg, A. (1974). *J. Biol. Chem. 249*, 3999.
22. Shlomai, J., and Kornberg, A. (1980). *Proc. Natl. Acad. Sci. U.S.A. 77*, 799.
23. Strathearn, M., and Ray, D.S. (1981). Submitted for publication.
24. Kaguni, J., LaVerne, L.S., and Ray, D.S. (1979). *Proc. Natl. Acad. Sci. U.S.A. 76*, 6250.
25. Hirota, Y., Yasuda, S., Yamada, M., Nishimura, A., Sugimoto, K., Sugisaki, H., Oka, A., and Takanami, M. (1979). *Cold Spring Harbor Symp. Quant. Biol. 43*, 129.
26. Kaguni, L.S., Kaguni, J., and Ray, D.S. (1981). *J. Bact. 145*, 974.
27. Barnes, W.M. (1979). *Gene 5*, 127.
28. Arai, K., and Kornberg, A. (1981). *Proc. Natl. Acad. Sci. U.S.A. 78*, 69.
29. Sanger, F., Nicklen, S., and Coulson, A.R. (1977). *Proc. Natl. Acad. Sci. U.S.A. 74*, 5463.

ESSENTIAL FEATURES OF THE ORIGIN OF BACTERIOPHAGE ØX174 RF DNA REPLICATION: A RECOGNITION AND A KEY SEQUENCE FOR ØX GENE A PROTEIN

P.D. Baas[1], F. Heidekamp[1], A.D.M. Van Mansfeld[1], H.S. Jansz[1], G.A. Van der Marel[2], G.H. Veeneman[2], and J.H. Van Boom[2]

[1]Institute of Molecular Biology and Laboratory for Physiological Chemistry, State University of Utrecht, Utrecht, The Netherlands, [2]Department of Organic Chemistry, State University of Leiden, Leiden, The Netherlands

ABSTRACT

This paper describes our current research on the structure-function relationship of the origin of bacteriophage ØX174 DNA replication. Results of four different approaches are reported.
1. Analysis of the nicking activity of ØX gene A protein on single-stranded synthetic and natural DNAs has identified the recognition sequence of ØX gene A protein within the decamer sequence CAACTTG↓ATA, corresponding to nucleotides Nr. 4299-4308 of the ØX DNA sequence.
2. Comparison of the DNA sequences around the cleavage sites of ØX gene A protein produced in RFI DNAs of bacteriophages ØX174, G4, St-1 and U3, has revealed a conserved region of 30 nucleotides. For St-1 DNA two nucleotide changes within this region were found.
3. Analysis of ØX mutants with preselected base changes in the origin region has indicated that the recognition sequence of ØX gene A protein is degenerated (CAACTPyG↓ATA or CAACTNG↓ATA).
4. Analysis of recombinant plasmid DNAs, containing ØX origin sequences has shown that the presence of the recognition sequence, CAACTTGATA, is not sufficient for the nicking activity of gene A protein on superhelical DNA, even if this sequence is followed by an AT-rich sequence. Also the presence of a 20 nucleotide long sequence corresponding to nucleotides Nr. 4299-4318 of the ØX origin was not sufficient.

These results indicate that for initiation of ØX RF DNA replication, besides the recognition sequence of the ØX gene A protein, a second specific nucleotide sequence, for which we propose the name key sequence, is required.

INTRODUCTION

The replication of bacteriophage ØX RF DNA is initiated by the endonucleolytic action of the viral gene A protein. The gene A protein nicks the viral strand of superhelical ØX RFI DNA, creating a free 3'OH-group at the G-residue at position 4305 of the ØX DNA sequence (1,2). After nicking, the gene A protein remains covalently bound to the 5'-end of the nick. Single-stranded ØX DNA is also nicked once by gene A protein at the same position as in double-stranded ØX RFI DNA (3). This is consistent with the observation that double-stranded ØX RF DNA is nicked by ØX gene A protein *in vitro* only if it contains superhelical turns, indicating that partial denaturation of the origin region is required for the nicking activity of ØX gene A protein (4). So in fact ØX gene A protein is a single-stranded specific endonuclease.

Recently we have shown that the synthetic decamer CAACTTGATA, corresponding to nucleotides Nr. 4299-4308 of the ØX DNA sequence, is cleaved by the gene A protein next to the G-residue (5). This indicates that the recognition sequence of ØX gene A protein lies within this 10-nucleotide sequence. Double-stranded superhelical RFI DNAs of the bacteriophages G4, St-1, U3, G14 and α3 are also nicked by ØX gene A protein (6-9). Comparison of the sequences around the nick sites in ØX, G4, St-1 and U3 DNA shows that these four phages have an uninterrupted stretch of 10 nucleotides, CAACTTGATA, the recognition sequence of ØX gene A protein, in common (Fig. 1). It is striking that, besides two nucleotide changes found in St-1 DNA (7), these phages also have in common a region of 30 nucleotides around the ØX gene A protein cleavage site. Outside this region the nucleotide sequences in the DNA of the four bacteriophages differ considerably. As discussed previously (7,9) this strongly suggests a special role for this conserved region during initiation, elongation and/or termination of ØX DNA replication.

In order to study which nucleotide sequences in bacteriophage ØX174 DNA are essential for origin function in the process of DNA replication, we have also constructed and analyzed ØX mutants with preselected base changes in the origin region and recombinant plasmid DNAs containing part of the conserved origin sequence.

The implications of the results obtained from these four approaches for our understanding of the initiation mechanism of bacteriophage ØX RF DNA replication are discussed.

14 ESSENTIAL FEATURES OF THE ORIGIN OF BACTERIOPHAGE   197

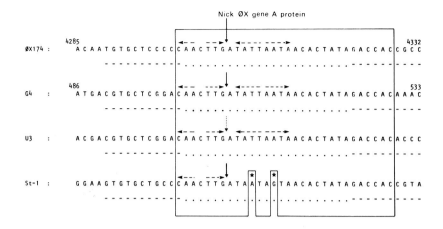

FIGURE 1. The nucleotide sequences in ØX (1,2), G4 (6,10), U3 (9), and St-1 (7) RF DNA in the region around the ØX gene A protein cleavage site. The site of the ØX gene A protein nick is indicated by vertical arrows. Horizontal arrows indicate short, self-complementary sequences. GC-rich tracts are indicated by broken lines, AT-rich tracts are marked with dots. Nucleotide sequences which are identical in all four phages are boxed.

RESULTS

The 5'-Cytosine Residue of the Decamer CAACTTGATA is an Essential Nucleotide in the Recognition Sequence of ØX Gene A Protein

In earlier studies with synthetic oligonucleotides of various lengths we have shown (5) that the decamer CAACTTGATA is still cleaved upon incubation with the ØX gene A protein. Further elimination of the 3'-terminal A-residue prohibits cleavage. These studies also indicate that the elimination of the 5'-terminal C-residue of these oligonucleotides greatly reduced their efficiency as a substrate for the gene A protein. In order to determine whether a specific nucleotide sequence or only the length of the substrate is important for the cleavage reaction, we have synthesized by means of the phosphotriester method (11) the decamer CAACTTGATA as well as the decamer *AAACTTGATA. Upon incubation with ØX gene A protein the decamer CAACTTGATA was cleaved efficiently, whereas almost no cleavage reaction was observed with the decamer *AAACTTGATA

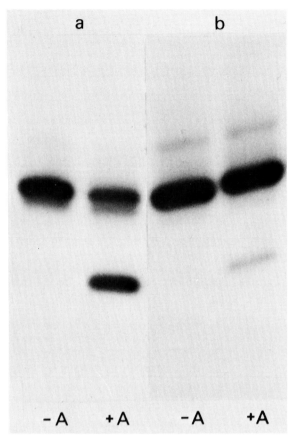

FIGURE 2. *Autoradiograph of a 25% polyacrylamide gel in 7 M urea, obtained after electrophoresis of synthetic decamers incubated with (+A) and without (-A) ØX gene A protein. The lanes represent from left to right* ^{32}*P-CAACTTGATA (a) and* ^{32}*P-AAACTTGATA (b).*

(Fig. 2). This indicates that the 5'-C residue is an essential nucleotide of the recognition sequence of ØX gene A protein.

The ØX Gene A Protein Recognition Sequence is Degenerated

Six different synthetic hexadecamers complementary to the origin region of bacteriophage ØX174, corresponding to nucleotides Nr. 4299-4314, except for one preselected base change, were used as primers for DNA synthesis on wild type ØX DNA as a template (Table I). DNA synthesis was performed with *E. coli* DNA polymerase I (Klenow fragment) in the presence of DNA ligase. Heteroduplex RFIV DNA was isolated and after limited diges-

TABLE I. *Nucleotide sequence of hexadecamers used as primers on wild-type ØX DNA as template*

		Mutation
	4314 ↓ 4299	
wild type	TATTAATAT CAAGTTG	–
m1	TATTAATA*CCAAGTTG	lethal
m2	TATTAAT*CTCAAGTTG	lethal
m3	TATT*GATATCAAGTTG	viable
m4	TATTAATC*GAGTTG	viable
m5	TATTAATA*ACAAGTTG	lethal
m6	TATTAATAT*TAAGTTG	lethal

The arrow ↓ indicates the position of the ØX gene A protein nick in the opposite viral strand. Nucleotides marked with an asterisk are changed nucleotides.

tion with DNase I the complementary strands containing the mutant primers were isolated by means of a poly (U,G)/CsCl gradient centrifugation (Fig. 3). The biological activity of these complementary strands was assayed in spheroplasts. Spheroplasts were made from *E. coli* K58 ung-(uracil-N-glycosylase) to prevent degradation of the complementary strands caused by uracil

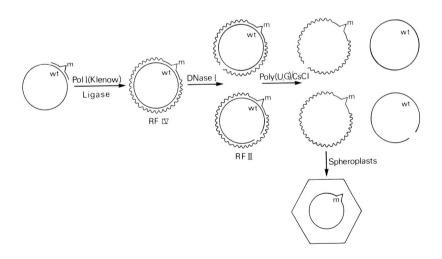

FIGURE 3. *General outline for the construction of ØX mutants in the origin region using synthetic oligodeoxyribonucleotides.*

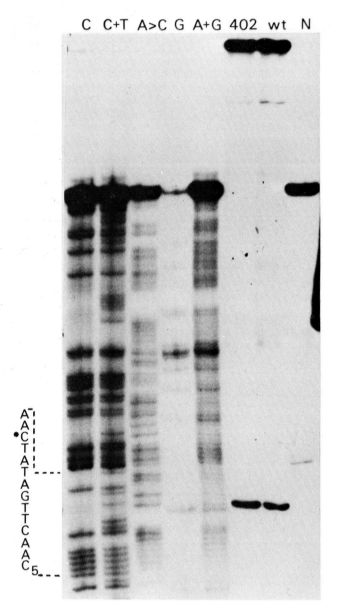

FIGURE 4. Autoradiograph of the sequence gel of the HaeIII Z6B-AluI A7C restriction DNA fragment, isolated from m316 RFI DNA. The lanes 402 and wt represent the HaeIII restriction DNA fragments Z6B, obtained from m402 RFI DNA and wt RFI DNA after incubation with ØX gene A protein, respectively. N refers to the non-modified m316 HaeIII Z6B-AluI A7C restriction DNA fragment.

incorporation (12). Viable ØX mutants were picked up with high frequency using a selection test which is based on the difference in UV sensitivity of homoduplex and heteroduplex ØX RF DNA (13). Phage DNA derived from single plaque lysates was annealed to wild-type complementary ØX DNA and from the resulting duplex DNA the UV survival curve was determined. We have been able to isolate in this way two different viable ØX origin mutants. The mutants dictated by the other four hexadecamer primers could not be detected. This indicates that priming with these hexadecamers resulted in the synthesis of complementary ØX DNA with lethal mutations (14).

The lethal mutations are all located within the recognition sequence of ØX gene A protein. From the two viable origin mutants one (m316) is located outside the recognition sequence. This ØX mutant contains a T → C transition at nucleotide Nr. 4310. However, the other viable mutant (m402), which contains a T → C transition at nucleotide Nr. 4304, is located within the recognition sequence. RFI DNA of these mutants is efficiently nicked by ØX gene A protein. Figure 4 shows part of the sequence gel of the HaeIII Z6B-AluI A7C restriction DNA fragment isolated from m316 RFI DNA. Onto this gel also the HaeIII Z6B restriction DNA fragments obtained after nicking of wt and m402 RFI DNA with ØX gene A protein respectively were loaded. The subfragments originating from the nicked Z6B restriction DNA fragments migrated in this gel at the same position as the A-residue at position 4306 of the ØX DNA sequence. This proves that ØX gene A protein nicks m402 RFI DNA in the same way as wt ØX RFI DNA, namely one nucleotide before the A-residue, i.e. after the G-residue at position 4305 of the ØX DNA sequence. Single-stranded DNA, which contains the sequence -CAACTCGATA- is also cleaved by ØX gene A protein. Figure 5 shows an autoradiograph of a polyacrylamide gel, obtained after incubation with ØX gene A protein of the single-stranded viral DNA, isolated from the Z6B restriction DNA fragment of m402 RFI DNA. A subfragment of approximately 100 nucleotides is visible, which is in good agreement with the expected length of 98 nucleotides from the $5'-{}^{32}P$ labelled end of the DNA to the G-residue of the recognition sequence.

These results indicate that the recognition sequence of ØX gene A protein is degenerated. Degeneration has also been observed in recognition sequences of various restriction endonucleases. It is not known if the nucleotide at position 4304 must be a pyrimidine base or that it can also be changed into a purine base.

The Presence of the Recognition Sequence in Superhelical DNA is not a Sufficient Condition for Nicking by ØX Gene A Protein

Inspection of the published nucleotide sequences of fd DNA

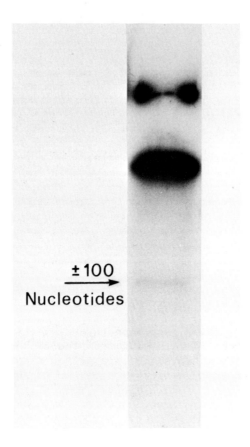

FIGURE 5. *Autoradiograph of a 25% polyacrylamide gel in 7 M urea, obtained after incubation of the viral strand of the restriction DNA fragment Z6B, isolated from m402 RFI DNA, with ØX gene A protein.*

(15), M13 DNA (16), SV40 DNA (17,18), BKV DNA (19), pBR 322 DNA (20) and polyomavirus DNA (21,22) revealed the presence of the decamer sequence CAACTTGATA in polyomavirus DNA (Nr. 1719-1710, ref. 21; Nr. 1737-1728, ref. 22). In order to investigate if the presence of the recognition sequence in a superhelical DNA is a sufficient condition for nicking by ØX gene A protein, polyomavirus DNA component I was incubated with ØX gene A protein. No conversion into polyomavirus DNA component II was observed. Control experiments with ØX RFI DNA and St-1 RFI DNA showed that the ØX gene A protein preparation which was used, was active. St-1 RFI DNA in the same reaction mixture as polyomavirus DNA component I was converted by ØX gene A protein into RFII DNA (Fig. 6). However, single-stranded polyomavirus

14 ESSENTIAL FEATURES OF THE ORIGIN OF BACTERIOPHAGE 203

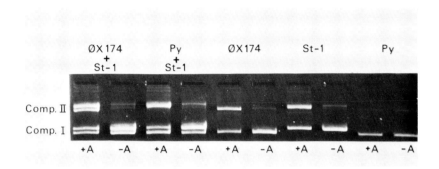

FIGURE 6. Analysis of the reaction products of ØX and St-1 RFI DNA, St-1 RFI DNA and polyomavirus (Py) DNA component I, ØX RFI DNA, St-1 RFI DNA, and polyomavirus DNA component I after incubation with ØX gene A protein (+A). The analysis was performed on horizontal 1% agarose slab gels containing 1 µg/ml ethidiumbromide. -A refers to the control experiment in which no ØX gene A protein was added.

DNA is cleaved by ØX gene A protein. After separation of the $5'-{}^{32}P$ labelled strands of the HaeIII 8 polyomavirus DNA restriction fragment, which contains the sequence CAACTTGATA, the individual DNA strands were incubated with ØX gene A protein. Analysis of the reaction products on a polyacrylamide gel in 98% formamide demonstrated cleavage of the DNA present in the slowest migrating band of the HaeIII 8 restriction DNA fragment (Fig. 7). A cleavage product of approximately 50 nucleotides was found. This is in good agreement with the expected length of 49 nucleotides from the $5'-{}^{32}P$ labelled end of the fragment to the G-residue of the recognition sequence.

Construction and Analysis of Recombinant Plasmid DNAs Containing ØX Origin Sequences

The synthetic hexadecamer CAACTTGATATTAATA, corresponding
                          GTTGAACTATAATTAT
to nucleotides Nr. 4299-4314 of the ØX DNA sequence, was cloned into plasmid pACYC177 ($Amp^R.Km^R$). Therefore pACYC177 DNA was linearized with the restriction enzyme HincII in the Amp^R gene or with the restriction enzyme SmaI in the Km^R gene, respectively (23). "Blunt end" ligation was performed with T4 DNA ligase and after transformation of *E. coli* strain K12-803 transformants with the antibiotic characteristics $Amp^S.Km^R$ and $Amp^R.Km^S$ respectively, were picked up. Restriction enzyme analysis of plasmid DNA isolated from several transformants showed the presence of an inserted piece of DNA. DNA sequence analysis of

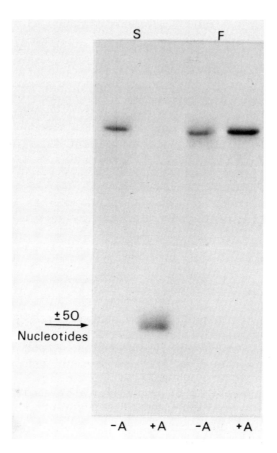

FIGURE 7. *Autoradiograph of a 10% polyacrylamide gel in 98% formamide of the separated DNA strands of the HaeIII 8 restriction DNA fragment of polyomavirus DNA incubated with (+A) and without (-A) ØX gene A protein. S and F indicate slow and fast migrating strand of the HaeIII 8 restriction DNA fragment of polyomavirus DNA.*

these recombinant plasmids around the original HincII or SmaI restriction site revealed the presence of ØX origin sequences. Figure 8 shows an autoradiograph of the sequence gel obtained from plasmid pFH 903, which contains the ØX DNA sequence Nr. 4299-4314, inserted into the SmaI restriction site. Besides this recombinant also plasmids were found, which contain the 16 b.p. DNA fragment in tandem. Moreover, sequence analysis of plasmid pFH 807 showed that in this plasmid a sequence of 20 nucleotides identical to the nucleotides Nr. 4299-4318 of the ØX origin sequence is present. However, neither of these plas-

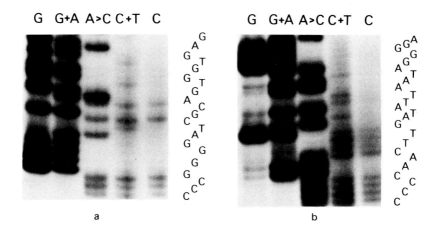

FIGURE 8. (a) Autoradiograph of part of the sequence gel of pACYC177 around the SmaI cleavage site with the deduced sequence. (b) Autoradiograph of part of the sequence gel of pFH 903 around the SmaI with the deduced sequence.

mid DNAs was converted into component II upon incubation with ØX gene A protein. This indicates that the presence in a plasmid of the ØX gene A recognition sequence, CAACTTGATA, followed by a long AT-rich stretch (tandem recombinants) is not sufficient for the nicking activity of ØX gene A protein. Also the presence in the plasmid pFH 807 of a 20-nucleotide long sequence corresponding to nucleotides Nr. 4299-4318 of the ØX origin was not sufficient for nicking (F. Heidekamp et al., manuscript in preparation).

DISCUSSION

The results obtained with the four different approaches as described in this paper are summarized in Figure 9.
Essential features of the origin of ØX RF DNA replication are:
1. The presence of the recognition sequence, CAACTTGATA, of ØX gene A protein. We have found that this decamer is cleaved by ØX gene A protein next to the G-residue (Fig. 2) and that natural single-stranded DNA, which contains this sequence in a surrounding different from ØX, e.g. polyomavirus DNA is also cleaved (Fig. 7). Figure 2 shows that the 5'-terminal C-residue is an essential nucleotide of the recognition sequence. Similar experiments in which nucleotides are changed at the 3' end of

Origin of ØX RF DNA replication

FIGURE 9. *Essential features of the origin of ØX RF DNA replication.* ✷ *indicates nucleotides which can be changed without effect on ØX gene A protein nicking or DNA replication ability.*
    At position 4304   T → C   (m402, 14)
    At position 4309   T → A   (St-1, 7)
    At position 4310   T → C   (m316, 14)
    At position 4312   A → G   (St-1, 7)
o *indicates nucleotides which cannot be changed without effect on ØX gene A protein nicking or DNA replication ability (lethal mutations).*
    At position 4305   G → A   (14)
    At position 4306   A → G or T (14)
    At position 4307   T → G   (14)

the decamer are in progress. Analysis of the mutants with pre-selected base changes in the origin region shows that only in one case we were able to isolate a viable ØX origin mutant with a nucleotide change within the recognition sequence. The other tested nucleotide changes within the recognition sequence have resulted in the synthesis of lethal mutations in the complementary ØX DNA strand. The finding that m402 contains a T → C transition at position 4304 indicates that the recognition sequence is degenerated. Other degeneracies have not been excluded.
2. The recognition sequence is followed by a stretch of non-essential nucleotides. This is based on the fact that in St-1 DNA two nucleotides, Nr. 4309 (T → A), and Nr. 4312 (A → G),

14 ESSENTIAL FEATURES OF THE ORIGIN OF BACTERIOPHAGE

are changed (7) and on the existence of a viable ØX origin mutant, m316, which contains a T → C transition at position 4310 (14). The nucleotide changes at position 4310 and 4312 convert an AT base pair into a GC base pair.

3. The non-essential sequence is followed by an essential nucleotide sequence for which we propose the name "key sequence". The following observations indicate that besides the recognition sequence of ØX gene A protein a second specific sequence is required in order to initiate ØX RF DNA replication. All sequenced isometric phages, whose RFI DNAs are nicked by gene A protein have, with exception of the two nucleotide changes in St-1 DNA, a region of 30 nucleotides in common (Fig. 1). This emphasizes the importance of a specific nucleotide signal longer than the recognition sequence of ØX gene A protein. Superhelical, double-stranded polyomavirus DNA, which contains the recognition sequence of the ØX gene A protein is not nicked by the A protein (Fig. 6). This indicates that for nicking more than the recognition sequence is required. That this extra requirement is an unspecific AT-rich stretch is highly improbable because two of the mutations found in the non-essential region change an AT base pair into a GC base pair. Moreover, recombinant DNA, which contains the 16 b.p. sequence (Nr. 4299-4314) in tandem, is not nicked by ØX gene A protein. In these molecules the recognition sequence is followed by an AT-rich stretch longer than in the ØX genome.

We envisage the following role for the key sequence. In superhelical covalently closed ØX RFI DNA the gene A protein recognizes and binds to the key sequence. After binding the gene A protein opens the DNA helix in such a way that the recognition sequence becomes exposed in a single-stranded form. Then the gene A protein is able to nick the viral strand next to the G-residue at position 4305 and initiates a cycle of DNA replication. The proposed way of action of the ØX gene A protein resembles the initiation of RNA synthesis by RNA polymerase. RNA polymerase first binds to the recognition sequence of the promoter site centered about 35 nucleotides from the mRNA initiation point and forms a closed complex. Then transition of the closed complex into an open complex takes place by binding of the RNA polymerase to the binding sequence of the promoter site centered about 8 nucleotides from the mRNA initiation point. During this transition RNA polymerase unwinds the DNA helix and this makes it possible to initiate an RNA chain (24). Also in this case superhelical turns in the template are important, because superhelical DNA is a better template for RNA polymerase than relaxed or linear DNA (25).

At this moment we do not know the end of the non-essential region nor the beginning or the end of the key sequence. An attractive candidate for the binding of ØX A protein is a nucleotide stretch which contains the sequence CACTAT. It is remarkable that this nucleotide sequence is found as part of the binding site of the Int protein of bacteriophage λ (26) as part of one of the promoters on fd DNA (P 0.64 nucleotide Nr. 4049-4054) (15) and that this sequence is also present in the origin region of bacteriophage λ (27,28).

REFERENCES

1. Langeveld, S.A., Van Mansfeld, A.D.M., Baas, P.D., Jansz, H.S., Van Arkel, G.A., and Weisbeek, P.J., *Nature (London)* 271, 417 (1978).
2. Sanger, F., Coulson, A.R., Friedmann, T., Air, G.M., Barrell, B.G., Brown, N.L., Fiddes, J.C., Hutchison, C.A. III, Slocombe, P.M., and Smith, M., *J. Mol. Biol.* 125, 225 (1978).
3. Langeveld, S.A., Van Mansfeld, A.D.M., De Winter, J.M., and Weisbeek, P.J., *Nucleic Acids Res.* 8, 2177 (1979).
4. Marians, K., Ikeda, J.E., Schlagman, S., and Hurwitz, J., *Proc. Natl. Acad. Sci. USA* 74, 1965 (1977).
5. Van Mansfeld, A.D.M., Langeveld, S.A., Baas, P.D., Jansz, H.S., Van der Marel, G.A., Veeneman, G.H., and Van Boom, J.H., *Nature (London)* 288, 561 (1980).
6. Van Mansfeld, A.D.M., Langeveld, S.A., Weisbeek, P.J., Baas, P.D., Van Arkel, G.A., and Jansz, H.S., *Cold Spring Harbor Symp. Quant. Biol.* 43, 331 (1979).
7. Heidekamp, F., Langeveld, S.A., Baas, P.D., and Jansz, H.S. *Nucleic Acids Res.* 8, 2009 (1980).
8. Duquet, M., Yarranton, G., and Gefter, M., *Cold Spring Harbor Symp. Quant. Biol.* 43, 335 (1979).
9. Baas, P.D., Heidekamp, F., Van Mansfeld, A.D.M., Jansz, H.S., Van der Marel, G.A., Veeneman, G.H., and Van Boom, J.H., in "Mechanistic Studies of DNA Replication and Genetic Recombination" (B. Alberts and C.F. Fox, eds.), ICN-UCLA Symp. on Molecular and Cellular Biology XIX, p. 267, Academic Press, New York (1980).
10. Godson, G.N., Barrell, B.G., Staden, R., and Fiddes, J.C., *Nature (London)* 276, 236 (1978).
11. Arentzen, R., Van Boeckel, C.A.A., Van der Marel, G.A., and Van Boom, J.H., *Synthesis* 137 (1979).
12. Baas, P.D., Van Teeffelen, H.A.A.M., Teertstra, W.R., Jansz, H.S., Veeneman, G.H., Van der Marel, G.A., and Van Boom, J.H., *FEBS Lett.* 110, 15 (1980).
13. Baas, P.D., and Jansz, H.S., *Proc. Roy. Neth. Acad. Sci. Ser. C* 74, 191 (1971).
14. Baas, P.D., Teertstra, W.R., Van Mansfeld, A.D.M., Jansz, H.S., Van der Marel, G.A., Veeneman, G.H., and Van Boom, J.H., *J. Mol. Biol.*, submitted.

15. Beck, E., Sommer, R., Auerswald, E.A., Kurz, C., Osterburg, G., Schaller, H., Sugimoto, K., Sugisaki, H., Okamota, T., and Takanami, M., *Nucleic Acids Res.* 5, 4495 (1978).
16. Van Wezenbeek, P.M.G.F., Hulsebos, T.J.M., and Schoenmakers, J.G.G., *Gene* 11, 129 (1980).
17. Reddy, V.B., Thimmappaya, B., Dhar, R., Subramanian, K.N., Zain, B.S., Pan, J., Gosh, P.K., Celma, M.L., and Weissman, S.M., *Science* 200, 494 (1978).
18. Fiers, W., Contreras, R., Haegeman, G., Rogiers, R., Van de Voorde, A., Van Heuverswijn, H., Van Herreweghe, J., Volkaert, G., and Ysebaert, M., *Nature (London)* 273, 113 (1978).
19. Yang, R.C.A., and Wu, R., *Science* 206, 456 (1979).
20. Sutcliffe, J.G., *Cold Spring Harbor Symp. Quant. Biol.* 43, 77 (1979).
21. Soeda, E., Arrand, J.R., Smolar, N., Walsh, J.E., and Griffith, B.E., *Nature (London)* 283, 445 (1980).
22. Deininger, P.L., Esty, A., LaPorte, P., Hsu, H., and Friedmann, T., *Nucleic Acids Res.* 8, 855 (1980).
23. Chang, A.C.Y., and Cohen, S.N., *J. Bacteriol.* 134, 1141 (1978).
24. Pribnow, D., *J. Mol. Biol.* 99, 419 (1975).
25. Hayashi, Y., and Hayashi, M., *Biochemistry* 10, 4212 (1971).
26. Hsu, P.L., Ross, W., and Landy, A., *Nature (London)* 285, 85 (1980).
27. Grosschedl, R., and Hobom, G., *Nature (London)* 277, 621 (1979).
28. Denniston-Thompson, K., Moore, D.D., Kruger, K.E., Furth, M.E., and Blattner, F.R., *Science* 198, 1046 (1977).

ACKNOWLEDGMENTS

The authors wish to thank Dr. A.J. Van der Eb and Dr. J.R. Arrand for gifts of polyomavirus DNA and Mrs. H.A.A.M. Van Teeffelen, Mrs. W.R. Teertstra and Mrs. J. Zandberg for expert technical assistance.

This work was supported in part by the Netherlands Organization for the Advancement of Pure Research (ZWO) with financial aid of the Foundation for Chemical Research (SON).

VIRAL DNA SEQUENCES AND PROTEINS IMPORTANT IN THE ØX174 DNA SYNTHESIS

Peter Weisbeek[1,2], Arie van der Ende[2], Harrie van der Avoort[1], Renske Teertstra[1], Fons van Mansfeld[2,3] and Simon Langeveld[1,2]

Department of Molecular Cell Biology[1], Institute of Molecular Biology[2] and Laboratory for Physiological Chemistry[3], University of Utrecht, Utrecht, the Netherlands

ABSTRACT Analysis of recombinant plasmids containing different fragments of the ØX174 DNA revealed that a DNA sequence containing the non-coding region between the genes A and H has a function in the *in vivo* viral DNA synthesis. This sequence, when cloned, interferes with the synthesis of parental RF DNA. The cloned (÷) origin interacts *in vivo* with the A protein of infecting phages, such that a single plasmid DNA strand is packaged into phagecoats. The resulting particle can transform ØX-sensitive cells.
In vitro analyses with the A* protein showed that this protein recognises separate parts of the sequence CAACTTGATATT. When, covalently bound to the DNA, this protein can still be enzymatically active on another DNA sequence.
Extrapolation of these results to the A protein gives new insights in the termination of DNA replication.

I. INTRODUCTION

The rolling-circle mode of DNA replication is initiated by the interaction of a unique nucleotide sequence (origin) with an initiator protein. The result of this reaction is the cleavage of one of the single DNA strands and the formation of a 3'-OH terminus (1,2,3). This 3' end can subsequently act as a primer for the synthesis of new DNA (1,4). Origin and initiator protein together determine the specificity of the initiation reaction. E.g. the origin of phage M13 can not be used by the initiator (A) protein of ØX174, although in closely related phages like ØX174 and G4 the two A proteins can recognize and cleave both origins (5,6).

After one round of DNA synthesis, the displaced strand has to be separated as a circular molecule from the replicating DNA (7). This requires the cutting off and the circularization of the displaced single strand. It is assumed that the initiator protein is involved in this termination reaction. Both the ØX A protein and the fd gene II protein possess a nuclease as well as a ligase activity (8,9).

The properties of the origin and the A protein of ØX174 have been studied *in vivo* and *in vitro* in order to determine the nature of their interactions during the replication process. In these studies often the A* protein has been used (10). This protein, a smaller A protein with most of its activities, has additional properties that are relevant for the understanding of the way the A protein functions (11).

In the course of these studies it became clear that next to the (+) and the (−) strand origins, there is a third DNA sequence that is important in the viral DNA synthesis.

II. RESULTS

Cloning of ØX-DNA fragments and selection of recombinant plasmids

The cloning of viral DNA sequences into a bacterial plasmid, followed by analysis of the recombinant plasmids, enables one to study these DNA sequences under conditions that differ considerably from those met by the phage during its normal life cycle. This approach makes it possible to study the viral genes and genproducts in the absence of other phage proteins, but also to analyse those regions that do not contain information to direct the synthesis of a protein but that have other genetic information. These regions e.g. intercistronic regions and replication origins are difficult to analyse in intact phages as their functions are often not revealed by standard genetic analyses.

We have cloned all parts of the ØX DNA separately or in combination with other genome parts into the bacterial plasmid pACYC177 (12). The resulting recombinant plasmids were analysed for the viral DNA inserts and for new properties acquired by the uptake of their viral DNA. The effects of the insertion of intact viral genes and combinations of genes on the plasmid and on infecting phages will be described elsewhere; in this paper two special types of recombinant plasmids will be described in more detail. One group of plasmids have obtained the phage DNA sequence containing the origin of (+) strand DNA synthesis together with varying lengths of other viral sequences. The other group of plasmids have in common a specific sequence with an untill now unknown function; this sequence interferes with the DNA synthesis of infecting phage particles.

Complete and partial digests of wild type ØX174 RF DNA incubated with the restriction enzymes $Hind$II or HaeIII were inserted into the single $Hind$II site of the plasmid pACYC177 (Amp^r Kan^r) by blunt end ligation with T4 DNA ligase. Insertion of a DNA fragment inactivates the plasmid gene determining the resistance against ampicillin. Transformed E. coli HF4704 (ØX-sensitive sup^-, rec^+) cells that were Kan^r Amp^s, were tested for their content of ØX DNA by infection with phage amber and temperature-sensitive mutants. When recombination between the inserted DNA and that part of the infecting phage DNA that contains the mutation is possible, than wild type phage particles will be formed and these are able to form plaques on this strain. Complementation between intact and expressed inserted genes and the mutant viral gene will also result in a (large) increase in plaque number. These plaques however, contain mainly mutant phages. An increase in the number of plaques of at least ten times was used as an indication for the presence of the corresponding phage DNA sequences. Relevant recombinant plasmids were than purified and further characterized by restriction enzyme analyses. In this way the size and the genetic content of the inserted ØX DNA fragments was determined. These results are given in Fig.1 for the plasmids discussed in this paper.

Plasmids that interfere with infecting phage particles

When wild type ØX phage is grown on HF4704 cells containing the plasmids shown in Fig.1 than part of these strains give a normal efficiency of plating (e.o.p.) and normal burst sizes as compared to a plasmid-less HF4704. Several strains however give largerly reduced plating efficiencies and phage yields. The reduction in the e.o.p.is even more pronounced when a recombination-deficient strain (E. coli C recA171, sup^-, ØX-sensitive) is used instead of HF4704. In the latter case 1% or less of the phage particles give rise to a plaque. Moreover the plaques found are much smaller and have fuzzy edges. This effect is ØX-specific because the closely related phages G4 and St-1 are not at all affected by these ØX-pAC177 recombinant plasmids.

Comparison of Table I with Fig.1 learns that all plasmids that cause this reduction in plating efficiency have in common the ØX-$Hind$II fragment R4. This fragment covers the end of gene H, the beginning of gene A and the non-coding region between these two genes. Analysis of the plasmid pPR256 (Fig.1) shows that the left part of R4 is not involved in this reduction as pPR256 does not reduce the e.o.p. whereas its insert contains part of R4. This means that the ØX-sequence that causes this

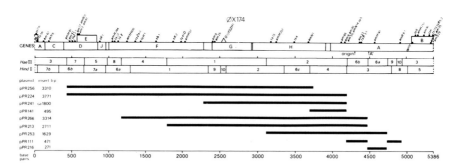

FIGURE 1. Size and genetic content of the ØX-DNA inserts in the recombinant plasmids. All ØX-DNA was inserted in the single HindII site in the β-lactamase gene of pACYC177.

TABLE I. The effect of ØX DNA insertions on phage production

HF4704 with plasmid	e.o.p. of phage [a]	relative phage yield of a single infection [b]	C recA171 with plasmid	e.o.p. of phage [a]
–	1.0	1.0	–	1.0
pAC177	1.0	1.0	pAC177	1.0
pPR256	1.0	1.0	pPR256	1.0
pPR224	0.035	0.024	pPR224	0.005
pPR241	0.008	0.010	pPR241	0.003
pPR141	0.13	0.12	pPR141	0.011
pPR266	0.010	0.012	pPR266	0.004
pPR213	0.16	0.17	pPR213	0.005
pPR253	0.010	0.010	pPR253	0.003
pPR111	1.0	1.0	pPR111	1.0
pPR216	1.0	0.9	pPR216	1.0

a) A fixed number of wild-type phages was titrated at $37^{o}C$ on the plasmid-containing cells. Efficiency of plating (e.o.p.) of 1 is the plaque-number on the plasmid-less strain. b) Cultures of the plasmid-containing HF4704 cells were infected with the lysis-defective am3 phage at a moi = 0.1. After 30' 37°, the cells were collected, lysed and titrated. The yield on the plasmid-less HF4704 was set to 1.

reduction is located in fragment R4 approx. between the nucleotides 3754 and 4202. This fragment does not code for a complete viral protein.
Analysis of the infection of these cells with a reducing plasmid showed a) that the adsorption of the phages to the cells is perfectly normal, 2) that the burst size of the infecting

15 VIRAL DNA SEQUENCES AND PROTEINS

cells is lowered six- to hundred-fold (depending on the particular plasmid used, (see also Table 1) and 3) that the ØX DNA synthesis is strongly affected. The latter aspect will be dealth with in more detail.
Plasmid-containing HF4704 cells were infected with wild-type phage in the presence of 35 µg/ml chloramphenicol and ^{14}C-thymidine was added to label new phage DNA. Under these conditions only RF DNA replication is possible in the cell, although at a lower rate; the ss DNA synthesis is completely blocked (13). The phage and plasmid DNA was then isolated together and analysed by electrophoresis on a 1.4% horizontal agarose slab gel in the presence of ethidiumbromide. The gels were photographed under ultraviolet light (to locate the plasmid and phage DNA bands) and subsequently cut in 1.5 mm slices. The gelslices were dissolved in scintillation liquid and counted. In this way the ethidiumbromide-stained bands can be quantitated for the amount of ^{14}C-thymine they contain. This is shown for the reducing plasmid pPR253 and the non-reducing plasmid pPR216 in Fig.2. Fig.2 shows that in the cells with the reducing plasmid pPR253 the amount of phage RF DNA (mainly RFI DNA) produced is decreased as compared to the cells with the non-reducing plasmid pPR216. Quantitatively the phage RF DNA has decreased approx. 3 times. The same holds for the other plasmids (data not shown). The conclusion of this experiment is that although there is still substantial RF DNA formation, it is decreased in HF4704 cells containing reducing plasmids. Under the conditions of this experiment, no or very little plasmid DNA is labelled.

A decrease in the amount of progeny RF DNA formed can mean two things: either the RF DNA replication itself is slowed down e.g. by a decrease in the efficiency of initiation, or the average number of parental RF DNA molecules per infected cell is lowered. When the infecting phage ssDNA is not or less efficiently converted into parental RF DNA, than less templates for RF DNA replication will be present. The efficiency of duplication of the ss DNA was determined by infection of plasmid-containing cells with ^{14}C-thymine - labelled phages in the presence of 150 µg/ml chloramphenicol.
No protein synthesis and therefore no RF DNA replication can occur under these conditions. The formation of parental RF DNA depends however only on already present host proteins and this process is therefore not hampered by the large concentration of chloramphenicol (13). The phage and plasmid DNA was isolated 30' after infection and again analyzed on agarose-gels as described above. Fig.3 shows the results of this analysis. In cells with the non-reducing plasmid pACYC177 90% of the ^{14}C-label recovered from the infected cells (after extensive washings) is found in the RFI and RFII fractions whereas approx. 10% of the label is recovered as single-stranded phage DNA.

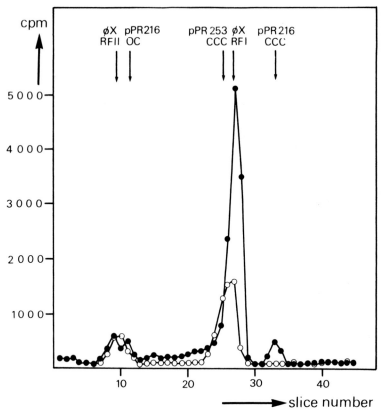

FIGURE 2. Electrophoresis of ^{14}C-labelled DNA on a 1.4% agarosegel. Plasmid-containing cells were infected with am3 phage (moi = 5) in the presence of 35 μg/ml chloramphenicol and ^{14}C-thymine. After 30' 37°, the DNA was extracted and run on gels. The gels were cut in 1.5 mm slices; each slice was dissolved and counted. —●— pPR216, —○— pPR253.

The ss DNA peak represents probably eclipsed but not properly penetrated ss DNA's (14).
In cells with the reducing plasmid pPR241 the results are completely different. The peak of single-stranded phage DNA is of about the same size as for pACYC177, but the amount of RFI and RFII DNA has decreased considerably, to less than 10% of the RFI and RFII DNA found in the cells with the non-reducing plasmid. The total amount of label recovered from each infected cell-culture was approximately the same, (indicating the same efficiency of adsorption), therefore much label must have been present as small non-discrete DNA fragments. In some experiments a very broad band of fast migrating labelled mate-

15 VIRAL DNA SEQUENCES AND PROTEINS

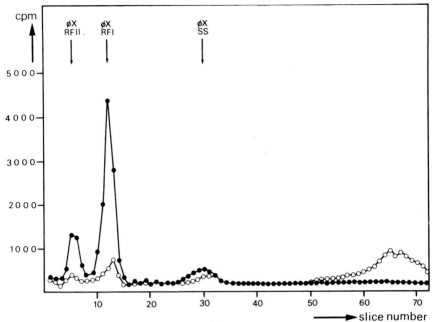

FIGURE 3. Electrophoresis of ^{14}C-labelled DNA on a 1.4% agarose gel. Plasmid-containing gels were infected with ^{14}C-labelled am3 phage (moi = 5) in the presence of 150 μg/ml chloramphenicol. After 60' 37°, the DNA was extracted and run on gels. The gels were cut in 1.5 mm slices. —●— pACYC177, —○— pPR241.

rial (much faster than ss DNA) was observed. Our interpretation of these experiments is that in cells with a reducing plasmid the infecting ssDNA is not (or very inefficiently) converted to parental RF DNA and that the non-converted ssDNA is gradually broken down to smaller DNA fragments. In those cells in which a parental RF molecule has been formed, about normal amounts of progeny RF DNA can be made (at lower chloramphenicol concentration). This can account for the much less pronounced reduction in progeny RF DNA, as seen in Fig.2.
The intracellular presence of a particular ØX DNA sequence influences strongly the ability of the cell to make parental RF DNA out of the infecting ssDNA. No phage proteins appear to be involved in this process. This interference therefore strongly suggests that the phage DNA sequence in the endogenous multicopy plasmid competes with the corresponding sequence in the infecting phage ss DNA for the interaction with some intracellular membrane structure or protein that is present in only limiting amounts.

Recombinant plasmids that contain the ØX-origin for (+) strand DNA synthesis

A large number of different plasmids has been isolated that contained the ØX-origin of RF DNA replication ((+) strand origin), either in a small fragment e.g. the *Hae*III fragment $Z6_B$ or in combination with other parts of the ØX genome. Fig.1 shows for 4 of these plasmids the size and genetic nature of the insertion. Plasmid pPR111, the best studied one, was constructed from a complete *Hae*III digest of ØX RF DNA and it contains not only the origin-fragment $Z6_B$ but also the fragments Z9 and Z10. The other plasmids were obtained from partial *Hae*III digests.

All plasmids that contain the ØX-origin of replication and that are in a superhelical form are *in vitro* a substrate for the ØX A protein. The origin sequence is nicked in the viral (+) strand at the same site as in the ØX RFI DNA and the A protein becomes also covalently attached to the 5' end of the nick site. Experiments performed in collaboration with Dr. K. Geider of the Max Planck Institut of Medizinische Forschung in Heidelberg showed that DNA synthesis can be initiated on such an A protein-nicked plasmid. Similar results have been reported by Zipursky et al. (15). The sequence on the *Hae*III fragment $Z6_B$ therefore contains all the information necessary for A protein-nicking and subsequent elongation of the 3' OH terminus of the (+) strand.

The presence of the ØX-origin in pACYC177 does not have any noticable effect on the viability of the recombinant plasmid. The extra origin apparently does not interfere with the plasmids own replication system. This is not very surprising as the ØX-origin may be expected to become activated only by the ØX A protein and this protein is not present in the cell. When however pPR111-containing cells are infected with phage particles, than a situation is created where the cloned origin and A protein is brought together in the same cell. Can these two components interact *in vivo* and will they initiate a rolling circle-type of DNA synthesis of the recombinant plasmid?

Analysis of the progeny of such an infection revealed the presence of two types of particles; the normal ØX particle that forms plaques and a particle that makes ØX-sensitive cells resistant to kanamycin. These Kan^r-cells have now obtained the plasmid pPR111. This new particle has been purified and analyzed. It can be separated from the ØX particle because it bands at a different, higher, position in a equilibrium-density gradient. The particle is resistant to nucleases and it can infect only ØX-sensitive cells. When the DNA is phenol-extracted from the particle and studied with an electronmicroscope, the structure shown in Fig.4A is found; a double-stranded

15 VIRAL DNA SEQUENCES AND PROTEINS   219

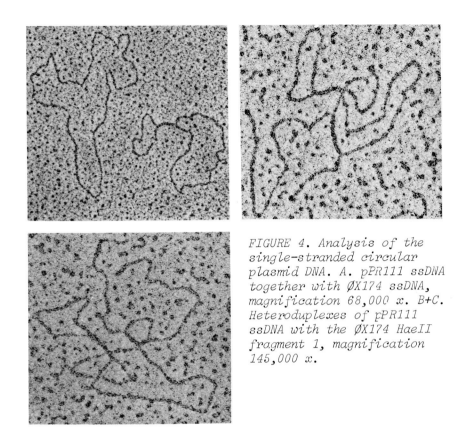

FIGURE 4. Analysis of the
single-stranded circular
plasmid DNA. A. pPR111 ssDNA
together with ØX174 ssDNA,
magnification 68,000 x. B+C.
Heteroduplexes of pPR111
ssDNA with the ØX174 HaeII
fragment 1, magnification
145,000 x.

stem with a small and a large single-stranded loop. Such a
structure is never found for ØX ssDNA. This structure agrees
with known properties of pACYC177. This plasmid has an inverted repeat (12,16) and its single strands therefore can form
the looped stem structure seen in Fig.4A. The DNA in the new
particle therefore is a single-stranded circular molecule. It
can easily be distinguished from ØX ssDND because of its typical form. The structure of a double-stranded stem with two
single-stranded loops of different size makes the orientation
of the molecule very easy. The small loop consists of the kanamycin -resistance gene and the larger loop contains the β-lactamase gene in which the ØX DNA is inserted (12,16).
Hybridization between the single-stranded plasmid DNA and
double-stranded ØX DNA fragments can be used to identify the
ØX DNA inserted and to determine its orientation in the plasmid. This is demonstrated for the single-stranded pPR111 DNA.
In Fig.4B and C hybridization with the HaeII fragment 1 shows

the presence of three different fragments. The size of the duplex region and the length of the single-stranded linear tails identify the fragments Z6B, Z9 and Z10 as the inserted ØX DNA. Hybridization with (+) strand and (-) strand ØX DNA separately (data not shown) proves Z6B to be inserted as (+) strand DNA and Z9 and Z10 as (-) strand DNA in the DNA strand of pACYC177 that does not code for β-lactamase (17). Length measurements on the single-stranded DNA also gives information about the size and position of the inverted repeat in pACYC177. The resulting structure of pPR111 is drawn in Fig.5.

The conclusion of this EM analysis is that the origin plasmid pPR111 is present in the phage coat as a single-stranded circle and that the ØX-origin sequence in this single-stranded circle is of the viral (+) strand type. This means that *in vivo* the cloned ØX-origin is indeed recognized and cleaved by the A protein produced by the infecting phage. It also means that after nicking at least strand displacement but most probably rolling-circle type DNA synthesis has taken place. The displaced strand is properly circularized and packaged in phage coats.

In this experimental set up, the A protein is acting in *trans*, although in genetic complementation tests, the A protein was shown many times to be *cis*-acting (18). Therefore there still must be a difference in the intracellular situation for the two types of experiments.

Stability of the plasmid-particles

The pPR111 plasmid-particle contains a single-stranded DNA molecule of about 4150 bases i.e. approx. 1200 bases less than the 5386 bases that are normally packaged in the phage coat. We have tested origin plasmids of different lengths for their capability of forming plasmid particles. For plasmids from 4150 basepairs up to 5500 basepairs there is no difference in the efficiency of the formation of plasmid phages and also the stability of the particles at $4^\circ C$ and $56^\circ C$ is not different from the stability of the ØX particle. There appears to be no relation between the amount of DNA packaged and the stability of the particle. This is different for phage lambda where deletion mutants are less stable than the wild type phage (19). For ØX the only limit seems to be the total amount of DNA that fits into the phage coat. These experiments show that the ØX DNA can have an insertion of approx. 120 bases without problems in the maturation. The next larger plasmids we used was 6400 basepairs and this one was not packaged. The figure of 120 extra basis agrees well with recent data of Müller and Wells (20) who showed that an insert of 160 bases in the region between the genes J and F does not affect the

15 VIRAL DNA SEQUENCES AND PROTEINS

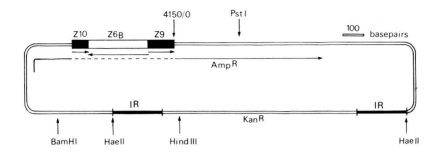

FIGURE 5. *Structure of the recombinant plasmid pPR111. For the pACYC177 part of this molecule the data of Chang and Cohen (12) was used. The direction of transcription of the Amp-gene was derived from the nucleotide-sequence of pBR322 (17).*

phage viability.

Infectivity of the plasmid-particle

The plasmid-particle portion of the total phage yield is quite variable. It ranges between 1 and 20% as judged by the intensity of the particle-bands in a RbCl-equilibrium density gradient. The reason for this large variation is as yet unclear to us. The ratio of plasmid particles to phage particles is much lower ($5 \times 10^{-5} - 10^{-3}$) when judged by infectivity; i.e. the number of Kan^r cells formed compared to the number of plaques formed. It was determined that approx. 1 in 500 plasmid-particles is able to succesfully transform the sensitive host cell, although the adsorption of the particle to the cell is the same as for a phage particle.

The ss DNA of the plasmid-particle is apparently blocked in its intracellular development. This inhibition can occur at the level of the conversion, e.g. when the infecting ss DNA does not have the proper signals for complementary strand synthesis or at the level of RF DNA replication when no progeny RF DNA can be made.
When the ØX (-) strand origin (21) is inserted in (+) strand origin plasmid such that both ØX-origins have the same orientation, than the resulting plasmid-particles are just as infective as the normal ØX particle. This result means that indeed a proper signal for conversion was absent on the single-stranded plasmid DNA. Moreover it also means that once the ss DNA is converted to ds DNA it can use the plasmid-replication system for its multiplication.

Properties of the A^x protein; Cleavage sequences in ØX ss DNA

Recently Langeveld et al. (22) found that the A^* protein is a very active endonuclease (see Table II). It cleaves single-stranded ØX-DNA at many different sites producing discrete DNA fragments.

The position of the cleavage sites was determined by digesting specific ss DNA fragments with A^* protein. Single-stranded ØX DNA isolated from phage particles was digested with the restriction enzyme *Hae*III which cleaves single-stranded DNA with the same sequence specificity as it does double-stranded DNA (22,23). The single-stranded DNA restriction fragments were ^{32}P labelled at their 5' termini with T4 nucleotide kinase and purified.

All restriction fragments were separately incubated with A^* protein, deproteinized with proteinase K and applied onto formamide-polyacrylamide gels. On each gel a set of *Hae*III restriction fragments of known sizes was coelectrophoresed to serve as length markers. When the *Hae*III fragments are plotted semilogarithmically versus their migration distance a good linear relationship is found for the fragments Z4 (603 nucleotides) through Z10 (72 nucleotides) on a 6% polyacrylamide gel made up in formamide. Therefore the sizes of the A^* protein produced subfragments can be deduced with reasonable accuracy from their relative positions on the gel. Since only the 5' termini had been labelled, the size of each subfragment determines the position of the cleavage site relative to the 5' end of the

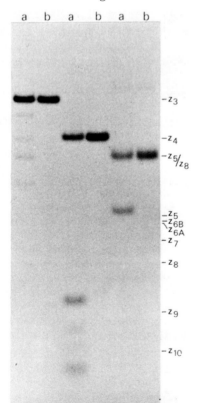

FIGURE 6. Formamide-polyacrylamide gelelectrophoresis of the HaeIII restriction fragments Z3, Z4 and the partial Z5-Z8 of ØX174 ssDNA. (a) after incubation with A^x protein, (b) control without A^x protein.

TABLE II. CLEAVAGE SITES OF THE A* PROTEIN IN ØX VIRAL SINGLE-STRANDED DNA

Fragment	Length of subfragment	Position of cleavage site	Fragment	Length of subfragment	Position of cleavage site	Fragment	Length of subfragment	Position of cleavage site
Z1	235 - 245	2016 ± 5	Z3	740 - 760	313 ± 10	Z5	180 - 205	862 ± 13
	190 - 200	1971 ± 5		580 - 600	153 ± 10	Z6A	165 - 170	4656 ± 3
	160 - 170	1941 ± 5		500 - 510	68 ± 5		130 - 135	4621 ± 3
	100 - 120	1886 ± 10		380 - 400	5339 ± 10		30 - 35	4521 ± 3
	90 - 95	1869 ± 3		285 - 295	5239 ± 5			
	84 - 86	1861 ± 1		195 - 200	5147 ± 3	Z6B	98*	4305
Z2	670 - 690	3809 ± 10		170 - 175	5122 ± 3	Z5-Z8	320 - 325	992 ± 3
	150 - 160	3284 ± 5		145 - 155	5099 ± 5	Z9	67*	4826
	95 - 105	3229 ± 5		90 - 100	5044 ± 5		29*	4788
	85 - 95	3219 ± 5		7*	4956			
	50 - 60	3184 ± 5	Z4	130 - 135	1306 ± 3	Z10	60*	4937
	< 30	$3159 \pm ?$		80 - 85	1256 ± 3			
				60 - 65	1236 ± 3			
				40 - 45	1226 ± 3			

Restriction fragments were incubated with A* protein as described in Materials and Methods. The restriction fragment Z7 was not cleaved under these conditions. The lengths of the subfragments derived from the formamide polyacrylamide gels are expressed in nucleotides. The values determined in different experiments fall in the range that is indicated for each subfragment. Asterisks mark the subfragments of which the length has been established by sequence analysing techniques.

fragment and so its position in the ØX DNA sequence (24). An example of such an A* protein digestion is given in Fig.6 for the fragments Z3, Z4 and the partial digestion product Z5-Z8. The cleavage sites determined in this way are listed in Table II; for a few cleavage sites (with asterisk in Table II) the exact cleavage site was determined by sequencing of the subfragment.

The efficiency with which the different cleavage sites are nicked by the A* protein varies greatly. The sequences of the best-cleaved sites are given in Table III; their homology to the origin sequence (Z6B) is evident. The exact position of the nick for five less efficient cleavage sites is given in

TABLE III. *Nucleotide sequences at the best cleavage sites*

Fragment	Sequence
Z6B	4293 - C T C C C C C A A C T T G↓A T A T T A A T A
Z8	978 - . . C C C C . . A C T T G A . . . T A A . .
Z2	3804 - . . . . T T G A T A T T . . T .
Z3	59 - . . . . A . C T T G A T A . . . . . .

The sequences at the best cleaved sites are compared with the origin sequence. Dots mark the position of the non-homologous nucleotides. The point of cleavage at the origin in the restriction fragment Z6B is indicated by an arrow.

TABLE IV. *Cleavage sites determined by sequence analysis*

Fragment	Sequence
Z6B	4299 - C A A C T T G↓A T A T T A A T - 4313
Z3	4950 - C c t g T T G A↓T g c T A A a - 4964
Z9	4781 - g c A C T T t A↓T g c g g A c - 4795
Z9	4821 - a A t t T T↓G g T c g T c g g - 4835
Z10	4932 - C g t t c T↓G g T g g t t g - 4946

Points of cleavage are indicated by arrows. Nucleotides which are homologous with the origin sequence are represented by uppercase characters.

Table IV. Comparison with the Z6B sequence shows the nick to be at approximately the same position as in the origin but with a small variation either to the left or to the right. Comparison of the cleavage sequences permits the following conclusions: 1) there is no common or unique sequence in which the A* protein nicks; 2) most cleavage-sites show homology with at least parts of the origin sequence -CAACTTGATATT-; 3) better homology results in more efficient cleavage; 4) entirely different parts of this sequence, e.g. CAACTTG and GATATT, are recognized and cleaved by the A* protein; 5) the nick is introduced within the cleavage sequence at approximately the same position as in the origin sequence.
One important consequence of these conclusions is that the A* protein possesses different DNA recognition domains which are able to recognize parts of the origin sequence separately. When only part of the protein interacts with a specific nucleotide-sequence, it is still able to cleave the DNA. The other part of the protein molecule might than still be able to interact with another sequence.

Nuclease-activity of the A protein on RFI DNA*

The A* protein can nick RFI DNA when Mn^{++} instead of Mg^{++} is present as cofactor during the incubation. The result of such an incubation is a mixture of RFII DNA, RFIV DNA (relaxed covalently closed) and linear RFIII DNA (see also 22). The RFII DNA molecules are mainly (approx. 80%) nicked in the (+) strand at the origin site, but EM-analysis showed that other sites in the RFI DNA are also substrates for the A* protein. The RFIII DNA has been analysed in more detail, mainly by electron microscopy. This DNA had a rather surprising structure; at one end it contained a covalently bound A* protein (Fig. 7A) whereas the strands at the other terminus are covalently linked to each other. After partial denaturation with formamide and fixation with glyoxal one end of the molecule showed a terminal single-stranded loop (Fig.7B). Electrophoresis on denaturing agarose gels confirmed the covalent linkage between the (+) and the (-) strand. Restriction analysis and electrophoresis on denaturing polyacrylamide gels resulted in the picture of this RFIII DNA as shown in Fig.8. One end of the linear molecule has the A* protein covalently bound to nucleotide 4306 of the (+) strand (the same position where the A protein is bound) and this terminus is not fully double-stranded; four bases at the end of the (-) strand are missing. These four bases are now found at the other end where the (+) and the (-) strands have become covalently linked together.
Inspection of the nucleotide sequence around the origin (Fig.9) reveals that it contains not only a cleavage-site for the A*

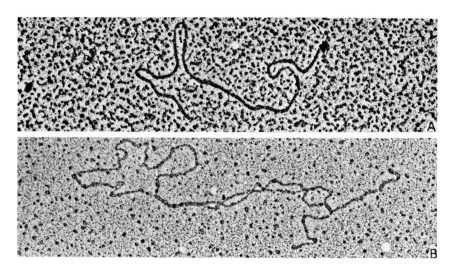

FIGURE 7. Linear double-stranded DNA formed by incubation of RFI DNA with A* protein. A) The RFIII DNA-A* protein complex. The protein has been visualized by the method of Wu and Davidson (25). B) After partial denaturation with formamide and fixation with glyoxal.

protein in the (+) strand (5' -ACTTGATAT) but also one in the (-) strand (5' -ATTAATAT). These sequences can be cleaved by the A* protein at the positions indicated by the arrows. Cleavage at these positions and ligation of the single-strands of the left end results exactly in the RFIII DNA structure that has been found (Fig.8).

The most simple model to explain the formation of the RFIII DNA would be that of two A* protein molecules nicking in the opposite strands followed by a ligation reaction at one end.

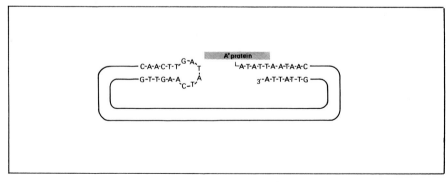

FIGURE 8. Structure of the RFIII DNA-A* protein complex

15 VIRAL DNA SEQUENCES AND PROTEINS

```
            4300          A*          4310
    5' - C C A A C T T G↓A T A T T A A T A A C A - 3'
    3' - G G T T G A A C T A T A↑A T T A T T G T - 5'
                              A*
```

FIGURE 9. *Nucleotide sequence around the (+) strand origin*

For several reasons however, this model seems incorrect to us. In the first place only superhelical DNA is a substrate for the A^* protein. The first nick would therefore inhibit the second nick. When RFII DNA made by A^* protein was incubated with extra A^* protein (even excessive amounts) no RFIII DNA formation could be observed. Moreover RFIII DNA is already formed at concentrations of A^* protein where less than 1 protein molecule per DNA is present. No RFIII DNA molecules with a protein attached to both ends were found against 100 RFIII DNA molecules with one protein knob. This would mean that the ligation step has to be very efficient. We know however that formation of RFIV DNA, that arises by a cleavage followed by a ligation, is not very efficient. Between 10 and 20% of the

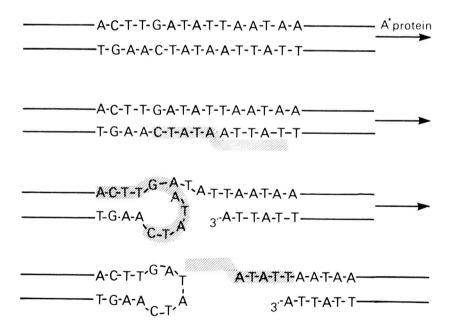

FIGURE 10. *Proposed pathway for the formation of the RFIII DNA-A^* protein complex.*

RFII DNA formed is converted into RFIV DNA.
Because of the reasons mentioned, we think that the model given in Fig.10 is the correct one. The RFI DNA is nicked with a low frequency in the (-) strand at the indicated position. The A^* protein covalently attached to the 5' end of the nick-site binds with part of its molecule to the sequence 5' -ATAT- but the rest of the protein has a low binding affinity for the other part of the cleaved sequence 5' -TATTA-. In the opposite (+) strand the optimal sequence (5' -ACTTG-) for this part of the protein is present. Partial separation at this end (maybe induced by the A^* protein itself) enables the bound A^* protein to interact with this sequence. A rearrangement of the phosphodiester bonds and the covalent bond between the protein and the terminal nucleotide i.e. cleavage and ligation by an already covalently bound A^* protein, results in the observed RFIII DNA structure.

III. DISCUSSION

   The insertion of a specific viral DNA fragment (between the nucleotides 3754 and 4202) into pACYC177 gives this plasmid the ability to interfere with the viral DNA synthesis. The inserted DNA contains parts of the genes A and H together with the non-coding region between these two genes. It does not code for an intact protein and its effect is maintained in both orientations with respect to the plasmid DNA. The interference is ØX-specific as the related phages G4 and St-1 are not affected.
   It was found that the conversion of the infecting phage ss DNA into parental RF DNA was drastically reduced in cells that contain this recombinant plasmid. This accentuates the fact that the inserted DNA does not code for a protein because the conversion of ss DNA to ds DNA does not require viral proteins (13). These results indicate that the interference with the viral DNA synthesis is the result of the interaction of the inserted viral DNA with a cellular structure that is present in limited amounts and that is needed for by the infecting phage DNA. In this competition for such a structure e.g. a membrane site, the multicopy plasmid will be largely favoured.
   The interference described here has many similarities with the process of superinfection exclusion described for ØX174 and S13 many years ago (26,27). In these experiments the DNA of the secondary infecting phage was also not converted to RF DNA. Moreover, the number of phage RF DNA molecules already present in the cell determined whether superinfection was succesful or not. This phenomenon might well have the

15 VIRAL DNA SEQUENCES AND PROTEINS

same molecular basis as the interference described above.

The insertion of the (+) strand origin in pACYC177 makes the resulting recombinant plasmid *in vitro* a good substrate for the A protein. *In vivo* this cloned origin can interact with the A protein produced by infecting phages. This results in particles that contain a single-stranded circular plasmid in a phage coat. The *in vivo* interaction of A protein and origin means that the so-called cis-activity of the A protein (18) is absent under these conditions. This stresses once more the special position of the parental RF DNA of the phage. Only on the parental DNA is this cis-activity of the A protein functional. The parental phage DNA can apparently not be reached by most of the intracellular proteins.

The infectivity of the plasmid phages is low: approximately 0.5% of these particles are able to make a cell resistant to kanamycin . This can be caused by problems with the conversion of the plasmid ss DNA into ds DNA or in the replication of the plasmid ds DNA. When also the viral (-) strand origin (21) is inserted in the same orientation as the (+) strand origin, than the efficiency of infection increases hundredfold and equals the efficiency of the ØX174 particle. We conclude from these experiments that the infecting plasmid ss DNA is dependent on the ØX174 mode of conversion. Eventual signals for complementary strand synthesis on the plasmid part of the ss DNA are apparently not effective under these conditions. Once the parental ds DNA is made, subsequent replication of this DNA is not restricted by the absence of viral A protein.

Analysis of the sequences found at the sites where the A^* protein can cleave ss DNA has shown that this protein can recognize and interact with sequences that have partial homology to the origin-sequence -CAACTTGATATT-. The position of the nick is approximately the same as in the origin. It is concluded from these results that separate domains of the A^* molecule bind to different parts of the origin sequence. When only part of the A^* protein is bound to a specific sequence, the protein still is able to cleave the DNA. Such a partially bound A^* protein still possesses the potential to interact with its unbound part with another nucleotide sequence. That this indeed can happen was shown by the analysis of the linear RFIII DNA made after incubation of RFI DNA with A^* protein. From the structure of this DNA molecule together with the known properties of the A^* protein, it is concluded that the RFIII DNA is formed according to the pathway drawn in Fig. 10. The important conclusion of this analysis is that indeed a part of the covalently bound A^* protein can specifically interact with another sequence. This bound protein can carry out a nicking-closing reaction with a concomittant change in the position of the bound protein.

TABLE V. *Properties of the A and A* proteins in vitro*

	A protein	A* protein	ref.
cleaves RFI DNA (Mg^{++})	+	−	(1,3,4)
cleaves RFI DNA (Mn^{++})	+	+	(8)
forms RFIII DNA (Mn^{++})	−	+	(8)
forms RFIV DNA (Mg^{++})	−	−	(1,8)
forms RFIV DNA (Mn^{++})	+	+	(8)
cleaves ss DNA at origin	+	+	(22)
cleaves ss DNA at other sites	−	+	(11,22)
attaches covalently to 5' end of the nick	+	+	(1,4,29)
binds to ds DNA	−	+	
binds to ss DNA	+	+	

A comparison of the nucleotidesequence of the A genes of ØX174 and G4 (24,28) learns that the A* part of these genes is much more conserved (21% amino acid changes) than the part that is found only in the A protein (51% amino acids changes). This suggests that most of the essential properties of the A protein will also be found in the A* protein. Table V shows that what the A and A* proteins have in common, are indeed such basic abilities as the binding of DNA, the breaking of phosphodiesterbonds and the resealing of such breaks. The difference between the two proteins lies mainly in the specificity of the reactions; e.g. the A protein binds only to single-stranded DNA and it requires a more specific nucleotide sequence for its nicking activity whereas the A* protein will nick only in superhelical DNA when Mn^{++} is present.
This conclusion makes it acceptable to expect that the described properties of the covalently bound A* protein will also be found in the A protein. The A protein fullfills many activities in the viral RF DNA replication, especially in the initiation and termination steps (7). With an A protein, covalently bound to the 5' end of the displaced strand, that can interact with the left part of the origin sequence (see Fig.11) and that is able to nick in that sequence and to ligate, many details of the termination reaction can very well be explained. In this way the displaced strand is cut off and circularized in one reaction, leaving the A protein at the 5' end of the new (+) strand.

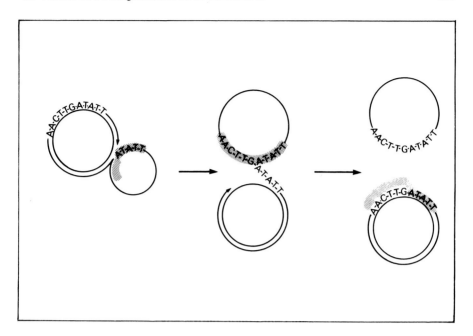

FIGURE 11. Interactions and enzymatic activities of the A protein during the termination step of the (+) strand DNA synthesis.

REFERENCES

1. Eisenberg, S., Scott, J.F. and Kornberg, A. (1976). *Proc. Natl. Acad. Sci. USA 73*, 1594.
2. Meyer, T.F., Geider, K., Kurz, C. and Schaller, H. (1979). *Nature (London) 278*, 365.
3. Langeveld, S.A., van Mansfeld, A.D.M., Baas, P.D., Jansz, H.S., van Arkel, G.A. and Weisbeek, P.J. (1978). *Nature (London) 271*, 417.
4. Ikeda, J., Yudelevich, A. and Hurwitz, J. (1976). *Proc. Natl. Acad. Sci. USA 73*, 2669.
5. Van Mansfeld, A.D.M., Langeveld, S.A., Weisbeek, P.J., Baas, P.D., Van Arkel, G.A. and Jansz, H.S. (1979). *Cold Spring Harb. Symp. Quant. Biol. 43*, 331.
6. Weisbeek, P.J., van Mansfeld, A.D.M., Kuhlemeier, C., van Arkel, G.A. and Langeveld, S.A. (1981). *Eur. J. Biochem., in press.*
7. Eisenberg, S., Griffith, J. and Kornberg, A. (1977). *Proc. Natl. Acad. Sci. USA 74*, 3198.
8. Langeveld, S.A., van Arkel, G.A. and Weisbeek, P.J. (1980). *FEBS Lett. 114*, 269.

9. Meyer, T.F. and Geider, K. (1979). *J. Biol. Chem.* 254, 12636.
10. Linney, E. and Hayashi, M. (1973). *Nature New Biol.* 295, 6.
11. Langeveld, S.A., van Mansfeld, A.D.M., Van der Ende, A., van de Pol, J.H., van Arkel, G.A. and Weisbeek, P.J. (1981). *Nucl. Acids Res, in press.*
12. Chang, A.C.Y.C. and Cohen, S.N. (1978). *J. Bacteriol.* 134, 1141.
13. Hutchinson III, C.A. and Sinsheimer, R.L. (1966). *J. Mol. Biol.* 18, 429.
14. Newbold, J.E. and Sinsheimer, R.L. (1970). *J. Mol. Biol.* 49, 49.
15. Zipursky, S.L., Reinberg, D. and Hurwitz, J. (1980). *Proc. Natl. Acad. Sci. USA* 77, 5182.
16. Sharp, P.A. Cohen, S.N. and Davidson, N. (1973). *J. Mol. Biol.* 75, 235.
17. Sutcliffe, J.G. (1979). *Cold Spring Harbor Symp. Quant. Biol.* 43, 77.
18. Tessman, E.S. (1966). *J. Mol. Biol.* 17, 218.
19. Weil, J., Cunningham, R., Martin, R., Mitchell, E. and Bolling, R. (1972). *Virology* 50, 373.
20. Müller, U.R. and Wells, R.D. (1975). *Nature (London)* 257, 421.
21. Shlomai, J. and Kornberg, A. (1980). *Proc. Natl. Acad. Sci. USA* 77, 799.
22. Langeveld, S.A., van Mansfeld, A.D.M., de Winter, J. and Weisbeek, P.J. (1980). *Nucl. Acids Res.* 7, 2177.
23. Blakesley, R.W. and Wells, R.D. (1975). *Nature (London)* 257, 421.
24. Sanger, F., Coulson, A.R., Friedmann, T., Air, G.M., Barrell, B.G., Brown, N.L., Fiddes, J.C., Hutchinson III, C.A., Slocombe, P.M. and Smith, M. (1978). *J. Mol. Biol.* 125, 225.
25. Wu, M. and Davidson, N. (1978). *Nucl. Acids. Res.* 5, 2825.
26. Hutchinson III, C.A. and Sinsheimer, R.L. (1971). *J. Virol.* 8, 121.
27. Tessman, E.S., Borrás, M. and Sun, I.L. (1971). *J. Virol.* 8, 111.
28. Godson, G.N., Barrell, B.G., Staden, R. and Fiddes, J.C. (1978). *Nature (London)* 276, 236.
29. Eisenberg, S. (1980). *Virology* 35, 409.

CONSERVATION AND DIVERGENCE IN SINGLE-STRANDED PHAGE DNA SECONDARY STRUCTURE: RELATIONS TO ORIGINS OF DNA REPLICATION

Thomas D. Edlind
Garret M. Ihler

Department of Medical Biochemistry
Texas A&M University
College of Medicine
College Station, Texas

ABSTRACT Using the same partial denaturing conditions employed for ϕX174 DNA (1), the secondary structure of G4 DNA was analyzed by electron microscopy. Both ϕX and G4 viral DNAs were folded into similar characteristic three-lobed structures, an observation consistent with their similar genetic maps and nucleotide sequences. The locations of stem and loop structures comprising the three lobes were mapped relative to the ends of *Pst*I-, *Eco*RI-, and *Sst*II-cleaved G4 replicative form DNA. It was found that the secondary structure surrounding the origin of viral strand replication was conserved in ϕX and G4 DNAs. Since ϕX and G4 have similar modes of viral strand replication, this conservation is consistent with our previous suggestion that secondary structure specifically folds viral DNA to promote its circularization. Two major differences between the secondary structures of ϕX and G4 DNAs, involving the smallest of the three lobes, were located in the region of the G4 origin of complementary strand replication. These structural changes can be correlated with the different modes of complementary strand replication utilized by ϕX and G4.

I. INTRODUCTION

The secondary structure of φX174 DNA has been characterized by electron microscopy (using partial denaturing conditions) and sequence analysis (1). This analysis revealed a characteristic three-lobed structure for φX viral DNA. The origin for viral strand DNA replication (ori_v), the site at which φX gene A protein specifically nicks viral (2) and replicative form(RF) (3) DNA, was located in a 92 nucleotide long single-stranded region. This region was situated between the arms of a Y-shaped structure consisting of two small loops and a large loop. This structure brings close together the 5' and 3' ends of displaced viral DNA cleaved at ori_v during rolling circle replication. This would promote ligation of the ends to regenerate circular viral strands.

The nucleotide sequences of G4 and φX DNAs have diverged 33% in coding regions and to a greater extent in intergenic regions (4,5). Nevertheless, their genetic maps are very similar, and in many respects these phage are also physiologically similar (6). One important difference between φX and G4 involves initiation of complementary strand replication, which requires for G4 DNA two host functions, primase and DNA binding protein (DBP) (7), and a unique site (ori_c) between genes F and G (8,9). Initiation of φX complementary strands requires an additional six host functions and does not occur at a unique origin (10).

We report here that G4 viral DNA, like φX, was folded into a characteristic three-lobed structure. Using restriction enzyme-cleaved RF DNA, the locations of these lobes and the structures comprising them have been mapped. Much of the secondary structure seen in φX DNA has been conserved in G4 DNA, including that located near ori_v. The structure surrounding G4 ori_c and the analogous region of φX have diverged. Base pairing models are presented and their implications to DNA replication and evolution of single-stranded DNA phage are discussed.

II. RESULTS

When spread for electron microscopy by the procedure of Davis et al. (11) modified by the addition of 0.5 M ammonium acetate (1), 30% of G4 viral DNA molecules were folded into a characteristic three-lobed structure connected at a central point (Fig. 1). An additional 30% of the molecules included two of the three lobes, while 27% included one. These lobes

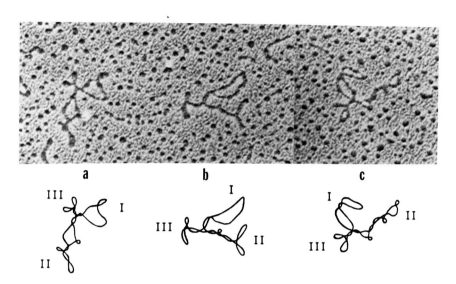

FIGURE 1. Electron micrographs and drawings of G4 three-lobed viral DNA (5577 nucleotides long), prepared using the partial denaturing conditions described for φX (1). Structural domains I, II, and III are indicated.

defined three structural domains on G4 DNA.

Domain I was 1780 ± 190 nucleotides long. At its base was a large loop, designated L_I, with an 80 nucleotide long stem. On 55% of the molecules domain I lacked additional structure, while a small loop (S_I) about 700 nucleotides long was observed on the remainder near the center of domain I (Fig. 1a,c).

Domain II was the largest of the three, 2480 ± 250 nucleotides long, and also the most complex in structure. At its base was loop $L1_{II}$ which had a 120 nucleotide long stem. Within domain II was a Y-shaped structure (Fig. 1a,c) consisting of a large loop $L2_{II}$ (1370 nucleotides long with a 170 nucleotide stem) and two small loops $S1_{II}$ (600 nucleotides) and $S2_{II}$ (400 nucleotides). A third small loop $S3_{II}$ (180 nucleotides) was observed between the stem regions of $L1_{II}$ and $L2_{II}$. The combined presence of these large and small loops with additional weak interactions folded domain II into an elongated Y-shaped structure with a small bulge in its stem, as in Fib. 1b.

Domain III was the smallest, 1310 ± 200 nucleotides long. It was frequently folded into a Y-shaped structure, consisting of loop L_{III} (140 nucleotide long stem) and loops $S1_{III}$ (420 nucleotides) and $S2_{III}$ (600 nucleotides). The structure of domain III thus resembled the Y-shaped structure within domain II.

The locations, relative to the G4 genetic map and sequence, of the domains and loops observed on viral DNA were determined from analysis of restriction enzyme-cleaved G4 RF DNA. Examples are shown in Fig. 2. The $PstI$ cleavage site was located within domain II, about 600 nucleotides from the base of $L2_{II}$. The $EcoRI$ cleavage site was located within domain I, about 300 nucleotides from L_{III}. The $SstII$ cleavage site was also within domain I, but on the other side of S_I and about 500 nucleotides from Ll_{II}. It should be noted that linearization of G4 DNA by restriction enzymes introduced considerably more variation in loop size and morphology than was observed on circular viral DNA. This variation was minimized by allowing sufficient time for the denatured RF DNA to self anneal prior to spreading for electron microscopy.

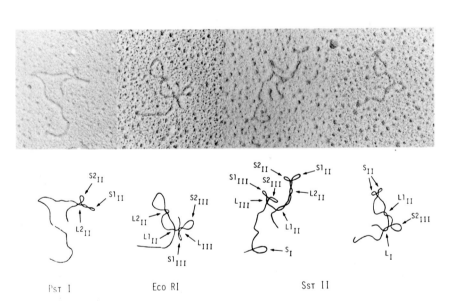

FIGURE 2. *Electron micrographs and drawings of restriction enzyme-cleaved G4 RF DNA, denatured and prepared as described for ϕX (1). Arrows indicate loops or loop stems, as described in the text.*

III. DISCUSSION

The secondary structure map as determined by electron

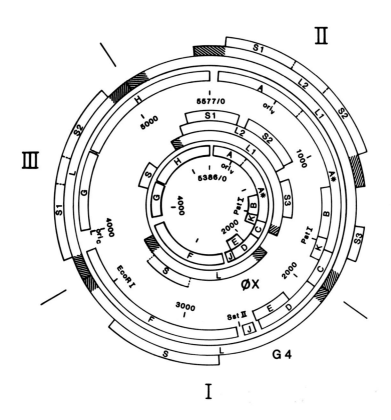

FIGURE 3. *Genetic maps (4,5) and DNA secondary structure maps as determined by electron microscopy (1, these results) for G4 three-lobed viral DNA (outer circles) and PstI-cleaved φXRF DNA (inner circles). Hatched areas of secondary structure maps represent loop stems. L and S refer to specific large and small loops within the indicated domains I, II, and III. Dotted lines in secondary structure maps indicate uncertain or alternative locations for the ends of loops.*

microscopy and the genetic map of G4 viral DNA are shown in Fig. 3. Alignment of the maps was based on analysis of restriction enzyme-cleaved RF DNA. The relative orders of small loops within the Y-shaped structures of domains II and III on G4 DNA are tentative as indicated by dotted lines, either of which may be correct. Data derived from electron microscopic analysis of partially denatured single-stranded DNA is subject to relatively large errors, due primarily to unresolvable (less than 100 nucleotides long) secondary structures. Therefore, precise alignment of the genetic and secondary structure maps requires identification of the actual sequences involved in the base pairing responsible for the observed structures. This identification can in principle be made using currently accepted procedures for evaluating base pairing (14,15) between regions of the DNA defined by the electron microscope data. Two of the ten loops mapped on G4 DNA have been analyzed in this way, as reported below. The procedures for evaluating base pairing are, however, somewhat crude, and verification of proposed base pairing by biochemical methods is clearly desirable.

The sequences of G4 and ϕX have diverged about 35% (4,5), including a net addition of 191 nucleotides to G4. Despite this variation there was at the level of resolution afforded by electron microscopy nearly complete conservation of secondary structure (compare Fig. 3 inner and outer circles). Paired regions involved in particular loop stems were in genetically similar or identical positions on the two DNAs. There were three resolvable differences in pairing interactions of ϕX and G4 viral DNAs, only one of which resulted in a real difference in gross structure. Stable pairing at the base of G4 domain III (loop L_{III}) was observed, while on ϕX DNA domain III appeared to be formed primarily by the juxtaposition of domains I and II. The third small loop ($S3_{II}$) of domain II was present on both ϕX and G4 DNAs in roughly analogous positions. On ϕX, however, $S3_{II}$ appeared to be mutually exclusive with $L2_{II}$, while on G4 these loops were clearly separated and frequently present on the same molecule. The only gross difference in the structures of ϕX and G4 resulted from the presence of an additional small loop ($S1_{III}$) within domain III of G4 DNA. On ϕX DNA domain III contained a single small loop (analogous to $S2_{III}$ of G4), and in its most folded state ϕX domain III appeared as a long, curved hairpin. The additional loop on G4 DNA folds domain III into a Y-shaped structure.

In light of the sequence divergence, the conservation of secondary structure implies it serves one or more important functions. Unfortunately, we do not yet know enough about these phage to positively identify a function for any one loop. There are, however, two contrasting observations regarding the replication of ϕX and G4 DNAs which appear to have bearing on their secondary structures. ori_v is located in equivalent positions on these DNAs, to the extent that a 30 nucleotide sequence surrounding that site is completely conserved (5). Based on the location of ori_v relative to the secondary structure of φX DNA, we suggested previously that viral strands may be specifically folded so as to promote the final step in viral strand replication - circularization of the displaced linear strands (1). The Y-shaped structure within which ori_v is located was conserved on φX and G4 DNAs (Fig. 3). This is consistent with their similar modes of viral strand replication (6), and provides additional support for the model in which structure promotes circularization. However, the location of φX ori_v within a 92 nucleotide single-stranded region between loops $S1_{II}$ and $S2_{II}$ (ref. 1) may be important for rejoining of the 5' and 3' ends, but an analogous location for G4 ori_v has not yet been determined. In addition to φX and G4, filamentous phage fd (12) and probably Col El plasmid (13) DNAs possess stable loops within which origins of rolling circle replication are found.

Loop $L2_{II}$ formed the base of the Y-shaped structure within which ori_v was located (Fig. 3). An extensive region of the G4 sequence was analyzed to locate the most stable base pairing corresponding to this loop. Our expectation was that the base paired sequences proposed for $L2_{II}$ would be conserved on φX and G4 DNAs, since the loop itself was conserved. This is apparently not the case. Assuming that the procedures used for evaluating base pairing are adequate to identify the specific nucleotides primarily responsible for loop formation, the results shown in Fig. 4 suggest $L2_{II}$ was conserved while the bases whose pairing is responsible for the loop differ between φX and G4. The proposed interactions for G4 and φX loop $L2_{II}$ have calculated negative free energies of -25 and -27 kcal and include 14 and 21 base pairs, respectively. The homologous regions of φX and G4 DNAs contain 5 and 9 substitutions, respectively, which reduce the negative free energies to -7 kcal in each case. The effect of this sequence divergence on loop $L2_{II}$ is surprisingly minimal. Not only are the calculated stabilities very similar, but also the locations of the paired sequences are similar, within 22 nucleotides of each other. Moreover, the lengths of the φX and G4 loops plus stems are 1377 and 1537 nucleotides, respectively. The length difference of 160 nucleotides corresponds closely to the 161 nucleotide net addition to the region of G4 DNA enclosed by $L2_{II}$.

In contrast to viral strands, initiation of complementary strand replication differs considerably between G4 and φX, both in terms of host function requirements (7, 10) and in unique versus multiple priming sites on the viral DNAs (8-10). As shown in Fig. 3, ori_c maps near the 5' end of domain III. This location corresponds to that of the major resolvable structural difference between G4 and φX DNAs, the presence of loop $S1_{III}$ as discussed above. It is also near the stem of

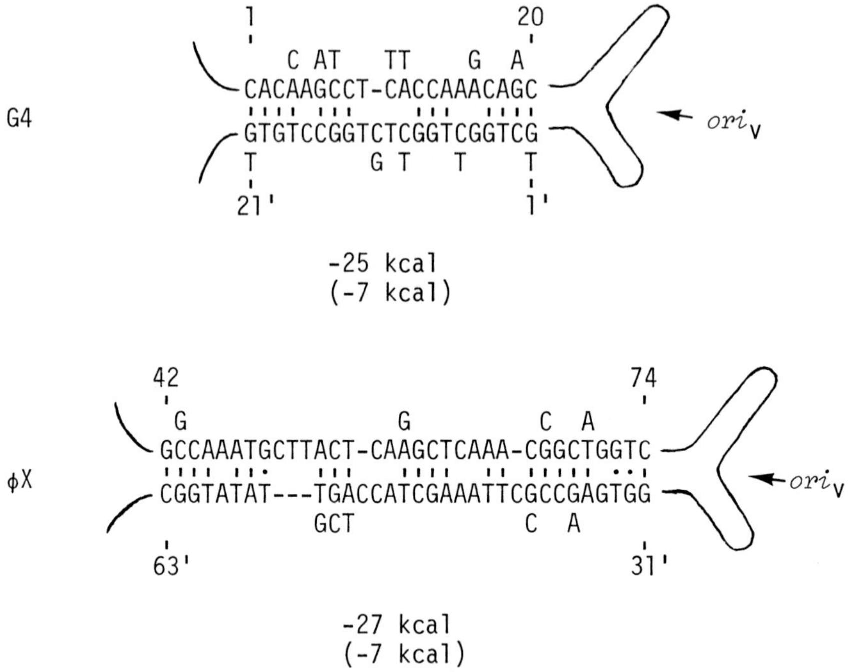

```
              1                    20
               '                    '
              C  AT  TT        G A
            CACAAGCCT-CACCAAACAGC
            ::::  :::   :::    ::::                    ← ori_V
            GTGTCCGGTCTCGGTCGGTCG
G4          T             G T T        T
            '                          '
           21'                         1'
```

                    −25 kcal
                    (−7 kcal)

```
              42                                    74
               '                                     '
              G              G            C A
            GCCAAATGCTTACT-CAAGCTCAAA-CGGCTGGTC
            ::::  ::: :    :::: :: :::::  :::                ← ori_V
            CGGTATAT---TGACCATCGAAATTCGCCGAGTGG
φX                       GCT            C A
             '                                       '
            63'                                     31'
```

                    −27 kcal
                    (−7 kcal)

FIGURE 4. *Base pairing models for loop L2$_{II}$ of G4 and φX DNAs. For G4, nucleotides 840-1440 were compared with 4940-120, according to numbering of ref. 5 (see also Fig. 3). Negative free energies were calculated (14, 15), but corrections for loop size were omitted. The φX base pairing was previously described (1). Nucleotides above and below base paired sequences are differences in the corresponding regions of φX and G4; calculated negative free energies for the substituted sequences are shown in parentheses. Numbering is relative to the ends of the G4 stem; the actual G4 stem location is nucleotides 5196-5215 paired with 1136-1156.*

L$_{III}$. While loop L$_{III}$ formed in G4 by pairing two regions at the base of domain III, on φX three-lobed DNA an analogous loop formed largely as the result of the L$_I$ and Ll$_{II}$ interactions. This again suggests, as with L2$_{II}$, that structure can be conserved while the base pairing responsible for it varies.

The base pairing responsible for loop L$_{III}$ was analyzed. The three individually most stable configurations identified are shown in Fig. 5. All three involve sequences between

FIGURE 5. Base pairing models for G4 loop L_{III}. Nucleotides 3660-4320 were compared with 4800-5400 (ref. 5) (see also Fig. 3). Negative free energies were calculated (14, 15), but corrections for loop size were omitted.

3890-3980 paired with 4860-4918. The region 3890-3980 is part of the spacer between genes F and G within which the primer for complementary strand replication is made (8, 9). In fact, base paired regions 1 and 3 shown in Fig. 5 are involved in the large downstream hairpin proposed for this region, and region 2 is involved in the proposed RNA primer hairpin. The downstream hairpin with a negative free energy of -40 kcal (omitting the loop size correction) is somewhat more stable than the most stable structure shown in Fig. 5 for loop L_{III}. Thus, the base pairing proposed for L_{III} may be incorrect. The primer hairpins in that case may correspond to the upper portion of loop Sl_{III}.

As an alternative, the base pairing analysis (Fig. 5) suggests a possible interrelationship between long and short range pairing involving the F-G spacer. Under certain conditions (i.e., those favoring additional weak interactions), long range pairing may be more stable than short range, or vice versa. For example, DNA binding protein may destabilize the three-lobed structure and loop L_{III} in particular. The intergenic hairpin structures would then be permitted to form, resulting in an active primase binding site (16). This model would account for the DBP requirement for initiation of complementary strand replication (7) and suggests a possible control mechanism for DNA replication based on interconversion of long and short range pairing. Conformational changes of this type may control various aspects of nucleic acid function, including mRNA translation.

ACKNOWLEDGMENTS

This research was supported by National Institutes of Health grants GM24432 and GM27727.

REFERENCES

1. Edlind, T. D., and Ihler, G. M. (1980) *J. Mol. Biol.* 142, 131.
2. Langeveld, S. A., van Mansfeld, A. D. M., de Winter, J. M., and Weisbeek, P. J. (1979) *Nuc. Acids Res.* 7, 2177.
3. Langeveld, S. A., van Mansfeld, A. D. M., Baas, P. D., Jansz, H. S., van Arkel, G. A., and Weisbeek, P. J. (1978) *Nature* 271, 417.
4. Sanger, F., Coulson, A. R., Friedmann, R., Air, G. M., Barrell, B. G., Brown, N. L., Fiddes, J. C., Hutchinson, C. A., Slocombe, P. M., and Smith, M. (1978) *J. Mol. Biol.* 125, 225.
5. Godson, G. N., Barrell, B. G., Staden, R., and Fiddes, J. C. (1978) *Nature* 276, 236.
6. Bass, P. D., Teertstra, W. R., van der Ende, A., and Jansz, H. S. (1980) *J. Mol. Biol.* 137, 283.
7. Zechel, K., Bouche, J. P., and Kornberg, A. (1975) *J. Biol. Chem.* 250, 4684.
8. Fiddes, J. C., Barrell, B. G., and Godson, G. N. (1978) *Proc. Natl. Acad. Sci. USA* 75, 1081.
9. Sims, J., and Dressler, D. (1978) *Proc. Natl. Acad. Sci. USA* 75, 3094.
10. McMacken, R., and Kornberg, A. (1978) *J. Biol. Chem.* 253, 3313.
11. Davis, R. W., Simon, M., and Davidson, N. (1971) *Methods Enzymol.* 21D, 413.
12. Shen, G. K. J., Ikoku, A., and Hearst, J. E. (1979) *J.*

Mol. Biol. 127,163.
13. Edlind, T. D., and Ihler, G. M. (1981) *J. Bacteriol.* (in the press).
14. Tinoco, I., Borer, P. N., Dengler, B. Levine, M. D., Uhlenbeck, O. C., Crothers, D. M., and Gralla, J. (1973) *Nature (New Biol.)* 246,40.
15. Salser, W. (1977) *Cold Spring Harbor Symp. Quant. Biol.* 42,985.
16. Sims, J., and Benz, E. W. (1980) *Proc. Natl. Acad. Sci. USA* 77,900.

PRIMARY AND SECONDARY REPLICATION SIGNALS IN BACTERIOPHAGE λ AND IS5 INSERTION ELEMENT INITIATION SYSTEMS

Gerd Hobom, Manfred Kröger, Bodo Rak

Institut für Biologie III,
Universität Freiburg

and Monika Lusky

Department of Molecular Biology,
University of California, Berkeley

ABSTRACT. A comparison of the available lambdoid origin sequences allows to point out several structural features in common with the DNA sequence of the phage G4 intercistronic region between genes F and G, which can be interpreted as recognition elements of a primase promoter sequence. These leading strand initiation signals are located in leftward orientation on the A-rich strand of all lambdoid origins, and by comparison with G4 require a single-stranded binding sequence, which appears to result from transcriptional activation. Binding of O protein to the origin interferes with transcription across *ori*, which therefore, has to precede these *ori/O* interactions.

A replication system which has many features in common with the λ-colE1 hybrid plasmid minimal replication system we have described earlier is located at both ends of the *E.coli* IS5 insertion element. According to a DNA sequence and functional analysis this 1195 bp element codes for two proteins in opposite, overlapping directions, with correlated, internal transcription signals attached to them in both orientations. A terminal inverted repeat sequence of 16 basepairs is the only structure remaining in common after both ends have been minimized for their replication signals. In comparison with the lambdoid *ice* signals this analysis suggests that both types of secondary replication signals become exposed in single-stranded conformation when either the leading strand DNA synthesis (maxi system) or transcription (mini system) is extended across their location. In the activated state they are expected to serve as primosome binding sites for initiation of lagging strand DNA synthesis.

INTRODUCTION

Replication of bacteriophage λ DNA is initiated from a unique site of its genome and proceeds bidirectionally around the circularized genome (1). Deletion mapping (2), isolation of cis-dominant mutations (3,4), comparative sequence analysis of wildtype and mutant phage DNAs (5-7), and a functional minimization in the cloned hybrid plasmid state (8) served to define a 2o2 basepairs origin of replication (*ori*) segment which is located close to the 80% position on the λ map (9).

Although at the resolution level of the electron microscope all four points of daughter strand DNA synthesis by back extrapolation appeared to have originated from the same *ori* segment (in most of the molecules) it is indeed extremely unlikely that the same small segment of DNA could have evolved to fulfill the requirements for each of the two leading *and* lagging strand initiation mechanisms, in bidirectional orientations. In a number of other systems including the human mitochondrial genome (10) initiation reactions have been observed that are separated both in space and in time for the two DNA strands. An extreme case of separately initiated plus and minus strand DNA synthesis is provided by the ØX174 RF→RF amplification mechanism (11), but also in plasmid ColE1 replication both strands are initiated at distinct positions on the DNA sequence and in specific sequential order (12,13). Both these systems belong to the class of unidirectional replication mechanisms, but upon minimization or cloning into hybrid plasmid combinations additional initiation signals have also been uncovered in a number of bidirectional replication systems. Here, the additional signals have been termed secondary origins of replication in plasmids F (*ori*II; 14,15) or R6K (*ori*α-*ori*γ; 16, 17), and inceptors of DNA replication in lambdoid phage DNAs (*ice*-signals; 8,18).

Usually the secondary replication signals are located at some distance from the primary *ori* signals and they are able to support replication of the minimized (or hybrid) plasmids in the absence of the primary *ori* signal, even without considerable change in the plasmid copy number (18,19). These results cannot be taken, however, to indicate a similar initiation mechanism for the integrated (complete) and the disintegrated initiation systems, and even the secondary signal itself may not function identically in both cases. Data that bear on the structure of these secondary replication signals and their function in both types of systems will be discussed in this article.

Besides the integrated primary and secondary signals which are directly and mechanistically involved in initiation of replication still another group of signals participates in the overall process, certainly in the case of bacteriophage λ. This class of several attached or tertiary replication signals can be exemplified by the λ transcription signals which are known to control initiation or attenuation of rightward transcription across *ori* required for activating the origin structure (20), and to indirectly regulate the overall initiation frequencies (21) through modulating *ori* activation rates.

RESULTS AND DISCUSSION

<u>Initiation of leftward leading strand DNA synthesis at the origin of replication signal.</u> In agreement with a template function for initiating only a single and not all four of the daughter strand growing points the DNA sequence of the lambdoid phage origins does not show an overall symmetrical organization, although individual symmetry elements are located in their left and right external sections (7). In our functional analysis in the hybrid plasmid system *ori* has also been shown to have a unidirectional leftward activity and to become inactive upon inversion relative to the conjugated *ice*-signal which is located to the left of *ori* in all these phage DNAs (8). Further data obtained in this system have confirmed the idea that the *ori* signal serves as a template for initiating the first leading strand DNA synthesis in leftward

direction, while inception of DNA synthesis at the remaining three growing points for an overall bi-directional replication may occur outside of *ori* in separate and consecutive reactions (see below).

In order to serve as a template for initiation of daughter strand synthesis it is necessary for the DNA sequence at *ori* to become strand separated and to expose a primase binding and promoting sequence, at least if primase is used for synthesizing the daughter strand primer RNA. As is known from analysing the phage G4 system in vivo and in vitro primase promoter sequences have to be single stranded (22). Strand separation at the primase promoter sequence is of course only the first step of parental DNA strand separation which continuously has to precede daughter strand RNA and DNA synthesis, since both primase and DNA polymerase III require single stranded templates, as is ultimately reflected in the overall result of semiconservative replication.

We have earlier proposed a strand separation model for the *ori* activated state as a result of transcriptional activation (1,23). Three different features of the *ori* DNA sequence can be pointed out to serve in its capacity of maintaining a metastabile strand separated structure long enough for primase to interact with its *ori* recognition sequence. One element of the *ori* structure important in this regard is its very A:T-rich central section, which is also highly asymmetric in the adenine and thymine distribution over both strands. A second property of the lambdoid origin structures is a very specific organization of direct and inverted sequence repetitions in the left A-sections which after strand separation has the potential of forming a large number of secondary structures (7) capable of decreasing the DNA renaturation rate in that region. And a third element involved in maintaining a metastabile strand separated structure at *ori* appears to be the organization of the right C-section, which alternatively allows the formation of one or the other of two larger DNA hairpin structures comparable in size to a common element in prokaryotic transcription termination signals. According to a more general theory of RNA polymerase and non-coding strand DNA hairpin interaction during transcription (23) a section C palindrome qualifies for an RNA polymerase pause site, and pausing of

RNA polymerase at *ori* section C would assist to maintain a strand separated state in nearby proximal sections A and B.

It is known from various *ori* cloning experiments (8,24) that deletion mutants in either of both external sections A and C will reduce but not eliminate *ori* activity, while deletion mutants in section B such as λr99 will block initiation of replication completely. These results argue that the *ori* primase promoter sequence should be located within section B. By comparing the DNA sequences of all available central origin sections of the lambdoid phages and bacterial chromosomes with the minus strand origins of G4 and related phages we arrive at a conclusion about the *ori* primase promoter consensus sequence which is presented in Fig. 1.

According to this comparison the primase promoter signal is located in lambdoid phage DNAs on the A-rich non-coding strand (relative to the transcriptional activation reaction), and it essentially covers all of the B section. This non-coding strand segment of the *ori* DNA sequence is also expected to be of rather low conformational flexibility even as a single stranded region. This is due to the high degree of purine (adenine) stacking interactions which may be important for the kinetics of the primase binding reaction. The observed unidirectional leftward activity of the lambdoid origins is in agreement with the conclusion drawn in Fig. 1 from our structural comparison with G4-like phage DNAs. No similar information is, however, available for any of the bacterial origins.

The leftward orientation of *ori* activity in λ together with a number of other results also excludes the rightward directed mRNA produced in transcriptional activation from further consideration as a primer RNA. And neither in RNA polymerase binding experiments in vitro, nor in *ori* cloning experiments in vivo has an *ori*-related RNA polymerase promoter activity been observed in any of the phage DNAs used (unpublished results). The mutants of *ori* section B studied so far, mainly λr99 and λr96 (4,5,8) are certainly in agreement with an absolute requirement for this segment, but many more mutants will have to be constructed before a more detailed analysis becomes possible.

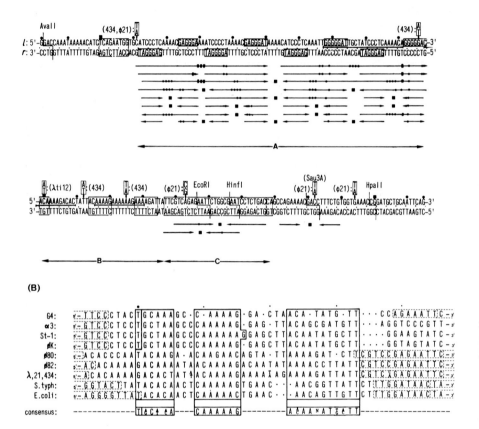

Fig. 1. Structural organization of the bacteriophage λ origin of replication (A), and comparison of DNA sequences located in the central section of the origin signal in G4-related, lambdoid and bacterial replicons (B). Nucleotide exchanges observed for the homologous 434 and 21 DNAs, and a selection of restriction sites are indicated above the λ sequence. Between both DNA strands horizontal bars refer to the λori deletion mutants r93, r99, and r96, in that order (5) and dots refer to the O gene reading frame. Double strand boxes mark segments of λ/ϕ80 DNA homology, single strand boxes point out a multiply repeated sequence element. Symmetry elements observed in the λ ori sequence are indicated in terms of potential secondary structures below the lines. Deviations from a perfect match are marked by dots on the arrows, while squares refer to inversion centers of inverted repeats. Three different sections of the λori structure are pointed out by double-headed arrows.

17 REPLICATION SIGNALS IN BACTERIOPHAGE λ

Besides facilitation of metastabile strand separation and exposure of a single stranded primase promoter sequence the origin structure also has to be able to protect the resulting primer RNA 5'-end from being exonucleolytically degraded or squeezed out of the template because of parental DNA rewinding. While at least part of all three *ori* reactions is directly supported by the (activated) origin DNA structure itself, some of them may also have to be assisted by an interaction with the phage specific initiator protein O or with host proteins involved in initiation. It had been concluded that the O protein directly acts at the origin structure (25,7), and we have found evidence that this interaction in vivo more specifically involves the A section of *ori* and the NH_2-terminal domain of the O protein (26). Based on these results we have earlier proposed a model with O involved in all three *ori* reactions pointed out here, by binding to one of the activated state secondary structures for stabilization of the strand separated structure in section A, and indirectly in section B (7). This structure would also be able to serve in protecting the 5'-end of the daughter strand (23). With the in vitro O protein footprinting data taken into consideration that are described in the preceding article (27) it might appear that the initial strand separa-

---

In order to maximize sequence similarities in the comparison of eleven origins (7,24,36,37) in the bottom part of the figure, a few gaps (dots) had to be inserted, equivalent to an extended rather than a regular stacked conformation in the (separated) DNA single strands. Regions of highest sequence homology have been boxed, and are summarized in a consensus sequence. The T residue boxed separately in G4 and related DNAs indicates the template position determining the 5'-terminal nucleotide of the resulting primer RNA (38). Dashed boxes to the left and right of the origin centers indicate boundaries of flanking *ori* sections, i.e. DNA segments that belong to inverted repeat sequences as explained in the top half of this figure.

tion reaction at *ori* does not require O interaction, while O binding to section A is part of the daughter duplex DNA protection system. In addition O/*ori* interaction is likely to be part of a larger protein complex in which a replicating phage DNA molecule is bound to a membrane 'replication site' through interaction with a larger number of other initiation factors (28,29), as is also suggested by λO^{ts} shift-up experiments (30,31).

In order to study whether O-binding would interfere with transcription across *ori* we have cloned the ori^+ containing *Sau*3A fragment of λdv*imm*21AB5 (32) and also a similar fragment containing the λr99*ori*⁻ mutation (isolated from pHL65 DNA;8) into a derivative of pBR322 which carries a p*lac*UV5 promoter fragment inserted in clockwise orientation near the *Eco*RI site (as described previously;see 26). Because *ori* is inserted between the *lac* promoter and the *tet* region of the vector DNA tetracycline resistance of the host cells can be used to monitor transcription across *ori*. Due to the absence of other λ replication signals and genes these hybrid plasmids are unable to replicate under *polA*⁻ conditions in the host cell, even if complemented for O from a second plasmid. The insert fragment of the second plasmid which is used for O complementation has been derived from pHL81 (phage 434 DNA;see 8) and because of its *Sal*I/*Hpa*I boundaries which closely bracket the O gene (and exclude *ice*, but also t_{R1}) it is also unable to replicate in *polA*⁻ cells.

When *E.coli*490A (*recA*⁻) cells containing a single plasmid with either the ori^+ or the *ori*99 allele as described above are plated on increasing concentrations of tetracycline colonies will be formed (in the absence of O) up to a 30µg/ml tetracycline limit, which has to be compared with similar limiting concentrations of 50-70 µg/ml for the vector DNA itself and for other related plasmids. Upon complementation of O from a second plasmid in the same cell the tetracycline concentration tolerated now drops to around 6µg/ml for both alleles, which is at the level of a control clone containing $\phi 80\text{-}t_O$ instead of *ori* (pHK6, constructed from a very closely related vector plasmid; M. Kröger and G. Hobom, submitted for publication). From these results we conclude that O in its *ori* binding reaction will block RNA polymerase transcription across *ori* and into the tetracycline region. And since O is bound to the proximal part of *ori*, section A, O is very likely to exclude any interaction of RNA polymerase with the distal parts of *ori*, sections B and C. Transcriptional activation of *ori* is, therefore, inhibited by the O protein binding to parental *ori* DNA (at higher concentrations of O) and has to occur prior to O binding. Consequently *ori*/O binding should be involved in a somewhat later step during initiation. However, this type of analysis does not allow to decide whether O binds to an activated state DNA conformation, immediately after transcriptional activation, or to the initial segment of daughter DNA resulting from the *ori* activated state's reactivity as a template. - The results obtained in this *ori* transcription analysis will be published in more detail elsewhere.

The secondary replication signals in λ and IS5 appear to control initiation of lagging strand DNA synthesis. Our analysis of the secondary replication signal, *ice*, which is located in the *c*II gene of λ and related phages had shown earlier that for initiation of replication in the hybrid plasmid system it must be attached to either an activated origin of replication (in the maxi system) or a promoter for transcription (in the mini system) with both of them oriented in right to left direction. Structural comparisons between the various lambdoid *ice* signals and an analysis of several *ice* mutants allowed to conclude that these signals consist of an inverted repeat sequence followed (with minor variations) by a hexa-G:C 'tail' sequence. A high degree of similarity between this group of structures and the larger class of transcriptional terminators with a hexa-A:T 'tail' sequence (or close variations) instead suggested to us that the *ice* signals might act as termination signals during primer RNA synthesis (8, 18,23).

In order to extend our analysis of secondary replication signals to some additional systems, and to arrive at more general conclusions about their structure and function(s), we have searched for substitute inceptor signals. This was done through cloning various fragments or mixtures of fragments into a *c*II position on hybrid vector plasmids containing *ori*, but not *ice*. While we have not been able to detect an additional signal elsewhere on λ^+DNA such signals have been found at both ends of the IS5 insert in λKH100 DNA. From the cloning results obtained we have concluded that each of the IS5 ends is able to sustain hybrid plasmid replication if transcribed from an external promoter in outside→in direction (equivalent to the λ mini system; 26).

For a structural comparison of the IS5 terminal or near-terminal replication signals with the lambdoid *ice* sites we decided to determine the IS5 DNA sequence. A complete sequence of the IS5 transposable DNA element is also required for an analysis of additional genes or signal structures that might be involved in IS5 hybrid plasmid replication.

From the DNA sequence of the 1195 basepairs element and from several series of subcloning experiments we arrived at the following conclusion about its genetic organisation: two genes with correlated transcriptional promoter and terminator signals are located on IS5 in opposite orientations with one gene totally contained within the other, as presented schematically in Fig. 2.

The terminator containing fragments (Fig. 2) have been cloned between a proximal *lac*UV5 promoter and the distal tetracycline region of pBR322 for monitoring the resulting level of tetracycline resistance, essentially as described above for the transcription analysis of λ*ori* fragments. The promoter containing fragments have been cloned in a proximal location relative to the galactokinase gene on plasmid pKO1 (33). Expression of the two IS5 genes from plasmids containing a complete IS5 sequence, and expression of only the IS5 small protein from plasmids deleted for the NH_2-terminal segment together with the promoter region of the IS5 large gene has been observed in a minicell protein synthesis system (see Fig. 3). This result excludes the possibility that the IS5 small protein might be a breakdown product of the large protein of IS5. For a further verification of the reading frames used for protein expression we have constructed an insertion variant of the IS5

17 REPLICATION SIGNALS IN BACTERIOPHAGE λ

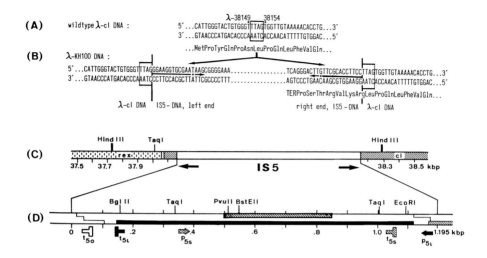

Fig. 2. (A) DNA and protein sequences of the wild type λcI region surrounding the IS5 target site in λKH100 DNA (boxed). Sequence positions refer to G=40000 which is located in the EcoRI site of ori section C, see Fig. 1. (B) DNA and protein sequences at the termini of the IS5 insert in λKH100 DNA. Boxes refer to the λcI target site duplication, the IS5 terminal inverted repeat is marked by two arrows. (C) Genetic map of the λ region close to the IS5 insert in λKH100 DNA indicating the location of the cI and rex genes and a few landmark restriction sites that have been used for cloning or subcloning. (D) Genetic map of insertion element IS5 as determined in DNA sequence and cloning analyses, and in protein expression from corresponding hybrid plasmids (see Fig. 3). The black bar and symbols refer to the IS5 large gene coding frame and correlated promoter and terminator sites (p_{5L}, t_{5L}), while the hatched bar and symbols indicate the oppositely oriented IS5 small gene coding region and transcription signals (p_{5S}, t_{5S}), respectively. A light symbol marks the position of a left end outside-in termination signal, and the location of the first near-terminal stop codons is also indicated for all six reading frames, one of which is actually used for cI termination. Some of the restriction sites used for IS5 subcloning are indicated above the map (M. Kröger and G. Hobom, submitted for publication).

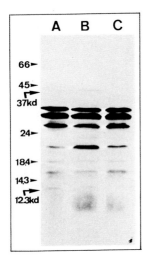
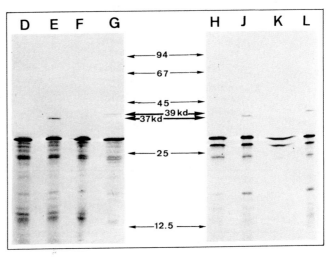

Fig. 3. Expression of proteins coded by IS5. The left panel shows the protein bands observed in a minicell system (39) after $[^3H]$-leucine labelling, separation on a 17% polyacrylamide/SDS gel and fluorography. The right panel gives the autoradiograms obtained after $[^{35}S]$-methionine labelling in a cell-free protein synthesis system (lanes D-G) and in the minicell system (lanes H-L). Lanes C,D,H: vector plasmid pML2 without insert. Lanes B,E,J: plasmid pML3 which contains a full size IS5 insert. Lanes A,F,K: plasmid pML4 with segment IS5:1-1097 inserted (p_{51} and NH_2-terminus of the IS5 large gene deleted). Lanes G,L: plasmid pHL128 (26) which has an IS5:154-1195 segment inserted and has an extended IS5 large gene reading frame because of in frame BglII/BamHI fusion at its penultimate codon. The positions of standard protein size markers are indicated, and the apparent molecular weights of the IS5 coded proteins has been calculated relative to these positions. The IS5 small protein is not visible in $[^{35}S]$-methionine labelling experiments, apparently due to its low methionine content. Further details will be described elsewhere (B. Rak et al., submitted for publication).

large gene. According to the DNA sequence of the fragment inserted into the penultimate codon the size of the protein product was expected to increase from 326 to 344 amino acids. A shift in the apparent molecular weight of the IS5 large protein from 37kd to 39kd was indeed observed (see Fig. 3).

Besides the general genetic organisation of IS5 the structure of both its left and right ends is particularly important in connection with the IS5 near-terminal replication signals. In order to preserve its state as an autonomously acting genetic unit IS5 or similar elements can be predicted to be organized in such a way as to avoid regulatory dominance exerted by adjacent genes or transcription units on the host DNA, in any location. As a necessary consequence we expected the presence of near-terminal stop codons in all six reading frames, which are indeed observed in IS5. Alternatively and in addition near-terminal termination signals for outside-in transcription, and therefore, polarity can be expected as a direct consequence of IS autonomy. An outside-in termination signal has been determined through subcloning in a near-terminal location at the left end of IS5, but no such signal could be detected at the IS5 right end. In this region there is essentially no space left between the terminal inverted repeat sequence and the -35 region of the IS5 large gene promoter (p_{51}, see Fig. 2). Therefore, the mechanism for protecting IS autonomy against right to left outside-in transcription is uncertain at present.

The most prominent feature of the IS5 left and right end sequence is a terminal inverted repeat of 16 basepairs with only a single mismatch (at position 13). Except for its terminal GG dinucleotide sequence the inverted repeat of IS5 is not related to any other such sequence determined for a number of different IS or Tn transposable elements (34). It has, however, been observed to bear similarity to a near-terminal sequence in phage mu DNA (35). The inverted repeat termini of IS5 are flanked by a directly repeated target site duplication of four basepairs (see Fig. 2).

When the left and right end IS5 sequence is analysed within the present minimization boundaries of the terminal or near-terminal replication signals (35; and unpublished results) no structure comparable

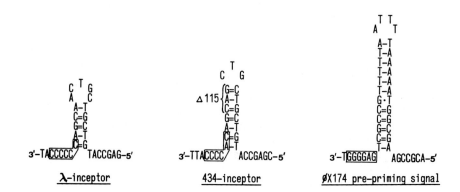

Fig. 4. Comparison of the lambdoid inceptor sites with the ØX174 pre-priming signal. The λ and 434 non-coding strand *ice* sequences as exposed during transcriptional activation or leading strand DNA synthesis in leftward direction are drawn in their secondary structure hairpin conformations, in analogy to the ØX174 pre-priming signal (4o). The 3'-hexa-C or hexa-G 'tail' sequences are indicated in boxes for all three signal structures. A three basepair deletion mutant with *ice*-phenotype, Δ115, is indicated for the 434 signal.

to the lambdoid *ice* signals can be detected. If the identical result obtained for both ends of IS5 is taken into consideration which should be reflected in an identical or nearly identical DNA signal sequence of opposite polarity, the only element of that character remaining in the close minimization brackets is the terminal inverted repeat sequence itself. This DNA sequence has no potential for forming a secondary structure (see Fig. 2), and its activity in initiation of DNA replication has to be attributed, therefore, to a single stranded conformation in the non-coding strand exposed as a consequence of transcriptional activation. Since the IS5 termini are unlikely to act as DNA signal structures on their own, this suggests that the inverted repeat sequences might expose single-stranded binding sites for an IS5 specific protein(s) with initiator activity.

17 REPLICATION SIGNALS IN BACTERIOPHAGE λ

The most likely candidate proteins for an inverted repeat binding activity are the two IS5 coded proteins. An analysis of the complementation requirement for the IS5 large or small protein(s) in initiation of IS5 plasmid replication is difficult, however, because the *E.coli* chromosome has been observed in Southern hybridization analyses to contain 11±1 copies of IS5 DNA (B. Rak et al., submitted for publication). A four-to sixfold reduction in copy numbers of incomplete IS5 plasmids as compared to complete IS5 plasmids both grown under $polA^-$ conditions is, however, suggestive in this regard, if the IS5 background copies in the chromosome are taken into consideration. In further experiments we will try to confirm this result and to reach a decision whether one or both of the IS5 coded proteins are involved in this reaction.

When the results obtained in the IS5 and λ hybrid plasmid initiation systems are compared with each other a conclusion about the λ initiation system and the λ*ice* signal is suggested that differs from our previous interpretation: the lambdoid secondary replication signals might   act as single stranded initiator protein binding sites after 'activation' by transcription or leading strand replication. Following this interpretation we would have to consider the similarity of the lambdoid *ice* signals with transcriptional terminators an accidental feature of these single-stranded DNA/protein recognition sites. A protein recognition site of some structural similarity with the lambdoid *ice* signals (see Fig. 4) has recently been detected in bacteriophage ØX174 DNA in the F-G intercistronic region. The ØX174 pre-priming signal is known to interact with a protein complex which is involved in initiating the lagging strand replication mode on the ØX174 single stranded template, and which has been termed the primosome (4o).

An analogous function for both the lambdoid *ice* sites and the IS5 termini as initiation signals (primosome recognition sites) for the lagging strand DNA synthesis would be in agreement with the data obtained for both systems and extend the interpretation given above for initiation of only the leading strand DNA synthesis at the lambdoid *ori* signal. A specific mode of local DNA replication is also expected to occur at some stage during transposition of an autonomous DNA element such as IS5, and may be more likely of lagging strand rather than leading strand DNA-synthesis character. Preliminary results obtained in a λori/ØX174 F-G fragment conjugated system are also in agreement with this general conclusion about the function of secondary replication signals such as the λice site (data not shown). The binding specificity of initiator protein/DNA signal sequence interactions may control replicon specific initiation, and may be used differently in different systems. The in vitro interactions of the λP protein with primosome proteins dnaB and dnaC (41, 42) may hint a λ specific primosome/ice recognition.

With leading and lagging strand DNA synthesis initiated in leftward direction at λori and λice, respectively, only two growing points of an overall bidirectional replication system have been installed. While the first fragment of the leftward lagging strand DNA synthesis may be extended backwards across ori and become converted into the leading strand DNA synthesis in rightward direction, the mechanism of initiating the rightward lagging strand DNA synthesis is still uncertain at present.

ACKNOWLEDGMENTS

We gratefully acknowledge the expert technical assistance of M. Hable, M. Sauter, G. Schlingmann and B. Traub. We thank M. Rosenberg and W. Szybalski for sending us plasmid or phage strains. This work has been supported by the Deutsche Forschungsgemeinschaft.

REFERENCES

1. Schnös, M., and Inman, R.B. (1970). J. Mol. Biol. 51, 61-73.
2. Stevens, W.F., Adhya, S., and Szybalski, W. (1971). In "The Bacteriophage Lambda" Cold Spring Harbor Laboratory (Hershey, A.D., ed.), Cold Spring Harbor, N.Y., p. 515-553.
3. Dove, W.F., Inokuchi, H., and Stevens, W.F. (1971). In "The Bacteriophage Lambda" Cold Spring Harbor Laboratory (Hershey, A.D., ed.), Cold Spring Harbor, N.Y., p. 747-771.

4. Rambach, A. (1973). Virology 54, 270-277.
5. Denniston-Thompson, K., Moore, D.D., Kruger, K.E., Furth, M.E., and Blattner, F.R. (1977). Science 198, 1051-1056.
6. Scherer, G. (1978). Nucleic Acids Res. 5, 3141-3156.
7. Grosschedl, R., and Hobom, G. (1979). Nature 277, 621-627.
8. Lusky, M., and Hobom, G. (1979). Gene 6, 173-197.
9. Szybalski, E., and Szybalski, W. (1979). Gene 7, 217-270.
10. Bogenhagen, D., Gillum, A.M., Martens, P.A., and Clayton, D.A. (1979). Cold Spring Harbor Symp. Quant. Biol. 43, 253-262.
11. Arai, K., Arai, N., Schlomai, J., and Kornberg, A. (1980). Proc. Natl. Acad. Sci. USA 77, 3322-3326.
12. Itoh, T., and Tomizawa, J. (1980). Proc. Natl. Acad. Sci. USA 77, 2450-2454.
13. Nomura, N., and Ray, D.S. (1981). this volume.
14. Kahn, M.L., Figurski, D., Ito, L., and Helinski, D.R. (1979). Cold Spring Harbor Symp. Quant. Biol. 43, 99-103.
15. Lane, D., and Gardner, R.C. (1979). J. Bacteriol. 139, 141-151.
16. Kolter, R., Inuzuka, M., Figurski, D., Thomas, C., Stalker, D., and Helinski, D.R. (1979). Cold Spring Harbor Symp. Quant. Biol. 43, 91-97.
17. Stalker, D.M., Kolter, R., and Helinski, D.R. (1979). Proc. Natl. Acad. Sci. USA 76, 1150-1154.
18. Lusky, M., and Hobom, G. (1979). Gene 6, 137-172.
19. Manis, J.J., and Kline, B.D. (1978). Plasmid 1, 492-507.
20. Dove, W.F., Hargrove, E., Ohashi, M., Haugli, F., and Guha, A. (1969). Japan. J. Genet. 44, Suppl. 1, 1-12.
21. Murotsu, T., and Matsubara, K. (1980). Molec. Gen. Genet. 179, 509-519.
22. Rowen, L., and Kornberg, A. (1978). J. Biol. Chem. 253, 758-764.
23. Hobom, G., Grosschedl, R., Lusky, M., Scherer, G., Schwarz, E., and Kössel, H. (1979). Cold Spring Harbor Symp. Quant. Biol. 43, 165-178.
24. Moore, D.D., Denniston-Thompson, K., Kruger, K.E., Furth, M.E., Williams, B.G., Daniels, D.L., and Blattner, F.R. (1979). Cold Spring Harbor Symp. Quant. Biol. 43, 155-163.

25. Furth, M.E., Yates, J.L., and Dove, W.F. (1979). Cold Spring Harbor Symp. Quant. Biol. 43, 147-153.
26. Hobom, G., and Lusky, M. (1980). In "Mechanistic Studies of DNA Replication and Genetic Recombination" (Alberts, B., ed.), Academic Press, New York, p. 231-255.
27. Matsubara, K., and Tsurimoto, T. (1981). this volume.
28. Tomizawa, J. (1971). In "The Bacteriophage Lambda" Cold Spring Harbor Laboratory (Hershey, A.D., ed.), Cold Spring Harbor, N.Y., p. 549-552.
29. Georgopoulos, C.P., Lundquist-Heil, A., Yochem, J., and Feiss, M. (1980). Molec. Gen. Genet. 178, 583-588.
30. Saito, H., and Uchida, H. (1977). J. Mol. Biol. 113, 1-25.
31. Klinkert, J., and Klein, A. (1978). J. Virol. 25, 730-737.
32. Matsubara, K., and Otsuji, Y. (1978). Plasmid 1, 284-296.
33. McKenney, K., Shimatake, H., Court, D., and Rosenberg, M. (1980). Federation Proc. 39, 2203.
34. Calos, M.P., and Miller, J.H. (1980). Cell 20, 579-595.
35. Lusky, M., Kröger, M., and Hobom, G. (1981). Cold Spring Harbor Symp. Quant. Biol. 45, 173-176.
36. Sims, J., Koth, K., and Dressler, D. (1979). Cold Spring Harbor Symp. Quant. Biol. 43, 349-365.
37. Zyskind, J.W., and Smith, D.W. (1980). Proc. Natl. Acad. Sci. USA 77, 2460-2464.
38. Bouché, J.P., Rowen, L., and Kornberg, A. (1978). J. Biol. Chem. 253, 765-769.
39. Trinks, K., Habormann, P., Beyreuther, K., Starlinger, P., and Ehring, R. (1981). Molec. Gen. Genet. in press.
40. Arai, K., and Kornberg, A. (1981). Proc. Natl. Acad. Sci. USA 78, 69-73.
41. Wickner, S. (1979). Cold Spring Harbor Symp. Quant. Biol. 43, 303-310.
42. Klein, A., Lanka, E., and Schuster, H. (1980). Eur. J. Biochem. 105, 1-6.

INTERACTION OF BACTERIOPHAGE LAMBDA O PROTEIN WITH THE LAMBDA ORIGIN SEQUENCE

Toshiki Tsurimoto
Kenichi Matsubara

Laboratory of Molecular Genetics
Osaka University Medical School
Kita-ku, Osaka, Japan

ABSTRACT

Initiation of bacteriophage lambda DNA replication requires two λ-coded proteins, O and P(1). Of these, the O protein is replicon-specific. Genetic studies (2,3) have suggested that the O protein will interact with a specific site on the λ genome, called ori (for origin), which is characterized by the presence of four tandemly repeated 19-base pair sequences (4,5). We have purified the O protein to near homogeneity (6) and examined its interaction with a cloned λ DNA fragment carrying the ori sequence by a protection assay against digestion by exonuclease III or DNase I. Our data demonstrate that at low concentration, the O protein binds to the inner two repeats of the ori sequence, while it binds to the outer two repeats only when its concentration is elevated. This unique molecular interaction may play a key role in activating the origin to initiate λ replication.

λ DNA fragment carrying the ori, when cloned in multicopy plasmid pBR322 or pACYC184, interferes with the replication of the complete λ genome. We inferred that this interference involves an in vivo interaction of λ O protein and λ ori, possibly a titration of O protein by multicopy ori sequence. Comparison of several deletion derivatives has revealed that the intensity of interference is inversely related to the number of repeats in the cloned fragment. The intensity of interference is greatly stimulated when the same cell carries another λ DNA fragment covering ice (for inceptor) and oop (an operon producing small RNA). These observations suggest that the O-ori interaction in vivo requires at least one additional λ-coded cytoplasmic factor.

INTRODUCTION

Intensive studies with bacteriophage λ on the mechanism of control in the process of DNA replication have established the following facts. a) All the genes and sites necessary for replication and its control are clustered in a region covering only one-tenth of the total phage genome (7). These elements are shown in Fig. 1. This region is transcribed as a unit from left to right in the figure. b) Transcription results in gene products

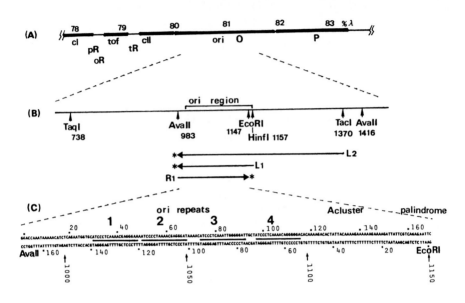

FIGURE 1. Bacteriophage λ genes and origin elements.
(A) A region of the λ genome covering 78 to 83 % of λ DNA. Thick lenes represent regions coding for polypeptides. Transcription starts from pR and proceeds in the direction tof→cII→O→P. Symbols for genes and sites are based on (19).
(B) Expanded map of the central region of the O gene. The ori region is defined according to Hobom et al. (4). Restriction endonuclease cutting sites used for this study are shown. Numbers represent the number of nucleotides from the transcription start point in pR. Horizontal arrows under the map indicate 5'-end-labeled fragments used for exoIII digestion assay and DNase footprinting. The labeled 5' ends are represented by (*) and the direction of exoIII digestion (3' → 5') on each labeled strand is expressed by the direction of the arrow.
(C) Nucleotide sequence of the ori region from AvaII (983) to EcoRI (1147) (refs. 4,5). Three unique features in this region are shown. They are the four repeating sequences (repeats 1 to 4, represented by lines drawn between the two strands), the adenine-rich region (A cluster) and a palindromic region (only a part of this sequence is shown). The upper and lower lines represent, respectively, the l- and r-strands. Noted on the strands are the number of nucleotides from the AvaII (l-strand) and EcoRI (r-strand) ends, respectively.

O and P that are needed for λ DNA synthesis (1). Genetical studies have shown that the O protein interacts with the P protein (8), and the P protein, in turn, interacts with several host enzymes, including dnaB (9). No other λ-coded proteins are needed for the initiation process. c) In the middle portion of the O gene is a unique DNA sequence called ori, consisting of four 19-base pair (bp) repeats(Fig. 1, refs. 4,5). It has been inferred that ori is the site for recognition by λ O protein (2, 3). λ DNA replication is thought to start at or very close to the ori repeats and the replication fork proceeds bidirectionally (10). d) The ori sequence or its vicinity must be "transcriptionally activated"; viz., this region must be transcribed prior to, or during, the process of replication initiation (11,12). However, the nature of this activation is not yet clear. e) A short leftward transcriptional unit which produces oop RNA is located next to the O gene (13). The role of this RNA synthesis is not yet clearly understood. f) A region called ice (for inceptor) has been defined to the left of the oop, and is argued to be a site for switching primer RNA to DNA (14). However, the exact function of ice is not clearly understood.

Despite many efforts, an in vitro replication system for λ is yet to be established. This may be because: g) The O protein, which is needed in λ-specific initiation of DNA synthesis, is unstable, and does not accumulate in quantity (15,16; Miwa and Akaboshi, personal communication). (On the other hand, there is evidence that a reasonable amount of P protein is accumulated under the same conditions.) The poor accumulation of the O gene product has prevented detailed analyses of molecular nature of this protein that plays a key role in λ DNA replication. h) The mechanism of transcriptional activation is not well understood. It is not clear if this mechanism functioned properly in the systems previously used for attempted in vitro λ replication. i) Similarly, a lack of knowledge about the role of ori-ice may have hindered proper activation of the origin region.

RESULTS AND DISCUSSION

λ-O protein specifically binds to the four repeating sequences in ori region

We have been able to purify λ O protein to near homogeneity (6). This was achieved by use of bacteria carrying mutant λdv, which accumulated a large number of copies of λdv plasmid and, therefore, a high level of O protein. The purified O protein had a molecular weight of 32,000. It bound to a DNA fragment carrying the λ ori sequence, but not to DNA fragments carrying ø80 ori or pBR322 ori, or to λ DNA fragments without the ori sequence.

We used purified O protein to study its mode of interaction with λ ori sequence, thought to be an early event and a key interaction in the initiation process of λ DNA replication.

Sufficiently strong binding of the O protein to a specific site(s) in the ori sequence would prevent the progressive degradation of the DNA beyond that interaction point by exonuclease III, an enzyme that digests DNA exonucleolytically from the 3' end. If the 5' terminus of this DNA molecule is labeled with ^{32}P (17), therefore, the interaction point should be accurately determinable from the size of the "protected fragment" carrying ^{32}P at the 5' end (18).

To test this possibility, we chose the R1 fragment, which has 5'-^{32}P in the 1-strand at the right end of the molecule, as shown in Fig. 1. The results are shown in Fig. 2. As expected,

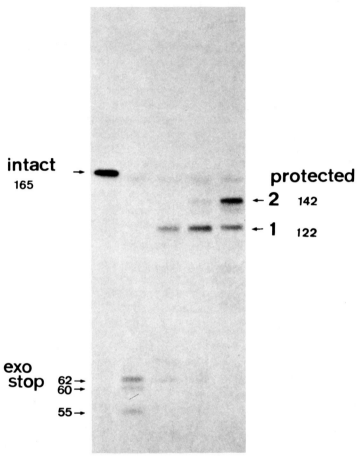

FIGURE 2. The O protein protects the R1 fragment against exo III disgestion. Approximately 3×10^{-15} moles of R1 fragment (^{32}P-end-labeled at EcoRI terminus: see Fig. 1) was incubated at 37° C with varying amounts of O protein for 12 min and then digested with exoIII (60 units/ml) at 37° C for 5 min according to (18). After digestion, the samples were electrophoresed in 10 % polyacrylamide and autoradiographed. Lane 1, undigested R1 fragment. Lane 2, digested with exoIII in the absence of O protein. Three exo stops are seen. Lanes 3,4,5, digested with the same amount of exoIII after incubation of the DNA with 3,6 and 12 ng of O protein, respectively. Two protected bands (protected 1 and 2) can be seen. Numbers represent fragment sizes expressed in nucleotides from the EcoRI terminus, as estimated from length standards.

the O protein prevented the degradation by exoIII from proceeding beyond a certain point in the R1 DNA. At low O concentration, the site of protection was 122 bp from the 5'-^{32}P, or 1029 position in the sequence map in Fig. 1. At high O concentration, the O protein protected a wider region of the R1 DNA, the protection site being 142 bp from 5'-^{32}P, or position 1009 in the sequence map.

We performed similar experiments with DNA fragment L1, which has ^{32}P at the 5' end of the opposite, l-strand. The results in Fig. 3 demonstrate that the O protein protected the DNA from processive degestion by exoIII at 96 bp and 119 bp from "left" 5'-^{32}P, at low and high O concentration, respectively. These sites are respectively positions 1078 and 1101 in the sequence map in Fig. 1.

We also carried out a protection assay against DNaseI, an enzyme that digests DNA endonucleolytically (21). At low concentration, it randomly incises DNA, but the region covered by the O protein should be protected from incision. The results in Fig. 4 demonstrate that the region in l-strand 50 through 96 bp from the 5' end of L2 DNA is protected by the O protein at low concentration, and the region 28 through 121 bp is protected at high concentration. These regions correspond in the sequence map to 1032 through 1078 and 1010 through 1103, respectively.
A similar footprinting experiment was carried out on the R1 fragment with high concentration of O protein, and a region 51 or 52 through 140 bp from the 5' end of R1 DNA was protected against DNaseI attack (data not shown). This region corresponds to 1011 through 1099 or 1100 in the sequence map (bracket 3 in Fig. 5).

These results are collected in Fig. 5. It is clear that at low concentration, O protein binds to the inner two repeats in the λ ori DNA sequence. At higher O concentration, the two outside repeats are bound. The molecular events involved in this sequential binding are not yet clear, though they apparently resemble the interaction of λ repressor (cI protein) or tof protein with three reiterated sequences in λ pR, a promoter-operator for transcription (22 - 24), and of SV40 T antigen with three reiterated sequences in the origin region for replication (25, 26). It may be reasonable to assume that the O-ori interaction activates the origin region in the process of initiation of replication. Examples of activation of a DNA region by its interaction with protein(s) include those of the oR2 sequence with λ repressor (22), of the cY region with λcII protein (27), and of several other operators with other proteins (28).

The O protein-ori interaction does not induce DNA strand-breaks (nicking activity) or modify strand helicity to a detectable extent (topoisomarase activity). The O protein is likely to act either by distorting the local DNA structure to induce the action of other proteins, or by binding directly to another replication factor, such as the λ P protein and/or a complex of host proteins.

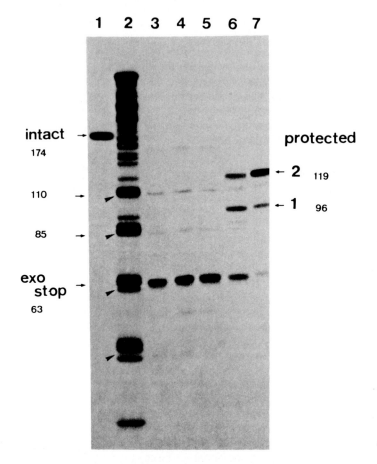

FIGURE 3. The $\underline{O}$ protein protection of the L1 fragment against exoIII digestion. Experiments were done as described in legend to Fig. 2, except that approximately 2×10^{-15} moles of L1 fragment (^{32}P-end-labeled at the AvaII terminus: Fig. 1) and 180 units/ml of exoIII were used. Lane 1, undigested L1 fragment. Lane 2, length standard: ^{32}P-end-labeled L2 fragment whose guanines were DMS-methylated and partially degraded according to Maxam and Gilbert (20). Four G clusters in repeats 1 to 4 can be seen (marked by arrow-heads). Lane 3, exoIII digestion of the L1 fragment without $\underline{O}$ protein. A strong exo stop and two weak exo stops can be seen. Lanes 4, 5, 6 and 7, exoIII digestions of L1 fragment after incubation with 1.5, 3, 6 and 12 ng of $\underline{O}$ protein, respectively. Two protected bands (protected 1 and 2) are seen. Arrows and numbers indicate band positions and their sizes in nucleotides (the 5' termini being always the $\underline{l}$-strand AvaII site labeled with ^{32}P).

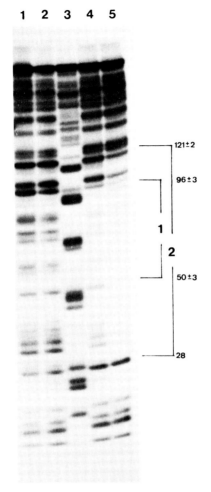

FIGURE 4. DNase footprinting reveals a region of protection by O protein. Approximately 2.5×10^{-15} moles of L2 fragment (An l-strand AvaII terminus was end-labeled with ^{32}P) was incubated at $20°C$ for 20 min with varying amounts of O protein. DNaseI was then added at 0.01 mg/ml and the mixture was incubated at $20°$ C for 2.5 min, followed by electrophoresis and autoradiography. Lane 1, DNaseI digestion in the absence of O protein. Lanes 2,4,5, DNaseI digestion in the presence of 3,6 and 12 ng O protein/0.1 ml, respectively. Brackets 1 and 2 at the side illustrate the regions protected against DNaseI attack by 6 ng and 12 ng of O protein, respectively. Numbers at the side of each bracket represent the 3' and 5' borders of the protected regions as estimated from size standard prepared similarly as in legend to Fig. 3 (in lane 3), expressed in number of nucleotides from the AvaII end.

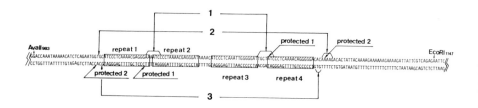

FIGURE 5. The ori region of λ DNA from AvaII (position 983) to EcoRI (position 1147) as protected by O protein against digestions by exoIII or by DNaseI. Upper and lower rows of nucleotide sequences represent the l- and r- strands of λ DNA, respectively. Protected positions against exoIII digestion (protected 1 and 2) were assayed with L1 and R1 fragments, and are shown with slanting arrows. Protected regions against DNaseI cleavage are indicated by brackets 1, 2 and 3. The results of the exoIII digestion assays and the DNaseI footprinting studies coincide almost perfectly. The four boxes containing the 19 bp repeating units (repeats 1 to 4) fit within the sites covered by the O protein, and match its borders quite well, as expected if one unit of O protein binds to each repeat.

Interference with λ replication by a cloned λ ori fragment

Is the unique interaction of λ O protein and the λ ori sequence also observable in vivo? If so, then we can imagine a situation where DNA fragment carrying the ori sequence (but not the complete O gene) perpetuated in a large number of copies in a cell "titrates" the O protein coded for by another complete, O-producing λ genome added in relatively few copies. The result will be interference with the replication of the λ genome due to short supply of the O protein. This prediction was tested with a cell carrying a λ ori DNA sequence cloned within a multicopy ColE1 derivative plasmid by measuring the growth of superinfecting λ phage added at low multiplicity. The cloned fragment originally used covered the ice-oop-ori sequences (see Fig. 6).

Fig. 6 demonstrates that the cloned ice-oop-ori strongly interferes with the replication of superinfecting λ phage. The cloned fragment does not carry the genes for cI, or tof, and therefore the superinfecting λ genome will not be affected by these repressors.

A series of deletion derivatives constructed from the parental recombinant plasmid pOA-4 supports the view that the λ ori sequence is responsible for the interference. In these derivatives, the vector region is identical, but the "right-hand-end

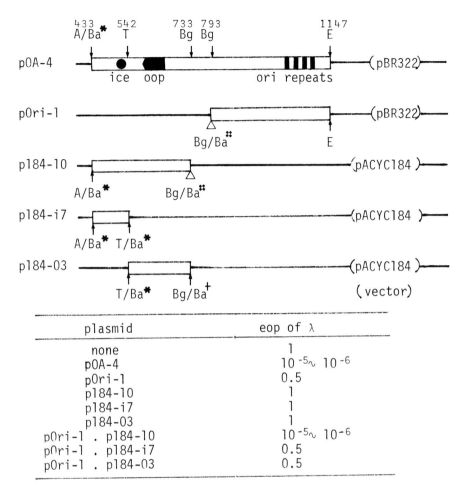

FIGURE 6. Recombinant plasmids carrying a portion of ice-oop-ori region and their ability to interfere with λ phage replication. Fragments of λ genome covering from nucleotide coordinate 433 in the sequence map (see ref.14) to 542, from 542 to 733, and from 733 to 1147 carry respectively ice, oop and the four repeats in ori. This entire segment was cloned in pBR322 by replacing the smaller BamHI-EcoRI segment of this vector DNA, and pOA-4 was the product of the recombination. The region from 733 to 793 is not correlated with any of these three functions.
Deleting the 433 to 793 segment from pOA-4 resulted in pOri-1, which carries only ori region. p184-10, p184-i7 and p184-03 are the pACYC184-based recombinant plasmids that carry respectively the ice and oop regions, the ice region and the oop region. All these fragments with BamHI termini were inserted in the unique BamHI site of the vector. Arrows with A, Ba, Bg, E and T represent cleavage sites for the restriction enzymes AluI, BamHI, BglII, EcoRI and TaqI, respectively. Some termini of the cloned

fragments were modified as follows: ⁂ Bg/Ba, BglII end joined to BamHI end; ✱ T/Ba or A/Ba, TaqI or AluI and BamHI ends joined by repair ligation; +Bg/Ba, repaired and ligated with BamHI linker. These recombinant plasmids were transformed into km723 (<u>E. coli</u> K12, <u>recA1,his,gal,rpsL</u>) individually or in combination. The resulting plasmid-carriers were grown overnight, and were used as indicator bacteria for scoring efficiency of plating (eop) of λ <u>cI</u>90<u>nin</u>$_5$ phage. Low eop represents strong interference by the plasmid.

FIGURE 7. Deletion derivatives missing a fraction of the four repeating units in the λ <u>ori</u> region, and their ability to interfere with λ phage replication. pOA-4 (see Fig. 6) DNA, cleaved at EcoRI site and digested to various extents by exonuclease BAL31 (29), was resealed into circular form with EcoRI- or BamHI-linker. Details of the construction of deletion derivatives will be published elsewhere (Tsurimoto and Matsubara, in preparation). Vertical lines represent the deletion endpoint coming in from the EcoRI cleavage site (right end in this figure), and $\overset{CCGGATCCGG}{GGCCTAGGCC}$ etc represents the joining of a BamHI linker at this deletion endpoint. All these derivatives have part of the <u>ori</u> region and an intact <u>ice-oop</u> region (to the left in the figure). The plasmids were transformed into km723, and the eop of the λ phage was scored as indicated in the legend to Fig. 6. Note that with the removal of 19-bp repeating units, the eop increased, <u>viz</u>, the ability to interfere with λ growth decreased.

portions" of the cloned λ DNA fragment have been deleted to various extents by progressive digestion with BAL31 exonuclease (29, Tsurimoto and Matsubara, in preparation). Fig. 7 shows the result. The derivative pOri-D1, in which the region lying to the right of the four repeats is deleted leaving the latter intact, interfered with λ growth to the same extent as the parental plasmid. But derivatives missing the rightmost one repeat (pOriD-4510), two repeats (pOriD-9010), and three repeats (pOriD-9011) showed progressively lower interference activity. This indicates that the four repeated <u>ori</u> sequences play an important role in the interference.

Dissection of the cloned λ DNA revealed the complex nature of the interference. This is illustrated in Fig. 6, which shows that fragments containing the ori repeats without the ice-oop region, such as pOri-1, interfere only weakly. The ice-oop region cloned together or separately showed no sign of interference (Fig. 6, pl84-10, pl84-17 and pl84-03; the ice-oop region in this set of experiments was cloned in pACYC184, a ColE1-type replicon, although replacing this vactor with pBR322 made no difference). Thus the coexistence of the ori and ice-oop regions appears to be necessary for the maximal interference to be expressed.

To see whether the stimulatory effect of the cloned ice-oop is also expressed in trans configuration, the ice-oop region cloned in pACYC184 and the ori region cloned in pBR322, which two vectors are compatible in a cell, were transformed into the same cell. The results (Fig. 6) demonstrate that the ori and ice-oop regions, whether in trans or cis configuration, exhibit strong interference. This effect was eliminated when the contiguous ice-oop sequence was replaced by ice or oop separately. These observations indicate, though they do not prove, that the cloned ori region primarily causes interference with growth of λ phage, and that this effect is stimulated by the simultaneously cloned intact ice-oop region. The action in trans configuration suggests that the stimulation by ice-oop may be mediated by a cytoplasmic material.

CONCLUSION

Lambda O protein has been shown to interact with the four 19-bp repeating sequences in the λ ori region. The inner two repeats are the preferred sites, and the outer two repeats are bound when the O concentration is elevated. The inferences drawn from genetic studies, that the ori sequence is the site of interaction with O protein, forming unique complex which acts in the process of initiation of DNA replication, have thus been confirmed on a molecular basis. However, our data do not indicate whether the ori-O complex is made first and then interacts with other components or, alternatively, other components interact with either ori or O first and then such complex(es) interact with each other through the specific interaction of ori and O.

A λ DNA fragment carrying the ori sequence but not coding for the complete O protein causes interference with λ growth when cloned in a multicopy plasmid pBR322. The interference does not involve the function of known repressor genes, but apparently involves the four 19-bp repeating sequences in ori. Deleting one, two and three of these four repeats progressively reduces the intensity of interference. Extrapolation from the in vitro observation suggests that this interference is likely to involve the ori-O interaction.

Our in vivo studies have demonstrated that the ori-O interaction is greatly stimulated by the presence of the ice-oop region. As this region is effective in trans configuration, the stimulation is likely to be mediated by a cytoplasmic factor(s).

And from the nucleotide sequence in the ice-oop region, this factor is probably not a protein.

The interference is directed toward not only λ phage but also the λdv plasmid carrying a complete set of replication genes and sites, and which is perpetuated in multicopies in the same cell (Miwa, personal communication). This type of interference may represent an important new element in autogeneous regulation in replication, because the appearance of excessive origins in such a system would hinder the burst of replication, and its reduction would allow the genomes to enter a round of replication. It should be noted that this regulatory system is independent of, and additional to, the autorepression system that also acts in autogeneous regulation of replication (30).

The exact mechanism of the interference remains to be elucidated in future, particularly its correlation with the λ O-ori interaction and the activation of λ ori region. Molecular studies on materials interacting in the process of initiation or inhibition of replication, in addition to the ori-O interaction, will provide clues to these problems.

ACKNOWLEDGMENTS

This work was supported by Scientific Research Grants from the Ministry of Education, Science and Culture of Japan.

REFERENCES

1. Ogawa, T., and Tomizawa, J., J. Mol. Biol. 38, 217 (1968).
2. Furth, M.E., McLeester, C., and Dove, W.F., J. Mol. Biol. 126, 195 (1978).
3. Furth, M.E., and Yates, J.L., J. Mol. Biol. 126, 227 (1978).
4. Hobom, G., Grosschedl, R., Lusky, M., Scherer, G., Schwarz, E., and Kössel, H., Cold Spring Harbor Symp. Quant. Biol. 43, 165 (1979).
5. Moore, D.D., Denniston-Thompson, K., Kruger, K.E., Furth, M.E., Williams, B.G., Daniels, D.L., and Blattner, F.R., Cold Spring Harbor Symp. Quant. Biol. 43, 155 (1979).
6. Tsurimoto, T., and Matsubara, K., Mol. Gen. Genet. in press (1981).
7. Matsubara, K., and Kaiser, A.D., Cold Spring Harbor Symp. Quant. Biol. 33, 769 (1968).
8. Tomizawa, J., in "The Bacteriophage Lambda" (A.D. Hershey ed.) p.549. Cold Spring Harbor Laboratory N.Y. (1971).
9. Georgopoulos, C.P., and Herskowitz, I., in "The Bacteriophage Lambda" (A.D. Hershey ed.) p.553. Cold Spring Harbor Laboratory N.Y. (1971).
10. Schnös, M., and Inman, R.B., J. Mol. Biol. 51, 61 (1970).
11. Dove, W.F., Hargrove, E., Ohashi, M., Haugli, F., and Guha, A., Japan J. Genet. 44 suppl. 1, 11 (1969).
12. Dove, W.F., Inokuchi, H., and Stevens, W.F., in "The Bacteriophage Lambda" (A.D. Hershey ed.) P.747. Cold Spring

Harbor Laboratory, N.Y. (1971).
13. Hayes, S., and Szybalski, W., Mol. Gen. Genet. 126, 275 (1973).
14. Lusky, M., and Hobom, G., Gene 6, 137 and 173 (1979).
15. Lipinska, B., Podhajska, A., and Taylor, K., Biochem. Biophys. Res. Commun. 92, 120 (1980).
16. Kuypers, S., Reiser, W., and Klein, A., Gene 10, 195 (1980).
17. Hirose, S., Okazaki, R., and Tamanoi, F., J. Mol. Biol. 77, 501 (1973).
18. Shalloway, D., Kleiberger, T., and Livingston, D.M., Cell 20, 411 (1980).
19. Szybalski,E.H., and Szybalski, W., Gene 7, 217 (1979).
20. Maxam, G., and Gilbert, W., Proc. Natl. Acad. Sci. USA 74, 560 (1977).
21. Galas, D.J., and Schmitz, A., Nucleic Acids Res. 5, 3157 (1978).
22. Ptashne, M., Jeffrey, A., Johnson, A.D., Maurer, R., Meyer, J.B., Pabo, C.D., Roberts, T.M., and Sauer, R.T., Cell 19, 1 (1980).
23. Johnson, A., Meyer, B.J., and Ptashne, M., Proc. Natl. Acad. Sci. USA 75, 1783 (1978).
24. Murotsu, T., and Matsubara, K., Mol. Gen. Genet. 179, 509 (1980).
25. McKay, R., and DiMaio, D., Nature 289, 810 (1981).
26. Myers, R.M., and Tjian, R., Proc. Natl. Acad. Sci. USA 77, 6491 (1980).
27. Shimatake, H., and Rosenberg, M., in EMBO workshop on protein-DNA interactions in bacteriophages (1980).
28. Taniguchi, T., O'Neill, M., and deCrombrugghe, B., Proc. Natl. Acad. Sci. USA 76, 5090 (1979).
29. Gray, H.B., Ostrander, D.A., Hodnett, J.L., Legerski, R.J., and Robberson, D.L., Nucleic Acids Res. 2, 1459 (1975).
30. Matsubara, K., Plasmid 5, 32 (1981).

ON THE ROLE OF RECOMBINATION AND TOPOISOMERASE IN PRIMARY
AND SECONDARY INITIATION OF T4 DNA REPLICATION[1]

Gisela Mosig, Andreas Luder, Lee Rowen,
Paul Macdonald and Susan Bock

Department of Molecular Biology
Vanderbilt University
Nashville, Tennessee 37235

ABSTRACT   We have shown that initiation of primary and of secondary replication forks of phage T4 requires some different functions. Primary initiation occurs in the absence both of recombination functions and of topoisomerase activity, while secondary intiation is largely dependent on both functions. In addition, we have shown that the preferred sites for primary initiation are different from initiation sites used later in the infection. We discuss these and other results within the framework of a model in which primary initiation from specific origin sequences is independent of recombination while secondary initiation requires formation of recombinational intermediates and is facilitated by topoisomerase activities.

Introduction
    There is ample evidence that replication of cellular chromosomes is tightly regulated to prevent or reduce amplification of specific DNA segments and to coordinate DNA replication with cell growth (22). The complex initiation mechanisms operating at origins of cellular replicons provide such control (25,63). Viral chromosomes, on the other hand, may not require all elements of this regulation; during lytic infections, when new replication forks accumulate rapidly, alternative mechanisms using different sites or structures may be used. Specifically, it has been postulated that in phages, recombination and initiation of DNA replication might

[1]This work was supported by NIH grants: GM 13221 to G.M. and a biomedical research support grant to Vanderbilt University.

be related (5,9,48, 49,58). Possible preferential replication of certain genome segments can subsequently be corrected and compensated for during packaging of the DNA into virions. Competition between alternative pathways of replication and maturation might introduce another level of regulation during viral growth.

Phage T4 infected cells are an excellent model system to investigate possible alternative initiation mechanisms, since genetics and biochemistry of the replication proteins are relatively well understood. T4 is particularly suited to investigate the possible role of recombination in replication, since T4 chromosomes are unusually recombinogenic (9,42). Approximately 20 genes are known to be involved in DNA replication in vivo (15,67). In contrast 7 purified proteins suffice to synthesize double-stranded DNA in vitro starting from nicks in the template (2,31,53,57). Five of these (DNA polymerase, gp 43; helix-destabilizing protein, gp 32; and three accessory proteins, gps 44, 62, and 45), the "five-protein-system", efficiently catalyze DNA elongation on primed single-stranded templates. Two additional proteins, gps 41 and 61,

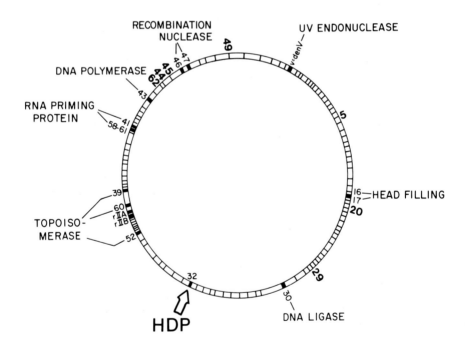

Fig. 1. Map of the phage T4 genome, showing the location of genes mentioned in this paper.

together catalyze synthesis of pentaribonucleotides (2,52,57) which resemble primers of Okazaki pieces found in vivo (29). (For map locations of T4 genes see Fig. 1).

De novo initiation at origin sequences in unnicked double-stranded DNA templates, however, has not yet been demonstrated in vitro. Presumably, initiation requires some or all of the more than 10 gene products which are involved in DNA replication in vivo but are not part of the present in vitro systems. It is noteworthy that most of the genes in the latter category are also involved in recombination (for review see ref. 9). To what extent are these processes related? Are recombination genes directly involved in initiation and, if so, are they required in all initiation? In fact, the numbers and locations of origin sequences that are used in vivo in the T4 genome are still a matter of controversy (11, 13,18,26,36,43,48).

To reconcile these apparently conflicting data and to explain a number of other observations, we have proposed a model in which T4 DNA replication is sequentially initiated by at least two different modes:

 1) from genetically determined bonafide origin sequences in duplex DNA and,
 2) from branched recombinational intermediates which are formed as soon as the first growing points have reached an end (49,50,50a). Our model is illustrated in Fig. 2.

The first round of DNA replication (primary replication) is initiated at an origin region (A). Because of the circular permutation of T4 DNA molecules (61), growing points initiated at any specific origin sequence must travel different distances before reaching an end. It has been proposed (7,64) that the terminus of the template for "lagging strand" synthesis at first remains unreplicated because of intrinsic priming problems (B). Our model proposes that such terminal single-stranded segments will invade homologous regions in other chromosomes to form branched recombinational intermediates (8,47a) (C). Because of circular permutation, such invasions will occur, in most cases, into the interior regions of other molecules. Rarely, an end may invade the terminal redundancy of the same molecule, generating a branched circle (C'), or a more complex structure if ends of other chromosomes also participate (see Figure 9). The branched recombinational intermediates can be resolved to (patch or splice) heteroduplexes (for review see Stahl, ref. 59). Alternatively, they can be converted to secondary replication forks (D and D'). (Both alternatives will finally yield genetic recombinants.) Reiteration of these processes will ultimately generate a highly branched network of DNA (20), whose complexity increases

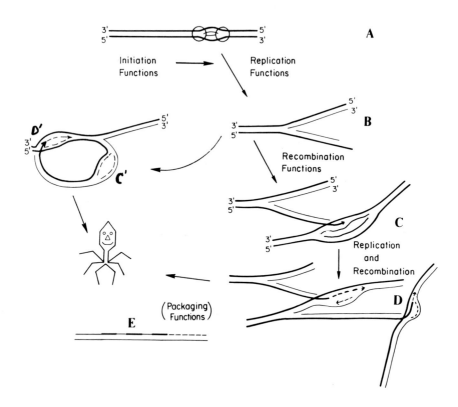

Fig. 2   A model for the pathway of T4 DNA after infection. The model postulates that primary initiation occurs at origin sequences, while secondary initiation can occur by conversion of recombinational intermediates to replication forks. Hot spots of recombination and preferred binding sites for T4 topoisomerase may behave as partial origins. Further details are explained in the text.

with increasing multiplicities of infection and with time (11,49). Late after infection, linear chromosomes are cut from such networks by the maturation process involving genes 16 and 17 (3,4,34,39) and the gene-49 maturation nuclease (14,23,35,40). Packaging restores a random population of circularly permuted and terminally redundant T4 chromosomes (61).

The key structures in this model are branched recombinational intermediates (Fig. 3) whose displaced single-stranded segments are mostly covered with gene-32 protein. Since gene-32 protein interacts both with DNA and with other proteins which act on DNA in different branches of this pathway, these interactions have pivotal regulatory roles at branch points of this pathway.

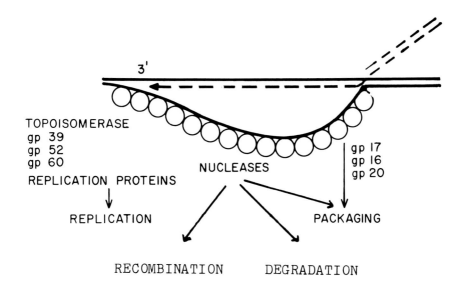

Fig. 3. A branched recombinational intermediate whose displaced single strand is covered with gene-32 protein (HDP).

Since several gene products compete for interactions with gene-32 protein on the recombinational intermediates (Fig. 3), interactions with certain products will facilitate while interactions with other proteins will prevent initiation of replication forks from such intermediates.

Allele-specific "conformational" suppression and "negative suppression" together with analysis of bypass mutations have revealed interactions between gene-32 protein and other proteins summarized in Fig. 4 (5,6,45 through 50a). Some of these interactions have also been observed in vitro (10,19,21).

We have tested several predictions of this model and we have asked which conditions selectively affect the primary or the secondary (recombinational) mode of initiation. To simplify the further discussion, we shall call all initiation (including reinitiation) from primary origins "primary (origin) initiation" and all other initiation "secondary (recombinational) initiation."

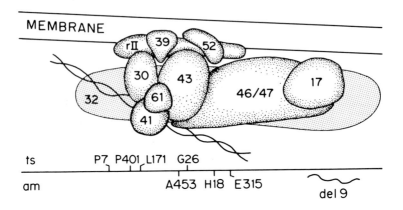

Fig. 4. Interactions of gene-32 protein with other proteins as deduced from genetic studies. A map of gene-32 mutations (<u>ts</u> mutations above and <u>am</u> mutations below the bottom line) is shown in the lower portion. Interactions of gene-32 protein with other gene products (indicated by corresponding gene numbers) are represented above that line. Fig. 1 shows the map positions of these genes. Gene 30: DNA ligase; gene 43: DNA polymerase; genes 47 and 47 control a recombination nuclease; genes 41 and 61: RNA priming protein; genes 39 and 52: T4 topoisomerase; genes rIIA and rIIB: membrane proteins; gene 17: DNA maturation function.

General Methods
    To arrest DNA replication and recombination at various intermediate states we have used appropriate single or multiple T4 mutants.
    To separate and analyze DNA arrested in these intermediate states we have used the following general protocol:
    1) infect with density and radioisotope labeled parental particles,
    2) permit replication in light medium containing a different radioisotope,
    3) lyse at different times after infection,
    4) separate, in appropriate density gradients, unreplicated T4 DNA from both progeny T4 DNA, which had or had not

recombined, and from bacterial DNA. In addition, we have
analyzed sedimentation properties of such DNA (5,6,36,44),

5) hybridize the separated DNA species to restriction
fragments of cytosine-containing T4 DNA by "Southern blotting"
(56). (Hydroxymethylcytosine-containing wild type T4 DNA
is not susceptible to most commonly used restriction enzymes;
cytosine-containing DNA produced by certain mutants, however,
is susceptible and restriction maps of such DNA have been
published, ref. 30.) In some experiments, T4 DNA was labeled
in vivo before isolation (50,50a). In other experiments
(e.g. Fig. 5) the T4 probe DNA was labeled by nick translation
(12) after it had been isolated,

6) as an alternative, we have viewed the DNA in the
electron microscope (11,33,48),

7) in addition, we have measured continuous (mainly
secondary) DNA synthesis by thymidine incorporation into
TCA precipitable material (46).

RESULTS

Our experiments ask three related questions: 1) Do
primary and secondary initiation use the same DNA sequences?
2) What is the role of T4 topoisomerase in initiation?
3) Are recombination functions required in primary or secondary
initiation?

1) <u>A preferred primary origin does not remain a preferred origin for subsequent initiations</u>.

Previous results had suggested that initiation of T4
DNA replication is poorly synchronized (50a) due to a slow
preinitiation step. We have tried to achieve better synchrony
by infecting with the gene-43 mutant P36 at the restrictive
temperature of 42°C, incubating for 15 minutes and then shifting the infected cells to the permissive temperature of 25°C.
This mutant synthesizes some DNA under restrictive conditions
(16) and we thought that it might accomplish, under these
conditions, a slow putative preinitiation step. Fig. 5C
shows that the DNA accumulated under restrictive conditions
hydridizes preferentially to a small Eco RI fragment. This
DNA is not covalently linked to parental strands and branch
migrates out of the template during the isolation procedure
(16). When replication is subsequently permitted for a short
time under permissive conditions, a few other Eco RI bands
are preferentially labeled (Fig. 5D). At least two of these
(1 and 2) are the same bands that are labeled first after
wild type infection (50). The sizes of these Eco RI fragments
are consistent with, but of course, do not unambiguously prove
their location on the T4 map (30) in the origin region that
we have found previously (36).

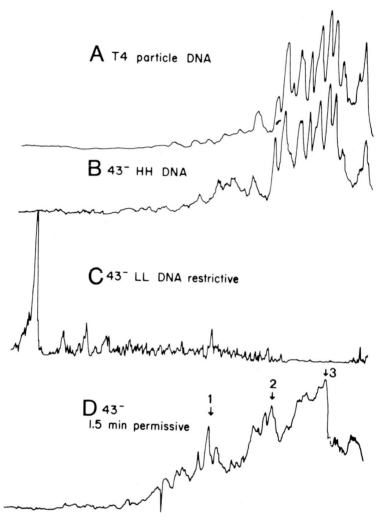

Fig. 5. Tracings of autoradiograms from Southern blots of DNA to T4 Eco RI restriction fragments. Replicating DNA was isolated from cells infected with the gene-43 mutant P36 and purified through Cs_2SO_4 gradients as described in the text. It was then nick-translated using ^{32}P-labeled dCTP (New England Nuclear). A) DNA extracted from wild type T4 particles. B) DNA banding at parental (heavy) density 15 minutes after infection with P36 at 42°C. C) Light DNA synthesized under restrictive conditions by this mutant. D) Replicated DNA of P36 infected cells after 15 min at restrictive, followed by 1.5 min at permissive temperature (25°C). Arrows 1 and 2 show peaks which are first labeled after infection with wild type T4 (50a).

Since T4 replication forks follow each other closely (65), the same genetic segments (i.e. Eco RI bands) should be preferentially labeled throughout the infectious cycle if all subsequent growing points were initiated at the same origins (62). Clearly, this is not the case. A few minutes after the onset of T4 DNA replication, the pattern of labeled intracellular DNA is virtually indistinguishable from the pattern of particle DNA shown in Fig. 5A.

2) <u>The T4 topoisomerase together with recombination functions facilitates initiation of secondary replication forks</u>.

T4 genes 39, 52, and 60, which together code for a type II topoisomerase (32,60) are involved in initiation of replication forks (37,38). Mutations in any of these genes delay, but do not eliminate rapid acceleration of T4 DNA replication without inhibiting primary replication (68). Primary replication occurs also (49) when these mutants infect E. coli S/6 under the most restrictive conditions for progeny production (51). We, therefore, asked if the corresponding gene products primarily facilitate the conversion of recombinational intermediates to replication forks. If so, secondary, but not primary initiation might be eliminated in these mutants if gene-32 protein or other recombination proteins were also mutationally altered. This is what we found with a gene-32 (HDP) mutation that allows some DNA replication in the single gene-32 mutant (45) (Figure 6). Additional gene-46 (recombination nuclease) mutations have similar effects (data not shown).

Conversely, our model (see Fig. 3) predicts that the topoisomerase functions become more dispensable when recombinational intermediates persist longer. Consistent with this prediction, mutations in the packaging gene 17 (Fig. 7) partially suppress a gene-39 mutation. The gene 17 <u>ts</u> mutation was originally selected as partial suppressor of a gene-32 mutation (50a), presumably because it reduces the affinity of gp 17 for gene-32 protein, a putative signal for packaging (see Fig. 3).

3) <u>Recombination functions together with topoisomerase functions can bypass gene 61 (priming protein) mutations</u>.

As mentioned above, the products of genes 61 and 41, together function as priming proteins (2,52,57) for initiation of Okazaki pieces <u>in vitro</u>. <u>In vivo</u>, gene 61 mutants initiate primary replication unidirectionally <u>de novo</u> (not from nicks) by displacement synthesis (33) (Fig. 8). These results show that the leading strand can be initiated in the absence of RNA priming protein, but that gp 61 is required for initiation of Okazaki pieces <u>in vivo</u>. In spite of this defect, however, gene 61 mutants eventually replicate both DNA strands repeatedly and produce substantial numbers of

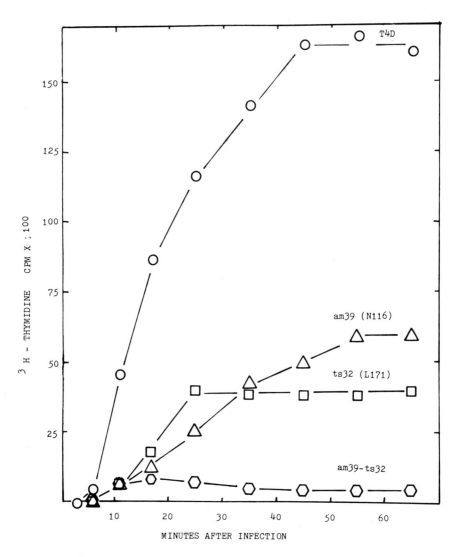

Fig. 6. ^{3}H-thymidine incorporation after infection of Su$^-$ bacteria with wild type T4 and the indicated mutants at 42°C. ^{3}H-thymidine was added 2.5 min after infection. The parental particles were labeled with ^{32}P, ^{13}C and ^{15}N. Final density shift of the parental DNA, measured in the same experiment was: in N116 120%, in L171 60% and in N116-L171 60% of the density shift under permissive conditions (Su$^+$ bacteria at 25°C). Fifty percent of the parental DNA of the double mutant was degraded to TCA soluble material, starting approximately 18 min after infection. Such degradation was not seen in the individual single mutants.

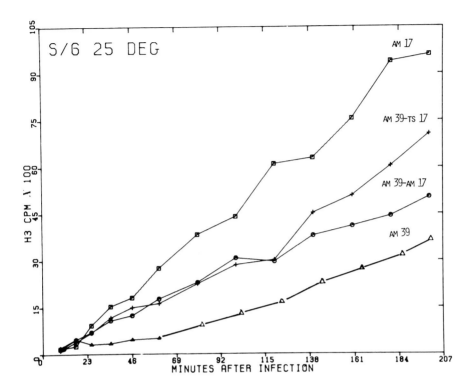

Fig. 7. ^{3}H-thymidine incorporation after infection of E. coli S/6 bacteria (Su$^-$) with gene 39 (N116) and with gene 39-gene 17 (N116-X6HE and N116 N56) double mutants. Note that these conditions are permissive for the gene-17 ts mutant X6HE, which was isolated at 25°C as a partial suppressor of a gene-32 mutant. Modified from Proceedings of the 7th Biennial Bacteriophage Assembly Meeting, Alan Liss, in press.

progeny particles. Neither the dnaG host primase nor the RNA polymerase substitute for the defective gp 61 in lagging strand initiation (33). How are the displaced template strands copied?

A modification of the model shown in Fig. 2 provides a ready explanation. When a growing point has reached the end of a DNA molecule, the displaced strand can invade another molecule to initiate a recombinational intermediate. From this intermediate both backward synthesis which copies the displaced strand in a continuous fashion and secondary initiation can occur. Again, as expected from this model, additional recombination deficient mutations completely abolish, and additional gene-52 (topoisomerase) mutations almost eliminate secondary replication, but do not interfere with primary synthesis of the leading strand in gene-61 mutants.

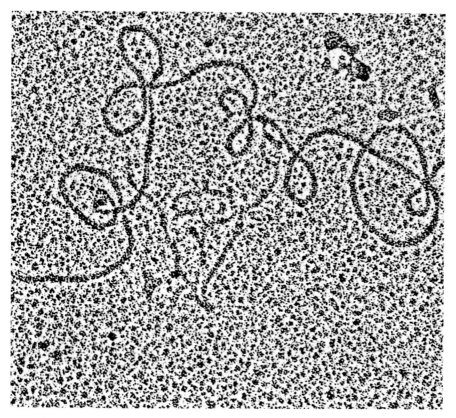

Fig. 8. Replicative intermediate isolated 10 min after infection of E. coli B with the gene-61 mutant E219. It shows a displacement loop.

DISCUSSION

We have shown that initiation of primary and of secondary replication forks requires some different functions. Primary initiation occurs in the absence both of recombination functions and of topoisomerase activity, while secondary initiation is largely dependent on both functions. Secondary initiation coincides with the appearance of highly branched networks of DNA. In addition, we have shown that the primary initiation sites do not remain the preferred initiation sites later in the infection (Fig. 5). Since T4 growing points follow each other in distances of approximately one tenth of a genome (65), our results suggest that most secondary growing points are initiated from different sites than primary growing points. This is consistent with the earlier observations that reinitia-

tion loops in replicative intermediates (11,13) are rare. The model shown in Fig. 2 readily provides a framework to understand all of our present results and other earlier observations. To simplify further discussion, we distinguish three steps in initiating DNA synthesis:
  I) Initiation of prereplicative synthesis,
  II) Primary initiation (recombination independent),
  III) Secondary initiation (recombination dependent).

Step I. Prereplicative synthesis (Fig. 2A, Fig. 5C) is at this time poorly understood. We do not know for sure whether this DNA is synthesized as part of the primary initiation process. Since prereplicative intermediates accumulate in certain gene 43 (DNA polymerase) mutants (Fig. 5C) and in certain gene 32(HDP)- gene 41(priming protein) double mutants (50a), it is, however, tempting to speculate that a short DNA segment can be copied without requiring proper interactions of DNA polymerase with all replication proteins. Subsequently, these interactions (e.g. between gps 43, 32, and 41) are required to open the template or to traverse a roadblock. It is presently unknown whether in this step gp 41 is required as priming protein or as a polymerase accessory protein (2).

Step II. Primary initiation presumably requires all replication proteins, but does not require functions or interactions with recombination proteins. In the absence of recombination functions or substrates, most, if not all chromosomes initiate replication at a single site and no complex branched structures appear (11,48). These results do not exclude the possibility that different chromosomes use different origins. We have, however, found that solitary incomplete T4 chromosomes from small T4 particles (41) initiate replication preferentially in a region counterclockwise from the DNA polymerase gene 43 (36,43). Consistent with this assignment, the DNA component of an RNA-DNA copolymer containing a putative primer of DNA replication, specifically transforms a gene 43 mutation (38a). On the other hand, multiple loops in replicating T4 chromosomes seen after multiple infection have been interpreted as multiple origins (13) and two origins have recently been mapped near genes 5 and 29 (18,26). It remains to be seen whether all of these origins can function in primary initiation (see discussion below).

Since T4 topoisomerase mutants initiate primary replication at the same time as wild type (48,68) but show delayed secondary initiation, we assume that the T4 topoisomerase is not required for primary initiation. We have been unable to block residual replication of topoisomerase mutants by inhibiting the host gyrase with novobiocin.

Step III. In contrast to primary initiation, secondary initiation requires recombinational intermediates and, in addition, it is facilitated by topoisomerase function. Topoisomerase mutants are recombinogenic (51). This suggests that topoisomerase is not required in the invasion step, but that it helps to channel recombinational intermediates into the replication pathway (Fig. 3). If recombinational intermediates are not made, no secondary initiation can occur. If their life time is prolonged (i.e. when packaging functions are defective), more secondary forks can accumulate even when T4 topoisomerase is defective (Fig. 7). As a first approximation, we consider the T4 topoisomerase as a facilitating protein in assembly of secondary replication forks analogous to the accessory proteins in phage morphogenesis (24). The cutting and unwinding functions of the T4 topoisomerase (Kreuzer and Alberts, this volume) are well suited for such a role. Any preferred binding sites for topoisomerase (as well as hot spots of recombination) would facilitate secondary initiation from preferred sites and might appear as "secondary" or "partial" origins. For all the reasons mentioned above, we believe that most, if not all multiple loops seen in wild type T4 chromosomes after multiple infection (11,14) (Figure 9) are initiated by the secondary (recombinational) mode.

All recombination deficient T4 mutants cease DNA replication prematurely and rather abruptly (9). This supports the earlier conclusion that there is little reinitiation from primary origin(s). Why is reinitiation later prevented? Luder's results (33) (Fig. 8) suggest one plausible explanation: Initiation of the leading strand at the primary origin is independent of the T4 priming proteins, and of the host dnaG protein. We assume it is primed by RNA polymerase. E. coli RNA polymerase is sequentially modified after T4 infection so that early promoters are not recognized later. The preferred origin region that we have identified is transcribed early and transcriptional activation and/or RNA priming from such regions are expected to cease at later times.

Secondary initiation involves most of the DNA-protein-protein interactions depicted in Fig. 4. In this context it is remarkable that most known gene-32 mutations, though they affect interactions with different proteins, map in the promoter proximal third of the gene (Fig. 4) (Mosig et al., 1977). In fact, they occur in the region which is also involved in binding to DNA (Williams et al., 1980; Krisch et al., 1980). This crowding of essential mutational sites in the gene and interaction sites on the protein implies that most of the potential interactions are competitive, and that they involve subtle allosteric modifications mediated by binding to DNA and

19 ON THE ROLE OF RECOMBINATION AND TOPOISOMERASE

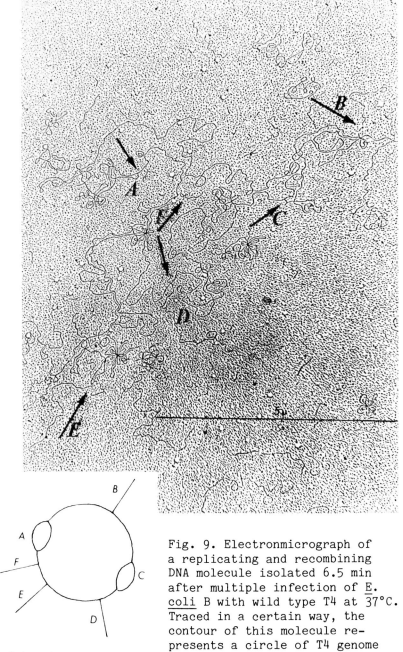

Fig. 9. Electronmicrograph of a replicating and recombining DNA molecule isolated 6.5 min after multiple infection of E. coli B with wild type T4 at 37°C. Traced in a certain way, the contour of this molecule represents a circle of T4 genome size (54 µm), with multiple branches which could have been formed when several other molecules invaded this circle (see Fig. 2C).

other proteins. Such subtle allosteric changes in response to different interactions together with variable concentrations of the participating components are exquisitely suited to coordinate the precise functioning, in time and space, of the many enzymes that have to act sequentially during initiation of replication forks.

We believe that secondary initiation from recombinational intermediates occurs in other systems as well. There is considerable circumstantial evidence that phages lambda (58) and T7 (54) might use this initiation mode. The recA dependent replication of E. coli (Lark et al. this meeting) may be related. In addition, current models of transposition of transposable elements invoke initiation of limited replication triggered by an intermediate in nonhomologous recombination (9a). Phage T4 might simply use recombinational initiation more frequently than other systems.

REFERENCES
(1) Alberts, B.M., and Frey, L.(1970).Nature 227, 1313.
(2) Alberts, B, Bedinger, B.P., Burke, L.R., Hibner, U., Liu, C.C., and Sheridan, R.(1980). In Mechanistic Studies of DNA Replication and Genetic Recombination, B. Alberts, ed. Academic Press, pp. 449.
(3) Black, L.W., and Silverman, D.J.(1978).J. Virol. 28, 643.
(4) Black, L.W., and Manne, V.(1980).J. Supramol. Structure, 4, 307.
(5) Breschkin, A.M., and Mosig, G.(1977).J. Mol. Biol. 112, 279.
(6) Breschkin, A.M., and Mosig, G.(1977).J. Mol. Biol. 112, 295.
(7) Broker, T.R.(1972).Ph.D. Thesis Stanford University, Stanford, California.
(8) Broker, T.R., and Lehman, I.R.(1971).J. Mol. Biol. 60, 131.
(9) Broker, T.R., and Doermann, A.H.(1975).Ann. Rev. of Genetics 9, 213.
(9a) Bukhari, A. (1981). Trends in Biochemical Sciences 6, in press.
(10) Burke, R.L., Alberts, B.M., and Hosoda, J.(1980). J. Biol. Chem. 255, 11484.
(11) Dannenberg, R.J.(1979).Ph.D. Thesis Vanderbilt University Nashville, Tennessee.
(12) Davis, R.W., Botstein, D., and Roth, J.R.(1980). Advanced Bacterial Genetics, Cold Spring Harbor Laboratory Press.
(13) Delius, J., Howe, C., and Kozinski, A.W.(1971).Proc. Natl. Acad. Sci. USA 68, 3049.
(14) Dewey, M.J., and Frankel, F.R.(1975).Virology 68, 387.

(15) Epstein, R.H., Bolle, A., Steinberg, C.M., Kellenberger, E., BoydelaTour, E., Chevalley, R., Edgar, R.S., Susman, M., Denhardt, G.H., and Lielaulis, A.(1963).Cold Spring Harbor Symp. on Quant. Biol. $\underline{28}$, 375.
(16) Garcia, G.M.(1978).Ph.D. Thesis Vanderbilt University Nashville, Tennessee.
(17) Gold, L. O'Farrell, P.Z., and Russel, M.(1976).J. Biol. Chem. $\underline{251}$, 7251.
(18) Halpern, M.E., Mattson, T., and Kozinski, A.W.(1979). Proc. Natl. Acad. Sci. USA $\underline{76}$, 6137.
(19) Hosoda, J., Burke, R.L., Moise, H., Kubota, I., and Tsugita, A.(1980). In Mechanistic Studies of DNA Replication and Genetic Recombination, B. Alberts, ed. Academic Press pp. 507.
(20) Huberman, J.A.(1969).Cold Spring Harbor Symp. Quant. Biol. $\underline{33}$, 509.
(21) Huberman, J.A., Kornberg, A., and Alberts, B.M.(1971). J. Mol. Biol. $\underline{62}$, 39.
(22) Jacob, F., Brenner, S., and Cuzin, F.(1963).Cold Spring Harbor Symp. Quant. Biol. $\underline{28}$, 329.
(23) Kemper, B., and Brown, D.T.(1976).J. Virol. $\underline{18}$, 1000.
(24) King, J.(1980). In Biological Regulation and Development, Vol. 2, Goldberger, ed. Plenum Press, pp. 101.
(25) Kornberg, A.(1980).DNA Replication, W.N. Freeman and Co., San Francisco.
(26) Kozinski, A.W., Ling, S.-K., Hutchinson, N., Halpern, M.E., and Mattson, T.(1980).Proc. Natl. Acad. Sci. USA $\underline{77}$, 5064.
(27) Krisch, H.M., Bolle, A., and Epstein, R.H.(1974). J. Mol. Biol. $\underline{88}$, 89.
(28) Krisch, H.M., Duvoisin, R.M., Allet, B., and Epstein, R.H.(1980). In Mechanistic Studies of DNA Replication and Genetic Recombination, B. Alberts, ed. Academic Press, pp. 517.
(29) Kurosawa, Y., and Okazaki, T.(1979).J. Mol. Biol. $\underline{135}$, 841.
(30) Kutter, E., O'Farrell, P., and Guttman, B.(1980). In Genetic Maps, S.J. O'Brien, ed., Vol. 1, Bethesda, Maryland, p. 33.
(31) Liu, C.C., Burke, L., Hibner, U., Barry, J., and Alberts, B.(1979).Cold Spring Harbor Symp. Quant. Biol. $\underline{43}$, 469.
(32) Liu, L.F., Liu, C.-C., and Alberts, B.(1979).Nature $\underline{281}$, 456.
(33) Luder, A.(1981).Ph.D. Thesis Vanderbilt University Nashville, Tennessee.
(34) Luftig, R.B., and Ganz, C.(1972).J. Virol. $\underline{10}$, 545.
(35) Luftig, R.B., Wood, W.B., and Okinaka, R.(1971).J. Mol. Biol. $\underline{57}$, 555.

(36) Marsh, R.C., Breschkin, A.M., and Mosig, G.(1971). J. Mol. Biol. 60, 213.
(37) McCarthy, D.(1979).J. Mol. Biol. 127, 265.
(38) McCarthy, D., Minner, C., Bernstein, H., and Bernstein, C.(1976).J. Mol. Biol. 106, 963.
(38a) McNicol, L.A.(1973).J. Virol. 12, 367.
(39) Minagawa, T.(1976).Virology 76, 234.
(40) Minagawa, T., and Ryo, Y.(1979).Mol. and General Genetics 170, 113.
(41) Mosig, G.(1963).Cold Spring Harbor Symp. Quant. Biol. 28, 35.
(42) Mosig, G.(1970).In Advances in Genetics 15, Academic Press, Inc., New York and London, pp. 1-53.
(43) Mosig, G.(1970).J. Mol. Biol. 53, 503.
(44) Mosig, G., and Werner, R.(1969).Proc. Natl. Acad. Sci. USA 64, 747.
(45) Mosig, G., and Breschkin, A.M.(1975).Proc. Natl. Acad. Sci. USA 72, 1226.
(46) Mosig, G., and Bock, S.(1976).J. Virol. 17, 756.
(47) Mosig, G., Berquist, W., and Bock, S.(1977).Genetics 86, 5.
(47a) Mosig, G., Ehring, R., Schliewen, W., and Bock, S. (1971). Molec. Gen. Genetics 113, 51.
(48) Mosig, G., Luder, A., Garcia, G., Dannenberg, R., and Bock, S.(1979).Cold Spring Harbor Symp. Quant. Biol. 43, 501.
(49) Mosig, G., Dannenberg, R., Ghosal, D., Luder, A., Benedict, S., and Bock, S.(1979).Stadler Symp. 11, 31.
(50) Mosig, G., Dannenberg, R., Benedict, S., Luder, A., and Bock, S.(1980). In American Society for Microbiology, Washington, D.C., **Microbiology 1980, pp.254.**
(50a) Mosig, G., Benedict, S., Ghosal, D., Luder, A., Dannenberg, R., and Bock, S.(1980). In Mechanistic Studies of DNA Replication and Genetic Recombination, B. Alberts, ed. Academic Press, pp. 527.
(51) Mufti, S., and Bernstein, H.(1974).J. Virol. 14, 860.
(52) Nossal, N.G.(1980).J. Biol. Chem. 255, 2176.
(53) Nossal, N.G., and Peterlin, B.M.(1979).J. Biol. Chem. 254, 6032.
(54) Powling, A., and Knippers, R.(1976).Molec. Gen. Genet. 149, 63.
(55) Rabussay, D., and Geiduschek, E.P.(1977). In Comprehensive Virology 8, Plenum Press, New York, pp. 1.
(56) Southern, E.M.(1975).J. Mol. Biol. 98, 503.
(57) Silver, L.L., Venkatesan, M., and Nossal, N.G.(1980). In Mechanistic Studies of DNA Replication and Genetic Recombination, B. Alberts, ed. Academic Press, New York, pp. 475.

(58) Skalka, A.(1974).In Mechanisms in Replication, Plenum Press, New York, pp. 421.
(59) Stahl, F.W.(1979). Genetic Recombination, W.H. Freeman and Co., San Francisco, CA.
(60) Stetler, G.L., King, G.J., and Huang, W.M.(1979). Proc. Natl. Acad. Sci. USA. 76, 3737.
(61) Streisinger, G., Emrich, J., and Stahl, M.M.(1967). Proc. Natl. Acad. Sci. USA. 57, 292.
(62) Sueoka, N.(1971).Genetics 68, 349.
(63) Tomizawa, J.I., and Selzer, G.(1979).Ann. Rev. Biochem. 48, 999.
(64) Watson, J.D.(1972).Nature New Biology 239, 197.
(65) Werner, R.(1969).Cold Spring Harbor Symp. Quant. Biol. 33, 501.
(66) Williams, K.R., LoPresti, M.B., Setoguchi, M., and Konigsberg, W.H.(1980).Proc. Natl. Acad. Sci. USA. 77, 4614.
(67) Wood, W.B., and Revel, H.R.(1976).Bacteriological Rev. 40, 847.
(68) Yegian, C.D., Mueller, M., Selzer, G., Russo, V., and Stahl, F.W.(1971).Virology 46, 900.

ANALYSIS OF SEVEN dnaA SUPPRESSOR LOCI IN ESCHERICHIA COLI

Tove Atlung

University Institute of Microbiology
University of Copenhagen
Copenhagen K, Denmark

ABSTRACT A genetic approach was used to study the function of the dnaA protein in initiation of chromosomal DNA replication. Spontaneous suppressor mutations for the temperature sensitive dnaA46 mutation were isolated to identify proteins with which the dnaA protein might interact and possibly the site of action of the dnaA protein in the oriC region. Seven loci for dnaA suppression were found by this selection. These loci have been mapped by Hfr and F⁻ matings and by Pl transductions. The dasC mutations have been mapped precisely by specialized transducing phages λilv and the rpoB mutations by specialized transducing phage λrifD18 and by complementation with plasmids carrying DNA segments derived from this phage. The rpoB mutations have been analysed in detail for allele specificity of suppression of seven different dnaA mutations.

INTRODUCTION

The replication of the bacterial chromosome is regulated at the step of initiation and requires de novo protein synthesis (Maaløe and Hanawalt, 1961; Lark et al., 1963) as well as transcription by the RNA polymerase (Lark, 1972). Initiation of replication takes place at oriC (v. Meyenburg et al., 1978; Yasuda and Hirota, 1977; v. Meyenburg and Hansen, 1980). Several components participating in the initiation process have been defined genetically by the isolation of dna mutants. These include at present mutations in the genes dnaA (Carl, 1970; Hirota et al., 1970; Abe and Tomizava,

1971; Wechsler and Gross, 1971; Beyersman et al., 1974), dnaB (Zyskind and Smith, 1977), dnaC (Carl, 1970; Wechsler and Gross, 1971), dnaP (Wada and Yura, 1974) and dnaI (Beyersman et al., 1974). There are strong indications that the dnaA protein participates in the regulation of initiation as well as in regulation of its own synthesis (Hansen and Rasmussen, 1977; Kellenberger-Gujer et al., 1978). The dnaA protein has been shown to act at an earlier step in the initiation proces than the dnaC protein and before or during the transcriptional step carried out by the RNA polymerase (Saitoh et al., 1975; Hannah et al., 1975; Zyskind et al., 1977) and RNA polymerase suppressor mutations of temperature sensitive dnaA mutations have been described by Bagdasarian and coworkers (1977, 1978). This work describes the isolation and mapping of suppressor mutations for the temperature sensitive dnaA46 mutation. It was carried out to define gene products and sites which might interact with the dnaA protein in initiation of chromosome replication.

TABLE 1
ESCHERICHIA COLI STRAINS

Strain	Genotype	Source/reference
FH116	K12 dnaA46 (thi thr leu thyA deoC lacY1 mal bglR)	derivative of CRT46 (Miller,1972)
TC152	K12 (thi argH metB his trp pyrE lac xyl tsx rpsL tna ilvY(b)) uhp	MM3ol(Masters,1977) mal+ (P1-)
TC187	K12 dnaA46 (a) (λ^+)	
CM734	K12 dnaA46 (thi metE his trp mtl ara gal lac tsx rpsL)	(Hansen and v. Meyenburg, 1980)
CM74o	K12 dnaA5 (c)	
CM746	K12 dnaA204 (c)	
CM748	K12 dnaA203 (c)	
CM782	K12 dnaA211 (c)	
CM9o3	K12 dnaA508 (c) (λ^+)	
CM9o5	K12 dnaA167 (b) (λ^+)	
WM448	B/r dnaA46 (leu proA trp his arg metB deoB lac gal HsdK12 rpsL)	(Beyersman et al. et al. 1974)
WM493	B/r dnaA5 (d) thyA	
WM433	B/r dnaA204 (d) thyA	
X10018	K12 rpoB2o3 rhoam trp tonB lacZam 8osupFTs	(Guarente and Beckwith, 1978)

(a) genotype of strain TC152 given in brackets
(b) The ilv mutation of TC152 was assigned to ilvY (Watson et al., 1979) by analysis of transduction with the specialized transducing phages λilv (see text).
(c) genotype of strain CM734 given in brackets
(d) genotype of strain WM448 given in brackets

RESULTS

The isolation of rpoB9o2 and rpoB917.

dnaA suppressor mutations - das (Wechsler and Zdzicnika, 1974) - were selected in an ilv⁻ bglR derivative of strain CRT46 (FH116, Table 1). The mutations giving rise to dnaA suppression in 8 out of 16 independent temperature resistant clones were shown to be 40 - 50 % cotransducible with argE. Two of these strains, TC2 and TC17 were analysed.

TC2 acquired rifampicin resistance (loo ug/ml) simultaneously with the dnaA suppression. These two phenotypes could not be separated in P1 transductions, they are more than 99% cotransducible, indicating that the suppressor mutation in TC2 is an rpoB mutation. It has been designated rpoB9o2.

The suppressor mutation in strain TC17 was also shown to reside in the gene for the beta subunit of the RNA polymerase. ColEl plasmids which carry segments of the rpoB rpoC region of the chromosome derived from λrifD18 (see fig. 1) reversed the dnaA suppression effect of the das mutation of TC17. The only common intact gene present on the chromosomal DNA on these plasmids is the rpoB cistron. The suppressor mutation of TC17 could therefore also be assigned to rpoB and was designated rpoB917.

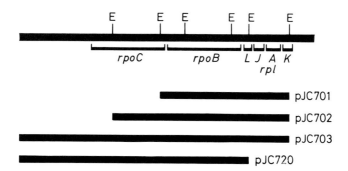

FIGURE 1. Plasmids carrying DNA segments from the rpoB region of the chromosome. The upper part shows the restriction enzyme EcoRI (E) site map of the chromosomal DNA carried on the plasmids and the location of the genes relative to this map (redrawn from Fiil et al., 1979). The lower part shows the extend of the chromosomal DNA carried on the plasmids which were derived from ColEl by insertion of EcoRI fragments of λrifD18 (Collins et al., 1976).

The second selection for dnaA suppressor mutations.

The strain - FH116 - used in the das selection described above was found to carry a cryptic dnaA suppressor mutation closely linked to ilv (see below). This mutation - designated dasC116 (see below) - suppresses the dnaA204 mutation and confers cold sensitivity of growth to dnaA46 carrying strains. To avoid any interference from this mutation I carried out a new selection for dnaA suppressor mutations using strain TC187 (table 1) which does not carry the dasC116 mutation. This strain also carries an appropriate set of markers for mapping by Hfr matings and is lysogenic for λcI^+ to allow mapping by specialized lambda transducing phages carrying cI857. Spontaneous temperature resistant colonies were selected at $42°C$ both on rich medium (LB) and minimal medium. Amongst 22o clones none was completely wild type with respect to growth and all were rifampicin sensitive. Twelve mutants with different growth properties (see Table 2) were chosen for mapping of the suppressor mutation.

TABLE 2
E. COLI STRAINS CONTAINING dnaA46 SUPPRESSORS (a)

Strain	Phenotypic characteristics	das allele	map position (b)
TC2	rifampicin resistant	rpoB9o2	89.5 min
TC17	rifampicin sensitive	rpoB917	89.5 min
TC388	rifampicin sensitive	rpoB908	89.5 min
TC379	cold sensitive	dasA379	80 min
TC387	rich medium sensitivity	dasB387	83.5 min
TC392	rich medium sensitivity	dasB392	83.5 min
TC382		dasC382	84.2 min
TC383		dasC383	84.2 min
TC371		dasG371	99 min
TC375	mucoid	dasF375	5 min
TC373	mucoid, slow growth	dasF373	5 min
TC387	mucoid, very slow growth	dasF387	5 min
TC377		dasF377	5 min
TC378		dasF378	5 min

(a) parental strain TC187 except for TC2 and TC17 where it was FH116 (Table 1)
(b) see text

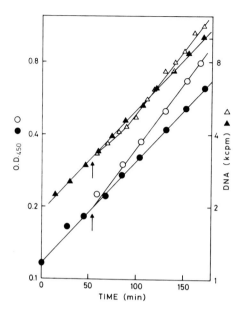
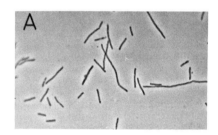
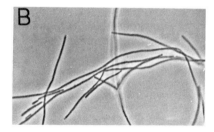

FIGURE2. Effect of a nutritional shift up experiment with the dasB392 strain. TC392 (λ^-) was grown exponentially at 33°C in AB glucose minimal medium (Clark and Maløe, 1967) supplemented with required amino acids and 10 ug/ml of ^{14}C-uracil (0.01 uCi/ug). At the time indicated by the arrow half of the culture was shifted up by addition of casamino acids (1% final conc.) OD450 and accumulation of DNA —measured as incorporation of ^{14}C-uracil into alkali stable TCA precipitable material (Hansen and Rasmussen) — was followed in the two parts of the culture. Samples for microphotography with a Zeiss light photomicroscope were also withdrawn from the cultures.
A: unshifted culture at 160 min
B: shifted culture 12o min after the shift.

Mapping of the dasA379 mutation.

Strain TC379 was found to be extremely cold sensitive; it did did not form colonies at temperatures below 39°C. However growth continued normally upon a shift down in temperature from 42°C to 33°C. DNA synthesis as well as cell division was stimulated by the shift but then reached a new balanced state with a slightly higher than normal DNA to mass ratio (data not shown). The cold sensitivity and the suppression phenotype is probably due to the same mutation as revertants which simultaneously acquired cold resistance and temperature sensitivity were frequently observed.

Hfr matings showed that the cold sensitivity of TC 379 is located in the 61 - 83 min region of the chromosome. F⁻matings narrowed the interval down to the xyl to pyrE region. The cold sensitivy is eliminated by both F⁻111 and F⁻14o (Low, 1973). A three factor cross with phage P1 showed the gene order to be dasA379 mtl pyrE. This allocates the dasA gene to 80 min on the genetic map, 30% cotransducibel with mtl.

Mapping of the dasB392 mutation.

Colony formation of the two suppressed strains TC392 and TC387 which had been selected on minimal medium was found to be inhibited by rich medium (YT). The sensitivity to rich medium appeared to be due to a division defect induced by nutritional shift up. Growth and DNA synthesis continued upon addition of casamino acids to a minimal glucose culture while long filaments with normal diameter were formed (see fig. 2). In the previous selection I had found a mutation conferring a similar phenotype which mapped between ilv and dnaA (data not shown).

The following P1 transductions were therefore carried out to map the das-392 mutation. A map according to Bachmann and Low (1980) of the ilv dnaA region is shown below:

uhp	dnaA tna	oriC	ilv
82	83		84 min

First a P1 lysate grown on TC392 was used to transduce the parental dnaA⁺ strain TC152 to uhp⁺; 60% of the transductants aquired the dnaA46 allele and 4% the phenotype of TC392, i.e. both the dnaA46 and das-392 alleles see below. In the next transduction an ilv⁺ wild-type strain was used as donor and TC392 as recipient; 68% of the ilv⁺ transductants were found to be temperature sensitive and rich medium resistant and 10% also recieved the dnaA⁺ allele. Last a P1 lysate grown on an ilv⁺ derivative of TC392 was used to transduce TC187 (the parental dnaA46 strain) to ilv⁺; 77%

of the transductants recieved the das-392 allele and became
temperature resistant and rich medium sensitive. Other P1
transductions had shown that uhp and ilv were less than 1%
cotransducible therefore the above transduction data show that
the das-392 mutation is located between ilv and dnaA in
the vicinity of oriC. It was designated dasB392.

The rich medium sensitivity conferred by dasB392 is
dependent on the presence of the dnaA46 mutation as TC392
was transduced to rich medium resistance by λtna phages
(Hansen and v. Meyenburg, 1979) only when they carried
the dnaA gene.

Mapping of the dasF locus.
Five dnaA suppressor strains (TC373, TC375, TC377, TC378
and TC387) yielded a high percentage of temperature sensitive
exconjugants in matings with HfrH selecting trp^+ (see fig. 3)
No Temperature sensitive arg^+ exconjugants were obtained
in matings with HfrR5. Thus the suppressor mutations of these
strains are located between the point of entry of HfrR5
at 3 min and trp at 28 min of the genetic map.

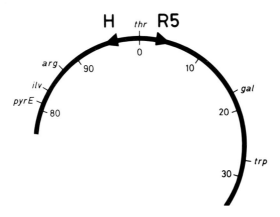

FIGURE 3. Map of the E. coli chromosome showing the
points of entry and the direction of transfer of HfrH (H)
and HfrR5 (R5) (according to Low, 1973). The position of a
few relevant markers are indicated (drawn according to Bachmann and Low, 1980).

P1 lysates were grown on these five strains and used to transduce the proA dnaA46 strain WM448 (see Table 1) selecting proA$^+$; in all cases 20 - 50 % cotransduction of temperature resistance with proA was observed, showing that the suppressor mutations of these five strains mapped close to proA. They have therefore all been assigned to a locus designated dasF.

Mapping of the dasG371 mutation and discovery of the cryptic dasE locus in strain TC187.

In Hfr matings with strain TC371 temperature sensitive exconjugants were obtained with high frequency both with HfrH selecting trp$^+$ and with HfrR5 selecting argH$^+$ indicating that the suppressor mutation was located in the 95 - 3 min interval defined by the points of entry of the two Hfr´s (see fig. 3).

In order to map this mutation more precisely derivatives of TC371 and its parent TC187 carrying selectable markers in the region (thr, pyrA, ara and leu) were constructed by P1 transduction. A map of the thr leu region of the chromosome according to Bachmann and Low (1980) is shown below:

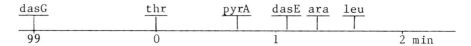

A P1 transduction with TC371 as donor and a thr::Tn5 derivative of TC187 as recipient (the parental dnaA46 strain) selecting thr$^+$ yielded 14% temperature resistant transductants showing 14% cotransducibility between thr and das-371 Another transduction was carried out in which a P1 lysate grown on a pyrA::Tn10 derivative of TC187 was used to transduce a leu::Tn5, ara derivative of TC371 to tetracycline resistance. No cotransduction of pyrA::Tn10 and the das$^+$ allele from the donor was found, while the expected cotransduction frequencies were found with ara (40%) and leu (17%). Other P1 transductions with these strains have shown 3% cotransduction of thr and ara and less than 0.3% cotransduction between thr and leu. As the das-371 mutation is 14% cotransducible with thr and less than 0.7% cotransducible with pyrA it must be located counterclocwise of thr at approximately 99 min. This locus was designated dasG.

Surprisingly I found that transduction of the above leu::Tn5 derivative of TC371 to leu$^+$ with P1 lysates grown on two other das K-12 strains gave rise to temperature sensitivity in 30 - 50 % of the leu$^+$ transductants. The genetic background therefore contains a mutation close to leu that is necessary for the dasG371 suppressor mutation to exert efficient suppression. This mutation was designated dasE187.

The P1 transduction shown in Table 3 was carried out to map this mutation more precisely. This transduction shows that the dasE locus is 30% cotransducible with leu and located between ara and pyrA as indicated by the absence of the three classes of transductants which would require a double crossover event.

That this mutation is a das mutation was shown when a P1 lysate grown on TC371 was used to transduce the dnaA46 strain FH116 to leu$^+$ and 35% of the leu$^+$ transductants were found to be temperature resistant.

TABLE 3
Mapping of the dasE locus by a four factor cross (a)

| | Genotypes | | | | no of trans- | % |
	leu	ara	dasE	pyrA	ductants	of total
Donor	+	+	+	−		
Recipient	−	−	−	+		
Transductants	1	0	0	0	56	49%
	1	1	0	0	23	20%
(b)						
	1	1	1	0	28	24%
	1	1	1	1	6	6
	1	0	1	1	0	0%
	1	0	0	1	0	0%
	1	1	0	1	0	0%
	1	0	1	0	0	0%
					115	100%

(a) A P1 lysate grown on NK6034 HfrH ΔlacproXIII pyrA::Tn10 was used to transduce a temperature resistant leu::Tn5 ara derivative of TC371 selecting leu$^+$ at 33°C. the transductants were screened for growth on arabinose as carbon source, temperature resistance of coloniformation at 42°C and tetracycline resistance.
(b) 1 indicates donor allele
     0 indicates recipient allele

Allocation of the suppressor mutation of TC388 to the rpoB region

One of the analysed twelve strains only TC388 became temperature sensitive when transduced to rifampicin resistance by λrifD18 (Collins et al., 1976), indicating that the suppressor mutation is in rpoB.

Precise mapping of the dasC locus by transducing phages λilv

Specialized transducing phages λilv carrying various segments of chromosomal DNA surrounding the ilvC gene (isolated by P. Jørgensen) have been used to map the suppressor mutations of TC382 and TC383. In a preliminary screening the twelve suppressor strains (Table 2) were transduced to ilv$^+$ by λilv28. The ilv$^+$ transductants were screened for temperature sensitivity due to introduction of the wild type allele of the suppressor mutation (the temperature sensitivity due to the dnaA46 mutation is reversible in contrast to the irreversible temperature sensitivity caused by induction of λcI857). Only the das mutations in TC382 and TC383 were transduced by λilv28. This suppressor mutation locus tightly linked to ilv was designated dasC.

The transducing phages λilv shown in figure 4 were used to map the dasC382 and dasC383 mutations more precisely. All the phages except λilv29 complement the ilvY mutation of TC382 and TC383. These two dasC strains were transduced with the λilv phages selecting ilv$^+$ and screened for reversible temperature sensitivity. The three phages λilv34, λilv41 and λilv29 all transduced the dasC mutations (20 - 50% cotransduction of ilv$^+$ and temperature sensitivity) while the other two phages λilv52 and λilv5 did not give any reversible temperature-sensitive transductants (less than 1% and 0.5% respectively).

Both dasC suppressor alleles are dominant over dasC$^+$ in a F´complementation test using F´14 (carrying ilv and metE but not dnaA) and recA derivatives of TC382 and TC383. The temperature-sensitive transductants obtained with the λilv phages must therefore represent reciprocal crossover events in the chromosomal DNA segment carried by the phages.

Thus the analysis of recombination of the λilv phages with the dasC mutations locates them to the left of the (common) point of insertion of λilv29 and λilv41 and to the right of the point of insertion of phages λilv52 and λilv5 (Fig. 4).

The dasC116 mutation mentioned above was also located to this region by transduction with the λilv phages.

20 ANALYSIS OF SEVEN dnaA SUPPRESSOR LOCI IN E. COLI

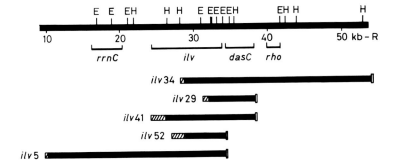

FIGURE 4. Specialized transducing phages λ carrying DNA from the ilv region of the E. coli chromosome.
The upper part shows the restriction enzyme cleavage map of the chromosomal DNA flanking the ilv genes. H: HindIII E: EcoRI cleavage sites. The kb scale has been adopted from v. Meyenburg and Hansen (1980) - kb-R indicates the position relative to oriC. The position of rrnC is according to v. Meyenburg et al., 1978 and of the ilv genes according to Watson et al., 1979 and Nargang et al., 1980 The lower part shows the extend of chromosomal DNA carried by the phages which was determined by analysis of restriction enzyme cleavage sites of HindIII, EcoRI and (not shown) XhoI, SalI, PstI and BglII. Bar (to the right) indicates the attachment site and the hatched area the region of crossover between chromosomal and lambda DNA.

Two relevant genes are located in this region of the chromosome: the rep gene which codes for one of the DNA helicases (Takahashi et al., 1977) and the rho gene which codes for the transcription termination factor rho (Roberts, 1969).

A detailed genetic and physical map of this region of the chromosome has been worked out by the construction of a set of pBR322 plasmids carrying restriction enzyme fragments from λilv34 and subsequent in vitro deletion analysis (T. Atlung and S. Brown, unpublished results). The rho gene was found to be located in the rigthmost part of the 6.9 kb HindIII fragment (35 kb-R to 42 kb-R) see Fig. 4, the N-terminal part of the gene being located at 4o kb-R (unpublished protein and DNA sequences, P. Klemm and S. Brown).

Thus the map position of the dasC suppressor mutations and the rho gene do not coincide.

TABLE 4
Range of temperature sensitivity of the dnaA strains (a)

Medium	Temp.	A46	A5	A204	A203	A211	A508	A167
				dnaA allele				
Min.(b)	37°C	+	+	+	+	+	+	+
	40°C	((+))	-	-	-	(+)	-	+
	42°C	-	-	-	-	-	-	((+))
YT (c)	37°C	+	+	+	+	+	+	+
	40°C	-	-	-	((+))	-	-	(+)
	42°C	-	-	-	-	-	-	-

(a) Ability to form colonies and the size of the colonies were scored after 2o hrs incubation at the indicated temperatures.
    +   normal sized colonies
  (+) small colonies
((+)) minute colonies
   -    no growth
(b) AB minimal medium (Clark and Måløe, 1967) containing o.2% glucose and required amino acids. (c) YT (Miller, 1972)

Allele specificity of rpoB suppressor mutations.

In order to determine whether the rpoB902 and rpoB917 could suppress other dnaA mutations than dnaA46, they were introduced into isogenic sets of dnaA strains by P1 transduction. This analysis required a thorough testing of the degree of temperature sensitivity of the recipient strains which is shown in Table 4. The order of the temperature sensitivity of the strains carrying the dnaA5, dnaA46 and the dnaA167 alleles was obtained by observing the ability to form colonies at different temperatures. It is in agreement with the results on the temperature sensitivity of DNA replication in strains carrying these three mutations (data not shown).

The rpoB902 mutation was found to suppress the temperature sensitivity of 4 of the tested 7 dnaA mutations (see Table 5) It gave the highest degree of suppression of the dnaA5 strain, which is the most temperature sensitive of the four suppresible strains and it only moderately raised the temperature resistance of the dnaA167 strain, the least temperature sensitive of the strains. Thus the differences in temperature resistance of the suppressed strains did not simply reflect differences in the temperature sensitivity of the dnaA strains. When spontaneous mutations to rifampicin resistance were selected in these dnaA strains, mutations

TABLE 5
Allele specificity of suppression of dnaA mutations
by rpoB mutations (a)

rpoB allele			dnaA allele				
	A46	A5	A2o3	A2o4	A211	A508	A167
rpoB9o2 (b)	+	++	−	−	+	−	(+)
rpoB917 (c)	+	+	NT	−	NT	NT	NT
rpoB2o3 (b)	−	−	−	−	−	−	−

(a) 4 isolates of each rpoB dnaA strain constructed as described below were screened for colony formation and growth as described in Table 4.
++, +, (+) and − indicates degree of temperature resistance:
++ > 4 C increase in nonpermissive growth temperature
+ 2-4 C increase in nonpermissive growth temperature
(+) 1-2 C increase in nonpermissive growth temperature
− no increase in nonpermissive growth temperature
NT Not tested
(b) an argH::Tn10 muation was introduced into the dnaA strains CM734 − CM9o5 (see Table 1). Subsequently the rpoB mutations were introduced by transduction with P1 lysates grown on TC2 and X10018 (Table 1) selecting arg and screening for rifampicin resistance.
(c) The rpoB917 mutation was introduced into rifampicin resistant derivatives of WM448, WM493 and WM433 (Table 1) by P1 transduction using TC17 as donor selecting metB and screening for rifampicin resistance.

were obtained which simultaneously conferred temperature resistance also in the three strains which could not be suppressed by rpoB902 (I found f. ex. that 20% of the mutations to rifampicin resistance obtained with the dnaA508 strain (CM903) conferred temperature resistance). This shows that all dnaA mutations can be suppressed by rpoB mutations.

The rpoB917 mutation also shows allele specificity like the rpoB902 mutation (Table 5). It suppressed the dnaA46 and the dnaA5 mutations but not thednaA204 mutation.

The rifampicin resistant rpoB203 rho suppressor allele (Guarente and Beckwith, 1978) was also introduced into this set of dnaA strains. I did not observe any suppression of the temperature sensitivity of any of the strains, but the presence of the rpoB203 mutation caused very slow growth in all of the dnaA strains, while it had no effect on growth of dnaA$^+$ strains.

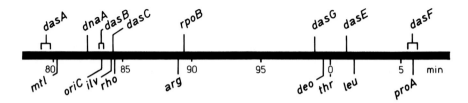

FIGURE 5. Map of the 80 to 5 min region of the chromosome summarizing the map positions of the das mutations and showing the position of a few other relevant genes (according to Bachmann and Low, 1980).

DISCUSSION

This study has lead to the identification of 7 genetic loci for spontaneous mutations resulting in suppression of the temperature sensitivity of a dnaA mutant (summarized in fig. 7). The selection for these suppressors was carried out in two strains with different genetic backgrounds both carrying the dnaA46 allele. No true revertants have been found in this work suggesting that the temperature sensitivity of the dnaA46 allele may not be a single-site mutation. Subsequent to the selections both genetic backgrounds were found to carry cryptic dnaA suppressor mutations – FH116 carries a dasC mutation and TC187 carries the dasE mutation. Amongst 12 dnaA suppressor mutant strains, preliminarily classified on the basis of growth characteristics, 7 different genetic loci, 3 of which were represented by a single mutation, were defined Additional genetic loci may turn out to be potential dnaA suppressor sites.

The selections have lead to the isolation of new rpoB suppressor mutations. The characterization of these mutations with respect to rifampicin sensitivity and allele specificity indicate that they are different from those described previously (Bagdasarian et al., 1978) In another study Wechsler and Zdzicnika (1974) have identified four classes of dnaA suppressor mutants. The published data on the map positions of their das mutations cannot easily be compared with my data, it seems likely however that some of the das loci defined in my work correspond to some of their classes of das mutations which are also clustered in the oriC region of the chromosome. The dnaA suppressor mutations - sin - isolated by G. Lindahl (unpublished results) map between metD and proA and are thus probably identical to the dasF mutations. described here (Fig. 6).

A comparison of the map positions of the das loci with genes and sites known to be involved in DNA replication shows that:
- the dasB locus might identify the site of action of the dnaA product within the origin; the present mapping data on dasB however do not allow any final conclusion at this point.
- dasE cannot be identical to dnaK or dnaJ as these these genes map at 0.3 min between thr and pyrA on the genetic map (Georgopoulos, 1977; Saito and Uchida, 1977) while dasE maps at 1.2 min between pyrA and ara.
- the dasG mutation could be a mutation in dnaC or dnaT (Bachmann and Low, 1980; Lark, 1978) which like dasG map at 99 min. The latter possibility should in particular be considered as one the other suppressor mutations (rpoB917) confers stable replication phenotype (unpublished results).
- the dasC mutations probably affect either the rep protein (Takahashi et al., 1977) or one of the other two gene products which are encoded by the DNA segment (unpublished results) where the dasC mutations are located.

The two RNA polymerase suppressor mutations rpoB902 and rpoB917 tested in this work exhibit allele specificity of suppression of dnaA mutations. The lack of suppression of some dnaA alleles by these rpoB alleles do not reflect an inability of a certain class of dnaA mutations to be suppressed by mutations in the beta subunit of the RNA polymerase since rifampicin resistant rpoB suppressor mutations could be found for all of the seven dnaA mutations.

The allele specificity of the rpoB suppressor mutations suggests that the suppression in this case is exerted by an interaction between the dnaA geneproduct - a 54 kD protein (Hansen and v. Meyenburg, 1979) - and the RNA polymerase, and not via bypassing of the requirement for dnaA protein in initiation of chromosome replication.

Allele nonspecific suppression of dnaA mutations by rpoB mutations has been suggested by Bagdasarian and coworkers (1978). The descrepancy between their conclusion and mine may only be an apparent one if the dnaA46, dnaA47 and dnaA177 alleles, which they have tested, are as similar in their behaviour as I found the dnaA46, dnaA5 and dnaA211 alleles to be. Alternatively there may exist both allele specific and nonspecific dnaA suppressor in rpoB, similarly to the described rho suppressor mutations in rpoB (Guarente and Beckwith, 1978; Guarente et al., 1979; Das et al., 1978).

The dnaA mutations affect expression of the trp operon; the degree of readthrough at the attenuator (trpAt) is increased in the dnaA mutants at elevated temperatures (unpublished results). The rpoB902 mutation also affects attenuation at trpAt giving decreased trp operon expression. This suggests that the dnaA protein plays a role as transcription termination factor at trpAt and that the rpoB902 mutation enables the RNA polymerase to terminate more efficiently at rho independent termination signals like the trpAt.

The rpoB203 mutation, which enables the RNA polymerase to terminate at rho dependent termination sites in the absence of active rho protein (Guarente and Beckwith, 1978) and endows the RNA polymerase with increased termination activity in vitro (Farnham and Platt, 1980), was found to be unable to suppress any of the dnaA mutations. The dasC suppressors are located apart from rho although close to this gene. Thus no indications for an involvement of rho in initiation of chromosome replication has been found in this study of dnaA suppressors.

The diversity of dnaA suppressor loci and the mutual interdependence of some of them (dasG and dasE) suggest that the dnaA protein is a constituent of a multicomponent complex - possibly located in the membrane - which interacts with the RNA polymerase in the transcriptional step of initiation of chromosome replication at oriC.

ACKNOWLEDGMENTS

I am very grateful to all those who have been helpful in supplying bacterial strains, advice and suggestions particularly Flemming Hansen, Kaspar von Meyenburg, Knud Rasmussen, Martin Pato Stanley Brown, Walter Messer. The expert technical assistance of Helle Frisk is gratefully acknowledged. This work was supported by a grant from the Danish Natural Science Research Council (J. no 511-20624).

REFERENCES

Abe, M., and Tomizawa, J. I. (1971). Genetics 69, 1.
Bachmann, B. J., and Low, K. B. (1980). Microbiol. Rev. 44, 1.
Bagdasarian, M. M., Izakowska, M., and Bagdasarian, M. (1977). J. Bacteriol. 130, 577.
Bagdasarian, M.M., Izakowska, M., Natorff, R. and Bagdasarian M. (1978) In "DNA Synthesis, Present and Future." Ed. by Molineux, I. and Kohiyama, M. 101.
Beyersmann, D., Messer, W., and Schlicht, M. (1974). J. Bacteriol. 118, 783.
Carl, P. L. (1970). Mol. Gen. Genet. 109, 107.
Clark, D. J. and Maaløe, O. (1967). J. Mol. Biol. 23, 99.
Das, A., Merril, C. and Adhya, S. (1978). Proc. Nat. Acad. Sci. 75, 4828.
Collins, J., Fiil, N. P., Jørgensen, P. and Friesen, J. D. (1976) in "Control of Ribosome Synthesis." Alfred Benson Symp. IX, Munksgaard, Copenhagen, 356.
Farnham, P. J. and Platt, T. (1980). Cell 20, 739.
Fiil, N. P., Bendiak, D., Collins, J. and Friesen, J. D. (1979). Molec. Gen. Genet. 173, 39.
Georgopoulos, C. P. (1977). Molec. Gen. Genet. 151, 35.
Guarente, L. (1979). J. Mol. Biol. 129, 295.
Guarente, L. and Beckwith, J. (1978). Proc. Nat. Acad. Sci. 75, 294.
Hanna, M. H., and Carl, P. L. (1975). J. Bacteriol. 121, 219.
Hansen, F. G., and von Meyenburg, K. (1979). Mol. Gen. Genet. 175, 135.
Hansen, F. G., and Rasmussen, K. V. (1977). Mol. Gen. Genet. 155, 219.
Hirota, Y., Mordoh, J., and Jacob, F. (1970). J. Mol. Biol. 53, 369.
Kellenberger-Gujer. G, Podhajska, A. J. and Caro, L. (1978). Molec. Gen. genet. 162, 9.
Lark, K. G. (1972). J. Mol. Biol. 64, 47.
Lark, K. G., Repko, T., and Hoffman, E. J. (1963). Biochim. Biophys. Acta 76, 9.
Lark, C. A., Riazi, J. and Lark, K. G. (1978). J. Bacteriol. 136, 1008.
Low, B. (1973). J. Bacteriol. 113, 798.
Maaløe, O., and Hanawalt, P. C. (1961). J. Mol. Biol. 3, 144.
Masters, M. (1977). Molec. Gen. Genet. 155, 197-202.
von Meyenburg, K., Hansen, F. G., Nielsen, L. D., and Riise, E. (1978). Molec. Gen. Genet. 160, 287-295.
von Meyenburg, K. and Hansen, F. G. (1980). In ICN-UCCLA Symp. on Molecular and Cellular Biology Vol. XIX., 137.

Miller, J. H. (1972). Cold Spring Harbor Laboratory, New York.
Nargang, F. E., Subramahnyam, C. S. and Umbarger, H. E. (1980) Proc. Nat. Acad. Sci. 77, 1823.
Roberts, J. W. (1969). Nature London 224, 1168.
Saito, H. and Uchida, H. (1977). J. Mol. Biol. 113, 1.
Saitoh, T. and Hiraga, S. (1975). Molec. Gen. Genet. 137, 249.
Takahashi, S. Hours, C and Denhardt, D. (1977). FEMS Microbiol Letters 2, 279.
Wada, C. and Yura, T. (1974) Genetics 77, 199.
Watson, M. D., Wild, J. and Umbarger, H. E. (1979). J. Bacteriol. 139, 1014.
Wechsler, J. A., and Gross, J. D. (1971). Mol. Gen. Genet. 113, 273.
Wechsler, J. A, and Zdzicnika, M. (1974). ICN-UCCLA Symp. on Molecular and Cellular Biology 3, 624.
Yasuda, S., and Hirota, Y. (1977). Proc. Nat. Acad. Sci. 74, 5458.
Zyskind, J. W. and Smith, D. W. (1977). J. Bacteriol. 129, 147
Zyskind, J. W., Deen, L. T., and Smith, D. W. (1977). J. Bacteriol. 129, 1466.

SUPPRESSION OF AMBER MUTATIONS IN THE dnaA GENE OF
Escherichia coli K-12 BY SECONDARY MUTATIONS IN rpoB.

Nancy A. Schaus[1], Kathy O'Day and Andrew Wright

Department of Molecular Biology
and Microbiology
Tufts University School of Medicine
Boston, Massachusetts

ABSTRACT

Three amber mutations, designated dnaA311, dnaA366, and dnaA91 have been isolated. These mutations were shown by both genetic and biochemical criteria to be amber mutations in the dnaA gene (Schaus etal., 1981). The ability of secondary mutations in rpoB to suppress the initiation defect of the amber mutant strains was examined. Of the three amber mutations tested, only dnaA311 can be suppressed by secondary mutations in rpoB under Su⁻ conditions. In addition, rpoB mediated suppression is clearly allele specific. These results suggest that the dnaA product is required for rpoB suppression.

INTRODUCTION

The initiation of DNA replication in Escherichia coli K-12 requires the product of the dnaA gene (Hirota, 1970; Wechsler, 1971). Conditional mutant strains, defective in dnaA function at the non-permissive temperature, are blocked in the initiation event, either before or during a rifampicin sensitive step (Zyskind, 1977; Kung, 1978). The initiation defect of dnaA mutant strains can be suppressed by the integration of a variety of replicons into the bacterial chromosome (termed integrative suppression), as well as by secondary mutations which map at several distinct loci on the

---

[1]Present address: Department of Molecular Biophysics and Biochemistry, Yale University, New Haven, Connecticut

E. coli chromosome (Nishimura, 1971; Nishimura 1973; Lindahl, 1971; Chesney, 1978; Wechsler, 1975; Bagdasarian, etal., 1977). One class of secondary mutations which suppress the initiation defect of several temperature sensitive dnaA mutant strains map in rpoB, the gene for the β-subunit of RNA polymerase (Bagdasarian, etal., 1977; Bagdasarian, etal., 1978; Felton & Wright, 1979). Several models have been proposed for the mechanism of suppression of the dnaA defect by secondary mutations in rpoB (Bagdasarian, etal., 1977; Bagdasarian etal., 1978). These models fall into two general classes in which the mutant RNA polymerse in one case bypasses the requirement for the dnaA product and in the other case, allows the mutant dnaA product to function.

We have recently described the isolation of amber mutations in the dnaA gene of Escherichia coli K-12 (Schaus etal., 1981). Thirty six initiation defective mutant strains were identified based on their dependence on P2sig5 integrative suppression. The defect in three of these mutant strains was shown to be due to an amber mutation. The amber mutations, designated dnaA311, dnaA366, and dnaA91 were identified by use of a strain, M7, which carries a temperature sensitive amber suppressor mutation, supF81(Ts). Spontaneous DnaA$^+$ revertants arose at frequencies between 10^{-6} and 10^{-7}, indicating that the initiation defect in the amber mutant strains most likely results from a single mutation. Mapping data, complementation analysis and the analysis of polypeptides produced by λ specialized transducing phages carrying the amber mutations provides strong evidence that the amber mutations are in the dnaA gene.

Amber mutations in dnaA may be useful to distinguish between the two basic models described above for rpoB suppression since at most only a fragment of the dnaA polypeptide will be present under non-suppressing conditions. We report here the results of experiments designed to determine whether the initiation defect in the amber mutant strains is suppressed by secondary mutations in rpoB.

MATERIALS AND METHODS

Bacterial Strains:

All strains are E. coli K-12 derivatives and are listed in Table 1.

TABLE I - BACTERIAL STRAINS

Strain	Chromosomal Markers	Source
148N	ΔlacX74 thi gyrA bfe rpoB148	M. Malamy
AW98	ΔlacX74 bglR thi gyrA dnaA46(Ts)	a
KO761	Δlac proA trp(Am) his thi tna::Tn10 supF81(Ts) rpsL tsx dnaA91(Am)	Schaus etal., 1981
CY15014	trpR rpoB2	Yanofsky and Horn, 1981
CY15015	trpR rpoB6	Yanofsky and Horn, 1981
CY15022	trpR rpoB7	Yanofsky and Horn, 1981
CY15023	trpR rpoB8	Yanofsky and Horn, 1981
NS387	same as KO761, except dnaA311(Am)	Schaus etal., 1981
NS388	same as KO761, except dnaA366(Am)	Schaus etal., 1981
NS389	Δlac gyrA supF58 tna::Tn10 dnaA366(Am)	a
NS449	lacY trpB9700 trp::Tn9 thr leu thi tna::Tn10 tonA21 supE44 dnaA508(Ts)	a
NS508	Δlac proA his thi bglR trpB9700 trp::Tn9 supF81(Ts) rpsL tsx argE::Tn10 dnaA311(Am)	a
NS509	same as NS508, except dnaA366(Am)	a
NS510	same as NS508, except dnaA46(Ts)	a
NS511	same as NS508, except dnaA91(Am)	a
NS520	same as AW98 plus rpoB321	a

a - Strains constructed in this laboratory.

Construction of rpoBdnaA mutant strains:

The rpoB148 mutation was introduced into the dnaA amber mutant strains and into the conditional dnaA mutant strains by P1vir transduction at 30°, using the donor strain 148N, and recipients NS387, NS388, NS389, KO761, AW98 and NS449 (Table 1). Selection was for rifampicin resistant growth. The presence of rpoB148 was verified by the cold sensitive phenotype (20°C) conferred by the rpoB148 mutation, and in some cases by cotransduction with a donor mutation which specifies resistance to bacteriophage BF23. This mutation (bfe) is approximately 60% cotransducible with rpoB148. Alternatively, the presence of the rpoB148 allele was confirmed by suppression of the T7 H3 gene1Am342 phenotype in suppressor minus, rifampicin resistant transductants (M. Malamy, personal communication),

The rpoB321, rpoB2, rpoB6, rpoB7, and rpoB8 mutations were introduced into the mutant strains at 30° using P1vir propagated on the donor strains NS520, CY15014, CY15015, CY15022 and CY15023 respectively and the recipient strains NS508, NS509, NS510, and NS511. The recipient strains carry Tn10 inserted in arg, a gene which is approximately 50% cotransducible with rpoB. Arg$^+$ transductants were selected and tested for rifampicin resistant growth on L medium plus rifampicin (50μg/ml).

RESULTS

Properties of the rpoB alleles

The rpoB alleles used in this study have several properties which could be relevant to their ability or lack of ability to suppress dnaA mutations. All are spontaneous mutations which confer rifampicin resistance on their host strains. The rpoB148 mutation suppresses the growth defect of a mutant of phage T7 (T7 H3 gene1Am342; Simon and Studier, 1973) which fails to produce the T7 - specific RNA polymerase. This suggests that the rpoB148 mutation alters the host polymerase specificity with respect to T7 transcription. Since the host RNA polymerase normally terminates transcription at the end of the T7 early genes, the rpoB148 mutation may alter the polymerase so that it no longer recognizes this termination signal. The rpoB7 and rpoB8 mutations lead to increased transcription termination at the tryptophan operon attenuator site (Yanofsky and Horn, 1981); like rpoB148, they also suppress the growth defect of T7 H3 gene1Am342. Mutations rpoB2 and rpoB6, in contrast to rpoB7 and rpoB8, lead to increased readthrough at the trp attenuator (C. Yanofsky, personal communication). Strains which carry these mutations are non-permissive for T7 H3 gene1Am342 growth (O'Connor and Malamy, personal communication). rpoB321 is a spontaneous mutation which was isolated by its ability to suppress the thermosensitivity of a dnaA46(Ts) mutant strain.

Allele specificity of rpoB suppression

In order to test the ability of the above rpoB alleles to suppress the dnaA mutations, they were introduced by P1 transduction into strains containing the amber mutations dnaA311, dnaA366, and dnaA91, and the temperture sensitive mutations dnaA46 and dnaA508 (Materials and Methods). Strains carrying the dnaA amber mutations contain a temperature-sensitive amber suppressor mutation, supF81(Ts), and are thus unable to grow at 40°. The dnaA rpoB mutant strains were tested for the ability to grow at the high temperature. As can be seen in Table 2, the presence of either the rpoB148 or the rpoB321 mutation suppresses the thermosensitivity of the conditional dnaA46(Ts) mutant strain. In contrast, the conditional dnaA508(Ts) mutant strain was not suppressed by any of the rpoB mutations tested (Table 2, line 2).

TABLE II

Suppression of dnaA Mutations by Mutations in rpoB

Relevant Genotype	Growth 40°					
	rpoB148	rpoB321	rpoB7	rpoB8	rpoB2	rpoB6
dnaA46(Ts)	+	+	-	-	0	0
dnaA508(Ts)	-	-	nt	nt	nt	nt
dnaA311(Am) supF81(Ts)	+	+	+	+	-	-
dnaA366(Am) supF81(Ts)	-	-	-	-	0	0
dnaA366(Am)supF58	+	nt	nt	nt	nt	nt
dnaA91(Am)supF81(Ts)	-	0	-	-	0	0

+, indicates growth on L medium; -, failure to grow on L medium; nt, not tested; 0, the indicated rpoB alleles could not be transduced into the indicated dnaA mutant strain.

The thermosensitivity of the dnaA311(Am) supF81(Ts) mutant strain was suppressed by rpoB148, rpoB7, rpoB8 and rpoB321. Suppression of the dnaA311(Am) defect by rpoB148 occurred even in the absence of the amber suppressor mutation supF81(Ts), since strain NS453 (NS387rpoB148) could be transduced to supF$^+$ (ie. Su$^-$) (data not shown). In contrast, the dnaA366 supF81(Ts) and dnaA91 supF81(Ts) mutant strains were not suppressed by any of the rpoB mutations tested.

Suppression of the dnaA366 amber mutation by the amber suppressing mutation supF58 (SuIII$^+$) leads to the production of a thermosensitive dnaA product (Schaus etal., 1981). As can be seen in Table 2, the thermosensitivity of a dnaA366(Am) supF58 mutant strain is suppressed by the rpoB148 mutation.

To determine whether mutations in rpoB could be isolated which would suppress the initiation defect in the dnaA366(Am) and dnaA91(Am) mutant strains, spontaneous rifampicin resistant derivatives of NS387 (dnaA311), NS388 (dnaA366) and KO761 (dnaA91) were isolated. Of 308 rifampicin-resistant derivatives of strain NS388 (dnaA366) and 520 rifampicin resistant derivatives of strain KO761 (dnaA91) none were able to grow at 40° (Table 3).

TABLE III

Frequency of Thermoresistant Colonies among Spontaneous Rifampicin Resistant Derivatives of dnaA Mutant Strains

Strain	Relevant Genotype	Experiment Number	No. of Colonies Rif^R	Rif^{R}40°^{R}
NS387	dnaA311(Am)supF81(Ts)	1	88	1[a]
		2	44	3
NS388	dnaA366(Am)supF81(Ts)	1	308	0
KO761	dnaA91(Am)supF81(Ts)	1	520	0

The dnaA mutant strains were grown to stationary phase in L medium at 30° and then plated on L medium plus rifampicin at 30°. Spontaneous rifampicin resistant mutants were tested for temperature resistant growth on L medium at 40°.

[a]The presence of the dnaA311(Am) mutation in 40°^{R} rifampicin-resistant derivatives of NS387 was verified by P1vir mediated transduction.

In the case of NS387 (dnaA311), 4 out of 132 rifampicin resistant derivatives grew at 40° (Table 3). One of these mutations (rpoB322) was transduced into strains NS373 (dnaA46(Ts)), NS449 (dnaA508(Ts)), and NS388(dnaA366(Am)); none of the resulting rpoB322 derivatives were able to grow at 40°. The other three rpoB suppressing mutations (rpoB323, rpoB324, rpoB325) also failed to suppress the thermosensitivity of dnaA46(Ts) and dnaA508(Ts) mutant strains (data not shown).

DISCUSSION

Our results indicate that suppression of the initiation defect in dnaA mutant strains by secondary mutations in rpoB is allele specific (Table 2 and above). In addition, of the three amber mutations studied, only one (dnaA311) is suppressed by mutations in rpoB under Su$^-$ conditions. Since the amber mutations dnaA366 and dnaA91 are not suppressed by secondary mutations in rpoB, it appears that they affect a function necessary for rpoB mediated suppression. These mutations may lead to the production of an unstable dnaA product or to the loss of a function essential for rpoB mediated suppression.

Our results are consistent with three different models in which there is a requirement for the dnaA product. The first model involves a direct interaction of the dnaA product with RNA polymerase during the initiation of DNA replication (Bagdasarian etal., 1977; Bagdasarian etal., 1978). According to this model, mutations in rpoB lead to the production of a mutant RNA polymerase able to interact with the mutant dnaA protein. In this case we expect that the polypeptide fragment produced as a result of the dnaA311 amber mutation is active in the initiation event due to its interaction with the mutant RNA polymerase. Since mutations dnaA366 and dnaA91 are not suppressed by rpoB mutations, they could give rise to products which are unable to form functional complexes with RNA polymerase. An alternative to the above model is that mutations in rpoB lead to an increased level of transcription of the dnaA gene, for example, by decreased termination at a hypothetical termination site. In this case, suppression of the initiation defect in the dnaA311(Am) mutant strain could result from an increased level of synthesis of a partially active fragment polypeptide. The dnaA366(Am) and dnaA91(Am) mutant polypeptides may not be active in initiation even when overproduced. A third model is that mutations in rpoB lead to the production of a mutant RNA polymerase able to bypass the requirement for one function of a multi-functional dnaA product. According to this model, the dnaA product is required and allele specificity could occur as a result of variations in the type of alteration in RNA polymerse required to bypass the particular dnaA defect.

Identification of the functional alterations in RNA polymerase which lead to suppression of the initiation defect in dnaA mutant strains may aid in defining the mechanism of rpoB mediated suppression. The rpoB mutations rpoB148, rpoB7 and rpoB8 each suppress the initiation defect in dnaA311(Am) mutant strains. Each of these rpoB mutations also suppresses the defect of the T7 H3 gene1Am342 mutant phage, a phenotype which suggests that the mutant RNA polymerase has altered specificity with respect to T7 gene expression. rpoB7 and rpoB8 are mutations which lead to increased termination at the tryptophan attenuator site as described above. It is not clear whether the mutant RNA polymerse affects the level of the dnaA product synthesized in the rpoB mutant strains. Further studies on the interaction between the dnaA product and RNA polymerase and on the regulation of dnaA expression should allow us to distinguish between these models and to elucidate the role of the dnaA product in the initiation of DNA replication.

ACKNOWLEDGMENTS

We thank Drs. Michael Malamy and Charles Yanofsky for providing strains and Dr. Malamy for helpful discussion. This research was supported by Public Health Service Grant GM 15837 from the National Institutes of Health.

REFERENCES

Bagdasarian, M. Izakowska, M., and Bagdasarian, M. (1977). J. Bacteriol. 130, 577.
Bagdasarian, M. Izakowska, M., Natorff, R., and Bagdasarian, M. (1978). In "DNA Synthesis Present and Future" (Molineux, I. and Kohiyama, M., eds.), Vol. 17, p. 101. Plenum, New York.
Chesney, R., and Scott, J. (1978). Plasmid 1, 145.
Felton, J., and Wright, A. (1979). Molec. gen. Genet. 175, 231.

Hirota, Y., Mordoh, J., and Jacob, F. (1970). J. Mol. Biol. 53, 369.
Kung, F., and Glaser, D. (1978). J. Bacteriol. 133, 755.
Lindahl, G., Hirota, Y. and Jacob, F. (1971). Proc. Natl. Acad. Sci. U.S.A. 68, 2407.
Nishimura, A., Nishimura, Y., and Caro, L. (1973). J. Bacteriol. 116, 1107.
Nishimura, Y., Caro, L., Berg, C., and Hirota, Y. (1971). J. Mol. Biol. 55, 441.
Schaus, N., O'Day, K., Peters, W., and Wright, A. (1981). J. Bacteriol, 145, 904.
Simon, M. and Studier, F. (1973). J. Mol. Biol. 79, 249.
Wechsler, J., and Gross, J. (1971). Molec. gen. Genet. 113, 273.
Wechsler, J. and Zdzienicka, M. (1975). In "DNA Synthesis and its Regulation" (Goulian, M., Hanawalt, P.C., and Fox, C.F. eds.) Vol. 3, p. 624. W.A. Benjamin, Menlo Park.
Yanofsky, C., and Horn, V. (1981). J. Bacteriol. 145, 1334.
Zyskind, J., Deen, L., and Smith, D.J. (1977). J. Bacteriol. 129, 1466.

EVIDENCE THAT THE *ESCHERICHIA COLI DNAZ* PRODUCT, A
POLYMERIZATION PROTEIN, INTERACTS IN VIVO WITH
THE *DNAA* PRODUCT, AN INITIATION PROTEIN[1]

*James R. Walker, William G. Haldenwang[2],
Joyce A. Ramsey, and Aleksandra Blinkowa*

Department of Microbiology
University of Texas
Austin, Texas

I. ABSTRACT

Temperature-sensitive (TS) *dnaZ* mutants of *Escherichia coli* are defective in chromosome polymerization at elevated temperatures. In an effort to identify other components of the chromosome replication mechanism, extragenic suppressors were sought. TS^+ "revertants" were screened for acquisition of cold-sensitivity. Approximately 1% of the "revertants" concomitantly became cold-sensitive (CS). Half of the CS revertants had acquired second mutations which were tightly linked to the original *dnaZ*(TS) mutation and might be in the *dnaZ* gene (near minute 10.5). Other suppressors mapped near minute 82 in the sequence *gyrB* CS *tna* by P1 transduction. Transduction experiments with λ transducing phages indicate that the CS mutations probably are located in *dnaA* - an initiation gene. The CS mutations may result in an altered *dnaA* protein which stabilizes the *dnaZ*(TS) protein.

To determine if the CS suppressor mutations would suppress other *dna*(TS) defects, P1 transduction was used to

---

[1]*Supported by American Cancer Society Grants NP169D and NP169E and, in part, by National Science Foundation Grant PCM78-078 08 and National Institutes of Health Grant AI08286.*

[2]*Present address: University of Kansas, Lawrence, Kansas.*

transfer CS into TS strains. Although CS could be transferred into wild-type of a *dnaZ*(TS) mutant, no CS transductants were isolated when *dnaB, C, E* or *G* recipients were used. The TS mutations might suppress the cold-sensitivity or the CS, TS double mutants could be lethal. In either case, the data suggest an interaction between *dnaA*(CS) and several *dna*(TS) proteins. Perhaps the *dnaA* protein has two functions - one in initiation and a second function as a matrix protein for binding the replication complex together.

II. INTRODUCTION

The *dnaZ* gene codes for a protein which is required for DNA polymerization, as shown by in vivo studies (1, 2, 3). Wickner and Hurwitz (4) isolated the *dnaZ* protein and showed that it is required in vitro for polymerization on a primed template. In Wickner's (5) model, *dnaZ* protein and Elongation Factor (EF) III transfer EFI to a primed template; DNA polymerase III then binds to this EFI-primed template complex. The *dnaZ* protein is the γ component of the Kornberg DNA polymerase III holoenzyme (6, 7).

We asked what components of the replication machinery does the *dnaZ* protein interact with in vivo. The approach was based on the Jarvik and Botstein (8) demonstration that extragenic suppressor mutations are often located in genes, the products of which interact with the product of the gene being suppressed. A second result of their experiment was the finding that extragenic suppressors of TS mutations sometimes were CS and maintained the cold sensitivity in the absence of the TS mutation. We used this approach by starting with a *dnaZ*(TS) mutant, selecting spontaneous TS$^+$ revertants, and scoring them for cold-sensitivity. Of about 6,000 TS$^+$ revertants, 54 independent strains were CS. About one-third of these were extragenic and probably are located with the *dnaA* gene-an initiation gene. This paper presents evidence that the *dnaA* protein interacts in vivo with the *dnaZ* and other replication factors.

III. METHODS

A. *Strains*

All suppressor mutations were selected in strain AX733 which is *dnaZ*2016(TS) *thyA deo lac rpsL* (9). Strains S1, S2, and S3 are spontaneous mutants of strain AX733 and carry

extragenic, CS suppressor mutations which probably are in
dnaA; they have the genotype *dnaZ*(TS) *dnaA*(SUZ2016, CS).
This designation indicates that a mutants allele of *dnaA* suppresses the *dnaZ*2016(TS) defect, but becomes CS. Strain JR8
contains the carbenicillin-resistance transposon, Tn406, in
the *rbs* gene, which is about 15% cotransducible with *dnaA*.
Its genotype is *dnaZ*(TS) *dnaA*(SUZ2016, CS) *rbs*::Tn406. Other
strains are described in figure and table footnotes.

B. *Media*

Yeast extract tryptone medium (YET) was supplemented with
0.5% NaCl. Minimal medium (10) was supplemented with glucose
(10 mg/ml), thiamin·HCl (5 µg/ml), amino acids (50 µg/ml),
glycerol (2 mg/ml), indole (2 µg/ml), and Casamino Acids (5
mg/ml), as desired. Thymine also was added to YET medium for
growth of *thy* mutants. Carbenicillin (500 µg/ml) and rifampicin (150 µg/ml) were added as desired.

C. *Transduction*

Spot tests for transduction by specialized λ transducing
phages to TS^+ or CS^+ were performed according to the procedure
of Walker et al. (11) on YET plates at 42°C or 20°C. Quantitative tests were by the Shimada et al. (12) procedure.
tna^+ transductants were selected by the method described by
Hansen and von Meyenburg (13).

D. *Mass and DNA Synthesis*

Culture absorbance at 540 nm was used as a measure of
bacterial mass. DNA synthesis was followed by the incorporation of ^{3}H-thymine into trichloroacetic acid-insoluble material. Cultures were grown in ^{3}H-thymine (1 µCi/2 µg/ml) at
least 9 generations before sampling began.

IV. RESULTS

A. *Isolation and Preliminary Mapping*

Independent cultures of a *dnaZ*(TS) mutant were grown at
30°C and plated at 39°C to select TS^+ revertants. Among 6,000
TS^+ mutants, 54 independent strains were cold-sensitive.
About one half of these mapped between the origins of Hfr
strains HfrH and KL226 in the *dnaZ* region (Fig. 1). Several
of these mutants were rendered CS^+ by λ*dnaZ*$^+$ transducing
phages and *dnaZ*$^+$ plasmids (11, 16); these suppressors might

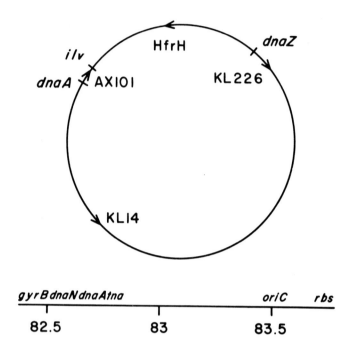

FIGURE 1. *E. coli* chromosome map from Bachmann and Low (14), Hansen and von Meyenburg (13), and Sakakibara and Mizukami (15). The region between minutes 82.4 and 83.8 is expanded at the bottom.

be in *dnaZ*.

About one-third of the suppressor mutations were located between the origins of KL14 and AX1001, i. e., between minute 66 and 83 on the chromosome map. This suggested that they might be in the minute 82 region near *gyrB dnaN dnaA*. P1 transduction demonstrated that the suppressors are near *dnaA* in the sequence *gyrB CS tna*. As will be shown below, the suppressors probably are in *dnaA*.

The remaining one-sixth of the suppressor mutations have not been mapped.

B. *Transduction of the Suppressor by Specialized λ Transducing Phages*

To position the extragenic suppressor mutations more precisely, a family of λ phages from Hansen and von Meyenburg

(13) was used. These phages carried *tna* and varying amounts of the *gyrB dnaN dnaA* region. All phages which carried an intact *dnaA*$^+$ transduced the CS suppressor recipients to CS$^+$ (Table 1). Any phage which did not carry *dnaA*$^+$ did not transduce CS$^+$. Two phages had host fragments which terminated between *dnaA* and *rim* and neither of these separated *dnaA* from the suppressor allele.

The conclusion that the suppressor mutations probably are in *dnaA* was made possible by the use of transducing phages which carry missense and nonsense mutations within *dnaA*. Schaus et al. (17) isolated a λ*dnaA*$^+$ *tna*$^+$ phage and derivatives which carry *dnaA*(amber) and *dnaA*(TS) mutations. The phage carrying the presumed missense mutation *dnaA*46(TS) did not transduce *dnaA*$^+$ or the wild-type CS$^+$ suppressor allele into *dnaA*(TS) or CS recipients (Table 2). The TS mutation on this phage is not expected to be polar; therefore, the failure of this phage to transduce the CS suppressor mutant to CS$^+$ indicates that the extragenic suppressor mutations probably are located in the *dnaA* gene. A λ*dnaA*(Am) phage did not transduce CS$^+$ into these strains, unless the recipients carried an amber suppressor. These mutant alleles of *dnaA* will be designated *dnaA*(SUZ2016, CS) to indicate that they suppress the *dnaZ*2016(TS) allele, but create cold-sensitivity of the *dnaA* protein. The symbol will be shortened to (SUZ, CS) in the remainder of this paper.

C. *The dnaA(SUZ, CS) Mutations Create a Defect in Initiation*

Mutants of genotype *dnaZ*(TS) *dnaA*(SUZ, CS) form colonies at 39 or 40.5°C but fail to grow at 20°C. This cold sensitivity is due to a defect specifically in DNA replication, as shown by measurements of mass, DNA and RNA synthesis. The extent of residual DNA synthesis was measured after shifting a culture to 19°C (Figure 2). Replication stopped gradually over a 5-hour period, during which time the total amount of DNA increased 60%. The immediate decrease in rate of synthesis after shifting to 19°C probably reflected the temperature effect on growth in general because the rate of mass increase decreased immediately also. The residual increase of 60% was compared to the residual increase when initiation was inhibited at the permissive temperature by rifampicin (18) (Figure 2). The residual increase was 60% also. When rifampicin was added to the parental strain (*dnaZ*(TS) *dnaA*$^+$) growing at 30°C, the residual DNA synthesis was 55%, approximately the same as observed in the suppressor-containing strain. These results are consistent with the interpretation that the *dnaA*(SUZ, CS) mutation creates an initiation defect

TABLE 1. Transduction of CS Suppressor Strain by λtna Phages

λtna phages	Markers transduced					
	$gyrB^a$	$dnaN^{a,b}$	$CS+^c$	$dnaA^d$	rim^a	Inner membrane protein^a	tna^e
λ216	−	−	−	−	−	−	+
λ18, 209	−	−	−	−	−	+	+
λ11, 530	−	−	−	−	+	+	+
λ7, 30, 45, 315, 624	−	−	+	+	+	+	+
λ406	−	+	+	+	+	+	+
λ330	+	+	+	+	+	+	+

^{a}Data from Hansen and von Meyenburg (13).
^{b}This gene was identified by Hansen and von Meyenburg (13) as a region that coded for a 45,000 dalton protein; Sakakibara and Mizukami (15) identified the gene as $dnaN$.
^{c}Strains S2(λ^+) and DK4(λ^+) were recipients. S2 is a $dnaZ$(TS) CS extragenic suppressor allele 72 strain. DK4 is a $dnaZ$ CS extragenic suppressor allele 71 strain.
^{d}Strains E177(λ^+) $dnaAI77$(TS), E508(λ^+) $dnaA508$(TS), and KH691(λ^+) $dnaA46$(TS) were used.
^{e}The recipient was strain BE280(λ^+) tna.

TABLE 2. Transduction of CS Suppressor Strain by λdnaA Missense and Amber Phages

Phage	Recipient			
	tna^a	$dnaA(TS)$, $supE44^b$	$dnaZ^+$, CS extragenic suppressor[c]	$dnaZ^+$, CS extragenic suppressor, $supE44^d$
λ425 tna^+ $dnaA^+$	+	+	+	+
λ423 tna^+ $dnaA46$ (TS)	+	−	−	−
λ366 tna^+ $dnaA366$ (AM)	+	+	−	+

[a] Strain WB566 tna.
[b] Strain E177 $dnaA177(TS)$ amber suppressor $supE44$.
[c] Strain JR24 carries CS extragenic suppressor allele 73.
[d] Strain JR16 carries CS extragenic suppressor allele 73 and amber suppressor $supE44$.

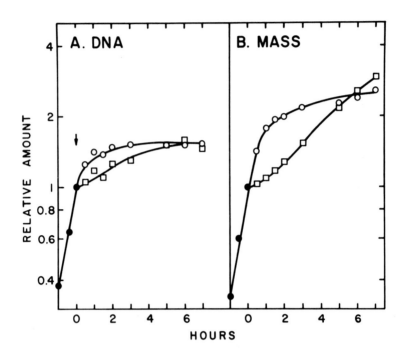

FIGURE 2. Residual DNA synthesis at 19°C and in the presence of rifampicin at 34°C in strain S1. This strain has the genotype dnaZ(TS) dnaA(SUZ, CS). A culture was grown at 34°C in Glucose-Casamino Acids medium containing ^{3}H-thymine and divided at 0 time (arrow) into two parts. One was shifted to 19°C (□). The second remained at 34°C but rifampicin was added (o). Relative amount 1 represents 2500 trichloroacetic acid-insoluble cpm/0.5 ml and an absorbance of 0.06. A, DNA synthesis; B, Mass increase.

at 19°C, as would be expected of an alteration within dnaA. In contrast, a strain with a CS suppressor mutation thought to be located within dnaZ stopped DNA synthesis immediately on a shift from 34° to 19°C, as expected of a polymerization defective mutant.

When the dnaA(SUZ, CS) mutations were moved into a dnaZ$^+$ background, the strains became cold-sensitive, which indicates that the expression of cold sensitivity does not depend on the presence of the dnaZ(TS) mutation. When a dnaZ$^+$ dnaA(SUZ, CS) strain was shifted from 34° to 19°C, the residual DNA synthesis was 50-60%, and the pattern of synthesis was similar to that in the dnaZ(TS) dnaA(SUZ, CS) double mutant.

D. Interaction of the dnaA(SUZ, CS) Allele with Other Replication Factors

It was of interest to determine if other dna(TS) mutants might be suppressed by the dnaA(SUZ, CS) mutations. Phage P1 was grown on a dnaA(SUZ, CS) rbs::Tn406 strain and used to transduce carbenicillin resistance and CS (15% cotransduction) into various recipients. The dnaA(SUZ, CS) mutation was transferred into $dnaA^+$ and dnaZ(TS) recipients with cotransduction frequencies of 22% (10/46) and 10% (6/59), respectively. However, no CS transductants could be isolated when the recipient was dnaB107(TS) (0/34), dnaB6(TS) (0/81), dnaB8(TS) (0/104), dnaC1(TS) (0/59), dnaE486(TS) (0/148), or dnaG3(TS) (0/52). These strains include ones with a mutation in an initiation gene (dnaC) and polymerization genes (dnaB, E, G). A control experiment using as a recipient a TS^+ revertant of dnaE486(TS) yielded CS transductants with a frequency of 12% (4/32). Explanations include (1) the possibility that the TS mutations suppress the CS defect, and (2) the possibility that combinations of dnaA(SUZ, CS) with TS mutants of dnaB, C, E, and G are lethal. In either case, the data suggests that the altered dnaA protein interacts with several replication proteins.

V. DISCUSSION

Extragenic suppressors of a dnaZ(TS) mutation have been isolated and localized in dnaA; these alleles are designated dnaA(SUZ, CS). The conclusion that they are alleles of dnaA depends on transduction by λ phages which carry $dnaA^+$ (13) and, principally, on the failure of a dnaA46(TS) phage (17) to transduce dnaA(SUZ, CS) strains to CS^+. The dnaA46(TS) phage is presumed to carry only one mutation and the 46(TS) mutations is presumed to be missense (i.e., non-polar). The suppressor mutations result in a cold-sensitive defect in initiation of replication, which is consistent with a location in dnaA.

Presumably, alterations within the dnaA product suppress the dnaZ(TS) defect by stabilizing the dnaZ(TS) protein at $39°-40°C$; as a result, however, the dnaA product becomes inactive at low temperature. This interpretation suggests that mutant dnaA protein and mutant dnaZ protein interact in vivo. If the mutant proteins interact, the wild-type proteins probably interact also. The failure to find double mutants of *dnaA*(SUZ, CS) and TS mutations in *dnaB*, *E*, *C*, and *G* suggests that the TS mutations suppress the CS phenotype or that the double mutants are lethal even at an intermediate permissive temperature. In the latter model, the mutant proteins would be inactive at the usually permissive 30°C.

These data suggest that the *dnaA* product, an initiation protein, interacts in vivo with the *dnaC* product (also an initiation factor) and the *dnaB*, *E*, *G*, and *Z* products, all of which are polymerization factors. Previously *dnaA* was thought to be required only in initiation and only once per replication cycle (19, 20, 21, 22). The polymerization proteins are required continuously during chromosome replication. How could the *dnaA*(SUZ, CS) protein suppress a *dnaZ*(TS) mutant polymerization protein? Presumably the *dnaA* protein must interact with the *dnaZ* protein continuously during polymerization. We propose that *dnaA* has two functions. First, there is the active, initiation function. Second, the *dnaA* protein might constitute the matrix which binds the replication complex together. In this way, *dnaA* can interact with several replication factors. Replication complexes have been proposed in several organisms, including T4 (23), *E. coli* (6), and two animal cells (24, 25).

This interpretation requires an explanation for why the usual *dnaA*(TS) mutants behave as initiation defective strains, rather than polymerization defective strains. Perhaps the usual TS mutations in *dnaA* affect one function but not the other. It should be recalled that the selection of TS^+ revertants of a *dnaZ*(TS) mutant is a powerful procedure and that the *dnaA*(SUZ, CS) mutants accounted for less than one per cent of all the revertants. This procedure might be expected to yield a special class of mutants.

Alternative explanations for the suppression of a *dnaZ* polymerization defect by a change in *dnaA* protein exist. The possibility that mutant proteins interact but wild-type proteins do not cannot be excluded. Possibly, the *dnaA*(SUZ, CS) mutations make repliction dependent on a polymerase (e. g., DNA polymerase I) which does not require *dnaZ* product.

ACKNOWLEDGMENTS

We thank Andrew Wright, Kaspar von Meyenburg, Richard Meyer, and Barbara Bachmann for strains.

REFERENCES

1. Filip, C.C., Allen, J.S., Gustafson, R.A., Allen, R.G., and Walker, J.R. (1974). *J. Bacteriol* 119, 443.
2. Haldenwang, W.G., and Walker, J.R. (1976). *Biochem. Biophys. Res. Commun.* 70, 932.
3. Haldenwang, W.G., and Walker, J.R. (1977). *J. Virol.* 22, 23.
4. Wickner, S., and Hurwitz, J. (1976). *Proc. Natl. Acad. Sci. USA* 73, 1053.
5. Wickner, S. (1976). *Proc. Natl. Acad. Sci. USA* 73, 3511.
6. McHenry, C., and Kornberg, A. (1977). *J. Biol. Chem.* 252, 6478.
7. Huebscher, U., and Kornberg, A. (1979). *Proc. Natl. Acad. Sci. USA* 76, 6284.
8. Jarvik, J., and Botstein, D. (1975). *Proc. Natl. Acad. Sci. USA* 72, 2738.
9. Chu, H., Malone, M.M., Haldenwang, W.G., and Walker, J.R. (1977). *J. Bacteriol.* 132, 151.
10. Howard-Flanders, P., Simson, E., and Theriot, L. (1964). *Genetics* 49, 237.
11. Walker, J.R., Henson, J.M., and Lee, C.S. (1977). *J. Bacteriol.* 130, 354.
12. Shimada, K., Weisberg, R.A., and Gottesman, M.E. (1973). *J. Mol. Biol.* 80, 297.
13. Hansen, F.G., and von Meyenburg, K. (1979). *Molec. Gen. Genet.* 175, 135.
14. Bachmann, B.J., and Low, K.B. (1980). *Microbiol. Rev.* 44, 1.
15. Sakakibara, Y., and Mizukami, T. (1980). *Molec. Gen. Genet.* 178, 541.
16. Henson, J.M., Chu, H., Irwin, C.A., and Walker, J.R. (1979). *Genetics* 92, 1041.
17. Schaus, N., O'Day, K., Peters, W., and Wright, A. (1981). *J. Bacteriol.* 145, 904.
18. Lark, K.G. (1972). *J. Mol. Biol.* 64, 47.
19. Hirota, Y., Ryter, A., and Jacob, F. (1968). *Cold Spring Harbor Symp. Quant. Biol.* 33, 677.
20. Fangman, W.L., and Novick, A. (1968). *Genetics* 60, 1.
21. Zyskind, J.W., Deen, L.T., and Smith, D. (1977). *J. Bacteriol.* 129, 1466.
22. Kung, F.-C., and Glaser, D.A. (1978). *J. Bacteriol.* 133, 755.
23. Tomich, P.K., Ciu, C.-S., Woocha, M.C., and Greenberg, G. R. (1974). *J. Biol. Chem.* 249, 7613.
24. Baril, E., Baril, B., Elford, H., and Luftig, R.G., (1973). in "Mechanism and Regulation of DNA Replication" (A. R. Kolber and M. Kohiyama, eds.), p. 275.

Plenum, New York.
25. Reddy, G.P.V., and Pardee, A.B. (1980). *Proc. Natl. Acad. Sci. USA* 77, 3312.

REC-DEPENDENT DNA REPLICATION IN *E. coli*: INTERACTION BETWEEN REC-DEPENDENT AND NORMAL DNA REPLICATION GENES[1]

Karl G. Lark
Cynthia A. Lark
Edward A. Meenen

Department of Biology
University of Utah
Salt Lake City, Utah

Rec-dependent replication is a form of error prone replication which enhances mutagenesis. It may enable bacteria to respond with higher mutation frequencies to variations in the environment. This form of replication depends for its formation on recA activity. RecA activity also is required to maintain this replication.

Wild-type cells require special treatments to induce the formation of this form of replication. Most of these treatments are conditions which induce S.O.S. recA-dependent repair. Mutants have been isolated ($dnaT$) in which, after inducing treatments, all replication is terminated in the absence of protein synthesis. Other mutants exist in which rec-dependent replication occurs as the only form of DNA replication ($sdrT_3$ and sdr_{Rec}). In the $sdrT_3$ mutant all replication is conditionally dependent on active recA function. When lysates are prepared from $sdrT_3$ mutants in which absence of recA function has inhibited replication *in vivo*, replication is also absent *in vitro*. However, replication can be restored *in vitro* by addition of purified recA protein. This restored activity is inhibited by ATP-γ-S, indicating that recA protein restores DNA replication by promoting strand recombination.

$DnaA_{ts}$ alleles suppress the $sdrT_3$ and $dnaT_{11}$ phenotypes.

[1]This work was supported by National Institutes of Health Grant AI10056 to K.G.L.

The phenotypes of these mutants can be restored *in vivo* or *in vitro* by raising the temperature to inactivate the $dnaA_{ts}$ gene product.

Using lysates of cells in an *in vitro* synthesis, we have developed an assay for the $dnaT^+$ gene product. These results indicate that dnaT has at least two activities; one involves regulating the initiation or termination of rec-dependent DNA replication, the other is required for chain elongation.

I. BACKGROUND

Under normal conditions of growth, DNA replication in *E. coli* ceases when protein synthesis is inhibited. However, a variety of treatments can result in the induction of an abnormal form of DNA replication which can continue in the absence of protein synthesis for long periods of time (see Figure 1) (1). Many of the conditions which give rise to this type of replication also induce a process referred to as error prone or S.O.S. repair (2). Thus, this form of replication may be essential for induced mutagenesis.

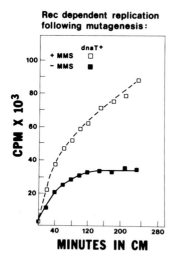

Figure 1. *Induction of rec-dependent replication by pretreatment with MMS. Cultures were exposed (or not exposed) to MMS and then incubated in the presence of chloramphenicol (150 μg/ml). DNA replication was measured as uptake of radioactive thymine (for details see (1) from which these data were taken).*

This type of replication cannot occur unless protein synthesis is allowed to proceed for some period after the inducing treatment (Figure 2) (3). Thus, a variety of seemingly different treatments appear to initiate a sequence of synthetic events which result in this form of replication. A considerable body of data has accumulated which indicate that this represents the synthesis of a special form of replication fork (4).

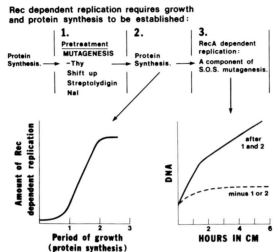

Figure 2. Representation of the conditions required for inducing rec-dependent replication. (Lower left) The period of growth determines the amount of rec-dependent replication. At least 2/3 of a generation of growth must elapse to synthesize the rec-dependent replication fork.

Under conditions allowing growth this replication fork is stable and remains present and active in the cell population, but is diluted by the increasing numbers of normal replication forks which are formed during subsequent growth (3). RecA protein is required both for initiation and during replication. It appears to be an essential part of the replication process (Figure 3) (1,5). Therefore, we shall refer to this as rec-dependent replication.

In other respects, rec-dependent replication resembles normal DNA replication, that is, it is semi-conservative, sequential and requires the gene products characteristic of the normal replication process (the products of *dnaB*, *E* and *G* as well as *C* (however, see below). At least 90% of the *E. coli* genome serves as a template for this replication (3).

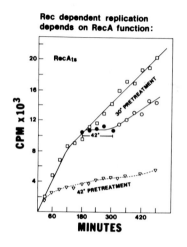

Figure 3. Rec-dependent replication in UT 3062 recA$_{200}$ (E. coli 15), a temperature-sensitive recA strain. Cultures were starved for thymine in the presence (□ - 30⁰) or absence (▽ - 42⁰) of recA function. They were then incubated in chloramphenicol at 30⁰ and replication measured as in Figure 1. An aliquot (●) of one was incubated at 42⁰ to inhibit recA function. Replication ceased but was restored when this aliquot was returned to 30⁰ (○): (Data taken from (1)).

II. MUTATIONAL VARIATION — A POSSIBLE SELECTIVE PRESSURE MAINTAINING REC-DEPENDENT REPLICATION IN BACTERIAL POPULATIONS

Rec-dependent replication appears to be a necessary part of recA-dependent mutagenesis (1). Thus, if rec-dependent replication is prevented, treatment with MMS or U.V. fail to induce mutations even though recA protein is synthesized. In contrast, cultures in which rec-dependent replication forks have been induced show a higher frequency of mutations during subsequent growth.

Rec-dependent replication is induced by a "shift-up" from nutritionally poor to nutritionally rich medium (Figure 4 (1)). We have measured mutation frequencies after a shift up into rich medium (Figure 5). In these experiments cells were grown in minimal media and transferred into broth. Mutation from streptomycin sensitivity to resistance, or from methionine auxotrophy to prototrophy was measured, both immediately after the "shift-up" and upon subsequent incubation in chloramphenicol. It is evident that mutations accumulate under these conditions. (No increase in mutation frequency was observed without the shift up pre-treatment (data not shown).)

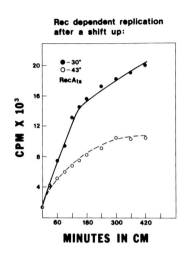

Figure 4. *A shift up induces rec-dependent replication. UT 3062 Rec$_{200}$ was shifted from glucose to broth and after 40 minutes placed in chloramphenicol. DNA synthesis was measured as in Figure 1. <u>RecA</u> function was inhibited by incubation at 43^0 (o). Data taken from (1).*

Figure 5. *Mutation frequencies in cultures of E. <u>coli</u> 15 or E. <u>coli</u> B after a shift up. A streptomycin-sensitive culture of E. <u>coli</u> B (UT 837) was grown in glucose -M9 minimal medium at 37^0. The cultures were transferred to L broth and after one hour chloramphenicol was added (to 150 µg/ml). The frequency of streptomycin-resistant colonies was measured by plating on agar which contained 2 layers. A 10 ml layer of streptomycin agar (20 µg/ml) over which 10 ml of nutrient agar was poured. The cells were plated within 6 hours. E. <u>coli</u> 15, UT 3062 metE$^-$ was grown in succinate M9 minimal medium and transferred to L broth. The culture was then treated in the →*

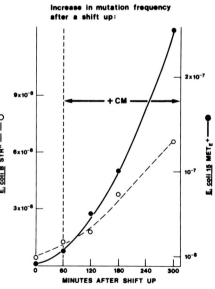

same manner as the *E. coli* B culture. The cells were plated on minimal M9 agar lacking methionine. A 2 ml of agar overlay containing 0.05 μg methionine was used in plating.

These data suggest a basis for the evolution of rec-dependent replication: In nature, bacteria encounter a succession of nutrient changes. Variation in the environment will result in periods of rapid growth or no growth, depending upon the genetic composition of the organism and its nutritional self sufficiency. Rec-dependent replication and the resulting increase in mutation frequency represents a source of variability in the genome which corresponds to the variation in the environment (Figure 6): When cells are shifted into a nutritionally rich environment, rec-dependent replication forks are formed. If essential nutrients are withdrawn, the cells can continue to synthesize DNA utilizing a replication system which favors mutagenesis. Thus, DNA accumulates which may contain favorable mutations. If the cells remain in a rich environment, the rec-dependent replication forks are diluted by the ever increasing proportion of normal replication forks. *Thus the effectiveness of rec-dependent replication in introducing mutants into the population is directly proportional to the variability of the environment.* Growth conditions with high variability will promote higher frequencies of mutations than those found in more constant environments.

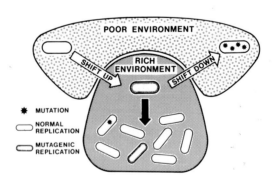

Figure 6. *A schematic representation of the effect of rec-dependent replication in increasing mutant frequency in cultures fluctuating between rich and poor environments on the one hand, or remaining in a rich environment on the* →

other. In the fluctuating environment few cells accumulate many mutated genomes. In the rich environment mutants are diluted out by the increasing number of normal organisms.

III. NATURE OF THE REC-DEPENDENT REPLICATION FORK

A. Working model of the rec-dependent replication mechanism

Previous data emphasized the ability of rec-dependent replication to proceed in the absence of protein synthesis. In particular, studies of the conditions under which replication forks were formed suggested that termination of normal replication resulted from the active destruction of the replication fork at the end of the replication cycle. Accordingly a model was proposed (Figure 7) (1) in which a protein component actively destroyed the complex of replication proteins rendering at least some of them unsuitable for further use in replication. According to this model, induced cells formed a replication fork in the presence of active recA protein. This cleaved, or destroyed, a terminator component (T). Moreover, it was suggested that recA protein played a continuing role in the replication process, either

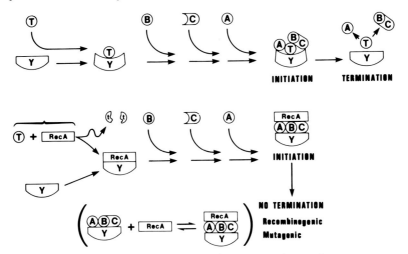

Figure 7. A working model for the formation of a rec-dependent replication fork. A termination protein, T, actively terminates replication. RecA protein cleaves this "terminator" and in addition acts continually during replication to promote rec-dependent replication.

as a part of, or as an adjunct to, the replication fork. Elaboration of this hypothesis required the isolation of mutants regulating or controlling rec-dependent replication.

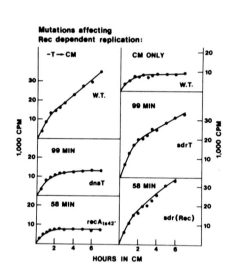

Figure 8. Mutations which affect rec-dependent replication: In each of the strains (mutant or wild type) DNA replication in chloramphenicol has been measured as in Figure 1. In one set of tests, cultures were starved for thymine for 60 minutes to induce rec-dependent replication. In the other set, exponential cultures were transferred directly into chloramphenicol. The isolation of dnaT (7) and sdrT (9) mutations have been described. The isolation of mutations linked to recA are described in the text. In the experiments in this paper, all of the mutations affecting rec-dependent replication have been studied in isogenic E. coli 15 strains derived from UT 3062 (7).

B. Mutants affecting rec-dependent replication

Figure 8 summarizes a variety of mutants which have been found to affect rec-dependent replication. Their identification has utilized the phenotype of DNA replication, or lack of replication, in the absence of protein synthesis. Two phenotypes have been observed, mapping in two regions of the chromosome. Two are near dnaC at 99 minutes on the E. coli genetic map (6), the other two are at, or near, the recA locus at 58 minutes.

1. *DnaT*. $DnaT_{11}$ was isolated from EMS treated cells using a form of "suicide" selection coupled with an autoradiographic screen for rec-dependent replication in chloramphenicol (7). Other dnaT mutants have been isolated by localized mutagenesis (8,9) or by making use of the fact that mutants in *dnaC* are *dnaC dnaT* double mutants which can be separated by P1 transduction (7).

$DnaT_{11}$ has been described in detail (1,7) and has the following phenotype:

A) Conditional lethality. DNA replication is temperature-sensitive under normal conditions. This indicates that the dnaT gene product may be essential for normal DNA replication.

B) Inability to continue rec-dependent replication at low or high temperatures (see Figure 8).

C) It does not synthesize high levels of recA protein except when exposed to S.O.S. inducing treatments.

D) It is an anti-mutator. That is, the ability of MMS to induce mutations is reduced in *dnaT* strains.

E) All of the above-mentioned mutant phenotypes appear to be transdominant in diploid heterozygotes containing $dnaT_{11}$ and $dnaT^+$.

F) *DnaT* interacts with other DNA replication genes (see below).

2. *SdrT*. *SdrT* mutants were isolated by localized mutagenesis using hydroxylamine-treated P1 grown on cells with a Tn*10* insertion at 99 minutes (8,9). Tetracycline-resistant transductants were screened for the *sdr* phenotype by autoradiography. This mutation can be cotransduced with *uxu* at a frequency of 30-50%. One mutant $sdrT_3$ has been studied in detail.

A) DNA replication occurs in the absence of protein synthesis without any inducing treatments.

B) It synthesizes high levels of recA protein only after S.O.S. inducing treatments.

C) All DNA replication in this mutant is dependent on recA function. This is shown in the experiment in Figure 10, which is discussed further below.

D) $SdrT_3$ increases the frequency of mutations obtained after MMS mutagenesis. This is seen in Figure 9 where the mutant and wild type are compared for the ability to revert uxu^- to uxu^+ at different MMS concentrations.

E) $SdrT_3$ interacts with *dnaA* (see below).

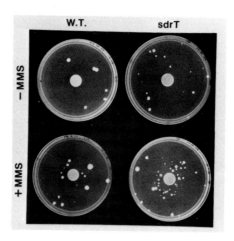

Figure 9. Mutation of uxu^- to uxu^+: Effect of $sdrT_3$. Cultures of UT 3062 uxu^- $sdrT^+$ or uxu^- $sdrT_3$ were plated on minimal agar containing glucuronic acid as an energy source (uxu^- cannot ferment glucuronic acid). One set of plates contained no mutagen, the other 2.5 µl of MMS. Colonies were scored after 3 days.

3. *recA* mutations. RecA mutants are unable to initiate rec-dependent replication. However, mutants similar to *sdrT* have been isolated which are closely linked to srl (sorbitol-utilization) at 58 minutes on the *E. coli* map. One class of these mutants were obtained by localized mutagenesis using hydroxylamine-treated P1 bacteriophage and selecting for the ability to grow on sorbitol. Another group of these were isolated by transduction of a sorbitol-negative strain with P1 phage grown on a sorbitol-positive *tif-1* strain (10). The lethality of *tif-1* mutations at elevated temperatures can be prevented by secondary mutations which suppress filamentation (11). Although strain 15 forms filaments (snakes) following a variety of treatments such as thymine starvation, no temperature sensitivity or filamentation has been observed in such transductants. Therefore, all sorbitol-positive transductants were screened for the *sdr* phenotype. These

mutants, referred to as *sdr rec* mutants, have not been studied in detail.

RecA dependence of growth and replication in sdrT recA$_{ts}$:

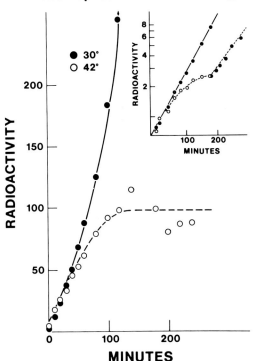

Figure 10. RecA dependence of DNA replication in UT 3062 sdrT$_3$ recA$_{200}$. A culture was grown at 30° or 42° to inhibit recA function. DNA replication was measured as in Figure 1. Data taken from a previous publication (1).

C. *RecA* function during rec-dependent replication

The *recA* gene product has been shown to have at least two activities: as a protease known to cleave specific proteins (the λ repressor (14) and the lexA protein (15); and as a promotor of synaptic activity on the molecular level, i.e., a DNA binding protein which promotes DNA-DNA interactions (16-18). These two activities can be distinguished *in vitro* by their interaction with ATP or its inhibitor ATP-γ-S. The latter enhances proteolysis (14) but inhibits resolution of synapsed DNA (19).

The *sdrT* mutant presented an opportunity to investigate the role of *recA* function during rec-dependent replication. Figure 10 presents data from an experiment in which an *sdrT$_3$ recA$_{ts}$* double mutant is grown at 30° and then placed at 42° to inactivate *in vivo* recA function. As can be seen, replication ceases at 42° but is immediately restored upon lowering the temperature. This suggested that *in vitro* studies of recA function would be possible using this mutant. This proved to be the case. Figure 11 presents a summary of work presented in detail in a previous volume of this series (9). Cultures of the *sdrT recA* double mutant were placed at 42° until replication had ceased. Cellophane disc lysates (12,13) were prepared from such cultures or from 30° control cultures. As can be seen, the lysate of the 42° culture failed to synthesize DNA. However, synthesis could be restored to such lysates by addition of purified recA protein and the restored synthesis was proportional to the amount of recA protein added (Figure 11 inset). By carefully adjusting the concentration of ATP-γ-S added to the system, it was possible to find a concentration which did not inhibit DNA synthesis (that is, the ratio of ATP to ATP-γ-S was such that synthesis proceeded at almost the full rate). This concentration of ATP-γ-S was then added to the system in which recA protein was restoring synthesis which had stopped *in vivo*. The restored synthesis was inhibited. We have concluded from these, and other experiments (9), that the active function of the recA product during rec-dependent replication is its synaptic or recombinational activity.

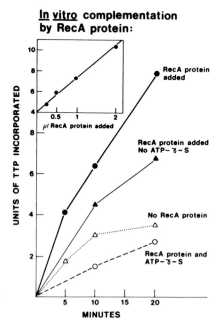

Figure 11. In vitro DNA replication in a lysate of UT 3062 sdrT$_3$ recA$_{200}$ts: effect of recA protein. Cultures were grown at 42° to inhibit recA function. After replication had ceased in vivo cellophan disc lysates were prepared and ^{3}H TTP incorporation measured as described previously (12,13). Replication was limited without recA protein (··△··) upon addition of recA protein replication was restored (-●-). The effect of recA protein was stoichiometric (inset): replication increases proportional to the recA protein added. In another series of experiments, the effect of ATP-γ-S was measured: ATP-γ-S did not inhibit normal repli- →

cation (9) data not shown here but identical to -▲-. ATP-γ-S did inhibit replication restored by recA protein (compare ▲ with ○.). These data are summarized from a previous publication (9). All incorporations were carried out in the presence of anti-polI antibody. Incorporations were completely inhibited by addition of N-ethyl maleiamide.

D. Interaction of dnaT and sdrT with other DNA gene products

1. Interaction of dnaT with dnaC. All *dnaC* mutants which have been isolated display the phenotype of *dnaT* shown in Figure 8. Using $dnaC_2$, it was possible to separate the *dnaT* phenotype from the temperature sensitivity of *dnaC* by P1 transduction (1,7). When this was done, the dnaC phenotype became much *more* severe (increased temperature sensitivity and poor growth at the lowest temperatures) but such "pure" *dnaC* strains had regained the ability to form rec-dependent replication forks. Temperature-resistant transductants were isolated from these same crosses. These possessed the dnaT phenotypes described in Figure 8 but were able to grow at 42°. Another indication of *dnaT dnaC* interaction was the observation that certain dnaC dnaT alleles, if combined, achieved a temperature-resistance greater than that possessed by either of the single temperature-sensitive mutants (1). These data led to the conclusion that the dnaT and dnaC gene products interact modifying each other's activity.

2. Interaction with dnaA. The requirement for dnaA function during rec-dependent replication was studied previously by examining a variety of existing *dnaA* mutants (20). These results were ambiguous because different dnaA alleles (or possibly the parent strains in which they were studied) behaved differently. A more consistent picture has been obtained recently, by transducing $dnaA_{508}$ or $dnaA_{N167}$ into *E. coli* 15 (strain 3062) and into the *dnaT* or *sdrT* mutants of this strain. Rec-dependent replication can proceed in the absence of $dnaA_{508}$ or $_{N167}$ function. This can be seen in the experiment in Figure 12, in which a culture of 3062 $dnaA_{508}$ was induced to initiate rec-dependent replication at 30° and replication in chloramphenicol was measured at 30 or 42°. It is clear that replication proceeds equally well at either temperature. It would appear that dnaA function is not required during rec-dependent replication.

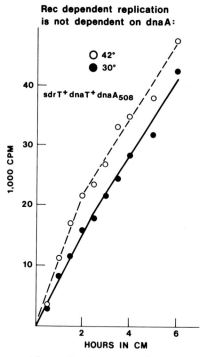

Figure 12. Rec-dependent replication at 30 and 42° in a $dnaA_{ts}$ strain. Rec-dependent replication was induced in UT 3062 $dnaA_{508}$ by starving for thymine for 90 minutes at the permissive temperature (30°). The culture was divided and incubated at 30 or 42° in chloramphenicol (150 µg/ml). Replication was measured as in Figure 1.

When $dnaA_{508}$ or $dnaA_{N167}$ were introduced into $dnaT_{11}$ or $sdrT_3$ mutants, a startling effect was obtained. At the permissive temperature, 30°, these dnaA alleles suppressed the sdrT or dnaT phenotypes. In the experiments in Figure 13 it can be seen that in the $sdrT_3$ $dnaA_{508}$ double mutant, rec-dependent replication is inhibited at 30°. Conversely, in the $dnaT_{11}$ $dnaA_{508}$ double mutant, rec-dependent replication proceeds at 30° (in contrast to the $dnaT$ mutant alone in which it is blocked (see Figure 8)). When the temperature is raised to 42°, the phenotype of the single $sdrT_3$ or $dnaT_{11}$ mutants again is expressed. Thus these two dnaA alleles appear to be suppressors of $sdrT$ or $dnaT$ and inactivation of the dnaA gene product reestablishes the original mutant phenotype.

The data in Figure 13 suggests that the $dnaA_{508}$ $dnaT_{11}$ double mutant ceases DNA synthesis immediately if placed at 42°. To test this further, we measured the *in vitro* synthesis of DNA by lysates prepared from cultures of a $dnaA$ $dnaT$ double mutant which had been growing at 42 or 30° in the presence of chloramphenicol (Figure 14). Lysates of cultures incubated in the presence of chloramphenicol at 30° could synthesize DNA *in vitro* at 30° but not at 42°. This supports the idea that, during rec-dependent replication,

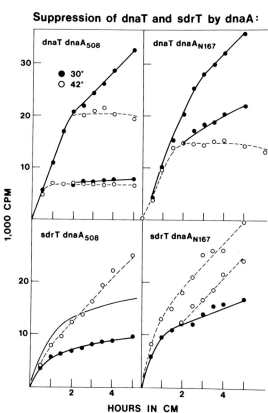

Figure 13. Rec-dependent replication in $\underline{dnaT}_{11}$ $\underline{dnaT}_{ts}$ or $\underline{sdrT}_3$ $\underline{dnaA}_{ts}$ double mutants. UT 3062 $\underline{dnaT}_{11}$ cultures containing the dnaA alleles shown, were starved for thymine for 90 minutes at $30°$. The cultures were divided and portions incubated in chloramphenicol at $30°$ (●) or $42°$ (○). Cultures of UT 3062 sdrT containing the $dnaA_{ts}$ alleles shown, were placed directly in chloramphenicol at $30°$ or $42°$:

In some experiments aliquots at 30 or $42°$ were returned to $42°$ or $30°$ respectively after DNA synthesis had been inhibited.

Cultures in these experiments and in Figure 12 were isogenic except for the dnaT, sdrT or dnaA alleles. The continuous line (lower left) represents synthesis at $42°$ in the absence of chloramphenicol.

chain elongation is inhibited in $dnaA_{508}$ $dnaT_{11}$ double mutants. Moreover, it is clear that lysates prepared from $dnaA_{508}$ $dnaT_{11}$ cultures grown at $42°$ had lost the ability to replicate DNA *in vitro* even at $30°$. The gene product of $dnaA_{N167}$ appears to be inactivated reversibly *in vivo* (see Figure 13 (21)). This is confirmed *in vitro* in that lysates of $dnaA_{N167}$ $dnaT_{11}$ cultures inhibited *in vivo* at $42°$ will replicate their DNA *in vitro* at $30°$ but not at $42°$.

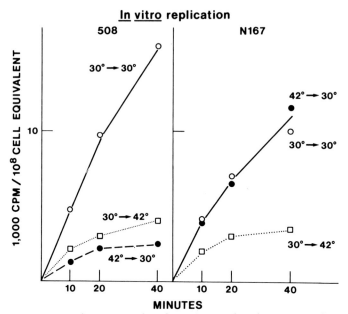

Figure 14. *In vitro* synthesis of DNA by lysates of UT 3062 $dnaT_{11}$ $dnaA_{508}$ (or $dnaT_{11}$ $dnaA_{N167}$) incubated in chloramphenicol at 30 or $42°$.

The cultures were starved for thymine at $30°$ for 90 minutes. They were then incubated at $30°$ (open symbols) or $42°$ (●) in chloramphenicol. After 150 minutes (when *in vivo* replication at $42°$ had ceased) cellophane disc lysates were prepared and replication measured *in vitro* as the incorporation of ^3H TTP. *In vitro* synthesis was measured at 30 or $42°$. All incorporations were carried out in the presence of anti-polI antibody. N-ethyl maleiamide completely inhibited *in vitro* replication.

3. *In vitro* complementation of $dnaT_{11}$ $dnaA_{508}$ lysates - an *in vitro* assay for $dnaT$. Lysates of $dnaT_{11}$ $dnaA_{508}$ cultures which have ceased rec-dependent DNA replication at $42°$ in chloramphenicol appear to be inhibited in the chain elongation step of replication. We have restored replication *in vitro* by use of complementing lysates (Figure 15). To do this, we initiated rec-dependent replication in the double mutant, $dnaT_{11}$ $dnaA_{508}$. The culture was incubated at $42°$ in chloramphenicol for 150 minutes. Lysates of this culture were mixed with lysates of complementing cultures which had

been treated with mitomycin C to prevent their DNA from acting as a template *in vitro*. Different complementation lysates were tried: $\underline{dnaT_{11}\ dnaA^+}$; $\underline{dnaT^+\ dnaA_{508}}$ or $\underline{dnaT^+\ dnaA_{N167}}$. Replication can be restored by lysates of the $\underline{dnaT^+}$ cultures but not by lysates of $\underline{dnaT_{11}\ dnaA^+}$.

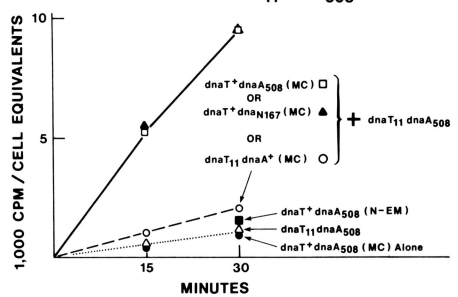

Figure 15. *In vitro* complementation of $\underline{dnaT_{11}\ dnaA_{508}}$. A recipient lysate was prepared as in the experiment in Figure 14. The culture has been incubated in chloramphenicol at 42° for 150 minutes. Cultures for donor lysates were incubated at 42° for 90 minutes and then with mitomycin C (●) for an additional 30 minutes. Incorporation was measured on cellophan discs for the recipient lysate alone ($\underline{dnaT_{11}\ dnaA_{508}}$); for two donors (only $\underline{dnaT^+\ dnaA_{508}}$ is shown) and for three complementing mixtures: $\underline{dnaT^+\ dnaA_{508}}$; $\underline{dnaT^+\ dnaA_{N167}}$; or $\underline{dnaT_{11}\ dnaA^+}$. Complementing mixtures were treated with N-ethyl maleiamide (■) to inhibit polymerase III. All incorporations were carried out in the presence of anti-polI antibody.

As a further test of whether the complementation observed restored replication in the $dnaT_{11}\ dnaA_{508}$ recipient lysate, we labeled the recipient cultures with ^{13}C-glucose ^{15}N ammonium chloride and analyzed the results of the complementation in a CsCl density gradient. The results in Figure 16 demonstrate that replication is occurring on the recipient ($^{13}C^{15}N$) template.

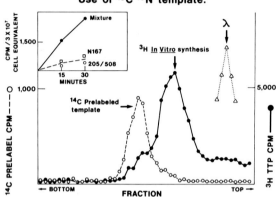

Figure 16. *CsCl density profile of DNA synthesized* in vitro *as a result of complementation. A complementing mixture of lysates was used (*dnaT$_{11}$ dnaA$_{508}$ *recipient +* dnaT$^+$ dnaA$_{N167}$ *donor). The* dnaT$_{11}$ *recipient was incubated in chloramphenicol at 42° and grown in* ^{13}C *glucose* ^{15}N *ammonium chloride medium (22). The* dnaT$^+$ *donor was treated with mitomycin C as in the experiment in Figure 15 but grown in regular (*^{12}C ^{14}N*) medium. The incorporation was measured (see inset). The bulk of the lysate mixture was harvested at 30 minutes, treated with pronase and centrifuged to equilibrium in CsCl.*

^3H labelled DNA synthesized in an *in vitro* complementation test was tested for its ability to hybridize with DNA from both the entire *E. coli* genome (Table 1) and to the replicative origin of the *E. coli* chromosome (data not shown) cloned in phage λ. The ^3H labelled DNA hybridized with the complete *E. coli* genome with the same efficiency as that with which ^{14}C labelled, template, DNA hybridized to *E. coli* genomic DNA. No hybridization could be detected with λ *oriC* DNA (23).

These results show that $dnaT_{11}\ dnaA_{508}$ mutants are inhibited in the process of DNA chain elongation when placed at 42°. Addition of wild type dnaT gene product can restore replication *in vitro*.

TABLE I. Hybridization to *E. coli* DNA of DNA Synthesized *in vitro* by Complementing Lysates

DNA on filter:	DNA Hybridized (cpm)		Input DNA cpm/ml
	E. coli 15		
	2.5 µg	1.25 µg	
^{3}H *in vitro* DNA	618 cpm	324 cpm	4580.0
^{14}C prelabel	59	31	452.5

DNA from E. coli 15 dnaT$_{11}$ was bound to nitrocellulose filters. (Another set of filters received λ DNA to measure non-specific binding.) These were hybridized (in (4.5 x SSC) + 45% formamide at 45° for 3 days) with DNA from an in vitro complementation reaction similar to that in Figure 16. Non-specific background binding to λ was: ^{3}H - 110 cpm; ^{14}C - 15 cpm. The counts in the table have been corrected for this background.

IV. DISCUSSION

A. Regulation of rec-dependent replication

Inducing treatments which lead to rec-dependent replication require participation of activated recA product during the synthesis of the new replication fork (1,5). This replication fork is stable in that it will participate in many successive cycles of replication without requiring additional protein synthesis. It is abnormal in the sense that it promotes mutagenesis and probably is a form of error prone replication. Two types of mutations at two loci affect the induction of the fork. One mutation, $dnaT_{11}$, allows a fork to be formed which is unstable and cannot continue replication in the absence of protein synthesis. Another $sdrT_3$ leads to rec-dependent replication without inducing treatments. Yet a third, sdr_{Rec}, results in a similar effect, but maps in the recA region. It is not yet conclusive that sdrT and dnaT are different genes, although transduction data suggest that they are separate. Two models (Figure 17) for the formation of rec-dependent replication can be proposed according to whether sdrT and dnaT are alleles of the same or different, genes. In one (Figure 17A) activated recA protease (14) modifies dnaT$^+$ to form an altered protein which consequently changes the stability and fidelity of the replication fork. According to this view, $dnaT_{11}$ protein cannot be changed by activated recA protease. SdrT protein, on the other hand, is an *already* altered form of dnaT protein which therefore does not require further alteration by recA protease. Mutants in the recA region, sdr $_{Rec}$, would already possess an active form of recA protein (as has been postulated for *tif-1* mutants) which therefore does not require activating treatments such as thymine starvation, mutagenesis, etc.

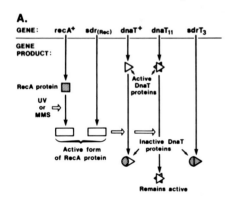

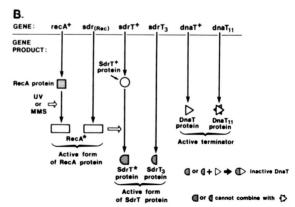

Figure 17. Two models to explain the interactions of dnaT and sdrT mutations (see text for details).

In the other model (Figure 17B), $sdrT$ and $dnaT$ are mutations in different genes which produce different products. Activated recA protein modifies sdrT product (sdrT*). SdrT* in turn interacts with dnaT⁺ product to modify its participation in the replication complex. $DnaT_{11}$ produces a protein which is resistant to alteration by sdrT*; whereas $sdrT_3$ product is already in an active state similar to sdrT*.

These two models present different experimental predictions. For example, in one variant of the second model, $sdrT_3$ should be dominant to sdrT⁺. This, and other predictions, should allow us to distinguish which is correct.

B. Role of recA and dnaA

Because $sdrT_3$ mutants require recA protein for DNA replication, it has been possible to test the function of recA protein $in\ vitro$. Complementation with recA protein restores replication. Such restoration is stoichiometric with the quantity of recA protein added but is inhibited at ATP-γ-S. These observations strongly suggest that recA protein is required as a recombination promoting agent.

During rec-dependent replication chromosome replication does not appear to be initiated by a dnaA-dependent initiation mechanism (Figure 12). An alternative mechanism may be the transfer of replication forks from one template to another by means of recombination. In the absence of replication fork assembly, dnaA product may not be needed to initiate chromosome replication, but it may be impossible to transfer an existing replication mechanism by "hopping" off from one template and on to another. Instead, recombination may be required to transfer the replication fork on a piece of DNA to another chromosome much like a train switching tracks. It should be possible to test this hypothesis experimentally by hybridizing the DNA synthesized to cloned fragments of the $E.\ coli$ chromosome.

C. Role of dnaT

A new finding from these experiments is that a form of dnaT product is required for rec-dependent replication and that this requirement can be supplied $in\ vitro$ by protein from a dnaT⁺ lysate. The fact that dnaT⁺ can supply the requirement for a $dnaT_{11}\ dnaA_{508}$ mutant recipient indicates that the lack of chain elongation is not due to inhibition of chain elongation by a dominant $dnaT_{11}$ product, but rather the absence of an essential factor. Thus $dnaT$ may function to regulate cycles of replication as well as to participate in chain elongation. It is interesting to note that such a dual function has been observed for the dnaC product (24), a protein with which $dnaT$ interacts (1,7). The existence of a complementation assay should facilitate purification of the dnaT product, after which it may be possible to distinguish the basis of these different dnaT activities.

D. Interaction with $dnaA_{ts}$ alleles

At permissive temperatures, $dnaA_{ts}$ alleles suppress the phenotypes of $sdrT_3$ or $dnaT_{11}$ (Figure 13). At non-permissive temperatures, this suppression ceases, indicating that the suppressor has become inactive. We can explain this behavior on the basis of interaction between $dnaT$ and $dnaA$ gene products. The $sdrT$ altered dnaT$^+$ protein interacts with dnaA$_{508}$ protein to form a product which is inactive or extremely inefficient in DNA replication. If $dnaA_{508}$ is inactivated the efficiency of the altered dnaT$^+$ product is restored. DnaT$_{11}$ product is stabilized by interaction with $dnaA_{508}$ hence replication can proceed in chloramphenicol. However, when $dnaA_{508}$ is heated $dnaT_{11}$ product is destabilized and replication ceases. Restoration of dnaT$^+$ protein during complementation restores activity. In connection with the stability of the $dnaT_{11}$ product it should be noted that $dnaT_{11}$ $dnaA_{508}$ double mutants are much more sensitive to temperature than either of the single mutants ($dnaT_{11}$ or $dnaT_{508}$).

E. Other rec-dependent replication genes: the complexity of the replication process

We have described and discussed results from our own laboratory. Earlier, Kogoma (25) isolated another mutation, sdrA, which maps at yet a different region of the *E. coli* chromosome from sdrT or Sdr$_{Rec}$. This mutant also continues to replicate DNA in chloramphenicol without any special induction treatments. Recently (26) it has been shown that with this mutant, as in sdrT, *all* replication is recA dependent. Moreover, a mutation which suppresses this recA dependence has been isolated. Thus at least three genes, sdrA, sdrT and dnaT, as well as components of the recA system, are required for the formation and maintenance of rec-dependent DNA replication.

The fact that rec-dependent replication appears to be error prone, indicates that the stringency of the replication process is relaxed. This may occur by interactions of rec-dependent gene products (sdrA, sdrT or dnaT) with other DNA replication genes such as dnaC or dnaA. Further studies of the effects of suppressing error prone replication are required to elucidate this process, but it appears evident that modifications in replication observed, are not simply due to addition or deletion of components but rather the result of gene product interaction.

REFERENCES

1. Lark, K. G., and Lark, C. A., *Cold Spring Harbor Symp. Quant. Biol. 43*, 537 (1978).
2. Witkin, E. M., *Bacteriol. Rev. 40*, 869 (1976).
3. Kogoma, T., and Lark, K. G., *J. Mol. Biol. 52*, 143 (1970).
4. Lark, K. G., *in* "Biological Regulation and Development" (R. F. Goldberger, ed.), vol. 1, p. 201. Plenum Pub. Corp., New York, (1979).
5. Kogoma, T., Torrey, T. A., and Connaughton, M. S., *Molec. Gen. Genet. 176*, 1 (1979).
6. Bachmann, B. J., and Low, K. B., *Bacteriol. Rev. 44*, 1 (1980).
7. Lark, C. A., Riazi, J., and Lark, K. G., *J. Bacteriol. 136*, 1008 (1978).
8. Hong, J., and Ames, B. N., *Proc. Natl. Acad. Sci. U.S.A. 71*, 3158 (1971).
9. Lark, K. G., and Lark, C., *in* "Mechanistic Studies of DNA Replication and Genetic Recombination" (B. Alberts and C. F. Fox, eds.), ICN Symposia on DNA Replication and Genetic Recombination XIX, p. 881. Academic Press, New York, (1980).
10. Goldthwait, D., and Jacob, F., *C. R. Acad. Sci. (Paris) 259*, 661 (1964).
11. George, J., Castellazzi, M., and Buttin, G., *Mol. Gen. Genet. 140*, 309 (1975).
12. Schaller, H., Bernd, O., Nüsslein, V., Huf, J., Hermann, R., and Bonhoeffer, F., *J. Mol. Biol. 63*, 180 (1972).
13. Lark, C., *J. Bacteriol. 137*, 44 (1979).
14. Craig, N. L., and Roberts, J. W., *Nature 283*, 26 (1980).

15. Little, J. W., Edmiston, S. H., Pacelli, L. Z., and Mount, D. W., *Proc. Natl. Acad. Sci. U.S.A. 77*, 3225 (1980).
16. Weinstock, G. M., McEntee, K., and Lehman, I. R., *Proc. Natl. Acad. Sci. U.S.A. 76*, 126 (1979).
17. Shibata, T., Das Gupta, C., Cunningham, R. P., and Radding, C. M., *Proc. Natl. Acad. Sci. U.S.A. 77*, 2606 (1980).
18. Cassuto, E., West, S. C., Mursalim, J., Conlon, S., and Howard-Flanders, P., *Proc. Natl. Acad. Sci. U.S.A. 77*, 3962 (1980).
19. McEntee, K., Weinstock, G. M., and Lehman, I. R., *Proc. Natl. Acad. Sci. U.S.A. 76*, 2615 (1979).
20. Kogoma, T., and Lark, K. G., *J. Mol. Biol. 52*, 143 (1975).
21. Fralick, J. A., and Lark, K. G., *J. Mol. Biol. 80*, 459 (1973).
22. Lark, K. G., and Renger, E. H., *J. Mol. Biol. 42*, 221 (1969).
23. Meyenberg, K. von, Hansen, F. G., Nielsen, L. D., and Riise, E., *Mol. Gen. Genet. 160*, 287 (1978).
24. Wechsler, J. A., *J. Bacteriol. 121*, 594 (1975).
25. Kogoma, T., *J. Mol. Biol. 121*, 55 (1978).
26. Torrey, T. A., Pickett, G. G., and Kogoma, T., *J. Supramolec. Struct. and Cell. Biochem.* Supp. 5, p. 340, Abstract 929, (1981).

AN ALTERNATIVE DNA INITIATION PATHWAY IN E. coli

Tokio Kogoma
Ted Albert Torrey
Nelda L. Subia
Gavin G. Pickett

Department of Biology, University of New Mexico
Albuquerque, New Mexico 87131

ABSTRACT   The sdrA102 mutation confers upon cells the ability to replicate DNA in the absence of protein synthesis. The mutation maps near metD. When this mutation was combined with the recA200 mutation which renders the recA protein thermolabile, it had little effect on normal replication. However, the sdrA102 recA200 double mutant exhibited temperature-sensitive stable DNA replication: it continuously replicated DNA in the presence of chloramphenicol at 30°C, whereas at 42°C the DNA content increased only 40-45% before DNA synthesis ceased. Suppressor mutations (rin; recA-independent) capable of stable DNA replication at 42°C were isolated in the double mutant. The rin mutations specifically suppressed the recA$^+$ dependence of stable DNA replication and did not alleviate the defective characteristics of of the recA56 mutants, i.e. the deficient general recombination, extreme UV sensitivity and inability of prophage induction. It was also found that the sdrA102 mutation is replicon-specific: i.e., the mutation allows only replication of the bacterial chromosome but not replication of F plasmid DNA in the absence of protein synthesis. A model is proposed in which the sdrA$^+$ gene product is viewed as a repressor of a switch from normal replication to an alternative replication pathway.

INTRODUCTION

In contrast to normal DNA replication, which requires concomitant protein synthesis for repeated initiation, stable DNA replication occurs for many hours despite the absence of protein synthesis (1, 2). This activity can be induced in wild-type cells by various treatments, including thymine starvation, exposure to nalidixic acid, and ultraviolet irradiation (2, 3). Stable DNA replication (SDR) represents several rounds of semi-conservative replication of the entire E. coli chromosome and requires most, if not all, of the gene products known to be involved in normal DNA replication.

One extraordinary aspect of this replication is the requirement for the $recA^+$ gene product in the induction process (3, 4) as well as during replication (4). Since the induction process also requires the $lexA^+$ gene product, we proposed that SDR is one of the so-called "SOS" functions (3). Other SOS functions similarly induced include: amplified synthesis of recA protein, enhanced mutagenesis, new DNA repair activity, and prophage induction in lysogens. Our demonstration that induced stable DNA replication is remarkably resistant to ultraviolet radiation is compatible with this proposal (3).

Mutants capable of SDR without an inducing treatment have been isolated in E. coli 15T⁻ strains and designated Sdr^c (constitutive stable DNA replication) mutants (6). This replication has been demonstrated, by DNA-DNA hybridization methods, to be semi-conservative replication of the entire E. coli chromosome (7). One of the mutant alleles (sdrA2) has been located near the pro-lac region of the E. coli genetic map and is recessive to the wild-type allele (6).

Recently, we have characterized another allele (sdrA102) also isolated in the E. coli 15T⁻ background which, when conjugated into E. coli K-12, maps very close to metD. In this study we have demonstrated the requirement of constitutive stable DNA replication for the $recA^+$ gene product and isolated suppressor mutations that specifically suppress the $recA^+$ dependence of this replication. We have also examined whether or not the sdr A102 mutation allows F plasmid replication in the absence of protein synthesis.

MATERIALS AND METHODS

The E. coli strains used are listed in Table 1. All media, chemicals and growth conditions used have been described previously (6). Mutagenesis and isolation of rin mutants by autoradiography was performed essentially as described by Kogoma (6). Separation of plasmid CCC DNA from linear DNA by centrifugation in CsCl-ethidium bromide was performed according to Collins and Pritchard (23).

Table 1 E. coli strains

Strain #			Genotype
AQ2	15T⁻	F⁻	thyA42 deoB20 metE101 arg-19 trp-68 thr-32
AQ145	15T⁻		as AQ2 except sdrA102
AQ146	15T⁻		as AQ145 except temperature resistant sdrA⁺
AQ300	15T⁻		as AQ145 except trp⁺ Hfr Cavalli
CB0129	K-12	F⁻	leu⁻ thi⁻ thy⁻ (25)
AQ331	K-12		as CB0129 except leu⁺ sdrA102
KL320	K-12	F⁻	proB48 his-29 metE90 trpA9605 trpR lacI22 lacZ118 rpsL171 azi-9 nalA18 (24)
AQ291	K-12		as KL320 except thy709 deo-29
AQ345	a		as AQ291 except pro⁺ sdrA102
AQ377	a		as AQ345 except his⁺ recA200
AQ527	a		as AQ377 except rin-15
AQ543	a		as AQ377 except srlC300::tn10 recA56
AQ547	a		as AQ527 except srlC300::tn10 recA56
AQ399	a		as AQ331 except carries F⁺zzz::tn10

a. These strains are hybrids containing a small portion of the 15T⁻ chromosome in a K-12 genetic background

RESULTS

Isolation and characterization of the sdrA102 mutation

Although most of the Sdr^C mutants could grow normally at 37°C, their growth rates were slightly lower than that of the parental strain. To determine whether or not complete loss of the sdrA⁺ gene activity is lethal for the cell, temperature-sensitive (Ts) Sdr^C mutants were isolated from AQ2, a derivative of 15T⁻, after mutagenesis with EMS. One of the mutants (AQ145, sdrA102) formed only minute colonies on nutrient agar at 42°C. The Ts phenotype of the mutant was cell-density dependent and also reversed by the addition of salts (NaCl, KCl, NH$_4$Cl, Na$_2$HPO$_4$) at concentrations higher than 0.5%. Figure 1A depicts ³H-thymine incorporation into DNA in the presence of the protein synthesis inhibitor, chloramphenicol (CM). The mutant continued to synthesize DNA in CM at 30°C although the rate declined somewhat at the later times of incubation. The ability of the mutant to replicate DNA in CM was enhanced after a period of incubation at the temperature restrictive for growth (Figure 1A).

A temperature-insensitive revertant which formed large colonies on nutrient agar at 42°C was isolated. It was un-

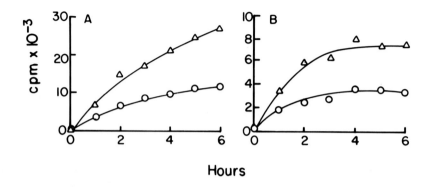

FIGURE 1. Presence and absence of stable DNA replication in AQ145 and AQ146. Cultures of AQ145(A) and AQ146(B) were grown at 30°C to 1 x 10^8 cells/ml. Aliquots (o) were added to ^{3}H-thymine (5μci/4μg/ml) and CM (150μg/ml). Another pair of aliquots (Δ) were incubated at 42°C for 40 min and then added to ^{3}H-thymine and CM. All subcultures were subsequently incubated at 30°, and samples (o.1 ml) were taken at time zero (the time of addition of CM) and at hourly intervals. The radioactivity in the acid-insoluble fraction was determined as described previously (6).

Table 2 Mapping of sdrA102 by conjugation

Exp.	Selected Marker	Unselected Marker			Total
			Sdr^c	Sdr^+	
I	Pro^+	Lac^+	42	16	
		Lac^-	29	13	100
II	Lac^+	Pro^+	58	15	
		Pro^-	0	5	78

Male cells (AQ300) and female cells (AQ291) were grown in nutrient broth to 3 x 10^8 cells/ml and mixed at a ratio of 1:2 (with female cells in excess). After 60 minutes aliquots of the mating mixture were plated on selective plates for Pro^+ Sm^r or Lac^+ Sm^r.

able to replicate DNA in CM (Figure 1B). Reversion frequency was approximately 10^{-7}. It appears, therefore, that a single mutation (sdrA102) is responsible for both the Ts phenotype and the constitutive stable DNA replication capability.

To map the sdrA102 mutation, an Hfr derivative of AQ145 (Table 1) was constructed and conjugated with an E. coli K-12 female strain. The results shown in Table 2 indicated that the mutation is located on the left side of proA. P1 transduction has indicated that the mutation is 4% linked to proA and about 33% linked to metD.

Despite the Ts phenotype of AQ145 and its Hfr derivative (AQ300), we have not been able to demonstrate the Ts phenotype in any of the sdrA102 conjugants or transductants in the K-12 background, although they exhibited strong Sdr^c phenotypes. It is possible that the mutation may require particular genetic backgrounds for the expression of the Ts phenotype.

$RecA^+$-dependent SDR and isolation of suppressor mutations, rin.

The strain AQ377 was constructed by mating the recA200 mutation into a K-12 sdrA102 strain, AQ345. The recA200 mutation renders recA protein thermolabile. The sdrA102 recA200 double mutant was found to grow normally in a temperature range of $30°$-$42°C$, indicating that the recA gene product is not required for normal DNA replication. However, when SDR in the double mutant was examined, it was found that both the establishment and the maintenance of SDR require functional recA product. Figure 2A indicates that while the sdrA102 recA200 double mutant was capable of SDR at $30°C$ (open circles), the SDR capability was abolished if the cells were incubated at $42°C$ for a period of time immediately prior to incubation in CM at $30°C$ (solid circles). The DNA synthesis in this heat-treated culture gradually ceased, reaching a total increase of only 40-45% of the amount present at the time of addition of CM. The residual amount of the DNA synthesized was very close to that expected for a culture growing in glucose medium if initiation of a new round of replication is blocked by the incubation at the restrictive temperature. A very similar result was obtained when a culture growing at $30°C$ was shifted to $42°C$ and incubated in CM (data not shown). When a culture exhibiting SDR at $30°C$ (e.g., open circles in Figure 2A) was shifted to $42°C$, slow shut-off of DNA synthesis was observed (data not shown). In all these experiments no significant amount of DNA degradation was detected. These results indicate that the functional recA protein is a specific requirement for constitutive SDR, possibly at the initiation step, although it is not required for normal DNA replication.

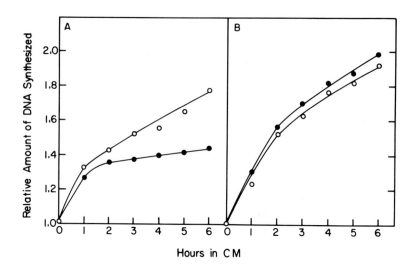

FIGURE 2. Effects of heat pre-treatment on the establishment of stable DNA replication in AQ377 and AQ527. Cultures of AQ377 (A) and AQ527(B) were grown at 30°C for 4 hours in the presence of 14C thymine (0.5 µCi/4 µg/ml) to a density of 2×10^8 cells/ml. 1 ml aliquots of the cultures were transferred to culture tubes and incubated at 30°C (o) or 42°C (●) for an additional 90 min. After the end of the incubation period, CM(150µg/ml) was added to all tubes, and incubation was continued at 30°C. Samples (0.1 ml) were taken at time zero and at hourly intervals. The data are presented as ratios of the later samples to the first sample at time zero.

The above observations prompted us to search for $recA^+$-independent mutants of AQ377 (sdrA102 recA200) insensitive to the loss of recA function at 42°C. The autoradiographic method previously described (Kogoma, 1978) yielded several such $recA^+$-independent (rin) mutants capable of SDR at 42°C. One rin mutant (sdrA102 recA200 rin-15) is shown in Figure 2B to continue SDR despite the heat pre-treatment. This strain (AQ527) also continued to exhibit SDR when cultures were shifted directly to 42°C upon addition of CM (data not shown). To test if the rin-15 mutation mapped within the recA locus, the recA56 missense allele was mated in, selecting for the closely linked srlC300::tn10 tetracycline resistance determinant (8). Figure 3 depicts DNA synthesis at 37°C in the presence of CM for AQ547 (sdrA102 recA56 rin-15) and for AQ543 (sdrA102 recA56 rin^+). Replacement of recA200 with the recA56 allele did not prevent the rin-15 mutant from expressing the Sdr^C phenotype (circles). The rin^+ control strain did not continue DNA synthesis in CM (triangles).

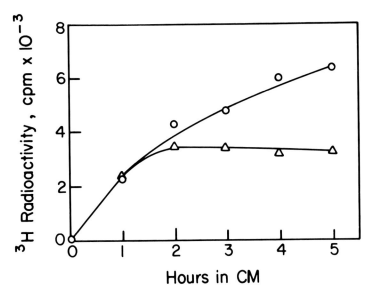

FIGURE 3. Presence and absence of stable DNA replication in AQ547 and AQ543. 1 ml cultures of AQ547 (o) and AQ543 (△) were grown at 37°C to cell density of $1-2 \times 10^8$ cells/ml. At time zero, ^{3}H-thymine and CM were added to final concentrations of 5μCi/4μg/ml and 150μg/ml, respectively.

Table 3. Recombination Frequency

Strain	Relevant Phenotype	Recombinants (a)	
		Met$^+$ (b)	Trp$^+$
AQ345	recA$^+$ sdrA102	5.1×10^{-1}	5.6×10^{-2}
AQ543	redA56 sdrA102	1.2×10^{-3}	$<2.9 \times 10^{-4}$
AQ547	recA56 sdrA102 rin-15	3.9×10^{-3}	$<2.9 \times 10^{-4}$

Male cells (CSH61 or CSH63) and female cells (AQ345, AQ543 or AQ547) were grown in nutrient broth to 3×10^8 cells/ml and mixed at a ratio of 1:50 (with female cells in excess). After 60 minutes aliquots of the mating mixtures were plated on selective agar plates. (a) recombinants/100 Hfr cells. (b) an average of two experiments.

Table 4. UV Sensitivity

Strain	Surviving fraction (N/No) at:			
	3.6	7.2	10.8	21.6 (J/m^2)
AQ345	—	—	9.9×10^{-1}	6.5×10^{-1}
AQ543	1.9×10^{-3}	1.5×10^{-4}	—	—
AQ547	2.0×10^{-3}	0.8×10^{-4}	—	—

Exponentially growing cells were placed in M9 buffer, irradiated with UV at various fluences and plated on nutrient agar plates. Colonies were counted after 18 hours of incubation in the dark.

Table 5. Prophage Induction

Strain	Induction frequency*
AQ345 (λ^+)	3.0×10^{-1}
AQ543 (λ^+)	5.7×10^{-7}
AQ547 (λ^+)	1.2×10^{-7}

Exponentially growing cells were placed in M9 buffer, irradiated with UV at 20 J/m^2 and plated with 5×10^8 cells of indicator strain (CSH25). *the number of plaques per the number of lysogens irradiated and plated.

The above results suggested to us that the rin mutations might be suppressor mutations specific for $recA^+$ dependence of SDR. Therefore, we examined AQ345 (sdrA102 $recA^+$), AQ543 and AQ547 strains for various measures of recA function. Tables 3, 4 and 5 summarize the results. The rin-15 suppressor mutation did not suppress the defective recA56 for general recombination, induction of new DNA repair activity (UV

24 AN ALTERNATIVE DNA INITIATION PATHWAY IN E. COLI

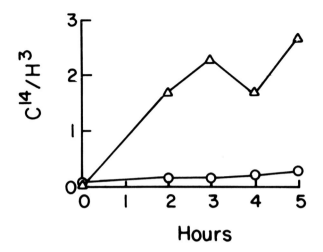

FIGURE 4. Replication of bacterial and CCC F DNA in CM treated AQ399 cells. A culture of AQ399 was labelled with 14C - thymine (1 μCi/4μg/ml) for 4 hours to a density of 2x10^8 cells/ml. The cells were then collected on filter, washed with warm M9 buffer and suspended in medium containing ^{3}H-thymine (5 μCi/2 μg/ml) and CM (150 μg/ml), and incubation was continued at 37°. 1 ml samples were taken at hourly intervals and analyzed for CCC DNA (O) and linear DNA (Δ) as described in Materials and Methods.

radiation sensitivity) nor prophage induction. Since the latter two are presumed measures of the proteolytic activity of the recA$^+$ gene product (9, 10), the rin-15 mutation does not complement either this function nor the recombination activity. This conclusion is supported by the observation that the lexB30 allele of recA which abolishes the proteolytic function of recA protein but not the recombinational function (11) did not affect the expression of the sdrA102 mutation (data not shown).

Introduction of F'133, which carries an 84-89 minute segment of the E. coli chromosome, into AQ527 (sdrA102 recA200 rin-15) abolished the ability to exhibit SDR at 42°C. The Rin phenotype was restored when the F' factor was cured by treatment with acridine orange. Hfr strains derived from AQ527 by integration of F'133 into the chromosome lost the Rin phenotype completely. These results indicate that the rin-15 mutation lies between 84 and 89 minutes on the E. coli genome and that the mutation is recessive to the wild type.

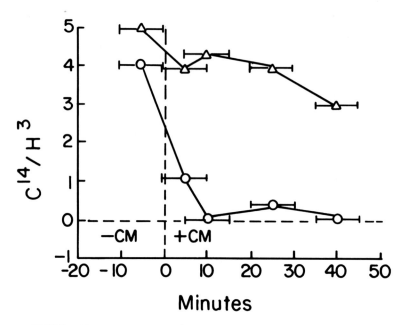

FIGURE 5. Incorporation of pulse-labelled 3H thymine into CCC DNA and bacterial DNA. A culture of AQ399 was labelled with 14C-thymine as described in the legend to Figure 4. After the 14C-thymine was washed out, 2 ml aliquots were pulse-labelled with 3H-thymidine (5μCi/0.5μg/ml) for 10 min before and after addition of CM (150μg/ml) as indicated by the horizontal bars. 3H/14C ratios were determined for CCC DNA (o) and linear DNA (Δ).

F plasmid replication in sdrA102 strains in the presence of CM.

F plasmid is normally unable to replicate the absence of protein synthesis (12). To determine whether or not the sdrA102 mutation allowed F plasmid replication in the presence of CM, a Tn10-containing derivative of F plasmid was introduced into an sdrA102 strain and the presence or absence of F replication was monitored by two methods. First, lysates from cells which had been labelled with ^{14}C-thymine before and ^{3}H-thymine after the addition of CM were centrifuged in CsCl-ethidium bromide gradients to separate covalently closed circular (CCC) plasmid DNA from linear DNA. The net increase in DNA synthesized during incubation in CM was determined from the ratio of ^{3}H and ^{14}C radioactivities. The results are shown in Figure 4, which indicated that F plasmid did not replicate in the sdrA102 mutant in the presence of CM despite the fact that bacterial linear DNA replicated continuously under the same conditions. A pulse labelling experiment (Figure 5) revealed that the F plasmid replication ceased within a few minutes after the addition of CM.

If F plasmid DNA replicated in CM in forms other than CCC, the first type of experiment described above would have underestimated the amount of F DNA. Therefore, in the second method F replication was monitored by measuring the copy number of the Tn10 sequence carried by the F plasmid, utilizing DNA-DNA hybridization. The result of a preliminary experiment indicated that there was little increase of the Tn10 sequence in CM (data not shown).

DISCUSSION

The fact that the $recA^+$ gene produce is required only for stable DNA replication (SDR) but not for normal replication of sdrA mutants suggests that SDR is distinct from normal replication and that the inhibition of protein synthesis triggers a switch from normal replication to SDR. The initiation mechanism operating during SDR may be an alternative to the normal initiation process. SDR may employ the $recA^+$ protein in place of some initiation factor thought to be unstable (2, 6). The $recA^+$ protein has been shown to be very stable in vivo. This would account for the $recA^+$ dependence of SDR as well as the independence from concomitant protein synthesis.

In this switch model, the product of the $sdrA^+$ gene may act as a repressor of the switch from the normal initiation pathway to the stable DNA replication alternative upon the inhibition of protein synthesis. In wild-type strains the $sdrA^+$ gene product prevents the switch when the normal mechanism fails to reinitiate due to the lack of the initiation factor. The $sdrA^+$ repressor can be inactivated after an inducing treatment, and the switch ensues, resulting in induced stable DNA replication (1, 2). This inactivation of the $sdrA^+$ repressor upon induction may be mediated by the $recA^+$ protease activity as proposed (3). In sdrA mutants, however, inactive sdrA repressor allows the switch to occur without an inductive treatment, confering upon the cells the phenotype of constitutive stable DNA replication. In the switch model, it is assumed that the unstable initiation factor is preferentially used as long as it is available. Only when the supply of the factor is interrupted, by inhibition of protein synthesis, does the $recA^+$-dependent alternative mechanism prevail.

Lark and Lark (13) isolated a mutant (sdrT), similar to the sdrA mutants, which is capable of DNA replication in the absence of protein synthesis. The sdrT mutation maps near dnaT but these two may be distinct loci (14). The sdrT seems to be mutagenic (14). In contrast to sdrA mutations, the sdrT mutation renders not only SDR but also normal replication dependent on the recA$^+$ gene product (13). In light of the switch model, the preferential usage of the unstable initiation factor may be affected in the sdrT mutant in such a way that the recA$^+$ protein is utilized even under normal growth conditions.

At present the role of the recA protein in SDR is not clear. In vitro, three enzymatic activities of recA$^+$ protein have been demonstrated: assimilation of single-stranded DNA into homologous duplex regions (15, 16), a DNA-dependent ATPase (17, 18), and a protease. The ATPase activity is closely associated with DNA strand assimilation. The protease function (presumably activated by some "SOS" signal) has been shown to cleave the λ repressor (9) and the lexA$^+$ protein (10). The latter is thought to be a repressor of the recA$^+$ operon and possibly other "SOS" operons. Besides the possible role of recA protein in inactivating the sdrA$^+$ repressor, direct involvement of the recA$^+$ protease activity in SDR is unlikely since the lexB30 mutation which inactivates the proteolytic function of recA (19) does not affect the SDR ability of the sdr102 mutant. Also supporting this conclusion is the observation that the rin mutation did not suppress in vivo the deficient proteolytic function of the recA56 protein (Tables 4 and 5).

Lark and Lark (13) reported that the addition of purified recA$^+$ protein restored in vitro replication to lysates prepared from sdrT cells which had ceased DNA replication in the absence of recA function. The restoration of replication activity was senstitive to adenosine 5- (γ-thio) triphosphate, an inhibitor of ATPases, suggesting to them involvement of the recombinational function of recA protein (13). Our in vivo results clearly rule out the general recombination function of recA protein as a role in SDR since the suppression by the rin mutation is specific to the requirement for the recA$^+$ product during SDR and does not alleviate the impaired general recombination capability of the recA56 product (Table 3). It is possible, however, that recA$^+$ protein binds DNA, utilizing ATP to stablilize a specific structure similar to an intermediate in recombination. If the requirement for recA$^+$ protein is at the initiation step as suggested by the residual amount of DNA synthesized in the absence of recA$^+$ function (Figure 2A), the proposed stabilized structure may function in the initiation of SDR. The product of the rin mutation may maintain this specific structure in the absence of recA$^+$ without restoring general recombination proficiency. It is important to point out that ssb$^+$, the structural gene for single-stranded DNA binding protein (20), lies outside the chromosomal region within which the rin-15 mutation appears to map. Recently, it has been reported that initiation of T4 DNA replication can occur from recombinational intermediates in later stages of infection (21).

Initiation of induced stable DNA replication after a period of thymine starvation occurs at the origin (1). However, it has not been proven whether SDR in the Sdr^C mutants is initiated at the oriC site. Recently, LaVerne and Ray (22) reported that DNA replication in dnaAl77 mutants integratively suppressed by F'128 becomes recA$^+$-dependent. The integration of the F' factor occurs at the oriC region in a site-specific reaction. It is of great interest to see whether the replication under the influence of an integrated F'128 factor is an SDR type and whether the rin mutation suppresses the recA$^+$ dependence of this replication.

We have shown that the effect of the sdrA mutation is replicon-specific: i.e., the mutation allows only replication of the bacterial chromosome but not replication of F plasmid DNA in the absence of protein synthesis. This observation suggests that stringent plasmids such as F and R factors have regulatory mechanisms similar to the bacterial sdr system. We anticipate isolation of mutations in stringent plasmids which specifically allow plasmid replication in the absence of protein synthesis.

ACKNOWLEDGMENTS

The excellent technical assistance of Marty Connaughton and Suchit Tusiri is greatly appreciated. We thank Drs. B. Bachmann, C. Lark and D. Mount for their strains. This work was supported by grants from NIH (GM22092), NSF (78-23222) and NIH (Minority Biomedical Support Program grant RR08139).

REFERENCES

1. Kogoma, T., Lark, K. G. (1970). J. Mol. Biol. 52, 143.
2. Kogoma, T., Lark, K. G. (1975). J. Mol. Biol. 94, 243.
3. Kogoma, T., Torrey, T. A., Connaughton, M. J. (1979). Molec. Gen. Genet. 176, 1.
4. Lark, K. G., Lark, C. A. (1978). Cold Spr. Habor Symp. Quant. Biol. 43, 537.
5. Witkin, E. M. (1976). Bacteriol. Rev. 40, 869.
6. Kogoma, T. (1978). J. Molec. Biol. 121, 5.

7. Subia, N. L. (1980). Thesis, University of New Mexico.
8. Csonka, L. N., Clark, A. J. (1980). J. Bacteriol., 143, 529.
9. Roberts, J. W., Roberts, C. W., Craig, N. L. (1977). Proc. Natl. Acad. Sci. USA, 75, 4714.
10. Little, J. W., Edmiston, S. H., Pacelli, L. Z., Mount, D. W. (1980). Proc. Natl. Acad Sci. USA 77, 3225.
11. Morand, P., Blanco, M., Devoret, R. (1977). J. Bacteriol. 131, 572.
12. Kline, B. C. (1974). Biochemistry 13, 139.
13. Lark, K. G., Lark, C. A. (1980). Mechanistic Studies on DNA Replication and Genetic Recombination. ICN-UCLA Symposia on Molecular and Cellular Biology, Volume XIX, 881. ed. Bruce Alberts, Academic Press, New York, N.Y.
14. Lark, K. G., Lark, C. A. (1981). Structure and DNA-Protein Interactions of Replication Origins. ICN-UCLA Symposia on Molecular and Cellular Biology, Vol. XXI. eds. Dan S. Ray and C. Fred Fox, Academic Press, New York, N.Y.
15. Shibata, T., DasGupta, C., Cunningham, R. P., Radding, C. M. (1979). Proc. Natl. Acad. Sci. USA, 76, 1638.
16. McEntee, K., Weinstock, G. M., Lehman, I. R. (1979). Proc. Natl. Acad. Sci. USA 76, 2615.
17. Ogawa, T., Wabico, H., Tsurimoto, T., Horri, T., Masukata, H., Ogawa, H. (1978). Cold Spr, Habor Symp. Quant. Biol. 43, 909.
18. Roberts, J. W., Roberts, C. W., Craig, N. L., Phizicky, E. M. (1978). Cold Spr. Habor Symp. Quant. Biol. 43, 917.
19. Kenyon, C. J., Walker, G. C. (1981). Nature 289, 808.
20. Meyer, R. R., Glassberg, J., Kornberg, A. (1979). Proc. Natl. Acad. Sci. U.S.A. 76, 1702.
21. Mosig, G., Luder, A., Bock, S., (1981). Structure and DNA-Protein Interactions of Replication Origins. ICN-UCLA Symposia on Molecular and Cellular Biology, Vol XXI. eds. Dan S. Ray and C. Fred Fox, Academic Press, New York, N. Y.
22. LaVerne, L. S., Ray, D. S. (1980). Molec. Gen. Genet. 179, 437.
23. Collins, J., Pritchard, R. H. (1973). J. Mol. Biol. 78, 143.
24. Birge, E. A., Low, K. B. (1974). J. Mol. Biol. 83, 447.
25. Caro, L. G., Berg, C. M. (1968). Cold Spr. Harbor Symp. Quant. Biol. 33, 559.

GENETIC ANALYSIS OF DNA REPLICATION IN SALMONELLA TYPHIMURIUM

Russell Maurer
Barbara C. Osmond
David Botstein

Department of Biology
Massachusetts Institute of Technology
Cambridge, Massachusetts

ABSTRACT

We have begun a systematic genetic analysis of replication-defective mutants of Salmonella typhimurium. The two goals of this work are the description in vivo of: 1. the protein-protein interactions involving replication proteins and 2. the temporal sequence of steps in the replication pathways (1,2). A necessary intermediate aim is the description and unambiguous gene assignment of replication-defective (dna⁻) mutants of Salmonella. We report here the isolation of some 60 new cold- or temperature-sensitive dna⁻ mutants and the analysis of some of these mutants by new, rapid recombination and complementation tests. The mutants analyzed map to 12 loci defined by transductional linkage (using phage P22) to reference Tn10 insertions. The new complementation procedure involves a large library of nearly random fragments of Salmonella DNA cloned in a lambda vector, and a Salmonella strain which adsorbs lambda. Plaques made by hybrid lambda phages containing a dna⁺ gene are overlaid with a lawn of lambda-sensitive Salmonella carrying a conditional-lethal mutation in that gene. The appearance of colonies above the plaque at the non-permissive temperature was shown to be the result of the formation of partial diploids for the dna gene, and thus to be a true indicator of complementation. In two cases tested using the new "plaque complementation" procedure, several mutations at a locus were found to be allelic. Intergeneric plaque complementation of dna genes between Salmonella and E. coli identified Salmonella genes functionally equivalent to dnaC and dnaE. These and other results support the idea that the replication proteins of E. coli and Salmonella are very similar in structure and function.

INTRODUCTION

An important question about DNA replication is whether the biochemical reactions and pathways observed in vitro correctly reflect the situation in vivo. This question arises particularly in the case of the replication of the bacterial chromosome because most in vitro studies have, of necessity, depended on more accessible viral templates. Two general approaches exist for dealing with this question. One approach, made possible by recombinant DNA technology, is the creation of template molecules more akin to the bacterial chromosome for use in biochemical studies. For example, several groups have cloned the oriC region from E. coli onto small plasmids (3,4). The other general approach, the one we have taken, is the study of bacterial mutants altered in some step(s) of DNA replication (for recent review see ref. 5).

The first dna$^-$ mutants in E. coli were isolated in 1965 (6), and since that report additional dna$^-$ mutants have been isolated by several different groups (references cited in 5). These mutants have contributed to the biochemical analysis of DNA replication principally by guiding the purification of dna gene products. Further exploitation of these mutants in vivo has been hampered by the complexities of dealing with them genetically. These complexities are:
1. The mutations exist in several non-isogenic E. coli backgrounds.
2. The genes map all over the E. coli chromosome, a fact which complicates strain construction, mapping. and complementation using F' factors.

We describe here improved genetic methods for Salmonella tymphimurium, a close relative of E. coli, which reduce to a manageable level the technical burden of the complexities referred to above. These methods make feasible the systematic investigation of gene function in bacterial DNA replication by genetic tests (1,2). Earlier investigations using the genetic approach have provided evidence suggesting that the dnaB and dnaC gene products interact (7-9), that the dnaA product and the beta subunit of RNA polymerase interact (10,11; T.Atlung, this volume), and that dnaA acts before dnaC in replication(12). Genetic analysis of dnaZ function is currently in progress (J. Walker, this volume). Clearly, the extension of this approach to the numerous as yet unexplored replication genes is likely to be rewarding.

Our results also demonstrate the feasibility of intergeneric complementation of dna genes between Salmonella and E. coli. It appears that the dna gene products of these two organisms are sufficiently similar in structure and function that the products of single Salmonella genes introduced into E. coli can function in the E. coli replication apparatus. Thus it is possible to identify functionally equivalent genes in the two organisms and to interpret the behavior of the Salmonella mutants in the context of previous work in E. coli.

METHODS

A. Bacteria

Salmonella typhimurium DB9005 (thyA deo) is free of the fels-1 prophage and the "cryptic" LT2 plasmid. The Salmonella strain DB4673 (mal val trp met galE hsdL hsdSA rpsL/F' mal^{B+} from E. coli) is an isolate of strain TS736 (13) obtained from S. Kustu. DB4641 (dnaA1 his3594 thy1152) is strain TB37 (14) and was obtained from the Salmonella Genetic Stock Center. DB4647 (dnaC1 met trp thy) is strain 11G (15) and was obtained from R. Rowbury. Salmonella strains carrying various Tn10 insertions were obtained from J. Roth, P. Osdoby, N. Kleckner, or were constructed in this laboratory. E. coli strains with dna ts mutations were obtained from J. Wechsler. DB5564 is E. coli K12 thr leu supE.

B. Bacteriophage

P22 int HT 12/4 (16) was used for generalized transduction. Two lambda-Salmonella libraries were used. An EcoRI library, carried in lambda gt7 (17), was obtained from R. Davis. A Sau3a partial-digest library derived from DB9005 was constructed following the procedure of Karn et al. (18).

C. Special Growth Conditions

DB4673 and its derivatives adsorb lambda only if grown in the absence of galactose (including galactose present in yeast extract). Therefore for complementation such cells were grown in lambda broth (17) + 0.2% maltose + 1 mM $MgSO_4$. For use in P22 transduction, such cells were grown in LB (17) + 0.05% galactose.

D. Mutagenesis

Bacteria were treated with EMS (19) and lambda phage with hydroxylamine (17) as described.

E. Isolation of Salmonella dna Mutants

Heat- or cold-sensitive mutants identified by replica-plating of EMS-treated DB9005 were screened for a Dna$^-$ phenotype by an adaptation of a published colony autoradiography procedure (20). Full details will be published elsewhere.

F. Transductional Mapping

P22 transducing lysates were prepared separately on each Tn10-bearing reference strain and adjusted to a concentration of 2×10^9 pfu/ml. With an inoculating device, a drop of each lysate was delivered onto a tetracycline agar plate seeded with 5×10^8 dna cells. After incubation at the non-permissive temperature the plates were scored for growth of cells in a patch, indicating cotransduction of dna$^+$ and Tn10 mediated by a particular lysate. Cotransduction was confirmed by repeating the transduction at low multiplicity, selecting tetracycline resistance, and scoring dna as an unselected marker.

G. Plaque Complementation

Salmonella tester cells for complementation were derivatives of DB4673 which had been lysogenized with lambda 112 (21) and made dna ts by cotransduction with an appropriate Tn10. Phage to be tested were allowed to develop overnight into plaques or spots (as appropriate) on a lawn of DB5564 using glass petri dishes. The plates were then sterilized by inversion over chloroform for 30 min. The plates were treated with UV to a dose of 340 ergs/mm^2, and were then overlaid with a mixture containing 1.5 ml top agar, 0.3 ml 20% maltose, 0.1 ml of 0.4% 2,3,5 triphenyl tetrazolium chloride and 0.3 ml of logarithmic phase tester cells. The plates were then incubated at the nonpermissive temperature overnight and scored.

RESULTS

A. Transductional Analysis

Our overall strategy is heavily dependent on the use of Tn10 insertions located near genes of interest. These insertions impose a fundamental uniformity on the manipulation of various loci, serving as selectable markers in strain construction, mapping, and local mutagenesis (22,23). In this context, a locus is defined by a Tn10 insertion and consists of all genes detectably cotransducible (using P22) with a particular Tn10 insertion. Roughly, each such locus corresponds to 1% of the Salmonella chromosome.

In assigning each dna^- mutation to a locus, the first step was to ascertain whether the mutation belonged to one of the loci defined by a reference collection of previously characterized Tn10 insertions. If none of the insertions was nearby, then a new Tn10 insertion was isolated on the basis of its proximity to the dna^- mutation (23), and this new Tn10 then became part of the reference collection.

In terms of the entire collection of mutants, the full application of this approach involves a separate transductional cross of each Tn10-bearing donor strain with each dna^- recipient strain, or a total of thousands, and eventually tens of thousands, of crosses. This approach is only practical with the use of the rapid transduction protocol described in Methods. The unique feature of this protocol is that 32 separate transductions (32 donors x 1 recipient) can be performed on a single agar plate. A similar rapid transduction protocol was developed independently by Rosenfeld and Brenchley (24).

The approach just outlined serves two important purposes. First, it results in the identification of a Tn10 insertion suitable for manipulation of each gene of interest.

Second, because all the mutations are tested with the same reference insertions, the results identify mutations which are linked to a common Tn10 and which therefore might be allelic. A new complementation method for testing allelism of linked mutations is described below.

The results of the transductional anlaysis are summarized in Table 1. Twelve loci containing dna genes have been defined so far. Slightly more than half of the mutants fail to map with the present reference collection of Tn10 insertions, and thus define additional dna genes.

B. Complementation Analysis

We have developed a general method for complementation among dna genes, or in fact any essential genes, of Salmonella. The test utilizes hybrid lambda phages carrying Salmonella DNA. In this test a positive complementation response is observed as a distinctive "plaque morphology" of the lambda hybrid. A valuable feature of the test is that noncomplementing mutant derivatives of the hybrids can be recognized easily. Previously this has been possible only with burdensome effort (25-27) or under unusually favorable circumstances unique to particular genes (28,29).

The method is an adaptation of that developed by Auerbach and Howard-Flanders (30) for the complementation of genes involved in UV resistance in E. coli. First, plaques of the lambda-Salmonella hybrids are allowed to develop normally on an E. coli host. Subsequently, the plate is overlaid with the Salmonella tester strain (i.e., dna ts) which is lysogenic for a lambda phage expressing the same immunity as the hybrid. The plate is then incubated at a temperature nonpermissive for the tester strain. Tester cells overlying a plaque may be infected by the lambda-Salmonella hybrid phage, and if the hybrid carries a gene which complements the defect in the tester, then the infected cells will be able to grow. Thus a positive response is manifested as a red colony of cells above a plaque.

C. Characterization of the Plaque Complementation Assay

An experiment was undertaken to determine whether cells growing above a plaque in a typical complementation assay were in fact diploid for the gene in question, as would be required for true complementation. As the presence of the ts^+

TABLE I

Locus assignment and complementation behavior of Salmonella dna mutants

Tn10 insertion	Map position	Allele numbers	λ hybrids isolated	Allelism results	E. coli locus
zei4::Tn10	48	ts98, ts554, ts469, nalR	yes	not tested	nrdA[a]
purFk741::Tn10	49	ts598	yes	----------	not identified[b]
zib6::Tn10	81	dnaA1c	no	----------	dnaA[c]
zjj1::Tn10	99	dnaC1[d], ts141, ts601, ts602	yes	all mutations are allelic	dnaC
zzz305-1::Tn10	--	ts229, ts305	yes	both mutations are allelic	dnaE
zzz653-3::Tn10	--	ts611, ts653	yes	not tested	not identified[b]
zzz288-2::Tn10	--	ts262, ts288	yes	----------	not identified[b]
zzz603-5::Tn10	--	ts603, cs638	yes	not tested	not identified[b]
zzz660-3::Tn10	--	ts660	no	----------	not identified
zzz645-1::Tn10	--	ts645	yes	----------	not identified[b]
zzz662-1::Tn10	--	ts662	yes	----------	not identified[b]
zzz663-1::Tn10	--	ts663	yes	----------	not identified[b]

[a]tentative; the clone which complements ts98 also complements E. coli nrdA101.
[b]Not dnaB, C, E, G, Z, lig, or nrdA.
[c]Assignment based on map position and mutant phenotype (10, 14, 40)
[d]Mutant allele of Spratt and Rowbury (15)

allele was implied by the growth at high temperature, the essence of the experiment was a demonstration of the presence of the ts⁻ allele. The key to the analysis was the presence of a Tn10 in the tester strain adjacent to the ts⁻ marker. Cells growing over twenty different plaques were repurified by single colony isolation at the permissive temperature (30°) before the analysis to permit the possible segregation of temperature sensitive cells. Isolated colonies were then tested for temperature sensitivity. and 46/170 colonies were in fact temperature sensitive at this point. Thirty-two of the remaining, temperature resistant, colonies were assayed for the presence of the ts⁻ allele by serving as donors of Tn10 in a P22 transduction with a wild-type (ts⁺) recipient. Twenty-eight of the thirty-two gave rise to both ts⁺ and ts⁻ transductants, while the other four gave rise only to ts⁺. Thus the vast majority of the cells examined in sufficient detail were shown to carry both the ts⁻ and the ts⁺ alleles. We conclude that true complementation is the basis of the plaque assay.

D. Allelism Tests and Intergeneric Complementation

The plaque complementation test can be used to determine whether two chromosomal mutations are allelic. The criterion of allelism is that the Salmonella tester strain bearing one of the mutations in question must respond the same as the other tester when challenged with wild-type or mutant lambda hybrids derived from the appropriate part of the chromosome. If an E. coli tester strain is used in place of one of the Salmonella strains, the same criterion defines functionally equivalent genes in the two organisms. These applications of the complementation method can be illustrated by the example of the dnaC locus.

In 1970, Spratt and Rowbury (15) isolated a dna⁻ mutation in Salmonella which we call dnaC1. This was first identified as a dnaC mutation mainly on the basis of mapping data which placed it at a position equivalent to the position of dnaC in E. coli (31). In our collection of mutations, there were three which, on the basis of transductional data, were found to map very close to dnaC1. The following experiments established that all four mutations in Salmonella are allelic and define a gene in Salmonella which is the functional homologue of the dnaC gene of E. coli. A lambda hybrid (lambda 4661A) which carries a 10.7 kb EcoRI fragment of Salmonella DNA, complements all four putative dnaC alleles and also complements a standard dnaC mutant of E. coli, PC2

(32). From a stock of lambda 4661A treated with hydroxylamine, we isolated 10 independent mutant derivatives which now fail to complement dnaC1 at $42°$. Every one of these mutant phages also fail to complement the other Salmonella dnaC mutants at $42°$. All but three of the mutant phages fail to complement the E. coli mutant PC2 at $40.5°$, the highest temperature at which the parent of PC2 can grow. The three exceptional phages apparently carry ts mutations in dnaC because these phages complement dnaC1 at $40.5°$. Complementation by the mutant phages is not restored by an amber suppressor introduced into the tester cells, and therefore it is unlikely that any of these phages carry a polar mutation which inactivates several genes at once. We conclude that lambda 4661A carries the Salmonella dnaC gene, that the phage mutants bear mutations in this gene as do the Salmonella mutants, and that this gene is the functional equivalent of the dnaC gene of E. coli.

A similarly rigorous analysis indicates that two other mutatations define the dnaE-equivalent gene of Salmonella. In addition, wild-type clones corresponding to virtually all the defined Salmonella loci have been isolated (Table 1). Preliminary data suggest that one Salmonella locus may prove to contain the nrdA (dnaF)-equivalent gene. No Salmonella locus has been identified as dnaB, dnaG, dnaZ or lig. Nonetheless, the Sau3a lambda-Salmonella library readily yielded positive responders when screened with E. coli tester strains mutant in these genes.

DISCUSSION

At this early stage in the genetic analysis of DNA replication in Salmonella, our aims are the unambiguous gene assignment of Salmonella mutations and the identification of functionally homologous genes in Salmonella and E. coli. To achieve these aims we have developed two new and generally useful tools.

One tool is the rapid transductional mapping technique. This technique localizes mutations to a small region of the chromosome and thereby identifies mutations that are closely linked to one another. The method is particularly useful in situations like DNA replication where mutations at numerous loci confer a similar phenotype. The underlying idea of a transposon (e.g , Tn10) as a universal linked marker can be applied in other situations, as well. Ultimately the reference collection of Tn10 insertions will grow to a size (about 150-200 insertions) sufficient to define overlapping loci in the Salmonella chromosome. When this is achieved, it will be possible to map any mutation, selectable or not, by a manageable number of standard transductions.

The other technique is the plaque complementation assay. In the present context we have used this assay to isolate lambda hybrids bearing Salmonella dna genes, to demonstrate allelism of chromosomal mutations, and to demonstrate functional homology of Salmonella and E. coli dna genes. We also have shown that noncomplementing mutants derived from the lambda hybrids can be readily isolated, a fact which makes the lambda hybrids immensely useful tools in the study of the structure and function of the dna genes and gene products.

Our results extend earlier observations supporting the close similarity of the replication apparatus in Salmonella and E. coli. Previous work has shown that the Salmonella chromosomal origin can function in E. coli (33). Also, various extrachromosomal elements dependent on host functions for replication can replicate in either organism, including F, colicin E1, and phages lambda, P22, ØX174, fd, and M13 (34-37; A. Wright, personal communication; D. Koshland, personal communication). Crude Salmonella extracts will carry out the ØX174 SS to RF conversion in vitro, and partially fractionated Salmonella extracts will complement at least some heat-inactivated E. coli extracts in this assay (S. Wickner, personal communication). All of these observations can be accounted for by the substitution of entire subsets of gene products of one organism for those of the other. In the intergeneric complementation tests (i.e., using E. coli tester strains) we have found that single Salmonella dna gene products can function in an otherwise all E. coli replication apparatus. We have observed intergeneric complementation with dnaC and dnaE, two genes whose products are known to function as part of multiprotein complexes (38,39). In preliminary experiments we have observed intergeneric complementation for dnaB, dnaG, and dnaZ, genes which also fit this description. Intergeneric compatibility of the dnaC function in vivo and in vitro has recently been demonstrated directly (J. Kobori, personal communication). Thus it appears that to a remarkable degree these dna gene products have preserved the structural elements that enable them to engage in specific protein-protein complexes. Comparisons of the dna genes and proteins of these two organisms may help to define such preserved structural elements.

ACKNOWLEDGMENTS

This work was supported by grants to D.B. from the National Institutes of Health (GM18973 and GM21253). R.M. was supported by Fellowship DRG-246-FT of the Damon Runyon-Walter Winchell Cancer Fund.

REFERENCES

1. Jarvik, J. and Botstein, D. (1973). Proc. Nat. Acad. Sci. USA 70, 2046-2050.
2. Jarvik, J. and Botstein, D. (1975). Proc. Nat. Acad. Sci. USA 72, 2738-2742.
3. Yasuda, S. and Hirota, Y. (1978). Proc. Nat. Acad. Sci. USA 74, 5458-5462.
4. Messer, W., Bergmans, H.E.N., Meijer, M., and Womack, J.E. (1978). Mol Gen. Genet. 162, 269-275.
5. Wechsler, J.A. (1978). In DNA Synthesis, Present and Future, I. Molineux and M. Kohiyama, eds., Plenum, New York, pp. 49-70.
6. Bonhoeffer, F. and Schaller, M. (1965). Biochem. Biophys. Res. Commun. 20, 93-97.
7. Schuster, H., Schlicht, M., Lanka, E., Mikolajczyk, M., and Edelbuth, C. (1977). Molec. Gen. Genet. 151, 11-16.
8. Sclafani, R.A. and Wechsler, J.A. (1981a). J. Bacteriol. 146, 321-324.
9. Sclafani, R.A. and Wechsler, J.A. (1981b). J. Bacteriol. 146, 418-421.
10. Bagdasarian, M.M., Izakowska, M., and Bagdasarian, M. (1977). J. Bacteriol. 130, 577-582.
11. Wechsler, J.A. and Zdzienicka, M. (1975). In DNA Synthesis and Its Regulation, ICN-UCLA Symp. on Molecular and Cellular Biol., vol. 3, M. Goulian and P. Hanawalt, eds., W.A. Benjamin, Menlo Park, Cal.
12. Kung, F.-C. and Glaser, D.A. (1978). J. Bacteriol. 133, 755-762.
13. Palva, E.T. and Liljestrom, P. (1981). Mol. Gen. Genet. 181, 153-157.
14. Bagdasarian, M., Zdzienicka. M., and Bagdasarian, M. (1972). Mol. Gen. Genet. 117, 129-142.
15. Spratt, B.G. and Rowbury, R.J. (1970). J. Gen. Microbiol. 64, 127-138.
16. Schmieger, H. (1972). Mol. Gen. Genet. 119, 75-88.
17. Davis, R.W., Botstein, D., and Roth, R. (1980. Advanced Bacterial Genetics, Cold Spring Harbor Laboratory, Cold Spring Harbor, N.Y.

18. Karn, J., Brenner, S., Barnett, L., and Cesarini,G. (1980). Proc. Nat. Acad. Sci. USA 77, 5172-5176.
19. Miller, J.H. (1972). <u>Experiments in Molecular Genetics</u>, Cold Spring Harbor Lab, Cold Spring Harbor, N.Y.
20. Wechsler, J.A., Nusslein, V., Otto, B., Klein, A., Bonhoeffer, F., Herrmann, R., Gloger, L., and Schaller, H. (1973). J. Bacteriol. 113, 1381-1388.
21. Maurer, R., Meyer, B.J., and Ptashne, M. (1980). J. Mol. Biol. 139, 147-161.
22. Hong, J. and Ames, B. (1971). Proc. Nat. Acad. Sci. USA 68, 3158-3162.
23. Kleckner, N., Roth, J. and Botstein, D. (1977). J. Mol. Biol. 116, 125-159.
24. Rosenfeld, S.A. and Brenchlev, J. (1979). J. Bacteriol. 138, 261-263.
25. Murakami, A., Inokuchi, H., Hirota, Y., Ozeki, H., and Yamagishi, H. (1980). Mol. Gen. Genet. 180, 235-247.
26. Sako,T. and Sakakibara, Y. (1980). Mol. Gen. Genet. 179, 521-526.
27. Kimura, M., Yura, T., and Nagata, T. (1980). J. Bacteriol. 144, 649-655.
28. Georgopoulos, C., Lam, B., Lundquist-Heil, A., Rudolph, C.F., Yochem, J., and Feiss, M. (1979). Mol.. Gen. Genet. 172, 143-149.
29. Georgopoulos, C., Lundquist-Heil, A., Yochem, J., and Feiss, M. (1980). Mol. Gen. Genet. 178, 583-588.
30. Auerbach, J. and Howard-Flanders, P. (1981). J. Bacteriol., in press.
31. Spratt, B.G. and Rowbury, R.J. (1972). Mol. Gen. Genet. 114, 35-49.
32. Carl, P.L. (1970). Mol. Gen. Genet. 109, 107-122.
33. Zyskind, J.W.,Deen, L.T., and Smith, D.W. (1979). Proc. Nat. Acad. Sci. USA 76, 3097-3101.
34. Zinder, N.D. (1960). Science 131, 924-926.
35. Chisolm, R.L., Deans, R.J., Jackson, E.N., Jackson, D.A., and Rutila, J.E. (1980). Virology 102, 172-189.
36. Botstein, D. and Herskowitz. I. (1974). Nature 251, 584-589.
37. Friedman, D.I. and Baron, L.S. (1974). Virology 58, 141-148.
38. Wickner, S. and Hurwitz, J. (1975). Proc. Nat. Acad. Sci. USA 72, 921-925.
39. McHenry, C.s. and Crow, W. (1979). J. Biol. Chem. 254, 1748-1753.
40. Bagdasarian, M., Hryniewicz, M., Zdzienicka, M., and Bagdasarian, M. (1975). Mol. Gen. Genet. 139, 213-231.

INITIATION OF BACTERIOPHAGE T7 DNA REPLICATION [1]

Stanley Tabor, Michael J. Engler,
Carl W. Fuller, Robert L. Lechner, Steven W. Matson,
Louis J. Romano, Haruo Saito, Fuyuhiko Tamanoi,
and Charles C. Richardson

Department of Biological Chemistry
Harvard Medical School
Boston, Massachusetts 02115

ABSTRACT   The reactions occurring at the replication fork of bacteriophage T7 DNA can be partially reconstituted using purified T7 DNA polymerase and T7 gene 4 protein. Lagging strand synthesis is initiated by the synthesis of RNA primers, a reaction catalyzed by the gene 4 protein. The gene 4 protein binds to single-stranded DNA and translocates along the DNA in a 5' to 3' direction. Upon reaching a specific recognition site, the gene 4 protein catalyzes the synthesis of tetraribonucleotide primers complementary to the template. Using DNAs of known sequence we have shown the predominant recognition sequence for the gene 4 protein to be 3'-CTGGG-5' or 3'-CTGGT-5'; the products of synthesis at these sites are RNA primers having the sequence pppACCC or pppACCA. The sequences, 3'-CTGGA/C-5' and 3'-CTGTN-5', are also used but less frequently. Extension of these primers by T7 DNA polymerase constitutes the initial stage of lagging strand synthesis.

Using a set of deletion mutants we have shown

---

[1] This work was supported by United States Public Health Service Grant AI06045 and American Cancer Society, Inc. Grant NP-1K.

that the primary origin of T7 DNA replication <u>in vivo</u> is located at position 15 on the physical map. A secondary origin is present at approximately position 4. We have determined the nucleotide sequence of the primary origin region; this region consists of two tandem T7 RNA polymerase promoters followed by an AT-rich region containing a single gene 4 protein recognition site.

Using purified proteins we have obtained initiation of DNA replication at the primary origin of T7 DNA as measured by the formation of replication bubbles. In addition to T7 DNA polymerase and gene 4 protein, T7 RNA polymerase and the four ribonucleoside triphosphates are required. No initiation occurs on the DNA of T7 LG37, a deletion mutant which lacks the primary origin. We have studied initiation of T7 DNA replication <u>in vitro</u> using a plasmid into which the primary origin region has been cloned. DNA synthesis on this plasmid is also dependent on T7 RNA polymerase and rNTPs. DNA synthesis is specific for plasmids containing the primary origin region provided the plasmids are first converted to linear forms.

INTRODUCTION

The chromosome of bacteriophage T7 is a linear duplex DNA molecule, 40,000 base-pairs in length. Two types of initiation of DNA synthesis occur during the replication of the T7 chromosome: (i) the initiation of DNA replication at the origin of the viral DNA molecule and (ii) the initiation of lagging strand synthesis at replication forks. <u>In vivo</u>, the earliest replication intermediates observed by electron microscopy are replication bubbles, the midpoints of which are located approximately 17% of the distance from the genetic left end of the molecule (position 17) (1,2). Replication proceeds bidirectionally from this origin, generating first Y forms and eventually progeny DNA molecules.

The movement of the replication fork, following initiation at the origin, requires a second type of initiation event, the initiation of lagging strand synthesis. <u>In vivo</u>, synthesis of the lagging strand is the result of the synthesis and processing of multiple Okazaki fragments, each 1000 to 2000 nucleotides in length (3,4). The 5'-termini of these DNA fragments bear oligoribonucleotides (5,6), thus suggesting a role of RNA in priming the synthesis of T7 Okazaki fragments.

In this paper we describe our studies of both types of initiation of T7 DNA synthesis, and our attempts to reproduce them using purified proteins and templates. First we describe our current understanding of the initiation of Okazaki fragments on the lagging strand, and then our studies on initiation at the replication origin.

THE REPLICATION FORK

The fundamental reactions occurring at the replication fork (Fig. 1) can be partially reconstituted using two phage proteins; T7 DNA polymerase and T7 gene 4 protein. (For review, see refs. 7-9.) Whereas T7 DNA polymerase is solely responsible for the polymerization of deoxyribonucleotides, the T7 gene 4 protein has a dual role at the replication fork. The gene 4 protein enables leading strand synthesis to proceed by facilitating the unwinding of the duplex, and it initiates lagging strand synthesis via the synthesis of RNA primers.

Leading Strand Synthesis

Extensive leading strand synthesis by T7 DNA polymerase only occurs in conjunction with the gene 4 protein (10-12). As synthesis proceeds, the gene 4 protein facilitates the unwinding of the DNA in a reaction that is coupled to the hydrolysis of nucleoside triphosphates (13). In fact, the

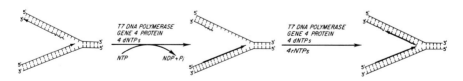

Figure 1. Schematic representation of T7 DNA replication by T7 DNA polymerase and gene 4 protein. (———) Template DNA; (———) Newly synthesized DNA; (∿∿) RNA primers.

gene 4 protein by itself is a single-stranded DNA-dependent nucleoside 5'-triphosphatase. DNA synthesis catalyzed by these two proteins can be quite extensive on circular DNA templates; product molecules with displaced strands up to 20 times the length of the template have been observed (14).

Lagging Strand Synthesis

The gene 4 protein initiates lagging strand synthesis by catalyzing the synthesis of RNA primers on the displaced strand (12,15-17) (Fig. 1). T7 DNA polymerase, like all DNA polymerases, is unable to initiate DNA synthesis de novo. However, it readily uses the RNA primers provided by the gene 4 protein as initiation sites. The RNA primers synthesized by the gene 4 protein in vitro are predominantly tetraribonucleotides having the sequences pppACCC or pppACCA (17,18).

Template Recognition Sequence For RNA Primer Synthesis By The Gene 4 Protein

One reasonable mechanism by which the gene 4 protein would catalyze the synthesis of RNA primers is through a template-directed reaction. Sequences in the template complementary to the primer sequence would direct the synthesis of the primer by a base-pairing mechanism. If this is the case, then the gene 4 protein should recognize the sequence complementary to the primer in the DNA template, or perhaps a larger recognition sequence. To investigate the nature of the recognition signal we have examined the sites of primer synthesis on the single-stranded circular DNA template of ϕX174 (19), whose entire 5386 nucleotide sequence is known (20).

The scheme for identification of gene 4 protein recognition sites is shown in Fig. 2. We have used ϕX174 DNA as a template for T7 DNA polymerase under conditions where DNA synthesis is dependent on RNA primer synthesis catalyzed by the gene 4 protein (15,16,18). Thus, all newly synthesized DNA will have an RNA primer at its 5' end. After removal of the RNA primers with alkali, the resulting 5'-hydroxyl ends of the DNA are labeled using polynucleotide kinase and [γ-^{32}P] ATP. After reannealing the end-labeled fragments to the template, the resulting duplex molecules are cut with an appropriate restriction enzyme. The DNA is denatured to release the radioactive fragments, which are then separated by gel electrophoresis. Each initiation site

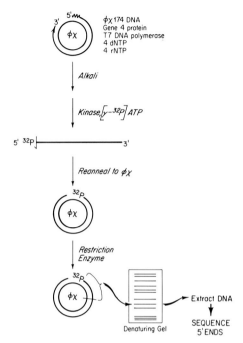

Figure 2. Scheme for the determination of gene 4 protein recognition sites for RNA primer synthesis.

on the template should provide a uniquely labeled fragment whose 5'-labeled end corresponds to the site of initiation.

The result of an experiment carried out as described above is shown in Fig. 3. Thirteen predominant bands are observed, demonstrating that the gene 4 protein recognized some sequence that occurs 13 times on the φX174 molecule. In order to identify the precise locations on the template where each initiation event occurred, the sequence of the 5'-terminus of the DNA of each of the 13 bands was determined by the procedure of Maxam and Gilbert (21).

The template sequences complementary to the 5' ends of all 13 predominant initiation fragments observed on φX174 are presented in Fig. 4. As expected, at every initiation site there is a sequence complementary to the primer sequences ACCC or ACCA. In addition, in every case there is a deoxycytidine residue adjacent to the 3' end of the sequence complementary to the primer. Of the 65 sequences on φX174 DNA complementary to gene 4 protein primers, only these 13 sites are used, and they alone have a deoxycytidine residue at this position.

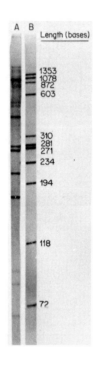

Figure 3. Site-specific initiation on φX174 DNA (19). Samples were run on a 6% polyacrylamide gel containing 7M urea. Lane A contains the RNA-primed molecules which have been treated with alkali to remove the RNA primers, followed by polynucleotide kinase and [γ-^{32}P] ATP to label the 5'-termini, prior to digestion with HaeIII. Lane B contains molecular weight markers: φX174 RFI was digested with HaeIII, and the 5'-termini labeled with ^{32}P. The number of nucleotides in each of the standards is shown on the right.

We conclude that the gene 4 protein recognizes the sequences 3'-CTGGG-5' and 3'-CTGGT-5' on a single-stranded DNA template, at which sites it catalyzes the synthesis of complementary pppACCC or pppACCA primers. Other sites, of the more general recognition sequence 3'-CTGGC/A-5' and 3'-CTGTN-5', can also be used for priming by the gene 4 protein, but less than 5% as efficiently as the predominant sites.

Position	Sequence	HaeIII Fragment (bases)
	3' 5'	
886	CTG|CTGGT|CCCG	216
952	CAA|CTGGT|GGAT	282
1069	AGA|CTGGT|CGTT	89
1078	CTG|CTGGT|TAGA	98
1585	CGT|CTGGG|TATT	411
1789	GTG|CTGGT|CTTT	12
2047	TGA|CTGGG|AGTC	270
2736	CCG|CTGGT|AAGT	959
2836	AAT|CTGGT|TTGG	1059
3650	TGA|CTGGT|CGGC	520
4664	ATG|CTGGT|TATA	175
4688	ACG|CTGGG|AGCC	199
4829	CTG|CTGGT|TTTA	69

Predominant Recognition Sequence 3'—CTGGG—5' (with T above middle G)

RNA Primer pppACCC–dN→ (with A below middle C)

Figure 4. Gene 4 protein priming sites on φX174 DNA. Shown are the nucleotide sequences of the template φX174 DNA in the region of each of the thirteen gene 4 protein initiation sites. The actual sequences analyzed were the thirteen unique 5' ends of newly synthesized DNA complementary to the sequences shown. The initiation fragments used to obtain the sequences were those shown in Fig. 3, lane A. The position refers to the actual site on the template DNA where the terminal ATP of the RNA primer is base-paired. The size of each HaeIII fragment refers to the nucleotide length from the terminal ATP of each RNA primer to the nearest HaeIII cut in the 3' direction.

Recently Okazaki's laboratory has mapped five in vivo initiation sites of Okazaki fragments isolated from newly replicated T7 DNA (22). From these studies they infer the same general recognition sequences (3'-CTGGN-5' or 3'-CTGTN-5'), for RNA primer formation in vivo as we have shown in our in vitro studies.

DNA Binding And Unidirectional Translocation Of The Gene 4 Protein

Comparison of the bands in the autoradiogram in Fig. 3, lane A shows that the gene 4 protein uses some primer sites more frequently than others. Four gene 4 protein recognition sites in a specific region of the φX174 molecule are listed in Table I, along with their relative frequencies of priming. The four sites are separated by 1014, 24 and 149 bases; the relative utilization of the three rightmost sites is 26:1:7.

The data presented in Table I and a similar analysis of all other primer sites on φX174 (19) show that the relative intensity of each band is directly proportional to the distance between the priming site and the next nearest site in the 5'-direction. We propose that the gene 4 protein binds to single-stranded DNA and then translocates in a 5' to 3' direction with regard to the template DNA until it finds a primer site. Based on such a model, the predicted ratio of utilization of the three sites in Table I would be 42:1:6. Thus site 4688 is rarely used because it is effectively masked as a result of exclusive priming at site 4664; while site 4664 is used extremely frequently since there are no competing primer sites on the 5' side of this site for over 1000 bases.

Table I. Spatial arrangment of three primer sites and their relative frequencies of utilization

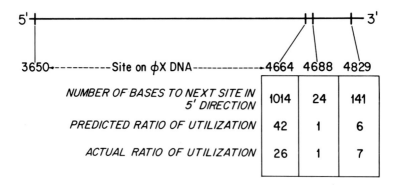

Predicted ratios of utilization are based on the model for polar, 5' to 3' translocation by the gene 4 protein. Actual ratios of utilization are based on the direct comparison of radioactivity in the initiation fragments (19).

Recently we have demonstrated stable binding of the gene 4 protein to single-stranded DNA. This gene 4 protein-DNA interaction requires the presence of a nucleoside triphosphate and is not dependent on the presence of T7 DNA polymerase. When a non-hydrolyzable nucleoside triphosphate (β, γ-methylene-dTTP) is substituted for the hydrolyzable nucleoside triphosphate, a more efficient binding is observed. This result suggests that the interaction of nucleoside triphosphate and gene 4 protein induces a conformational change in the enzyme which contributes to the stability of the enzyme-DNA complex. Non-hydrolyzable nucleoside triphosphates inhibit the priming activity of the gene 4 protein as well as the single-stranded DNA-dependent nucleoside triphosphatase activity. This result is consistent with an essential role of nucleoside triphosphate hydrolysis on unidirectional movement of the gene 4 protein to find primer recognition sites.

Our current concept of the interactions between the gene 4 protein and T7 DNA polymerase at a replication fork is summarized in Fig. 5. The gene 4 protein migrates 5' to 3' on the displaced strand, facilitating the unwinding of the duplex, a reaction coupled with the hydrolysis of nucleoside triphosphates. When a gene 4 protein recognition site occurs on the displaced strand, the gene 4 protein binds to this sequence, synthesizes a complementary tetranucleotide RNA primer, and lagging strand synthesis is initiated by another T7 DNA polymerase molecule.

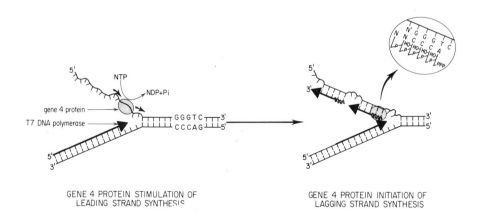

Figure 5. Model of gene 4 protein-T7 DNA polymerase interactions at a replication fork.

INITIATION OF DNA REPLICATION AT THE T7 REPLICATION ORIGIN

Mapping Of The Primary Origin

Dressler et al. (2), using electron microscopy, showed that during the first round of T7 DNA replication, DNA synthesis proceeds bidirectionally from an initial position located approximately 17% of the distance from the genetic left end of the T7 molecule. However, viable mutants of T7 exist that carry deletions between positions 14.55 and 21.85 (23,24). Using a set of these deletion mutants we recently physically mapped an essential component of the primary origin to a 100 base-pair region between positions 14.75 and 15.0 (25). When the primary origin is deleted, T7 DNA replication is initated at secondary origins, of which the predominant one is located at position 4. The nucleotide sequence of the primary origin region has been determined (26), and it is shown in Fig. 6.

Genetic Organization Of The Primary Origin Region

On the basis of functional mapping of the primary origin, sequence analysis of this region, and in vitro studies to be described here, we tentatively define the primary origin region as the region extending from position 14.51 to position 14.95. By this definition, the primary origin region lies between gene 1 (T7 RNA polymerase) on the left, and two small genes, genes 1.1 and 1.2, on the right. While the functions of the protein products of genes 1.1 and 1.2 are not yet known, we have recently shown that the gene 1.2 protein is required for T7 DNA replication, at least in some E. coli mutants (28).

A conspicuous feature of the primary origin region is the two tandem T7 RNA polymerase promoters, $\phi 1.1A$ and $\phi 1.1B$, located at the extreme left. Like all seventeen known T7 RNA polymerase promoters (27), these two consist of a highly conserved 23 base-pair sequence (29,30). However, these two are unique with regard to their close proximity to one another and the lack of any gene between them. Since this region of T7 is also transcribed by the E. coli RNA polymerase in vivo, there is no apparent requirement for these two T7 RNA polymerase promoters for the synthesis of messenger RNA. We speculate that these promoters, and therefore T7 RNA polymerase, are involved in the initiation

26. INITIATION OF BACTERIOPHAGE T7 DNA REPLICATION 397

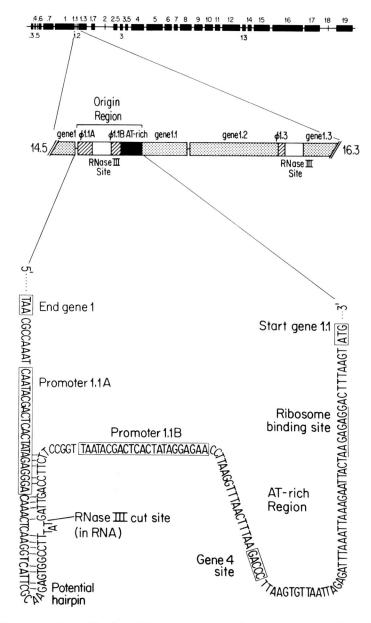

Figure 6. Nucleotide sequence of primary origin of T7 DNA replication. At top is the genetic map of T7 (adapted from ref. 27). Inserts show detailed genetic organization of the primary origin region, and the nucleotide sequence of the primary origin (26).

of DNA replication.

Between the second T7 RNA polymerase promoter, φ1.1B, and the start of gene 1.1, is a 61 base-pair region devoid of any genes or promoters. This region is extremely AT-rich (78% A+T). At its center, in the strand complementary to that shown in Fig. 6, is a single gene 4 protein recognition site, 3'-CTGGG-5'. Primer synthesis at this site would direct DNA synthesis rightward.

In Vitro Initiation Of DNA Replication At The Primary Origin

One approach we have taken to determine the enzymatic reactions involved in initiation has been to study initiation in vitro using T7 RNA polymerase, T7 DNA polymerase, and gene 4 protein (31). When these purified proteins are incubated with wild-type T7 DNA for one min, 6.9% of the DNA molecules contain replication bubbles (eye forms) as observed in the electron microscope (Fig. 7). The remainder are either unreplicated, or have short, randomly located branches characteristic of synthesis at nicks. Analysis of the eye forms reveals that all of the replication bubbles are located between 16 and 24% of the distance from an end of the T7 DNA molecule. In Fig. 8 the molecules have been oriented by denaturation mapping. All of the bubbles are located in the left half of the DNA molecules at a position similar to that observed in vivo.

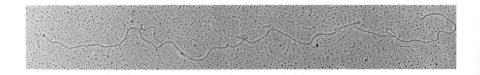

Figure 7. Electron micrograph of a replicating T7 molecule incubated for one minute in the presence of T7 DNA polymerase, T7 RNA polymerase and gene 4 protein.

26 INITIATION OF BACTERIOPHAGE T7 DNA REPLICATION

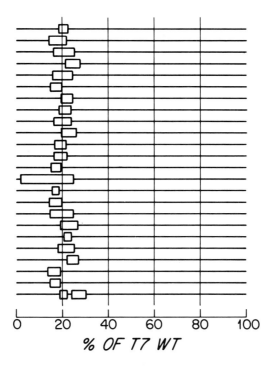

Figure 8. Line diagram of partially replicated T7 DNA molecules isolated from a reaction containing T7 RNA polymerase, T7 DNA polymerase, and gene 4 protein. After incubation for one min the DNA was analyzed by electron microscopy. Random replicating molecules were photographed and measured (31).

The requirements for initiation as measured in this assay are shown in Table II. In the complete system 6.9% of the molecules contain replication bubbles all of which are located near the primary origin. No initiation is observed in the absence of either T7 RNA polymerase or T7 DNA polymerase. In the absence of gene 4 protein the number of replication bubbles is reduced by approximately one-half. RNA synthesis is required for initiation; no eye forms are observed in the absence of rNTPs, and hybridization studies using restriction fragments of T7 DNA show that RNA synthesis is not restricted to the promoters located near the primary origin. When wild-type T7 DNA is replaced by the deletion mutant LG37, which lacks the primary origin, no replication bubbles are observed.

Table II. Requirements for initiation on T7 DNA

Conditions	Eye forms %
Complete system	6.9
- T7 RNA polymerase	0
- T7 DNA polymerase	0
- gene 4 protein	3.0
- 4 rNTPs	0
- wild type T7 DNA + LG37 DNA	0

The complete system contained wild-type T7 DNA, T7 RNA polymerase, T7 DNA polymerase, gene 4 protein, and the four ribo- and four deoxyribonucleoside triphosphates. After incubation for 1 min at 30°, the samples were analyzed by electron microscopy. In each analysis at least 200 full-length T7 molecules were randomly scored for the presence of replication bubbles (31).

In Vitro Initiation On A Plasmid Containing The T7 Primary Origin

A major difficulty in using the entire T7 chromosome to study initiation is that one must rely on electron microscopy to score specific initiation events, as differentiated from synthesis at nicks. To circumvent this problem, we have used a plasmid containing the T7 primary origin (31). As shown in Fig. 9, the plasmid pAR111 consists of pBR322 into which the T7 fragment from position 14.1 to 18.2, and thus including the primary origin region, has been cloned. In addition to the AT-rich region, this recombinant molecule contains the three T7 RNA polymerase promoters $\phi 1.1A$, $\phi 1.1B$, and $\phi 1.3$. As a control we have used the plasmid pDR100, which contains a fragment of T7 DNA of similar size (position 67.5 to 74.1). This fragment contains a single T7 RNA polymerase promoter, $\phi 13$.

In the absence of T7 RNA polymerase, no DNA synthesis is catalyzed by T7 DNA polymerase and gene 4 protein on the

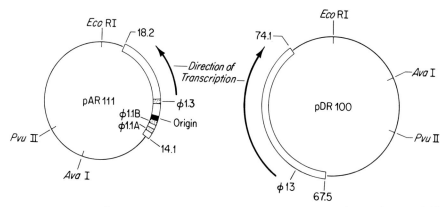

Figure 9. Schematic representation of plasmids pAR111 and pDR100. The boxed segments represent T7 DNA, and the single lines, pBR322 DNA. T7 RNA polymerase promoters are designated by ϕ.

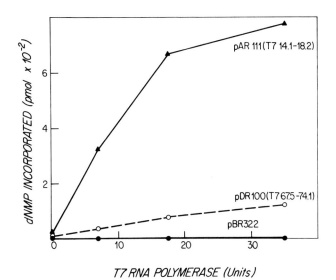

Figure 10. DNA synthesis on cloned T7 DNA. The plasmid DNAs were converted to linear form by digestion with PvuII prior to use as templates. After incubation for 5 min with the indicated amounts of T7 RNA polymerase, the acid-insoluble radioactivity was determined.

plasmids pBR322, pDR100, or pAR111 which have been converted to linear molecules with PvuII (Fig. 10). Upon addition of T7 RNA polymerase, DNA synthesis occurs specifically on the plasmid (pAR111) containing the primary origin. The initiation specificity shown in Fig. 10 requires that the plasmid templates be first converted to linear molecules. When supercoiled templates are used T7 RNA polymerase stimulates DNA synthesis to the same extent on both the plasmid containing the origin (pAR111), and the control plasmid (pDR100), but not on the vector pBR322. At present we do not know why supercoiling eliminates the specificity for initiation. Supercoiling is known to affect transcription from certain E. coli RNA polymerase promoters (32).

Electron microscopic analysis shows that 15% of the pAR111 molecules in the reaction mixture are converted to replicating forms during the 5-min incubation. DNA synthesis initiates specifically within the cloned T7 DNA fragment containing the primary origin since all of the replication bubbles are located near the center of the linear molecules. (PvuII cleaves the plasmid such that the T7 origin region lies near the center of the linear molecule, see Fig. 9.) A typical replicating pAR111 molecule is shown in Fig. 11.

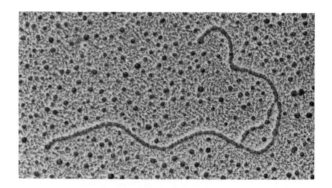

Figure 11. Electron micrograph of a replicating linear plasmid pAR111 molecule incubated for five minutes in the presence of T7 DNA polymerase, T7 RNA polymerase and gene 4 protein.

DISCUSSION

Transcription By T7 RNA Polymerase Is Required For Initiation At The Primary Origin

We have provided evidence for an essential role of T7 RNA polymerase in the initiation of T7 DNA replication at the primary origin. The combination of T7 DNA polymerase, gene 4 protein, T7 RNA polymerase, rNTPs, and dNTPs leads to the formation of replication bubbles on wild-type T7 DNA at a site similar to that observed in vivo. Furthermore, DNA synthesis occurs specifically on a linear plasmid, pAR111, harboring the primary origin region and is dependent on T7 RNA polymerase and rNTPs. We (9,26), as well as others (2,12,33-36), have previously suggested an involvement of T7 RNA polymerase in initiation, but the only direct evidence in vivo is the cessation of T7 DNA synthesis after inactivation of a temperature-sensitive T7 RNA polymerase (34). Studies by others (12,35) have attempted to mimic initiation in vitro using T7 RNA polymerase but no specificity for initiation at the primary origin was observed.

Comparison Of In Vivo And In Vitro Initiation At The Primary Origin

In our present studies the initiation event in vitro resembles that in vivo in that it is site specific, and the replication bubbles are located near the primary origin. However, the replication bubbles do appear to lie further to the right, with their centers having a wide spread around position 19. Although this shift to the right could be due to statistical error in the electron microscopic analysis, examination of the line diagram of the partially replicated molecules shown in Fig. 8 reveals an interesting pattern. With only one exception, the left fork of each replication bubble never extends beyond position 15. At least a third of the left forks of the bubbles are located at precisely position 15. Although not presented in this paper, we have obtained similar evidence for unidirectional replication rightward from position 15 on the cloned origin region using purified proteins.

We suspect that leading strand synthesis proceeds rightward from the primary origin at position 15 until a gene 4 protein recognition site is exposed on the lagging strand (see Fig. 12). Lagging strand synthesis then proceeds leftward in a normal manner until the primary origin is reached. At this point, what was initially lagging strand synthesis now represents leading strand synthesis leftward. However, there is no displaced lagging strand at the left fork, and hence the gene 4 protein cannot bind and translocate in a 5' to 3' direction to facilitate unwinding of the duplex. Thus the leftward synthesis cannot proceed past position 15. In vivo this would probably represent a delay rather than a barrier to the movement of the leftward fork. Removal of the RNA primer by the T7 gene 6 exonuclease which generates a single-stranded region at the fork is one of many mechanisms capable of overcoming this barrier. Such a delay in movement of the fork leftward, combined with a delay in initiation of lagging strand synthesis, may explain the observation that the replication bubbles found in vivo appear at position 17 (2,25) whereas the primary origin, as defined by deletion mapping (25,26), is located at position 15.

Possible Mechanisms Involved In Initiation At The Primary Origin

We do not yet know the precise mechanism by which transcription initiates replication at the primary origin. However, among the several possibilities two likely models are shown in Fig. 12. In Mechanism A, T7 RNA polymerase initiates transcription at one of the two tandem T7 RNA polymerase promoters upstream from the origin. Transcription through the AT-rich region melts apart the DNA strands, exposing the single gene 4 protein recognition site, 3'-CTGGG-5'. The gene 4 protein binds to this site and synthesizes the tetranucleotide RNA primer ACCC, which is then extended by T7 DNA polymerase, initiating DNA synthesis rightward. At a later time a gene 4 protein recognition site is exposed on the displaced strand, and lagging strand synthesis is initiated in a leftward direction by the mechanism described above. In this model, T7 RNA polymerase is serving only an activating role; its movement through the origin region, and not the transcript, is essential. This mechanism is similar to that of transcriptional activation proposed for the initiation of phage lambda DNA replication (37).
In Mechanism B, T7 RNA polymerase again initiates transcription at one of the two promoters. However, after T7 RNA polymerase terminates to make a unique RNA transcript, or the transcript is processed, T7 DNA polymerase uses the transcript itself as the primer for initiation of DNA

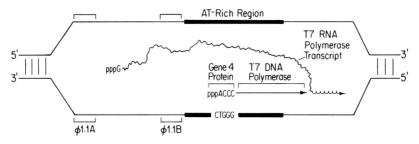

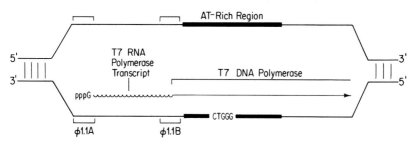

Figure 12. Two models for initiation of T7 DNA replication at the primary origin.

synthesis in a rightward direction. In this mechanism the gene 4 protein is not directly involved in the initiation process, but obviously does play a role in facilitating movement of the fork and synthesizing RNA primers for lagging strand synthesis. Such a mechanism, in which an RNA polymerase transcript is used as the primer, is responsible for the initiation of replication of the single-stranded DNA of phage M13 (37) and of the double-stranded plasmid ColE1 (38).

The possibility exists that both mechanisms depicted in Fig. 12 operate, at least under some conditions. The appearance of replication bubbles at the primary origin of wild-type T7 DNA in vitro in the absence of added gene 4 protein suggests, of course, that initiation can occur without the synthesis of a primer by the gene 4 protein. However, it is difficult to demonstrate conclusively the absence of trace amounts of gene 4 protein contaminating one of the purified enzymes.

One approach we are taking to determine definitively the role of each of the two tandem promoters, and the single gene 4 protein recognition site, is to mutagenize each of these genetic elements in vitro and then to analyze the effects of these lesions on initiation both in vivo and in vitro. We have cloned a 600 base-pair fragment containing the primary origin region into the single-stranded vector M13mp2. The gene 4 protein recognition sequence, 3'-CTGGG-5', has been specifically mutated to the sequence 3'-CCGGG-5', a sequence that is not a recognition site for the gene 4 protein. Mutagenesis was achieved by controlled DNA synthesis up to the gene 4 protein recognition site, and then site-specific misincorporation using reverse transcriptase. We are currently in the process of introducing this mutated fragment into wild-type T7 DNA. In addition we are altering the two T7 RNA polymerase promoters, ϕ1.1A and ϕ1.1B, by using chemically-synthesized mutant promoter fragments as primers for DNA synthesis on wild-type T7 DNA.

We are still some distance from understanding the detailed enzymatic mechanisms responsible for initiation at the primary origin. Regardless of the mechanism by which transcription initiates DNA replication, a number of intriguing questions remain. For example, what unique feature of the primary origin region results in initiation at this particular site? Numerous other T7 RNA polymerase promoters and gene 4 priming sites exist throughout the T7 genome. Although our in vitro initiation system, using only a few highly purified proteins, mimic in vivo initiation with regard to several important criteria, other proteins, perhaps yet unidentified, are likely to be involved. For example, other proteins, such as the T7 gene 6 exonuclease and DNA binding protein, may be necessary to obtain bidirectional replication from the primary origin. Hopefully, the in vitro initiation system can be used to develop assays with which to identify and to characterize these additional conponents of the initiation system.

REFERENCES

1. Wolfson, J., Dressler, D., and Magazin, M. (1972) Proc.Nat.Acad.Sci.U.S.A. 69, 499-504.
2. Dressler, D., Wolfson, J., and Magazin, M. (1972) Proc.Nat.Acad.Sci.U.S.A. 69, 998-1002.
3. Masamune, Y., Frenkel, G.D., and Richardson, C.C. (1971) J.Biol.Chem. 246, 6874-6879.
4. Sternglanz, R., Wang, H.F., and Donegan, J.J. (1976) Biochemistry 15, 1838-1843.
5. Seki, T., and Okazaki, T. (1979) Nucleic Acid Res. 7, 1603-1619.
6. Ogawa, T., and Okazaki, T. (1979) Nucleic Acid Res. 7, 1621-1633.
7. Richardson, C.C., Romano, L.J., Kolodner, R., LeClerc, J.E., Tamanoi, F., Engler, M.J., Dean, F.B., and Richardson, D.S. (1979) Cold Spring Harbor Symp.Quant.Biol. 43, 427-440.
8. Hillenbrand, G., Morelli, G., Lanka, E., and Scherzinger, E. (1979) Cold Spring Harbor Symp.Quant.Biol. 43, 449-459.
9. Tamanoi, F., Engler, M.J., Lechner, R., Orr-Weaver, T., Romano, L.J., Saito, H., Tabor, S., and Richardson, C.C. (1980) in Mechanistic Studies of DNA Replication and Genetic Recombination, ed. Alberts, B. (Academic Press, New York) pp. 411-428.
10. Kolodner, R., Masamune, Y., LeClerc, J.E. and Richardson, C.C. (1978) J.Biol.Chem. 253, 566-573.
11. Hinkle, D.C. and Richardson, C.C.(1975) J.Biol.Chem. 250, 5523-5529.
12. Scherzinger, E. and Klotz, G. (1975) Molec.Gen.Genet. 141, 233-249.
13. Kolodner, R. and Richardson, C.C. (1977) Proc.Nat.Acad.Sci.USA 74, 1525-1529.
14. Kolodner, R. and Richardson, C.C. (1978) J.Biol.Chem. 253, 574-584.
15. Scherzinger, E., Lanka, E. and Hillenbrand, G. (1977) Nucleic Acid Res. 4, 4151-4163.
16. Romano, L.J. and Richardson, C.C. (1979) J.Biol.Chem. 254, 10476-10482.
17. Romano, L.J. and Richardson, C.C. (1979) J.Biol.Chem. 254, 10483-10489.
18. Scherzinger, E., Lanka, E., Morelli, G., Seiffert, D., and Yuki, A. (1977) Eur.J.Biochem. 72, 543-558.
19. Tabor, S., and Richardson, C.C. (1981) Proc.Nat.Acad.Sci.U.S.A. 78, 205-209.

20. Sanger, F., Coulson, A.R., Friedmann, T., Air, G.M., Barrell, B.G., Brown, N.L., Fiddes, J.C., Hutchison, C.A., III, Slocombe, P.M., and Smith, M. (1978) J.Mol.Biol. 125, 225-246.
21. Maxam, A.M., and Gilbert, W. (1980) Methods in Enzymology 65, 499-560.
22. Fujiyama, A., Kohara, Y., and Okazaki, T. (1981) Proc.Nat.Acad.Sci.U.S.A. 78, 903-907.
23. Simon, M.N. and Studier, F.W. (1973) J.Mol.Biol. 79, 249-265.
24. Studier, F.W., Rosenberg, A.H., Simon, M.N., and Dunn, J.J. (1979) J.Mol.Biol. 135, 917-937.
25. Tamanoi, F., Saito, H., and Richardson, C.C. (1980) Proc.Nat.Acad.Sci.U.S.A. 77, 2656-2660.
26. Saito, H., Tabor, S., Tamanoi, F., and Richardson, C.C. (1980) Proc.Nat.Acad.Sci.U.S.A. 77, 3917-3921.
27. Studier, F.W., and Rosenberg, A.H. (1981) J.Mol.Biol. (in press).
28. Saito, H., and Richardson, C.C. (1981) J.Virol. 37, 343-351.
29. Rosa, M.D. (1979) Cell 16, 815-825.
30. Panayotatos, N., and Wells, R.D. (1979) Nature 280, 35-39.
31. Romano, L.J., Tamanoi, F., and Richardson, C.C. (1981) Proc.Nat.Acad.Sci.U.S.A. (in press).
32. Yang, H.-L., Heller, K., Gellert, M., and Zubay, G. (1979) Proc.Nat.Acad.Sci.U.S.A. 76, 3304-3308.
33. Fischer, H. and Hinkle, D.C. (1980) J.Biol.Chem. 255, 7956-7964.
34. Hinkle, D.C. (1980) J.Virol. 34, 136-141.
35. Wever, G.H., Fischer, H., and Hinkle, D.C. (1980) J.Biol.Chem. 255, 7965-7972.
36. Knippers, R., Stratling, W., and Krause, E. (1973) in DNA Synthesis In Vitro, eds. Wells, R.D. and Inman, R.B. (University Park Press, Baltimore) pp. 451-461.
37. Tomizawa, J. and Selzer, G. (1979) Ann.Rev.Biochem. 48, 999-1034.
38. Itoh, T. and Tomizawa, J. (1980) Proc.Nat.Acad.Sci.U.S.A. 77, 2450-2454.

THE PRIMOSOME IN φX174 REPLICATION[1]

Peter M. J. Burgers, Robert L. Low, Joan Kobori, Robert Fuller, Jon Kaguni, Mark M. Stayton, Jim Flynn, LeRoy Bertsch, Karol Taylor and Arthur Kornberg

Department of Biochemistry
Stanford University School of Medicine
Stanford, California

ABSTRACT

In vitro conversion of φX174 viral single-stranded (SS) DNA to the parental duplex replicative form, RF, requires at least eleven purified Escherichia coli proteins. Among these, protein n', a highly specific DNA-dependent ATPase (dATPase), guides proteins n, n", i, dnaC, dnaB and primase to the intragenic locus between structural genes F and G to assemble a mobile priming system, the primosome. Remarkably, the primosome, having served in the SS→RF reaction, is still physically and functionally present in nearly intact form following isolation by sucrose gradient centrifugation. During the next replicative stage (RF→RF), the conserved primosome facilitates cleavage, by the φX-encoded gene A protein, of the viral strand, thereby initiating continuous elongation of this strand. When fully restored by the addition of protein i, the primosome repeatedly primes the discontinuous elongation of the complementary strand through several cycles of progeny RF production.

INTRODUCTION

Replication of the E. coli chromosome requires a large number of replication proteins as predicted by the numerous genes necessary for this process (1). In fact the task of

---

[1] This work was supported in part by grants from the National Institutes of Health and the National Science Foundation.

reconstituting the replication machinery in vitro is so formidable that a simpler approach to dissect the basic features of E. coli replication had to be sought. Our approach lies in the study of the in vitro DNA replication of single-stranded phages M13, G4 and ϕX174 that depend heavily upon the host E. coli for their DNA replication proteins. ϕX174 replication has been especially useful since it requires most, if not all, the replication proteins that operate at the E. coli replication fork.

The replication of ϕX174 DNA can be divided into three stages (2). In the first stage, eleven purified E. coli replicative proteins serve to convert the single-stranded viral circle to the parental duplex, replicative form (SS→RF). In the second stage, multiple copies of progeny RF are produced using the parental RF as a template (RF→RF); in addition to the eleven proteins required in the first stage, this process requires the phage-encoded gene A protein and host rep protein (3,4). In the final stage, production of progeny RF ceases and only the viral strand circles are synthesized for encapsulation into progeny phage (RF→SS) (5).

In this paper two key aspects of this process will be discussed. First, we report that the prepriming proteins n, n', n", i, dnaC and dnaB proteins and primase necessary for stage I (SS→RF) and required in stage II (RF→RF) for strand synthesis assemble into a mobile, functional priming complex, the primosome, on the viral single-stranded DNA coated with single strand binding protein (SSB). Secondly, this primosome assembled in stage I is remarkably conserved and active in stage II to help produce multiple copies of progeny RF.

RESULTS AND DISCUSSION

Proteins Required for the Conversion of SS ϕX DNA into the Duplex Form

Synthesis of the complementary strand on the single-stranded circle of small DNA phages (M13, G4 and ϕX174) is an example of discontinuous replication. The various stages of this process are: prepriming, priming, elongation, gap filling and ligation. These stages have been elucidated and reconstituted in vitro for the three phages (Table 1). Although all three use DNA polymerase III holoenzyme for elongation (6), they differ in the mechanism used to synthesize the primer RNA that holoenzyme can elongate (7).

RNA polymerase recognizes and primes SSB-covered M13 DNA (8); primase recognizes and primes SSB-covered G4 DNA (9). SSB-covered ϕX DNA, however, is not recognized by primase;

TABLE 1. Protein Requirements for in vitro Conversion of Phage Single-stranded DNA to Duplex Closed Circular Form

Stage	M13	G4	φX174
Prepriming	SSB	SSB	SSB protein n protein n' protein n" protein i <u>dnaB</u> protein <u>dnaC</u> protein
Priming	RNA polymerase	Primase	Primase
Elongation	DNA polymerase III holoenzyme	DNA polymerase III holoenzyme	DNA polymerase III holoenzyme
Gap filling & ligation	DNA polymerase I ligase	DNA polymerase I ligase	DNA polymerase I ligase

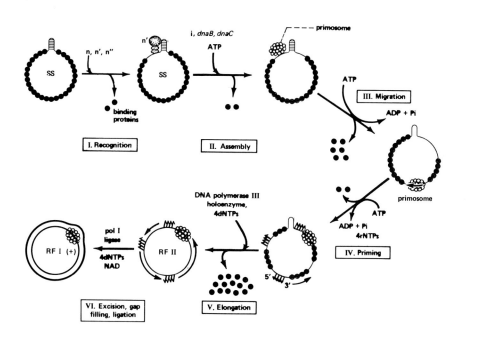

FIGURE 1. Various steps in the conversion of φX174 SS DNA into the duplex replicative form.

unless the DNA is preprimed by six prepriming proteins: proteins n, n', $_2$n", i, dnaB and dnaC (Fig. 1)(10).
Proteins n^2, n' (11), dnaC^3 and dnaB (12) have each been purified to near homogeneity and protein n$_4$" has been partially purified$_3$ through six fractionation steps^4. The dnaB (12,13) and dnaC^3 genes, as well as the dnaG (primase) gene^5 (13) have been cloned into high-copy-number plasmids to overproduce the proteins 20- to 200-fold, considerably simplifying the task of purification and providing material needed for more extensive characterization. The genes coding for proteins n, n', n" and i have not yet been identified.

Irrespective of the mechanism of primer formation used by the different single-stranded phage DNA's, chain elongation of the primer in each case is carried out by DNA polymerase III holoenzyme, a multi-polypeptide replication complex (6). Although the exact subunit structure of holoenzyme is not yet known, several subunits have been extensively characterized and their structural genes determined (Table 2) (14-20). Reconstitution experiments have shown that the form of holoenzyme used in the replication of each phage DNA may differ (Table 2) (18). A new subunit, ζ, has been recently identified as necessary for G4 and M13 replication^6. The requirement for this subunit had not been appreciated before because of its presence as a minor impurity in primase (18). Overproduction of primase (dnaG) eliminated this impurity^5 and thus provided an assay for ζ activity. From SDS polyacrylamide gel electrophoresis, a molecular weight of 48,000 D is tentatively assigned to the ζ subunit.

Roles of Protein n' and dnaB in the Primosome

The φX174 primosome, defined as the active priming complex of the prepriming proteins and primase, contains the two DNA-dependent ATPases, n' protein and dnaB protein, that have been studied in detail.

Protein n' (factor Y of Wickner and Hurwitz (21)) contains an intrinsic, DNA-dependent ATPase (dATPase) activated by a locus on φX DNA between structural genes F and G (11). This site is preserved in an exonuclease VII digest of the

---

^{2}Low, R. L., Shlomai, J. and Kornberg, A., manuscript in preparation.
^{3}Kobori, J., and Kornberg, A., manuscript in preparation.
^{4}Low, R. L., Shlomai, J. and Kornberg, A., unpublished results.
^{5}Stayton, M., Yasuda, S., and Kornberg, A., unpublished results.
^{6}Burgers, P. M. J. and Kornberg, A., unpublished results.

TABLE 2. DNA Polymerase III Holoenzyme Subunits

Subunit		Mw (x 10^3)	Locus	Required for SS→RF			Reference
				G4	φX	M13	
core	α	140	dnaE				(14,15)
	ε	25					
	θ	10		+	+	+	
	β	37	dnaN				(16,17)
	γ	52	dnaZ				(18)
	δ	32	dnaX	−	+	−	(19)
	ζ	(48)[a]		+	n.d.	+	
	τ	83		−	−	−	(20)

[a]Burgers, P. M. J. and Kornberg, A., unpublished results; n.d. = not determined.

HaeIII Z-1 fragment, a 55-nucleotide long fragment from residue 2301 to 2354 that contains the potential for a 44-nucleotide hairpin structure (22). Unlike the SSB-coated SS φX DNA, SSB-coated SS G4 and M13 DNA show negligible ATPase (dATPase) effector activity (23). Furthermore, this HaeIII Z1 fragment of φX174 uniquely promotes assembly of the φX priming system complex, the primosome, indicating that protein n' initiates the assembly of proteins n, n", i, dnaC, dnaB and primase at this site (24).

Protein n' effector sites have also been identified at the pBR322 (25) and ColE1[7] lagging-strand origin sites. A separate site on ColE1 is also present upstream on the opposite (L) strand[7]. The ColE1 lagging-strand origin site (H strand of the HaeII E fragment) when cloned into M13 SS DNA supports an in vivo rifampicin-resistant, dnaG-dnaB dependent conversion of the chimeric SS DNA to the RF (26). Recently, in collaboration with Nomura and Ray, we have found that this chimeric SS DNA in vitro promotes a rifampicin-resistant SS→RF conversion dependent on each of the purified φX replicative proteins[7].

The dnaB protein DNA-dependent ATPase (GTPase) differs significantly from the protein n' activity (27-30). The dnaB protein ATPase (GTPase) can use a wide variety of single-stranded DNAs, but ATP hydrolysis is completely suppressed by SSB. Studies with the dnaB-primase general priming system (31) and experiments with nonhydrolyzable ATP analogs suggest that the dnaB protein complexed with ATP helps the primosome identify sites where primers can be started; the dnaB protein:ATP complex induces DNA conformational changes required for primase initiation (30).

[7]Low, R. L, Kornberg, A., Nomura, N. and Ray, D. S., unpublished results.

Protein n' and dnaB are Stably Bound to φX174 DNA

To determine the stoichiometry of proteins n' and dnaB in the prepriming intermediates and their fate at the different stages of the conversion of the single-stranded circle into the replicative form, these two proteins were isotopically labeled by reductive methylation with tritiated sodium borohydride (31,32). At any given stage, proteins bound to the DNA can be separated from unbound proteins by Bio-Gel A-5m gel filtration or by sucrose density centrifugation. The stoichiometry of labeled, bound protein can be determined by the radioactivity associated with the DNA.

These experiments show that in the different stages of the φX SS→RF reaction, even up to the covalently closed duplex form, proteins n' and dnaB (expressed as a hexamer of the 50,000 D monomer) each remain bound in a stoichiometry of one molecule per replicated circle (Table 3) (33). Thus, these key primosomal proteins are fully retained in the completed parental RF. Moreover, the dnaB binding is highly specific as indicated by isotope exchange studies (30). A stable non-exchangeable dnaB-DNA complex can be obtained only if all the prepriming proteins are present in the prepriming intermediate. Furthermore, only φX, not G4 or M13 DNA, can support this specific complex (Table 4) (30).

Functional Retention of the Primosome on Covalently Closed Duplex φX DNA.

Recently, we have been able to reconstitute replication of the φX duplex form (RF→RF) from purified enzymes by combining the SS→RF with the RF→SS system (34). This reaction requires all the prepriming proteins, primase, gene A protein, rep protein and DNA polymerase III holoenzyme. Many RFII molecules are made from one parental gene A-RFII molecule. Since proteins n' and dnaB are physically retained on the φX duplex form, we determined whether other prepriming proteins are functionally present. The RF→RF reaction provides such a functional test for the presence of an active primosome. The synthetic RFI isolated by sucrose density centrifugation, when supplemented with the RF→SS enzymes (gene A, rep, holoenzyme), produced no RFII, but only SS φX DNA (33). However, when the reaction was further supplemented with protein i, the primosome became fully functional and many copies of RFII DNA were produced per input parental RFI (Fig. 2). On the other hand, when the synthetic RFI was isolated by gel filtration, the primosome had to be supplemented with prepriming proteins n + n", i and dnaC (33). These results indicate that the primosome is unstable and gentler methods are needed to isolate φX RFI DNA in an intact and stable form.

TABLE 3. Retention of Proteins n' and dnaB with synthetic RFI

Product	Molecules per replicated circle	
	Protein n'	dnaB protein
Prepriming intermediate	1.0	1.2
RFII (priming, DNA synthesis)	1.0	1.1
RFI (gap-filling, ligation)	0.9	0.9

The prepriming intermediate was formed in a 150-μl reaction mixture containing: 90 μl of buffer A (20 mM Tris-HCl pH 7.5/20 mM KCl/5% sucrose/ 8 mM MgCl$_2$/0.1 mM EDTA/5 mM dithiothreitol/0.2 mM ATP/0.1 mg/ml bovine serum albumin), 2.2 nmol (as nucleotide) of ϕX SS DNA, 6.6 μg of SSB, 0.15 μmol of ATP, 4.7 μg of dnaC protein, 0.4 μg of protein i, 0.5 μg of proteins n + n" mixture, and either 0.7 μg of ^{3}H-labeled dnaB protein (1.5 x 10^5 cpm/μg) and 43 ng of protein n' or 0.7 μg of dnaB protein and 43 ng of ^{3}H-labeled protein n' (5 x 10^4 cpm/μg). After a 30-min incubation at 30°C, the intermediate was filtered through a 5-ml Bio-Gel A-5m column (0.5 x 26 cm) equilibrated at room temperature with buffer A. Void-volume fractions (25-μl aliquots) were assayed for DNA replication by using [α-^{32}P]dTTP (10 cpm/pmol). The remainder of each fraction was assayed for labeled n' or dnaB protein. For synthesis of RFII, the reaction mixture was supplemented with rNTPs, unlabeled dNTPs, [α-^{32}P]dTTP (20 cpm/pmol), primase (0.3 μg), and pol III holoenzyme (0.25 μg) and incubated an additional 20 min; for RFI, DNA polymerase I (10 ng), NAD (7.5 nmol), and DNA ligase (1 μg) were also added and the incubation was extended 40 min.

TABLE 4. Requirement for the formation of dnaB protein·DNA complex

DNA	SSB	n + n"	n'	i	dnaC	dnaB complexed with DNA (molecule/circle)
φX174	+	−	−	−	−	<0.05
	+	+	−	−	−	<0.05
	+	−	+	−	−	<0.05
	+	−	−	+	−	<0.05
	+	−	−	−	+	<0.05
	+	+	+	−	+	<0.05
	+	+	−	+	+	0.07
	+	+	+	+	−	<0.05
	+	+	+	+	+	<0.05
	+	+	+	+	+	1.13
	−	+	+	+	+	0.73
M13	+	−	−	−	−	<0.05
	+	+	+	+	+	<0.05
G4	+	−	−	−	−	<0.05
	+	+	+	+	+	<0.05

The reaction mixture contained in 25 μl: 5 μl of buffer A (Table II), 15 nmol of ATP, 750 pmol (as nucleotide) of φX, G4 or M13 SS DNA, 0.5 μg of [^{3}H]dnaB protein, and, when present, 90 ng of protein n', 0.3 μg of proteins n + n" mixture, 90 ng of protein i, 1.2 μg of dnaC protein, and 2 μg of SSB. Incubation was at 30°C for 20 min. The reaction mixtures were filtered on a Bio-Gel A-5m agarose column (0.2 x 20 cm) equilibrated with buffer A. The [^{3}H]dnaB protein associated with DNA in the void-volume was measured in a liquid scintillation spectrometer.

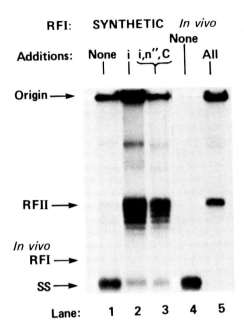

FIGURE 2. Autoradiogram of products from RF→RF reaction with synthetic RFI DNA isolated by sucrose-gradient. Reaction mixtures (80 µl) contained RFI (234 pmol as nucleotide), 1 mM ATP, rNTP's, unlabeled dNTP's, [α-^{32}P]dTTP (600 cpm/pmol), gene A and rep proteins, SSB, pol III holoenzyme, and prepriming proteins as indicated. Reaction mixtures with in vivo RFI contained 240 pmol (as nucleotide) in 80 µl of buffer A (Table III) plus 0.8 mM ATP, the above mentioned RF→SS DNA components and either all or none of the prepriming proteins plus primase.

Retention of the primosome renders the synthetic RFI a far more efficient substrate for gene A cleavage than supercoiled RFI extracted from infected cells. The initial rate of DNA synthesis in the RF→SS reaction with the synthetic RFI is 20 times faster than that obtained with in vivo supercoiled RFI (Fig. 3) (33). The need for gyrase catalyzed supercoiling, previously considered essential, can be dispensed with (35).

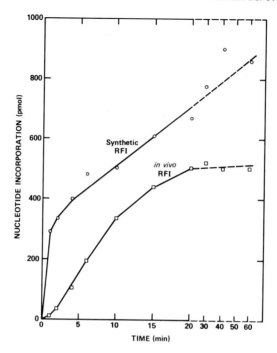

FIGURE 3. Kinetics of the RF→SS DNA reaction. The reaction mixture (120 µl) contained 360 pmol (as nucleotide) of either Bio-Gel-purified synthetic RFI (o) or in vivo RFI (□) and the RF→SS DNA system (Fig. 1). Incorporation of [^{32}P]dTTP into DNA was determined by precipitation of 5 µl aliquots with 10% trichloroacetic acid.

The Primosome Requires ATP.

The stages in which the primosome requires the energy of ATP hydrolysis have been analyzed. Protein n' and dnaB protein ATPases can easily be distinguished since the former has dATPase and very low GTPase activity (11), while the latter has GTPase and negligible dATPase activity (29). A stable prepriming replication intermediate (primosome minus primase) was assembled on the SSB-coated φX circle and the complex isolated by gel filtration in the presence of ATP (36). In the absence of ATP, no stable primosome could be retained on the DNA. Studies with nonhydrolyzable analogs of ATP indicate that ATP hydrolysis is not necessary for stable binding of the primosome (36). Although the primosome assembles at or near the n' site, movement of the primosome

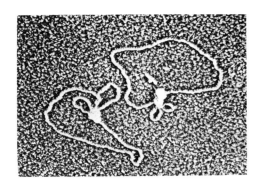

FIGURE 4. Electron micrograph of synthetic RFI DNA.

around the chromosome in the anti-elongation direction (37) requires the energy of ATP hydrolysis (24,36). A comparison of the ATPase, dATPase and GTPase activities of the prepriming replication intermediate (10 nmol, 6 nmol, and 1 nmol, respectively, of triphosphate hydrolyzed per pmol of complex per min at 30°C) is one indication that the n' ATPase likely fuels movement along the DNA (36).

The Primosome Freely Moves on the φX RFI DNA

SS φX DNA was converted to RFI DNA by the action of the SS→RF enzymes and the RFI DNA was isolated by sucrose density centrifugation in the presence of Mg-ATP (33). Electron microscopy of the DNA showed that a large protein complex was present on the DNA (Fig. 4)[8]. The size of the complex agrees with the expected mass of the primosome (ca. 0.5×10^6 D). To determine if the primosome was located at a specific site, the synthetic RFI was cut by the restriction endonuclease PstI and the linear molecules examined by electron microscopy. The primosome was randomly distributed on the DNA (Fig. 5). However, when the synthetic RFI was first cut by gene A protein and then by PstI endonuclease, most of the molecules had the primosome located at 20% from the nearest end of the linear molecule, exactly the distance between the gene A and PstI sites (38)) (Fig. 5). Thus the primosome appears to move freely along the DNA, using the n' ATPase (dATPase) as an energy source, until it encounters the site of gene A cleavage where further movement is restricted. Indeed, the ATPase (dATPase) activity associated with the primosome-RFI complex is abolished by gene A cleavage[8].

---

[8] Low, R. L., Kornberg, A. and Griffith, J., unpublished results.

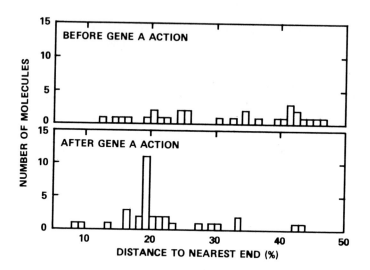

FIGURE 5. Length histogram of primosome location on synthetic RFI DNA cut by Pst1 without (upper) and with (lower) prior cleavage by gene A protein.

During the φX DNA replicative cycle, conservation of the primosome has important advantages. First, retention of the primosome provides a DNA topology that favors efficient cleavage by gene A protein, independent of supercoiling by gyrase. Secondly, conservation of an intact primosome on the DNA may provide a more efficient mechanism for fork movement in duplex DNA replication (Fig. 6). Traveling along the lagging strand of the replication fork in the direction of fork movement (37) without the necessity for repeated dissociation and reassociation, the primosome may, together with holoenzyme and rep protein, constitute part of a larger replisome and promote efficient duplex replication (Fig. 6).

27 THE PRIMOSOME IN θ/X174 REPLICATION

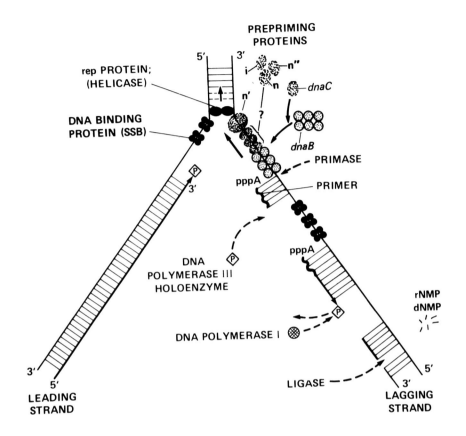

FIGURE 6. A scheme of the structure of the replication fork of the E. coli chromosome.

REFERENCES

1. Kornberg, A. (1980). "DNA Replication." W. H. Freeman and Co., San Francisco, p. 376.
2. Denhardt, D. T. (1975). CRC Crit. Rev. Microbiol. 4, 1611.
3. Tessmar, E. S. (1966). J. Mol. Biol. 17, 218.
4. Denhardt, D. T., Iwaya, M., and Larison, L. L. (1972). Virology 49, 486.
5. Fujisawa, H. and Hayashi, M. (1976). J. Virol. 19, 416.
6. McHenry, C. and Kornberg, A. (1977). J. Biol. Chem. 252, 6478.

7. McMacken, R., Bouché, J. P., Rowen, S. L., Weiner, J. H., Ueda, K., Thelander, L., McHenry, C., and Kornberg, A. (1977). RNA Priming of DNA Replication. In "Nucleic Acid-Protein Recognition" (H. J. Vogel, ed.) p. 15. Academic Press, New York.
8. Geider, K. and Kornberg, A. (1974). J. Biol. Chem. 249, 3999.
9. Zechel, K., Bouché, J. P. and Kornberg, A. (1974). J. Biol. Chem 250, 4684.
10. Meyer, R. R., Shlomai, J., Kobori, J., Bates, D. L., Rowen, L., McMacken, R., Ueda, K., and Kornberg, A. (1978). Cold Spring Harbor Symp. Quant. Biol. 43, 289.
11. Shlomai, J. and Kornberg, A. (1980). J. Biol. Chem. 255, 6789.
12. Arai, K., Yasuda, S. and Kornberg, A., submitted for publication.
13. McMacken, R. (1981). J. Supramol. Str. Cell. Biochem. Supplement 5, 343.
14. McHenry, C. S. and Crow, W. (1979). J. Biol. Chem. 254, 1748.
15. Gefter, M. L., Hirota, Y., Kornberg, T., Wechsler, J. A., and Barnoux, C. (1971). Proc. Natl. Acad. Sci. USA 60, 3150.
16. Johanson, K. O., and McHenry, C. S. (1980). J. Biol. Chem. 255, 10984.
17. Burgers, P. M. J., Kornberg, A., and Sakakibara, Y. Proc. Natl. Acad. Sci. USA, in press.
18. Hübscher, U., and Kornberg, A. (1980). J. Biol. Chem. 255, 11698.
19. Hübscher, U., and Kornberg, A. (1980). Proc. Natl. Acad. Sci. USA 76, 6284.
20. McHenry, C. S. (1980). In "Mechanistic Studies of DNA Replication and Genetic Recombination" (Alberts, B. and Fox, C. F., eds.) Academic Press, New York, p. 569.
21. Wickner, S. and Hurwitz, J. (1975). Proc. Natl. Acad. Sci. USA 72, 3342.
22. Shlomai, J. and Kornberg, A. (1980). Proc. Natl. Acad. Sci. USA 77, 799.
23. Shlomai, J. and Kornberg, A. (1980). J. Biol. Chem. 255, 6794.
24. Arai, K. and Kornberg, A. (1981). Proc. Natl. Acad. Sci. USA 78, 69.
25. Zipursky, S. L. and Marians, K. J. (1980). Proc. Natl. Acad. Sci. USA 77, 6521.
26. Nomura, N. and Ray, D. S. (1980). Proc. Natl. Acad Sci. USA 77, 6566.
27. McMacken, R., Ueda, K., and Kornberg, A. (1977). Proc. Natl. Acad. Sci. USA 74, 4190.

28. Ueda, K., McMacken, R., and Kornberg, A. (1978). J. Biol. Chem. 253, 4051.
29. Reha-Krantz, L. and Hurwitz, J. (1978). J. Biol. Chem. 253, 4051.
30. Arai, K., and Kornberg, A. (1981). J. Biol. Chem. 256, 5247; and subsequent papers (pp. 5247-5280).
31. Arai, K., and Kornberg, A. (1979). Proc. Natl. Acad. Sci USA 76, 4308.
32. Rice, R. H. and Means, G. E. (1971). J. Biol. Chem. 246, 831.
33. Low, R. L., Arai, K. and Kornberg, A. (1981). Proc. Natl. Acad. Sci. USA, 78, 1436.
34. Arai, K., Arai, N., Shlomai, J. and Kornberg, A. (1980). Proc. Natl. Acad. Sci. USA 77, 3322.
35. Henry, T. J. and Knippers, R. (1974). Proc. Natl. Acad. Sci. USA 71, 1549.
36. Arai, K., Low, R. L., and Kornberg, A. (1981). Proc. Natl. Acad. Sci. USA 78, 707.
37. Arai, K and Kornberg, A. (1981). Proc. Natl. Acad. Sci. USA 78, 69.
38. Eisenberg, S., Griffith, J. and Kornberg, A. (1977). Proc. Natl. Acad. Sci. USA 74, 3198.

ROLE OF THE β SUBUNIT OF THE ESCHERICHIA COLI DNA POLYMERASE III HOLOENZYME IN THE INITIATION OF DNA ELONGATION

Kyung Oh Johanson and Charles S. McHenry

Department of Biochemistry and Molecular Biology
The University of Texas Medical School
Houston, Texas 77025

ABSTRACT The β subunit of the DNA polymerase III holoenzyme has been purified 10,000-fold to homogeneity from E. coli HMS-83. The native and denatured molecular weights of β are 72,000 and 37,000, respectively, as determined by equilibrium sedimentation. Thus, β appears to exist as a dimer when in a state free of other holoenzyme components. An antibody directed specifically against the β subunit has been prepared and used to probe the structure of the DNA polymerase III holoenzyme and the mechanism of its reaction. This antibody has been shown to inhibit only those reactions for which holoenzyme is required and the polymerase III core alone will not suffice. Furthermore, the β antibody inhibits formation of an ATP (or dATP) requiring initiation complex with primed DNA, but not subsequent elongation. Using the β antibody to block reinitiation, we have shown that holoenzyme can replicate most of the G4 genome (>5,000 nucleotides) without dissociating.

INTRODUCTION

The DNA polymerase III holoenzyme has been implicated by both biochemical and genetic evidence as a true replicative enzyme responsible for the synthesis of most of the E. coli chromosome (for a review, see 1). It is a complex multisubunit enzyme composed of a DNA polymerase III core and several auxiliary proteins (2). The polymerase III core, which only catalyzes limited synthesis on nuclease treated duplex DNA (3,4), is composed of three subunits α, ε, and θ of 140,000, 25,000 and 10,000 daltons, respectively (4,5,6). The holoenzyme contains at least four auxiliary subunits (β, γ, δ and τ) which permit it to function as a replicative enzyme (2,5,7). Other laboratories have isolated three proteins, Factor I, dnaZ protein, and Factor III, which permit polymerase III to catalyze synthesis on a long primed single-stranded template (8,9,10). These proteins probably correspond to β, γ and δ (2).

The specific role of holoenzyme auxiliary proteins is not clear. However, preliminary work indicates a possible cycling of holoenzyme subunits during the replication process (10,11). Holoenzyme, in the presence of ATP (or dATP) can form an isolable initiation complex upon incubation with a primed single-stranded DNA template. Upon formation of this complex, elongation proceeds in the presence of an antibody which contains IgG directed against copol III*. β was probably the activity contained in copol III* preparations (2,12). Furthermore, it has been suggested that when isolated separately, replication elongaton Factor III and the dnaZ protein can transfer replication elongation Factor I to the DNA template in an ATP (or dATP) requiring process (10).

In this article we will discuss a portion of our ongoing research concerning the DNA polymerase III holoenzyme, confining ourselves to the β subunit. This protein appears to serve a role in the initiation of DNA elongation—not initiation at a unique chromosomal origin, but the beginning of Okazaki fragment synthesis from a primer.

RESULTS AND DISCUSSION

<u>Purification of β</u> - As we have reported (13) β has been purified 10,000 fold to homogeneity from E. coli HMS-83, a polA, polB mutant. The enzyme was purified using as an assay its requirement to reconstitute holoenzyme activity on a G4 template (Fig. 1). Holoenzyme was resolved into two proteins, β and DNA polymerase III* (pol III*) by phosphocellulose chromatography. These two proteins are alone inert in the G4 assay but together reconstitute holoenzyme activity. β was further purified by gel filtration, cation and anion exchange chromatography (13). An SDS acrylamide gel of pure β and our most highly purified holoenzyme is shown in figure 2.

<u>Molecular Weight of β</u> - SDS gel electrophoresis indicates that β has a molecular weight of 40,000 or greater, but all other methods suggest this value to be anomalously high (13). Equilibrium sedimentation in guanidine·HCl indicates β to be 37,000 daltons. β exhibits a native molecular weight of 72,000 in similar runs performed under nondenaturing conditions. The method of Siegel and Monty yields a value of 70,000 daltons for the native protein (13). Thus, β appears to exist as a dimer in its native state when dissociated from other holoenzyme subunits.

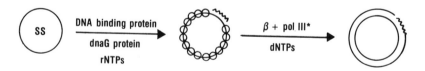

Figure 1. Assay for β subunit. β and pol III* together reconstitute holoenzyme.

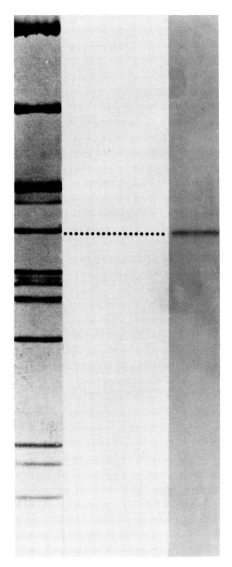

Figure 2. SDS acrylamide gel of purified DNA polymerase III holoenzyme (left) and β (right).

Table 1

Antibody Effects upon Holoenzyme on Various Templates

DNA Template	Antibody	DNA synthesis (% of antibody reaction)
G4	none	100[a]
G4	control IgG	96
G4	anti-β IgG	1
G4	anti-β IgG + excess β	94
Salmon Sperm	none	100[b]
Salmon Sperm	control IgG	77
Salmon Sperm	anti-β IgG	85

[a] Total incorporation was 68 pmol in the standard G4 assay (13).
[b] Total incorporation was 130 pmol in the standard pol III Salmon Sperm DNA assay (5).

<u>Inhibition of Holoenzyme by β Antibody</u> - Anti-β IgG inhibits the holoenzyme reaction on the natural G4 template (Table 1). However, this antibody is without effect in artificial nuclease-activated duplex DNA assays which do not require β (13).

Our anti-β IgG preparation even shows specificity within the holoenzyme reaction; it inhibits initiation but not elongation (Fig. 3, Table 2). Once an initiation complex is formed and isolated by Biogel A5m filtration; elongation is not affected. The inhibition observed is caused by a block in reforming initiation complex from components which have dissociated subsequent to isolation. To support this conclusion we note that (i) adding more antibody does not result in further inhibition; (ii) N-Ethylmaleimide (NEM), an agent which blocks reinitiation by attacking pol III, causes similar inhibition but does not further inhibit reactions to which anti-β IgG has been added; (iii) only the slow kinetic phase of the elongation reaction is blocked by anti-β IgG. Reactions beginning from initiation complex proceed to completion within 15 seconds under optimal conditions; reactions proceeding from dissociated components require several minutes.

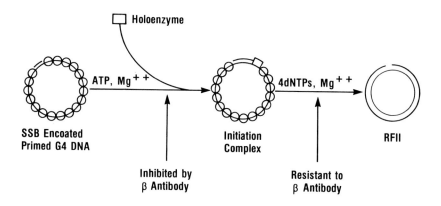

Figure 3. β antibody inhibits initiation but not elongation.

Table 2

DNA Synthesis in the Presence and Absence of Initiation Inhibitors[a]

		DNA synthesis (pmol)
A.	Initiation Complex + dNTPS	1300
B.	Initiation Complex + anti-β IgG + dNTPS	600
C.	Primed G4 + Holoenzyme + dNTPS	1500
D.	Primed G4 + anti-β IgG + holoenzyme + dNTPS	40

[a] Assay conditions were as previously described (15).

Processivity of the Holoenzyme Reaction - The availability of an agent which blocks reinitiation but not elongation provides a useful tool to determine the processivity of a polymerase. The Hae III restriction map of G4 (14) indicates that fragment Z4 should be made early and fragments Z1 and Z6 late relative to the unique G4 origin (Fig. 4). Thus, the Z4/Z1 or Z4/Z6 ratio could be used as an index of the portion of G4 circles completed. A time course of our in vitro reaction verifies that these fragments are synthesized in the proper order (Fig. 5). Fragment Z4 is completed before Z1 and Z6 are begun (15).

Next, the processivity of the holoenzyme was analyzed using this methodology (Fig. 6). The Z4/Z1 and Z4/Z6 ratios are the same for both unihibited reactions and reactions to which anti-β IgG was added subsequent to initiation complex formation (15). Addition of IgG to reactions prior to initiation complex formation results in complete inhibition.

To verify these results, reinitiation was blocked using an alternative method that is not dependent upon the inhibition of β (15). The β protein is resistant to NEM(2); the core polymerase III is very sensitive (3). Once the holoenzyme binds to DNA to form a tight initiation complex, it is protected from high concentrations of this reagent (Fig. 6). Thus, NEM can be used, like β antibody, to block reinitiation. The addition of this reagent prior to the addition of holoenzyme, results in complete inhibition. If the elongation reaction proceeds from initiation complex in the presence of NEM, the ratio of Z4 to Z1 or Z6 is the same as a completely replicated control (Fig. 5).

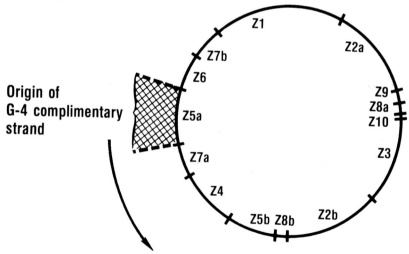

Figure 4. Hae III restriction map used to analyze processivity (14,15).

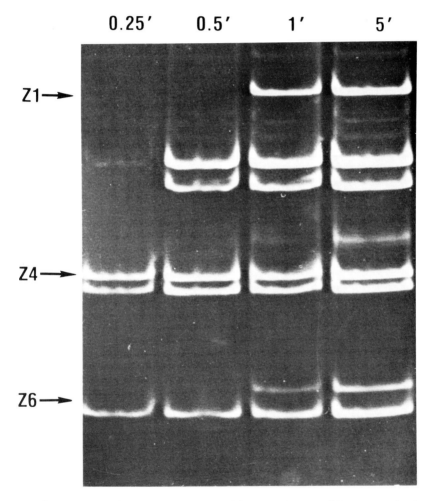

Figure 5. Time course of G4 reaction. Aliquots of the initiation complex were incubated with 0.9nmol of dTTP and 2.4nmol each of dATP, dGTP and dCTP in a total volume of 25 µl at 22°C. The conditions used were different from our standard reaction so that the reaction rate would be decreased, permitting time for appropriate sampling.

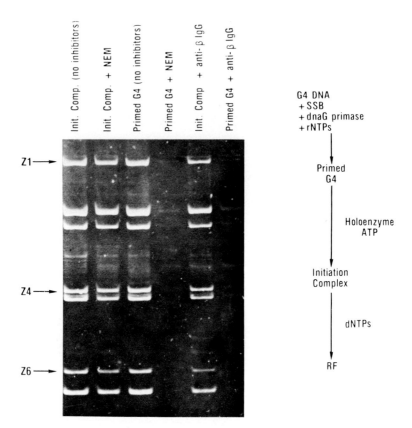

Figure 6. Reaction products in the presence and absence of initiation inhibitors.

Thus, holoenzyme can replicate the entire 5500 nucleotide G4 molecule without dissociating. This processivity is adequate for one holoenzyme to synthesize an entire Okazaki fragment processively. This conclusion indicates that an additional function of the holoenzyme auxiliary subunits is to increase the processivity of pol III. We have previously determined, in collaboration with the Bambara laboratory, that the processivity of the pol III core is 10→15 nucleotides.

From the aforementioned results it was tempting to speculate that β may function to initiate nucleic acid synthesis, then dissociate like the subunit of RNA polymerase. However, this creates a paradox. Pol III* cannot utilize a long single strand efficiently as a template—it can initiate but cannot elongate in a highly processive manner. If β dissociates after initiation, how does pol III* "remember" that it came from holoenzyme and replicate the entire G4 molecule processively? β-mediated conformational changes or chemical alterations of pol III* could be invoked, but we decided to first ask whether β is present in the elongation complex.

<u>Presence of β in Elongation Complexes</u> - To assay for the presence of β at various reaction stages we developed a sensitive immunodetection method for β. A complex is isolated by Biogel A5m filtration (13,15) subjected to SDS gel electrophoresis, blotted onto nitrocellulose sheets (16), reacted with rabbit anti-β IgG (17), then peroxidase conjugated goat anti-rabbit IgG (18). The position of β can be located and quantitated by a peroxidase-mediated colorimetric reaction (18). Standards ranging from 20ng to 500pg are shown in

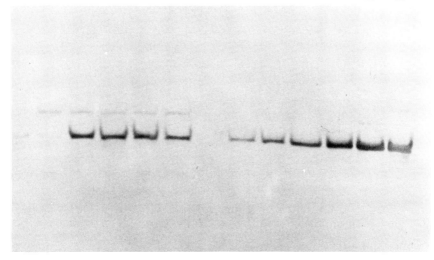

Figure 7. Detection of β in initiation and elongation complexes. Lanes A-G contained approximately 600 pmol (total nucleotide) of active G4 DNA. Complete details of this immunodetection technique will be published (K. Johanson and C. McHenry, in preparation).

figure 7 (lanes H →M). Using single stranded DNA binding protein encoated primed G4 DNA it can be seen that little β (<0.03/circle) is bound to the DNA when it is omitted (lane G), added without pol III* (lane A) or if pol III* is added without β (lane B). However, stoichiometric quantities of β are present in both the initiation complex (lane F) and even the completed RFII (lane C). The experiments reported in lanes E and D are of partially elongated G4. This was accomplished by removal of dCTP from the elongation mixture or by addition of dideoxythymidine triphosphate (90% inhibition). Thus, under noninvasive reaction conditions where β is present in stoichrometric quantities it remains bound to G4 DNA. Since β is an acidic protein which has no affinity for DNA by itself, it presumably interacts with pol III-bound 3'-OH termini.

To investigate whether β or pol III* can cycle in the replication reaction we added excess primed G4 DNA to isolated initiation complex which contained stoichiometric quantities of holoenzyme. It was observed (Table 3) that both pol III* and β cycle and that addition of either holoenzyme component to the reaction does not increase synthesis if the reactions are allowed to proceed to completion. Therefore, pol III* and β do not remain immobilized on the termini of

Table 3

Cycling of Holoenzyme Components in G4 Reaction

	DNA Synthesis (pmol) Component Added			
	none	β	pol III*	β+pol III*
Initiation Complex	107	103	96	88
Primed G4	0	0	3	144
Initiation Complex + Primed G4	205	225	207	205

Reactions were performed under standard conditions (13) except the designated reactions contained approximately 100 pmol (total nucleotide) of isolated initiation complex (13,15) or 200 pmol of primed single stranded DNA binding protein encoated dnaG primase primed G4 DNA (15).

RFII; they can dissociate to other DNA molecules if they are present.

The question still remains as to whether β can be stripped from pol III once initiation has occurred resulting in an efficient holoenzyme-like replicative complex or whether it is required as an essential elongation component. The latter possibility could be rationalized with the antibody resistance of elongation if inhibitory anti-β IgG is sterically blocked from contacting β in the elongation complex. Present experiments are directed toward answering this question.

ACKNOWLEDGMENT

This work was supported by grants from the American Cancer Society and National Institutes of Health.

REFERENCES

1. McHenry, C. and Kornberg, A., "DNA Polymerase III Holoenzyme" In: The Enzymes, Nucleic Acids, Vol. 14, Pt.A (ed. P. Boyer) Academic Press, New York, in press.
2. McHenry, C. and Kornberg, A. (1977) J. Biol. Chem. 252, 6478.
3. Kornberg, T., and Gefter, M.L. (1971) Proc. Natl. Acad. Sci. U.S.A. 68, 761.
4. Livingston, D., Hinkle, D. and Richardson, C. (1975) J. Biol. Chem. 250, 461.
5. McHenry, C. and Crow, W. (1979) J. Biol. Chem. 254, 1748.
6. Otto, B., Bonhoeffer, F. and Schaller, H. (1973) Eur. J. Biochem. 34, 440.
7. McHenry, C. (1980) In: "Mechanistic Studies of DNA Replication and Genetic Recombination" (ed. B. Alberts) pp. 569-577, Academic Press, New York.
8. Wickner, S. and Hurwitz, J. (1976) Proc. Natl. Acad. Sci. U.S.A. 73, 1053.
9. Hurwitz, J. and Wickner, S. (1974) Proc. Natl. Acad. Sci. U.S.A. 71, 6.
10. Wickner, S. (1976) Proc. Natl. Acad. Sci. U.S.A. 73, 3511.
11. Wickner, W. and Kornberg, A. (1973) Proc. Natl. Acad. Sci. U.S.A. 70, 3679.
12. Wickner, W., Schekman, R., Geider, K. and Kornberg, A. (1973) Proc. Natl. Acad. Sci. U.S.A. 70, 1764.
13. Johanson, K. and McHenry, C. (1980) J. Biol. Chem. 255, 10984.

14. Fiddes, J., Barrell, B. and Godson, N. (1978) Proc. Natl. Acad. Sci. U.S.A. 75, 1081.
15. Fay, P. Johanson, K., McHenry, C. and Bambara, R. (1981) J. Biol. Chem. 256, 976.
16. Towbin, H., Staehelin, T. and Gordon, J. (1979) Proc. Natl. Acad. Sci. U.S.A. 76, 4350.
17. Renart, J., Reiser, J. and Stark, G. (1979) Proc. Natl. Acad. Sci. U.S.A. 76, 3116.
18. Avrameas, S. and Guilbert, B. (1971) Eur. J. Immunol. 1, 394.

PRIMING OF PHAGE Ø29 REPLICATION BY PROTEIN p3, COVALENTLY LINKED TO THE 5' ENDS OF THE DNA[1]

Margarita Salas, Miguel A. Peñalva, Juan A. García
José M. Hermoso, José M. Sogo

Centro de Biología Molecular (C.S.I.C.-U.A.M.)
Universidad Autónoma. Canto Blanco
Madrid, Spain

ABSTRACT. Electron microscopy of protein-containing replicative intermediates of Ø29 DNA has shown the presence of a high proportion of replicating molecules that circularize due to protein-protein interaction. The results obtained suggest the presence of the terminal protein at the ends of the parental and daughter DNA strands and support the model on the role of the terminal protein, p3, as a primer in the initiation of Ø29 DNA replication. Length determination of the circular replicating Ø29 DNA molecules are in agreement with previous results using replicative intermediates treated with pronase which indicated that Ø29 replication starts at the ends of the DNA and proceeds by a mechanism of strand displacement. Extracts from Ø29-infected cells synthesize <u>in vitro</u> unit length Ø29 DNA. Replicative rather than repair synthesis takes place in the <u>in vitro</u> system. The DNA synthesized <u>in vitro</u> contains covalently linked protein. <u>In vitro</u> experiments using extracts from cells infected with mutant <u>ts</u>3(132) indicate that a functional protein p3 is required.

[1]*This work was supported by research Grant 1 R01 GM27 242-01 from the National Institutes of Health and by Grants from Comisión Asesora para el Desarrollo de la Investigación Científica y Técnica and Comisión Administradora del Descuento Complementario (INP).*

INTRODUCTION

Bacteriophage Ø29 which infects <u>Bacillus subtilis</u>, has a linear, double-stranded DNA of molar mass 11.8×10^6 g mol^{-1} (1) with a protein, p3, prduct of cistron 3, covalently linked to the two 5' termini (2-5). The linkage between protein p3 and Ø29 DNA is a phosphoester bond between the OH group of a serine residue and 5' dAMP (6). A six nucleotides long inverted terminal repetition has been found at the ends of Ø29 DNA (7,8). Ø29 DNA replication starts, non simultaneously, from either DNA end and proceeds by a mechanism of strand displacement (9,10). Protein p3 has been shown to be involved in the initiation of Ø29 replication (11). A role for the parental protein p3 in replication has also been demonstrated (2). A model has been proposed in which protein p3 acts as a primer in the initiation of replication by interaction with the parental protein p3 and possibly also with the inverted terminal repetition; reaction with the 5' terminal nucleotide, dATP, to form a protein-dAMP covalent linkage would provide the 3'OH group needed for elongation (9,11).

The protein p3 molecules, located at each 5' end of Ø29 DNA, interact with each other giving rise to circular DNA molecules (2). Similarly, protein p3 at the ends of two or more DNA molecules interact to form dimers or higher concatemers (2). If the protein were present at the ends of the parental and daughter DNA strands in replicating molecules (9), different kinds of circular structures, dimers and higher concatemers should be obtained. The results presented in this paper are consistent with the presence of protein p3 at the ends of the parental and daughter DNA strands in replicating molecules and, therefore, they support the role of protein p3 as a primer in the initiation of Ø29 DNA replication.

An <u>in vitro</u> system for DNA replication has been obtained from Ø29-infected B. subtilis. The <u>in vitro</u> system uses endogenous DNA as template and synthesizes replicative, unit-length Ø29 DNA with protein covalently linked to nascent DNA chains. A requirement for protein p3 in the <u>in vitro</u> system has been shown.

RESULTS

a) Sedimentation Analysis of Ø29 DNA Replicating Molecules.

B. subtilis infected with the Ø29 delayed-lysis mutant sus14(1242) at 30 °C was pulse-labeled at 50 min post infection for 1.5 min with ^{3}H-thymidine, and after treatment with lysozyme the lysates were incubated for 1 h at 37 °C in the presence of 1% sodium dodecylsulfate to release the DNA from the bacterial membrane and sedimented through a sucrose gradient containing 0.1% sodium dodecylsulfate. Figure 1a shows the presence of two radioactive peaks, one of them (peak A) with a sedimentation rate similar to that of mature Ø29 DNA used as marker. The radioactive peak with a lower sedimentation rate (peak B) dissappeared after a chase with an excess of cold thymidine (Fig. 1b). Uninfected cells, pulse-labeled under the same conditions as the infected cells, had essentially no radioactive material (Fig. 1a).

b) Electron Microscopy of the Ø29 DNA Replicating Molecules Present in Peak A.

The fractions from peak A were precipitated with ethanol and further sedimented through a sucrose gradient in the absence of sodium dodecylsulfate. After dialysis, the fractions containing the radioactive material were analyzed with the electron microscope using the BAC spreading technique (12,13). Figure 2 shows the four different kinds of circular type I molecules (double-stranded DNA with single-stranded tails located at a random position (9)) which can be produced by protein-protein interaction. Figure 2a shows a type I molecule in which the protein at the end of the daughter duplex-strand interacts with the protein at the end of the displaced parental single-strand. A type I molecule in which the protein at the end of the displaced parental strand has interacted with the protein at the end of the parental duplex strand is shown in Fig. 2b. In Fig. 2c, a type I molecule in which the proteins at the ends of the parental and daughter duplex-strands have interacted can be seen. Finally, Fig. 2d shows a molecule in which the two parental proteins and the daughter protein have interacted. The interaction of the protein from the parental single-strand and that from the daughter duplex-strand in type II molecules (DNA partially double-stranded and partially single-stranded, the transition point being random) gives rise to circular molecules (Fig. 3,a and b). Circular type I/II molecules which are a combination of type I and type

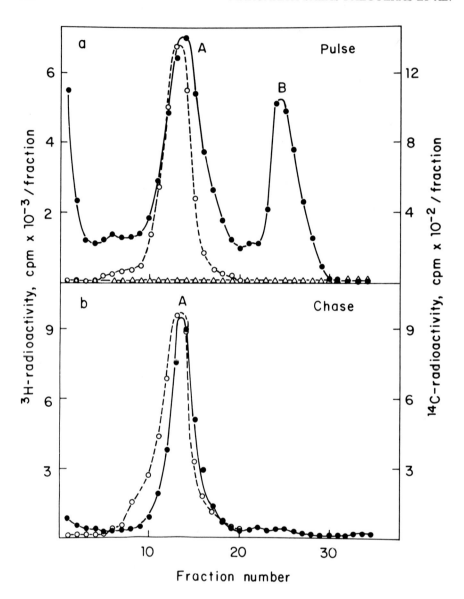

FIGURE 1. Neutral sucrose gradient of the DNA synthesized in B. subtilis infected by phage Ø29, pulse-labeled and chased.

a. B. subtilis was infected and labeled with ^{3}H-thymidine from 50 to 51.5 min (14). b. At 51.5 min a 200-fold excess of cold thymidine was added and the incubation continued for 20 min. Sedimentation is from right to left. ●——●, infected; ▲---▲, uninfected; o---o, ^{14}C-Ø29 DNA marker.

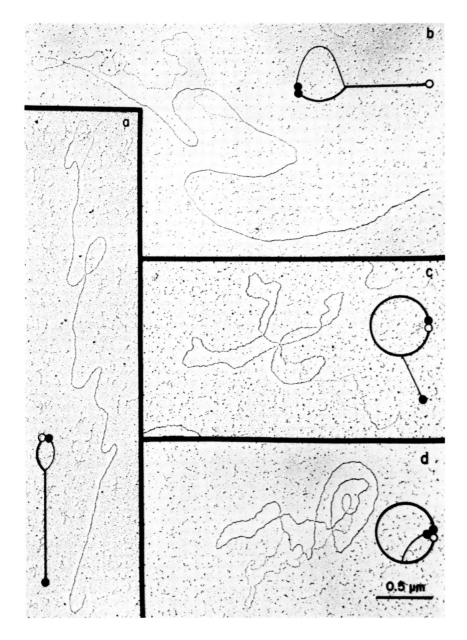

FIGURE 2. Electron micrographs of circular type I replicating ⌀29 DNA molecules. The heavy line represents double-stranded DNA and the light line single-stranded regions. Solid circles represent the parental protein and open circles newly synthesized protein. Bar = 0.5 μm.

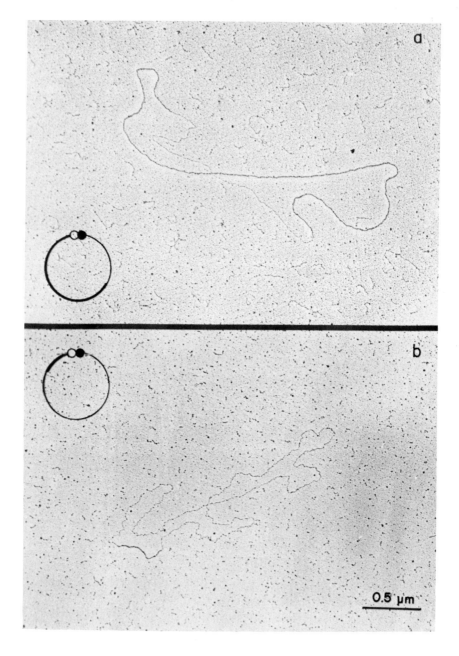

FIGURE 3. *Electron micrographs of circular type II replicating Ø29 DNA molecules.*
 Bar = 0.5 μm.

II molecules were also seen (not shown). Figure 4 shows two dimers, each formed by a circular double-stranded DNA molecule and a type II molecule interacting either through the protein at the end of the parental single-strand or through the protein at the end of the daughter duplex-strand. Other dimers consisted in two circular molecules, in two linear molecules or in one circular and one linear molecule, at least one of them being a replicating type I or type II molecule (not shown). Aggregates higher than dimers were also seen but, taking into account the difficulties for their analysis, they were not further studied.

The percentage of circular type I (and type I/II) and type II molecules relative to the linear forms is shown in Table 1. It can be seen that the percentage of circular replicating molecules is similar, or even higher, to that of circular unit length double-stranded DNA. Table 1 also shows that the amount of type I (and type I/II) molecules (13%) is lower than that of type II molecules (73%). When the replicating molecules were isolated by sucrose gradient sedimentation after treatment with 2.6% glyoxal for 1 h at 37 °C in the presence of 1% sodium dodecylsulfate, the amount of type I (and type I/II) molecules increased to 36% and that of type II molecules decreased to 59% (14) (see later).

c) Nature of the DNA Molecules Present in Peak B.

About 90% of the radioactivity present in peak B was degraded by nuclease S1, suggesting that it contained essentially single-stranded DNA. The buoyant density in CsCl of the material in peak B was 1.714 g cm^{-3}, being that of the DNA in peak A 1.705 g cm^{-3}, also consistent with the single-stranded nature of the DNA in peak B. Electron microscopy of the material in peak B confirmed the above conclusion. Single-stranded DNA molecules were mainly present in peak B. The length distribution of these molecules was rather heterogeneous, ranging from 0.8 µm to 4 µm.

When treatment with 2.6% glyoxal was included, the radioactivity present in peak B was drastically reduced (14) and, as indicated above, the percentage of type I (and type I/II) molecules greatly increased. From these results it seems likely that the heterogeneous single-stranded DNA present in peak B is derived from breakage of the single-stranded tails of type I molecules, the breakdown of these molecules being reduced after treatment with glyoxal.

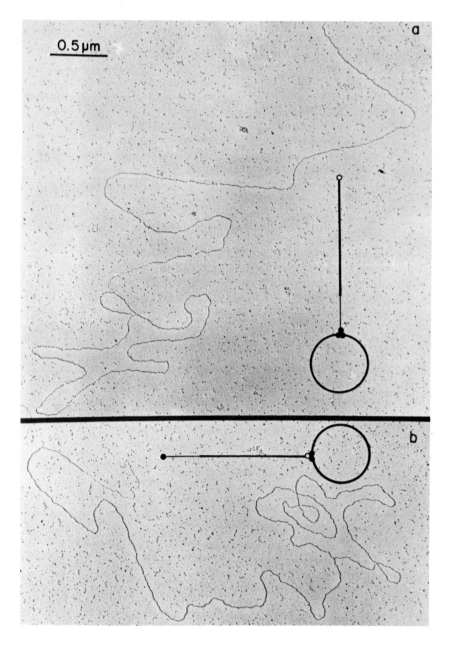

FIGURE 4. *Electron micrographs of dimers formed by a circular double-stranded DNA molecule and a type II molecule. Bar = 0.5 μm.*

TABLE I. Frequencies of the Different Types of Molecules Seen in the Preparation of Replicating Intermediates of Ø29 DNA

DNA molecule	Number	Circular molecules,%[b]	Replicating molecules,%[c]
Double-stranded			
Circular	176	34	-
Linear	338		
Type I and I/II			
Circular	27	60	13
Linear	18		
Type II			
Circular	125	48	73
Linear	136		
Type III			
Linear	43	-	12
Type IV			
Linear	6	-	2
Dimers[a]	130	-	-

[a] *The dimers have been considered as units.*
[b] *Calculated for each different kind of DNA molecule.*
[c] *The percentage was determined with respect to the total number of replicating molecules. Dimers have not been included.*

d) *Length Determination of Circular Type I and Type II Molecules.*

Length determination of the double-stranded DNA in circular type I molecules gave an average value of 7.01 ± 0.42 μm. This length is very similar to that obtained for double-stranded circular DNA molecules (7.07 ± 0.15 μm). Determination in type I molecules of the ratio between the length of the double-stranded region with the value more similar to the single-stranded one and the length of the total double-stranded DNA showed a random distribution, which indicates the existence of molecules at different stages of replication. The ratio between the length of the single-stranded region in type I molecules and the corresponding double-stranded region showed a maximum at a value of 1.

The length distribution of circular type II molecules gave an average value of 7.23 ± 0.34 μm. The ratio between the length of the double-stranded region in these molecules and the total length of the DNA (double- and single-stranded) gave a random distribution, indicating the existence of molecules that have replicated to a different extent.

e) *In vitro Replication of ∅29 DNA*

Although many E. coli in vitro systems are known to replicate exogenously added DNA (15), no similar systems were reported for B. subtilis. Extracts were prepared from ∅29-infected B. subtilis F25-1, pol I⁻ (kindly given by Dr. N. Cozzarelli (16) by centrifugation of the lysates for 30 min at 25 000 rpm in a Ti65 rotor and the incorporation of ^{3}H-TMP was followed relative to an uninfected control. A concentration of 0.35 M KCl was used to inhibit the host DNA polymerases II and III (16). As shown in Figure 5, ∅29-infected extracts incorporate ^{3}H-TMP to a much higher extent than uninfected extracts, the amount of ^{3}H-TMP incorporated after 20 min of incubation at 37 °C being 7.1 pmoles per 0.05 ml of reaction mixture. Neutral sucrose gradient of the material labeled in the ∅29-infected extracts under these conditions showed that the radioactivity cosedimented with a ∅29 DNA marker (Fig. 6). Alkaline sucrose gradient showed that the material labeled during a 2.5 min incubation of the infected extracts was shorter than unit-length DNA and was chased into unit-length DNA (not shown).

To show whether semiconservative replication rather than repair synthesis was occurring in the in vitro system, ∅29 DNA was synthesized in a 20 min of incubation at 37 °C in the presence of 5-bromodeoxyuridine triphosphate and the product was analyzed by CsCl gradient centrifugation to equilibrium. A single radioactive peak was obtained in neutral CsCl, with a density higher than that of light ∅29 DNA used as marker (Fig. 7a). When the centrifugation was carried out in alkaline CsCl, the radioactive peak was more heterogeneous but, nevertheless, its density was higher than that of light ∅29 DNA (Fig. 7b).

The DNA synthesized in vitro was heated for 10 min at 68 °C after addition of sodium dodecylsulfate to 1% and EDTA

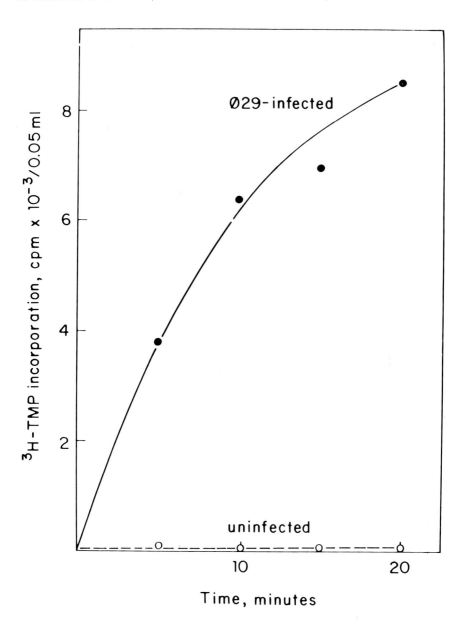

FIGURE 5. DNA synthesis by extracts of B. subtilis infected by phage ø29. The reaction mixture contained: 50 mM Tris-HCl, pH 7.5, 5 mM $MgCl_2$, 2.5 mM 2-mercaptoethanol, 0.35 M KCl, 10 μM dATP, dCTP and dGTP, 10 μM ^{3}H-dTTP (1200 cpm/pmol), 0.15 mM UTP, CTP and GTP, 1 mM ATP, 8 mM creatine phosphate, and 160 μg creatine phosphokinase/ml.

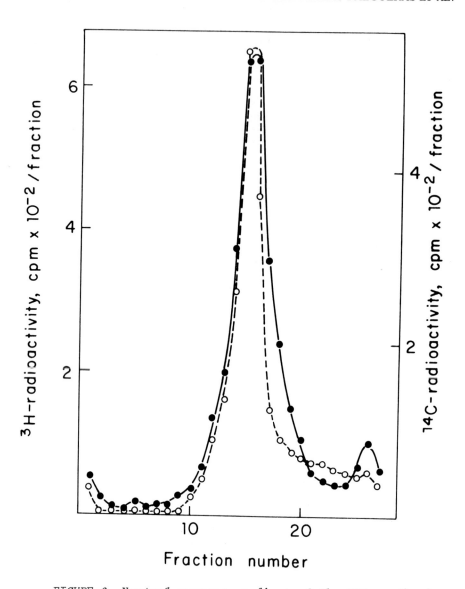

FIGURE 6. *Neutral sucrose gradient of the DNA synthesized by extracts of* B. subtilis *infected by phage ∅29.*
The DNA synthesized *in vitro* was treated with proteinase K (170 µg/ml) in the presence of 0.5% SDS and centrifuged in a 5-20% sucrose gradient in 50 mM Tris-HCl, pH 7.8, 10 mM EDTA, 1 M NaCl for 2.5 h at 50 000 rpm in a SW65 rotor at 20 º C. ●——●, DNA synthesized *in vitro*; o---o, ^{14}C-∅29 DNA marker.

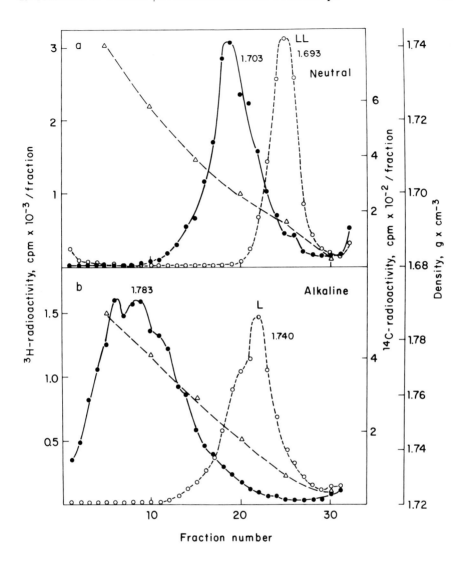

FIGURE 7. CsCl gradient of the DNA synthesized in the presence of 5Br-dUTP by extracts of B. subtilis infected by phage ∅29.
The DNA synthesized in vitro in the presence of 10 μM 5Br-dUTP and 0.2 μM ^{3}H-TTP was treated with proteinase K as indicated in Fig. 6 and centrifuged in a neutral (a) or alkaline (b) CsCl gradient to equilibrium for 90 h at 30 000 rpm at 15 °C in a Ti65 rotor. Light ^{14}C-labeled ∅29 DNA was used as a marker. ●——●, ^{3}H-radioactivity; o---o, ^{14}C-radioactivity.

to 20 mM and purified by chromatography on a Sepharose 2B column equilibrated with 10 mM Tris-HCl, pH 7.5, 50 mM NaCl, 1 mM EDTA, 0.1% sodium dodecylsulfate to eliminate protein non-covalently linked to the DNA (17). The purified DNA, both native and heat denatured, was retained to nitrocellulose filters to an extent similar to protein p3-containing Ø29 DNA isolated from phage particles, suggesting the presence of protein p3 in the product synthesized in vitro (not shown).

f) Requirement for Protein p3 in the In vitro Replication of Ø29 DNA

A way to show a requirement for protein p3 in the in vitro system for Ø29 DNA replication could be to study the DNA synthesized at 30 °C and 45 °C by extracts from B. subtilis infected with mutant ts3(132) at 30 °C. Figure 8 shows that the incorporation at 45 °C in wild-type-infected extracts was not decreased to a high extent relative to that obtained at 30 °C. On the contrary, the incorporation at 45 °C in ts3(132)-infected extracts stopped quickly compared to that obtained at 30 °C.

DISCUSSION

Ø29 DNA molecules isolated from viral particles, containing protein p3 at the 5' ends, circularize in vitro due to protein-protein interaction (2). Replicating Ø29 DNA molecules isolated without treatment with proteolytic enzymes can also form in vitro circular structures, dimers or higher aggregates. Different kinds of circles are found in the case of type I molecules, which can be accounted for by interaction of the protein at the end of the two parental strands and the protein at the end of the daughter duplex strand. Circular type II molecules could be formed by interaction of the protein at the end of the parental single strand and that at the end of the daughter duplex strand. Dimers of replicating molecules are also found, which can be formed by interaction of the protein at the ends of monomers. Taking into account that all possible interactions are formed in vitro and the finding of a high proportion of circular molecules, the results strongly suggest the presence of protein at the ends of all DNA strands, parental and daughter. Polyacrylamide gel electrophoresis of in vivo isolated ^{35}S-labeled DNA molecules has shown that the protein present in replicating molecules is p3 (14).

Length determinations of circular type I and type II molecules are in agreement with previous results obtained with

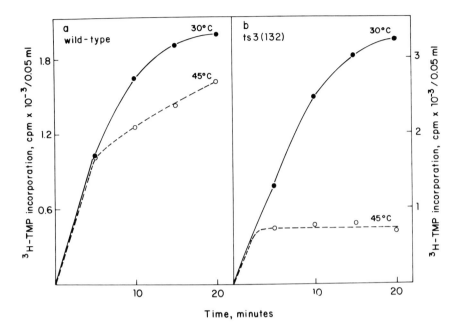

FIGURE 8. DNA synthesized in vitro at 30 ºC or 45 ºC by extracts of B. subtilis infected at 30 ºC with wild-type phage ∅29 or mutant ts3(132).
The incubation conditions were as described in Fig. 5.

linear type I and type II molecules, produced by treatment of the replicative intermediates with pronase (9) and support the conclussion that ∅29 replication starts at the ends of the DNA and proceeds by a mechanism of strand displacement.

The results obtained for ∅29 DNA replication are very similar to those obtained in the case of adenovirus replication where circular structures were found (18,19) probably due to the interaction of the protein linked at the 5' ends of the DNA (20,21). Evidence that protein is associated with the ends of parental and daughter strands in replicating adenovirus DNA has been also obtained using other techniques (22-24). Both, the results obtained with phage ∅29 and with adenovirus, which suggest the presence of the terminal protein at the ends of the parental and daughter DNA strands, support the model proposed for the initiation of ∅29 (9,11) and adenovirus (20) replication by which a newly synthesized molecule of the terminal protein would act as a primer by reaction with the dNTP corresponding to the 5' end and formation of the protein-dNMP covalent linkage, thus providing the 3'OH group needed for elongation by the DNA polymerase.

An in vitro system for Ø29 replication has been obtained from phage-infected cells. The system replicates endogenous DNA giving rise to the formation of unit-length Ø29 DNA in neutral sucrose gradient. When the in vitro-labeled DNA was analyzed by alkaline sucrose gradient, shorter than unit-length DNA was obtained after 2.5 min of incubation and it was chased into unit-length DNA. This result is in agreement with those obtained in vivo which suggested that there is some kind of discontinuity in Ø29 DNA replication (11).

When the DNA synthesized in the presence of 5-bromodeoxyuridine triphosphate was analyzed by neutral CsCl gradient a single peak with a density higher than that of light DNA was obtained suggesting that replicative synthesis rather than a repair reaction is taking place in vitro. Analysis of the above DNA by alkaline CsCl gradient showed a more heterogeneous peak, but with a density higher than taht of light Ø29 DNA, again suggesting that replicative DNA synthesis is occurring in vitro. The heterogeneity of the peak obtained in the alkaline gradient may be due to in vitro elongation on DNA chains already initiated and to the presence of single-strand nicks.

The DNA chains labeled in vitro contain protein as shown by its binding to cellulose nitrate filters even after heat denaturation. These results suggest the presence of protein covalently linked to the in vitro synthesized DNA, most probably protein p3.

The requirement for protein p3 in the in vitro replication of Ø29 DNA was shown since, when extracts from bacteria infected at 30 °C with mutant ts3(132) were incubated at 45 °C, DNA synthesis stopped quickly. These results could indicated that initiation, and very little elongation of initiated chains, is mainly occurring in the in vitro system described here. However, ^{32}P-labeled ser-P was not detected from DNA synthesized in vitro in extracts from wild-type infected cells labeled with dATP-α-^{32}P after HCl hydrolysis, under conditions in which ser-^{32}P was obtained from p3-{^{32}P}DNA complex labeled in vitro (6). This result suggests that no initiation occurs in this in vitro system and therefore the results obtained with mutant ts3(132) could indicate that protein p3 is also involved in an elongation process.

We are presently working out an in vitro replication system dependent on exogenously added Ø29 DNA to try to obtain a direct evidence for the priming of Ø29 replication by protein p3.

REFERENCES

1. Sogo, J.M., Inciarte, M.R., Corral, J., Viñuela, E., and Salas, M. (1979). J. Mol. Biol. 127, 411.
2. Salas, M., Mellado, R.P., Viñuela, E., and Sogo, J.M. (1978). J. Mol.Biol. 119, 269.
3. Harding, N., Ito, J., and David, G.S. (1978). Virology 84, 279.
4. Ito, J. (1978). J. Virol. 28, 895.
5. Yehle, C.O. (1978). J. Virol. 27, 776.
6. Hermoso, J.M., and Salas, M. (1980). Proc. Natl. Acad. Sci. U.S.A. 77, 6425.
7. Escarmís, C., and Salas, M. (1981). Proc. Natl. Acad.Sci. U.S.A., in press.
8. Yoshikawa, H., Friedman, T., and Ito, J. (1981). Proc. Natl.Acad. Sci. U.S.A., in press.
9. Inciarte, M.R., Salas, M., and Sogo, J.M. (1980). J. Virol. 34, 187.
10. Harding, N.E., and Ito, J. (1980). Virology 104, 323.
11. Mellado, R.P., Peñalva, M.A., Inciarte, M.R., and Salas, M. (1980). Virology 104, 84.
12. Vollenweider, H.J., Sogo, J.M., and Koller, Th. (1975). Proc. Natl. Acad. Sci. U.S.A. 72, 83.
13. Sogo, J.M., Rodeño, P., Koller, Th., Viñuela, E., and Salas, M. (1979). Nucleic Acids Res. 7, 107.
14. Sogo, J.M., García, J.A., Peñalva, M.A., and Salas, M. Submitted for publication.
15. Wickner, S.H. (1978). Ann. Rev. Biochem. 47, 1163.
16. Gass, K.B. and Cozzarelli, N.R. (1973). J. Biol. Chem. 248, 7688.
17. Challberg, M.D., Desiderio, S.V. and Kelly, T.J.,Jr. (1980). Proc. Natl. Acad. Sci. U.S.A. 77, 5105.
18. Girard, M., Bouché, J.P., Marty, L., Revet, B., and Berthelot, N. (1977). Virology 83, 34.
19. Kelly, T.J.,Jr. and Lechner, R.L. (1978). Cold Spring Harbor Symp. Quant. Biol. 43, 721.
20. Rekosh, D.M.K., Russell, W.C., and Bellet, A.J.D. (1977). Cell 11, 283.
21. Carusi, E.A. (1977). Virology 76, 390.
22. Stillman, B.W., and Bellett, A.J.D. (1979). Virology 93, 69.
23. Robinson, A.J., Bodmar, J.W., Coombs, D.H., and Pearson, G.D. (1979). Virology 96, 143.
24. Van Wielink, P.S., Naaktgeboren, N., and Sussenbach,J.S. (1979). Biochim. Biophys. Acta 563, 89.

FUNCTIONAL COMPONENTS OF THE SACCHAROMYCES CEREVISIAE
CHROMOSOMES - REPLICATION ORIGINS AND CENTROMERIC SEQUENCES

Clarence S. M. Chan
Gregory Maine
Bik-Kwoon Tye

Department of Biochemistry, Molecular and Cell Biology
Cornell University
Ithaca, New York

I. ABSTRACT

We have isolated replication origins and centromeric sequences from chromosomes of Saccharomyces cerevisiae. The replication origins were isolated by selecting for DNA sequences that allow autonomous replication of plasmids (1) in yeast. Two classes of replication origins were found, those which are unique and those which are reiterated in the genome. These repetitive replication origins may be flanked by single copy DNA sequences or by sequences which themselves are repetitive. Some of these repetitive replication origins are arranged in a tandem array interspersed by repeated flanking sequences. The centromere-like sequences were isolated by screening for yeast DNA sequences which are capable of conferring mitotic as well as meiotic stability to autonomously replicating plasmids.

II. INTRODUCTION

In order to understand how complex eukaryotic chromosomes replicate, we want to dissect the chromosomes into their simple functional parts and study each part indepen-

Abbreviations: $A = AvaI$, $B = BglII$, $Bm = BamHI$, $H = HpaI$, $H3 = HindIII$, $K = KpnI$, $P = PstI$, $PvI = PvuI$, $Pv = PvuII$, $R = EcoRI$, $S = SacI$, $S2 = SacII$, $S1 = SalI$, $X = XhoI$, $Xb = XbaI$, $kb = kilobase$.

dently before we try to understand how they function together as a whole. Two important functional components of the chromosomes, as we understand chromosomes today, are the replication origins - the sites where DNA replication is initiated, and the centromeric regions - the parts of the chromosomes responsible for the segregation of homologous chromosomes. We have chosen to analyse how chromosomes replicate by isolating individual replication origins and centromeric sequences. We have previously reported the isolation of replication origins from Saccharomyces chromosomes (1). They are found to fall into two categories; those which are unique and those which are repeated many times (15-20 times) in the yeast genome. In this paper, we will describe the organization of one family of such repetitive replication origins.

The isolation of centromeric sequences from yeast have been described by a couple of laboratories. Clarke and Carbon have described the isolation of centromeres from chromosome III (2) and chromosome XI (3). Stinchcomb et al. (4) have also described the isolation of the centromere of chromosome IV. The isolation of these centromeric regions required the cloning of genetic markers closely linked to the centromere. This method is cumbersome, sometimes requiring the isolation of several overlapping clones in order to reach from the centromere-linked markers to the centromere. This method is also not feasible for the isolation of centromeres which have no tightly linked genetic markers. We have isolated centromere-like sequences from yeast (called STB for stabilizing sequences) which confer stability to autonomously replicating yeast plasmids, both mitotically and meiotically. This method is also described by Carbon et al. (3) and Stinchcomb et al. (4) in the same volume. One of these sequences has been analysed extensively. Unlike the CEN3 sequence isolated by Clarke and Carbon, this sequence (STB 3) is not associated with a replication origin.

III. RESULTS

A. *The 131 Family*

We have previously described the isolation of a family of repetitive replication origins which we named the 131 family (1). There are seventeen independent clones isolated which belong to this family, all sharing sequence homology

30 FUNCTIONAL COMPONENTS OF THE *S. CEREVISIAE* CHROMOSOMES

with the prototype YRp131 plasmid clone. Four of the clones, YRp131, YRp120, YRp206 and YRp282, were isolated in the screen for autonomous replicating sequences by virtue of their high frequencies of transformation in yeast (5). These clones show homology with one another. The YRp131 insert can be cleaved into two halves, fragments A and B, by the HindIII enzyme. Figure 1 shows the restriction map of the plasmid YRp131. Fragment A contains the replication origin as well as the repetitive sequence. Fragment B does not contain

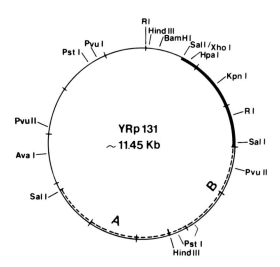

FIGURE 1. Restriction map of the plasmid YRp131. SalI fragment from yeast (--------------) is inserted into the SalI site of the vector which consists of pBR322 DNA (————) carrying the yeast LEU2 gene (▬▬). Fragment A contains the 131 prototype sequence which carries the replication origin and the repetitive sequence.

either the replication origin or the repetitive sequence. Fragment A was used as a probe to screen an E. coli library of SalI-restricted yeast DNA clones. Thirteen clones hybridized to the YRp131 A fragment. All 13 clones (YRp131A, B, C, F, J, L, M, N, O, P, Q, R, S) were capable of autonomous replication in yeast. Thus seventeen independent plasmid clones were isolated, all of which shared sequence homology with the YRp131 prototype sequence and they are all capable of autonomous replication in yeast. Since only about

1/5 - 1/4 of randomly selected plasmid clones carrying SalI-restricted yeast DNA fragments are capable of autonomous replication (1), this finding suggests a tight association between the repetitive sequence and the ability of autonomous replication. These sequences are termed repetitive replication origins. In order to understand the structural organization of these repetitive replication origins, extensive restriction mapping of these clones was carried out.

B. Members of the 131 Family with Unique Flanking Sequences

Figure 2 shows the restriction maps of the SalI inserts from YRp131 and YRp131J. The YRp131 SalI fragment can be cleaved into two halves by the HindIII enzyme. When used as hybridization probe for SalI-restricted total yeast DNA separated on an agaorse gel, fragment A hybridizes to multiple bands, whereas fragment B hybridizes to a single band. The YRp131J SalI insert can be divided into two parts at the PvuII site. The left half (hatched bar) hybridizes to the SalI insert from YRp131. The right half hybridizes to a unique band in a genomic DNA transfer experiment (6). Thus these two clones are similar in that the repetitive YRp131 prototype sequence present in each is flanked by unique sequences in the genome. Subcloning and transformation experiments showed that the YRp131 prototype sequence (hatched bar) contains a replication origin in each case. The YRp131J SalI fragment is about 11kb in size. Subcloning and transformation experiments also indicate that it contains a second replication origin which lies within the HindIII-SalI fragment on the right side (thick solid bar). Since the YRp131J fragment was isolated as a single SalI fragment, we must have isolated two contiguous replication origins from the chromosome. Using these naturally occurring contiguous replication origins, we may be able to study the temporal order of their activation of the cell cycle.

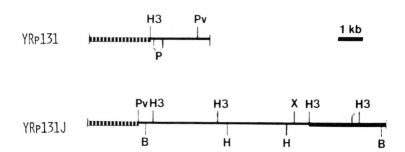

FIGURE 2. *Restriction maps of the SalI inserts of YRp131 and YRp131J.* ▪▪▪▪▪▪▪▪▪▪ *represents the YRp131 prototype sequence carrying the replication origin and the associated repetitive sequence.* ▬▬▬▬▬ *represents unique flanking sequence.* ▬▬▬▬▬ *represents DNA segment carrying a replication origin.*

C. *Members of the 131 Family with Repetitive Flanking Sequence X*

The restriction maps of another set of members of the 131 family are shown in Figure 3. There are seven clones which belong to this set, they all contain the YRp131 prototype sequence, each of which is flanked by the same repetitive sequence which we call X. Three (YRp131C, 131M and 131R) of the seven independent clones isolated have identical restriction maps. All of the cloned sequences, with the exception of the one from YRp131N, can be cleaved by the BglII enzyme to yield a fragment (hatched region) which contains the repetitive YRp131 prototype sequence and the associated replication origin. The other fragment does not hybridize to the YRp131 prototype sequence, but cross hybridize with the repetitive sequence X located in the region indicated as thick solid bar. These clones have very similar restriction sites. They all contain a HpaI site which separates the SalI-BglII fragment (hatched region) into two segments, both of which hybridize to the YRp131 prototype sequence. Thus, the HpaI site must lie within the YRp131 prototype sequence. Subcloning and transformation experiments show that the replication origin is located between the HpaI and BglII sites. The YRp131N clone is classified in this set because the small BglII-AvaI fragment (thick solid bar) of YRp131C, which contains the repetitive flanking sequence X, hybridizes to this clone. The leftmost SalI-HindIII fragment (hatched region) of 131N contains the 131 prototype repetitive sequence as well as the replication origin.

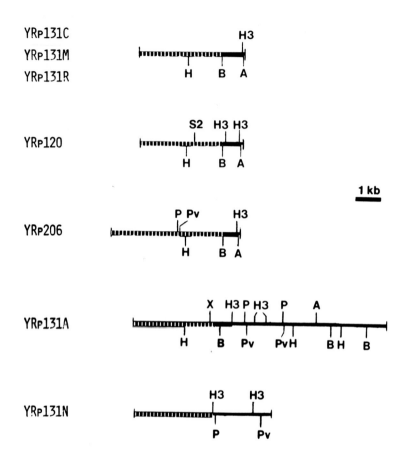

FIGURE 3. Restriction maps of the SalI inserts of members of the 131 family with repetitive flanking sequence X. ▮▮▮▮▮▮▮▮▮▮ represents DNA segment carrying the YRp131 prototype sequence. The replication origin as well as the associated repetitive sequence maps within this region. ▬▬▬▬ represents DNA segment which carries the repetitive flanking sequence X. ─── represents unique sequence.

D. Members of the 131 Family with Repetitive Flanking Sequence Y

Eight members of the 131 family belong to yet another set. They are YRp131B, 131F, 131L, 131O, 131P, 131Q, 282 and 131S. Seven of these independent clones contain inserts

30 FUNCTIONAL COMPONENTS OF THE S. CEREVISIAE CHROMOSOMES

which have identical restriction maps (Figure 4). This repeated occurrence suggests that there are probably multiple copies of these SalI fragments in the yeast genome. YRp131S, unlike the other members of this set, contains a smaller SalI fragment. However, comparison of their restriction maps shows remarkable similarities. All members in this set contain a XhoI site which divides the cloned fragments into two halves. One half contains the YRp131 prototype sequence with the associated replication origin. The other half shows homology with the other members in this set, but has no homology with the YRp131 prototype sequence. We call this repetitive flanking sequence Y, since it has no homology with the repetitive sequence X.

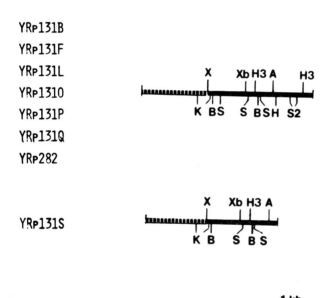

FIGURE 4. Restriction maps of the SalI inserts of members of the 131 family with repetitive flanking sequence Y. ▪▪▪▪▪▪▪▪▪▪▪ *represents DNA segment carrying the YRp131 prototype sequence. The replication origin as well as the associated repetitive sequence maps within this region.* ━━━━ *represents DNA segment which carries the repetitive flanking sequence Y.*

E. *Copy Number of the 131L Sequence in the Yeast Genome*

The fact that seven independent clones isolated from the yeast genome contains identical SalI inserts suggests that this SalI fragment may occur more than once in the yeast genome. Calibration study was carried out to determine the copy number of the YRp131L insert in the yeast genome (Figure 5). Five µg of SalI-restricted total yeast DNA was loaded in lane 2 of the 0.7% agarose gel. Various amounts of SalI-restricted YRp131L plasmid DNA were loaded in the other lanes such that the amounts of 131L sequence loaded were equivalent to 0.5X, 1X, 2X, 5X, 10X and 20X the amounts of a 6.8kb DNA fragment of the total yeast DNA loaded in lane 2. The digested DNA was separated by agarose gel electrophoresis and transferred to nitrocellulose filter (6). The BglII fragment of the YRp131L insert, which does not show homology to the YRp131 prototype sequence or the repetitive sequence X, was labeled with ^{32}P and used as probe for hybridization. Comparison of the intensity of the 6.8kb band (size of the YRp131L SalI fragment) in each of the lanes in the autoradiogram suggests that there are at least five copies of the YRp131L SalI fragment in the yeast genome.

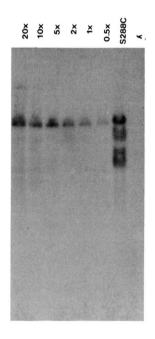

FIGURE 5. *Calibration of YRp131L copy number in the yeast genome. One μg of HindIII-restricted lambda DNA and 5 μg of SalI-restricted total yeast DNA were loaded in lane 1 and lane 2 respectively of the 0.7% agarose gel. Various amounts of SalI-restricted YRp131L plasmid DNA were loaded in the other lanes such that the amounts of 131L sequence loaded were equivalent to 0.5X, 1X, 2X, 5X, 10X and 20X the amounts of a 6.8kb DNA fragment of the total yeast DNA loaded in lane 2. The digested DNA was separated by agarose gel electrophoresis and transferred to nitrocellulose filter. The BglII fragment of the YRp131L insert was labeled with ^{32}P by nick-translation with DNA polymerase I and used as probe for the hybridization.*

F. *Arrangement of the Multiple Copies of the YRp131L Sequence*

The multiple copies of the YRp131L sequence could be arranged in various ways in the genome. They could be dispersed in different locations in the genome, arranged as tandem repeats, or occur in other arrangements. Analysis of the cleavage products of total yeast DNA digested by restriction enzymes which cut the YRp131L sequence once should yield information about their arrangement. Total yeast DNA was restricted by SalI, BamHI, HpaI, XhoI, EcoRI and XbaI respectively, separated by agarose gel electrophoresis, transferred to nitrocellulose filter, and hybridized with the P^{32}-labeled SacII-SalI fragment of the YRp131L sequence as probe. If the YRp131L sequence is tandemly repeated, and if it represents one complete repeat unit, then one would expect to see a 6.8kb band lit up by this probe. The autoradiogram (Figure 6) shows that a prominent 6.8kb band was lit up in each of these cases. Other bands of different sizes can also be seen. When total yeast DNA was digested by enzymes which do not cleave the YRp131L sequence, such as SmaI and PvuII, only DNA bands corresponding to large molecular weight DNA was lit up. The enzyme SacII cleaves the YRp131L sequence at two sites 0.25kb apart. When it was used for cleaving total yeast DNA, a prominent 6.55kb band, 0.25kb smaller than the 6.8kb repeat unit, was lit up. All these together support

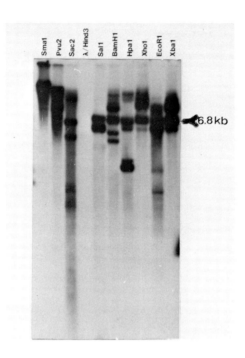

FIGURE 6. *Hybridization of restricted total yeast DNA with YRp131L SacII-SalI fragment. Genomic yeast DNA was digested by restriction enzymes which do not cleave the YRp131L SalI insert, such as SmaI and PvuII, enzyme which cleaves the YRp131L SalI insert twice such as SacII and enzymes which cleave the YRp131L SalI insert once such as SalI, BamHI, HpaI, XhoI, EcoRI and XbaI. The digested DNA was separated by agarose gel electrophoresis, transferred to nitrocellulose filter and hybridized with the P^{32}-labeled SacII-SalI fragment of YRp131L sequence as probe. HindIII digested lambda DNA was used as a control.*

the notion that some, but not all, of the YRp131L repetitive sequences are arranged as multiple tandem 6.8kb repeats in the yeast genome. This was confirmed by the isolation of BamHI clones from yeast, using the BglII fragment of the YRp131B sequence as probe. Four independent clones were isolated. These again were shown to have identical restriction maps corresponding to a circular permutation of the YRp131L sequence restriction map (Figure 7).

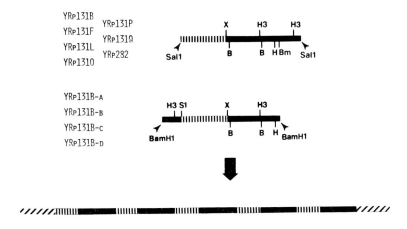

FIGURE 7. Some of the 131L repetitive sequences are arranged in a tandem array. Restriction maps of the SalI inserts of YRp131B, 131F, 131L, 1310, 131P, 131Q and 282 and the BamHI inserts of YRp131B-a, 131B-b, 131B-c and 131B-d. The restriction map of the first set is a circular permutation of that of the second set.

G. *Isolation of DNA Sequences which Stabilize Autonomously Replicating Plasmids*

All of the autonomously replicating plasmids which we have isolated share a common phenotype, they are not stably maintained in the cell, presumably due to non-uniform segregation of the plasmids during mitotic division. Clarke and Carbon (2) reported the isolation of the first centromere of chromosome III, CEN3, in yeast. The CEN3 sequence has the following properties; when inserted into autonomously replicating plasmids in yeast, the plasmids become mitotically stable, furthermore, the plasmids behave like chromosomes during meiosis, i.e., they segregate 2:2 in meiotic divisions. Using these properties of the known centromere, we proceeded to isolate other centromeric sequences by screening for DNA sequences isolated from total yeast DNA which are capable of stabilizing unstable autonomously replicating plasmids.

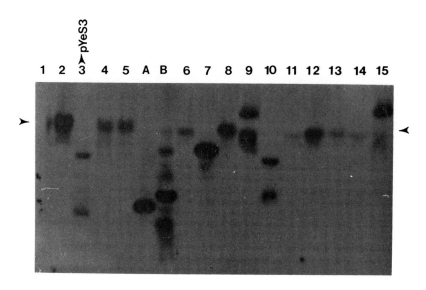

FIGURE 8. *Autoradiogram to show the location of pBR322 sequences in various stable yeast transformants. DNA from a number of Ura$^+$ stable transformants was extracted by the rapid procedure (7). DNA was separated on 0.7% agarose gel, transferred to nitrocellulose filter and hybridized to P^{32}-labeled pBR322 DNA probe. Arrow indicates position of chromosomal DNA. Transformants #3, 7, 9, 10 and 15 indicate the presence of plasmids containing pBR322 sequence. Transformants #3 (pYeS3) was further shown to contain a centromere-like sequence. Controls used are shown in lanes A (YRp10 plasmid DNA digested with BamHI) and B (transformant containing the plasmid YRp10).*

Total yeast DNA isolated from S288c strain was partially digested with MboI restriction enzyme and cloned into the BamHI site of YRp10 (pBR322 plasmid carrying ARS1 and URA3 gene, gift from Dr. Stinchcomb). This library of MboI DNA clones was transformed into a Ura$^-$ strain of yeast, selecting for Ura$^+$ transformants. Transformants were pooled and grown non-selectively for 16 generations and then plated on selective media. Stable and unstable transformants can be distinguished by their colony morphology under a dissecting scope. Stable transformants are large with a smooth appearance whereas unstable transformants are smaller and serrated. The URA3 marker in these stable transformants can be maintained

either by integration of the plasmid into the chromosome or
on the autonomous plasmids. These two possibilities can be
distinguished by analysis of the pBR322 DNA sequence in the
cell. Figure 8 shows the analysis of total yeast DNA isolated from each of these transformants. Those which were
shown to carry plasmids were further analysed by amplification
of the plasmids in E. coli. One of the transformants, pYeS3,
was shown to carry a plasmid with an insert of 2.8kb in the
BamHI site of YRp10. The insert is chromosomal in origin and
the restriction map of the insert is shown in Figure 9.

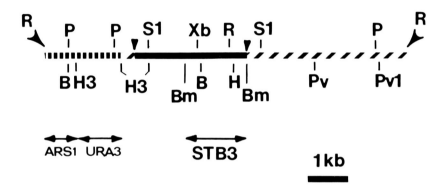

FIGURE 9. Restriction map of pYe3S. ▬▬▬▬ represents
yeast DNA sequences containing the ARS1 and URA3 gene. ▬▬
represents the MboI insert carrying the STB3 sequence. ////.
represents pBR322 DNA.

H. The STB3 Sequence is a Centromere-Like Sequence

The stabilizing sequence, STB3 was identified to be located within the 1.5kb BamHI fragment of the pYeS3 plasmid
(Figure 9). This was carried out by subcloning the 1.5kb
BamHI fragment into the BamHI site of YRp14 (pBR322 carrying
ARS1 URA3 LEU2). The mitotic stability of this plasmid,
pGM3, was compared with that of pGM1 (pBR322 carrying ARS1
CEN3 LEU2 URA3) and YRp14. Yeast transformants carrying
each of these plasmids were grown non-selectively for 23
generations and the fraction of cells which retained the
plasmids were compared (Table 1). Transformants harboring
pGM3, which carries the STB3 sequence, and pGM1, which carries CEN3, were stable for the URA3 marker, where as transformants harboring YRp14 have completely lost the URA3 marker.
This mitotic stability was also observed when the 1.5kb BamHI

STB3 sequence was subcloned into two autonomously replicating plasmids YRp121 and YRp131 (1). Centromeric sequences are

TABLE 1. Mitotic Stability of Various Plasmids in Yeast

Plasmid	Yeast Strain	% Stable
pGM1(CEN3)	TD4	96[a]
pGM3(STB3)	TD4	87[a]
YRp114	TD4	0[a]
pGM1(CEN3)	TD4	100[b]
pGM3(STB3)	TD4	95[b]
YRp114	TD4	2[b]
pGM3(STB3)	A2	96[b]
YRp121	A2	0[b]
YRp121(STB3)	A2	98[b]
YRp131	A2	1[b]
YRp131(STB3)	A2	91[b]

[a] Scored after 23 generations of non-selective growth in YEPD.

[b] Single colonies grown on selective medium were resuspended and plated on YEPD plates. Colonies were replica-plated on leucine-deficient medium to score for Leu$^+$ colonies.

TD4=a his4-519 ura3-52 leu-3 leu2-112 trp1-289
A2=α leu2-2 leu2-112 his3-11 his3-15

also known to confer meiotic stability to the plasmids. The behavior of the pGM3 plasmid was examined through meiosis and sporulation (Table 2). Yeast strain a his4-519 ura3-52 leu2-3 leu2-112 trp1-289/ARS1 LEU2 URA3 STB3 was crossed with α leu2-3 leu2-112 ura3-52 and the diploid was allowed to go through meiosis and sporulation. Eighteen tetrads were analysed. Twelve of them (66%) showed 2+:2-segregation for the LEU2 URA3 plasmid-linked markers. Of these 12 tetrads, there were 6PD and 6NPD but no TT with respect to the centromere-linked trp-1 marker of chromosome IV. Thus, the plasmid in at least 66% of the asci segregates in the first meiotic division as chromosomes as observed in pYE(CEN3)41 (2).

30 FUNCTIONAL COMPONENTS OF THE S. CEREVISIAE CHROMOSOMES

TABLE 2. Meiotic Segregation of pGM3

Cross:	TD4(pGM3)	X	AGH5

a his4-519 ura3-52 leu2-3 leu2-112   α leu2-3 leu2-112
       trp1-289/ARS1 LEU2 URA3 STB3   ura3-52

Distribution in tetrads of LEU2 and URA3 markers on pGM3

4+:0-	3+:1-	2+:2-	1+:3-	0+:4-
4	1	12	0	1
22%	6%	66%	0%	6%

LEU2 and URA3 markers are linked in every tetrad analysed.

In addition to the mitotic and meiotic stability of the pGM3 plasmid, we have also shown that the plasmid exists as an autonomous plasmid in the yeast cells through mitosis and meiosis and was never found integrated in the chromosome. Furthermore, we have shown that the STB3 sequence, unlike the CEN3 sequence, is not associated with an autonomously replicating sequence.

IV. DISCUSSIONS

From the restriction mapping analysis of the repetitive replication origins, some very interesting structural organization of the 131 family of repetitive replication origins has emerged. This family of homologous replication origins is found to be flanked by unique DNA sequences, or sequences which are repetitive themselves. There are at least two sets of flanking repetitive sequences found adjacent to this 131 family of replication origins. These flanking sequences seem to contain extensive homologies of at least 1 to 3kb to one another. They are only repeated several times in the genome (data not shown) and some of them are arranged in a tandemly repeated array, interspersed with the repetitive replication origins. Several questions are raised from these findings. Is this family of replication origins co-ordinately controlled? Are they activated by the same initiation factor? What is the significance of the structural organization of these repetitive replication is the significance of the structural organization of these repetitive replication origins? We do not have answers to these questions yet. One postulate is that these repetitive replication origins may serve a controlling function. Sets of genes that are expressed at the same time during the cell cycle may be located next to homologous replication origins that are co-ordinately activated. Initiation of DNA replication may be required for the transcription of neighboring genes. There are precedents for the coupling of DNA replication and late gene transcription in phage T4 (8) and adenovirus (9). There are also numerous examples of the close proximity of replication origins to promoters as illustrated in the tandemly repeated ribosomal RNA gene clusters in yeast (10), the non-tandemly arranged repetitive histone gene clusters, H2A and H2B, in yeast (Hereford, personal communication), and the well-documented work on SV40 (11) E. coli (12) and Salmonella typhimurium (13) etc. Our first step is to find out if this 131 family of repetitive replication origins are activated co-ordinately during the cell cycle. We would also like to know if the flanking sequences adjacent to this set of replication origins are in fact transcribed genes. If so, then we would like to find out if these sets of genes are co-ordinately expressed. Experiments to investigate these questions are in progress.

The isolation of centromeres by screening for sequences which stabilize autonomously replicating plasmids is a simple and feasible method. One stabilizing sequence, STB3, isolated in this manner has been shown to have the characteristics of a centromere. Autonomously replicating plasmids carrying this STB3 sequence are stable during mitotic divisions and behave as minichromosomes, segregating in a Mendelian fashion, during meiotic divisions. Stinchcomb et al. (4) and Carbon et al. (3) have reisolated the CEN3, CEN4 and CEN11 sequences using this procedure. We are presently mapping the chromosomal location of the STB3 sequence. It is conceivable that all seventeen centromeres of the seventeen chromosomes of yeast can be isolated in this way. A comparison of the detailed study of a large number of centromeres may yield important information about the structure and function of centromeres in yeast.

ACKNOWLEDGMENTS

We are grateful to Dr. John Carbon for his gift of pYe(CEN3)41 plasmid used in our comparison studies. We would like to thank Kim Muzzy for her help in the preparation of this manuscript. This work was supported in part by research grant from the National Institute of Health (AI/ GM 14980) and in part by grant from the U.S. Department of Agriculture (Hatch - NYC - 181412).

REFERENCES

1. Chan, C.S.M., and Tye, B.-K., Proc. Natl. Acad. Sci. 77, 6329-6333 (1980).
2. Clarke, L., and Carbon, J., Nature 287, 504-509 (1980).
3. Carbon, J., Clarke, L., Fitzgerald-Hayes, M., Hsiao, Chulai, and Buhler, J.-M., this volume (1981).
4. Stinchcomb, D.T., Mann, C., Thomas, M., and Davis, R. this volume, (1981).
5. Stinchcomb, D.T., Struhl, K., and Davis, R.W., Nature, Vol. 282, 39-43 (1979).
6. Southern, E., J. Mol. Biol. 98, 503-577 (1975).
7. Struhl, K., Stinchcomb, D.T., Scherer, S., and Davis, R.W., Proc. Natl. Acad. Sci. 76, 1035-1039.
8. Rabussay, D., and Geiduschek, E.P., Comprehensive Virology 8, H. Fraenkel-Conrat and R.R. Wagner, eds. (New York: Plenum Press), pp. 1-196 (1977).
9. Thomas, G.P., and Mathews, M.B., Cell 22, 523-533 (1980).
10. Szostak, J.W., and Wu, R., Plasmid 2, 536-554 (1979).
11. Seidman, M.M., Levine, A.J., and Weintraub, H., Vol. 18, 439-449 (1979).
12. Hirota, Y., Oka, A., Sugimoto, K., and Takanami, M., this volume, (1981).
13. Zyskind, J.W., Harding, N.E., Takeda, Y., Cleary, J.M., and Smith, D., this volume, (1981).

DNA SEQUENCES THAT ALLOW THE REPLICATION AND SEGREGATION OF YEAST CHROMOSOMES

Dan T. Stinchcomb
Carl Mann
Eric Selker[1]
Ronald W. Davis

Department of Biochemistry
Stanford University School of Medicine
Stanford, California

ABSTRACT

By introducing clonally isolated DNA molecules into the yeast Saccharomyces cerevisiae, we have identified sequences that may control DNA replication and segregation. One class of sequences transforms yeast at high frequency and allows autonomous replication of all colinear DNA (1). Such fragments (termed ars for autonomously relicating sequences) have been isolated from yeast and five other eukaryotes (2). Their properties are reviewed. The sequences responsible for the replication of ars1 have been studied in detail. Two functional domains are distinguishable. Sequence comparison of ars1 and the origin of replication of the endogenous yeast plasmid, Scp1, reveals only limited homologies. However, the homologies contain a core sequence, TAAAPyAPyAAPu, that is also present in ars3, a yeast ars linked to SUP11 on chromosome IV. Most ars-containing hybrid molecules are mitotically unstable, yet ars plasmids are present at high average copy number (approximately 60 copies/cell). Segregation of the plasmids must be nonrandom. A small plasmid (1.4 kb) escapes this aberrant segregation and is mitotically more stable. Its increased stability is not solely due to the absence of

---

[1] Present Address: Goethe Institute, Wasserburgerstraβe 54, 8018 Grafing, Federal Republic of Germany.

bacterial sequences nor to its sequence arrangement. A DNA fragment (CEN4) near the centromere-linked trp1 gene also alleviates mitotic instability. Circular molecules of 5-50 kilobase pairs carrying ars1 and CEN4 replicate and segregate properly during mitosis and meiosis; in meiosis the transformed phenotype primarily segregates $2+:2^0$. CEN4 like its predecessor, CEN3 (3), behaves as a centromere upon yeast transformation. The yeast transformation system provides an in vivo assay that functionally defines the DNA sequences which control DNA replication (ars) and segregation (CEN).

INTRODUCTION

The DNA transformation of yeast has revolutionized yeast molecular biology. Transformation was first demonstrated by Hinnen, Hicks and Fink (4). Using a procedure based on Ca^{2+} and polyethylene glycol treatment of yeast leu2 mutants, they obtained 2-10 Leu+ transformants per µg of hybrid plasmid DNA containing the yeast gene, LEU2. Hinnen, Hicks and Fink analyzed the genetic behavior and DNA structure of the transformants and found that the LEU2 gene and the colinear bacterial plasmid DNA were integrated into the yeast genome (4). In yeast, this integration event occurs by homologous recombination (4,5).

We have been interested in exploring other modes of yeast transformation. In transformation by integration, the foreign DNA sequences only replicate and segregate by virtue of their position in the host chromosome. A second type of yeast transformation can be imagined where foreign DNA can independently replicate without integrating into the host genome. We have asked whether a replicating mode of transformation is possible, and if so, what are its requirements? Finally, one might imagine transformation mediated by the introduction of functional yeast chromosomes that replicate and segregate independently. Again, what are the minimal DNA sequences required for this behavior? In this fashion, we have been using yeast transformation as an in vivo assay for sequences that control chromosome replication and segregation.

MATERIALS AND METHODS

I. BACTERIAL AND YEAST STRAINS

All strains used in this work as well as their storage and growth conditions have been described (2,6).

II. DNA AND ENZYMES

Bacterial plasmid DNA was isolated by repeated CsCl/ethidium bromide centrifugation (6). All enzymes used were purchased from commercial suppliers with the exceptions of DNA polymerase I and T4 DNA ligase which were generously provided by Stewart Scherer. Reaction conditions are outlined in reference 6.

III. CONSTRUCTION OF HYBRID DNA MOLECULES

EcoRI/BglII-generated fragments of Sc4101. 2 μg of the plasmid YRp12 (which carries the 1.4 kb TRP1 ars1 fragment inserted into the EcoRI site of YIp5, a vector that does not autonomously replicate, see reference 7) was digested with sufficient EcoRI, BglII and BamHI to ensure complete cleavage in 50 mM NaCl, 50 mM Tris-HCl (pH 7.5), and 7 mM MgSO$_4$. The DNA was then heated to 68° for 5 minutes and ligated by the addition of ATP to 1 mM, DTT to 10mM and T4 DNA ligase. After incubation at 4° C for 90 minutes, the ligation mixture was incubated again with BamHI. This second cleavage linearized any circles with reconstructed BamHI sites; the desired plasmids had a BglII site ligated to the BamHI site and were not cut. This DNA was used to transform BNN45 to ampicillin resistance. Colonies that were also tetracycline-sensitive (and thereby contained inserts into the EcoRI/BamHI sites of YIp5) were screened for the correct plasmid DNA structure by rapid DNA preparation and restriction endonuclease digestion. YIp5-Sc4105 contained the 0.85 kb EcoRI/BglII-generated fragment and YIp5-Sc4106, the 0.6 kb fragment.

Circularization of Sc4101 and Sc4128. 2 μg of YRp7 (pBR322 containing the EcoRI-generated 1.4 kb TRP1 ars1 fragment) was digested with EcoRI and 1.25 μg of λgt30-Sc4127 (a λ hybrid phage containing 11.2 kb of DNA near the centromere of chromosome IV including TRP1 and ars1, reference 8) was cleaved with both XhoI and SalI. After digestion, both samples were incubated for 5 minutes at 68°C, diluted to 10μg/ml in 100 mM NaCl, 50 mM Tris-HCl (pH 7.5), 7 mM MgSO$_4$, 10 mM DTT, 1 mM ATP and ligated at 4°C for 16 hours by T4 DNA ligase. These DNA samples were used to transform the trp1 yeast strain, YNN27, to Trp+. The frequency of Trp+ transformants was quite high - approximately 10^5 colonies/μg Sc4101 DNA. This efficiency is reduced 10-fold if the ligation step is deleted. All Trp+ transformants screened contained autonomously replicating molecules of the size expected for a circular 1.4 kb EcoRI fragment (Sc4101) or 8.0 kb XhoI/SalI fragment (Sc4128).

Construction of YR^{2}p12. 0.2 µg samples of YRp12 were digested with 1:1 serial dilutions of EcoRI. Concurrently, 10 µg of YRp7 (Sc4101 inserted into pBR322) was cut to completion. All samples were electrophoresed in a 1% agarose, Tris-acetate gel at 2.5 V/cm (6). Linears produced by partial EcoRI digestion of YRp12 and the small EcoRI-generated Sc4101 fragment were sliced out of the agarose gel. The gel slices were dissolved in 6M NaClO$_4$ and the DNA was bound to glass fiber filters as described by Chen and Thomas (9). The DNA (approximately 0.3 µg YRp12 linear and 3.0 µg Sc4101) was eluted, ethanol precipitated, and resuspended in 10 µl 100 mM NaCl, 50 mM Tris-HCl (pH 7.5), 7 mM MgSO$_4$, 1 mM ATP, 10 mM DTT and ligated by T4 DNA ligase at 4°C for 16 hours. This DNA was used to transform BNN72 (referred to as KM601 in ref. 1) to tetracycline resistance. Rapid plasmid DNA preparations from cultures grown from individual colonies were digested with HindIII. The presence of a 1.45 kb HindIII digestion product upon agarose gel electrophoresis was diagnostic of YR^{2}p12.

IV. ANALYSIS OF YEAST TRANSFORMANTS

The growth rates and mitotic stabilities of yeast transformants were measured as described previously (2). Autonomous replication of hybrid molecules was demonstrated in all cases by agarose gel electrophoresis of DNA from transformed cells and nitrocellulouse filter hybridization as outlined in the text and described in detail in reference 2.

IV. COPY NUMBER DETERMINATION

Estimates of plasmid copy number are often inaccurate due to differential extraction of covalently closed circles relative to linear, chromosomal DNA. To circumvent this difficulty, we used glass beads and phenol to extract total nucleic acid from ars transformants. Three Trp+ transformants of YNN27 were grown in 10 ml of yeast minimal medium. The three strains were transformed by YRp12, λgt30-Sc4127 (a λ/ars1 hybrid, see ref. 8), and λgt9-Sc4130 (a λ/ars1,CEN4 hybrid, see ref. 8). The cultures were harvested at 2×10^7 cells/ml by centrifugation, resuspended in an SDS lysis solution, 0.25-0.3µ glass beads, and phenol-chloroform and extracted as previously described (10). One tenth of each nucleid acid preparation was denatured in 0.1 N NaOH for 30 minutes at 25°C, neutralized with 2 M Tris-HCl (pH 7), and spotted onto four strips of nitrocellulose. Each strip also contained spots representing nine 1:1 serial dilutions of DNA standards. The four standards were preparations of YRp12, λgt9-Sc4130, sucrose gradient-purified HIS3 DNA (gift of Judith Jaehning), and YEp6 (a yeast/bacterial vector containing Scp1 DNA sequences). Each standard was linearized by restriction enzyme cleavage where necessary, denatured and neutralized. The four strips were baked and then hybridized to 10^6 cpm of the appropriate ^{32}P-labeled probe (11): labeled pBR322 was used to measure YRp12 sequences, λ DNA measured the amount of λ hybrid present, a labeled BamHI-generated DNA fragment of the HIS3 gene quantitated chromosomal DNA, and an XbaI/EcoRI fragment of Scp1 DNA was used to measure the endogenous yeast plasmid. After hybridization in 50% formamide, 0.9M NaCl, 50 mM NaPO$_4$ (pH 7), 5 mM EDTA, 100 µg/ml salmon sperm DNA, 0.2% SDS at 42°C for 20 hours, the strips were washed extensively in 10 mM NaPO$_4$ (pH7) and 0.2% SDS, air-dried, and autoradiographed at -70°C with flashed Cronex-4 (DuPont). Each spot was then cut out, placed in scintillation fluid and counted. The hybridization efficiencies as determined by the standards were linear over the entire 500-fold concentration range (4-0.008 ng DNA). The data are presented as ng DNA/10^7 cells. The number of copies per cell was calculated using the molecular weight of each DNA species and the total DNA content of a yeast cell. The copy numbers for HIS3 and Scp1 varied from 1.3-2.0 and 43-180, respectively.

V. DNA SEQUENCING

Sucrose gradient-purified DNA fragments containing Sc4101 was a gift of Judith Jaehning. The fragments were digested with HindIII and BglII restriction endonucleases and the 3' ends were labeled by incorporation of [α-^{32}P]-dGTP, -dCTP, -dATP, -dTTP (25 µCi each; 400 Ci/mmole; Amersham) as described previously (12). The end-labeled fragments were digested with EcoRI and separated on a 5% polyacrylamide gel. Similarly, the 0.6 Kb EcoRI/BglII fragment was 5' end-labeled using T4 polynucleotide kinase (13). AvaII was used to cleave this fragment to 0.56 kb such that it contained only one labeled 5' end. After elution from the polyacrylamide gel (14), the DNA fragments were sequenced by the methods of Maxam and Gilbert (13). A 528 nucleotide sequence confirmed on both strands is presented.

RESULTS AND DISCUSSION

I. AUTONOMOUSLY REPLICATING SEQUENCES

To explore alternative modes of yeast transformation it is first necessary to reduce the background of transformation by integration. This is achieved using the vector YIp5. YIp5 consists of the well-known pBR322 sequences and a small 1.2 kb yeast fragment bearing the URA3 gene. When this plasmid is used to try to transform a yeast strain that carries a ura3 chromosomal rearrangement, no Ura+ transformants are obtained. This lack of transformation is presumably due to extremely low rates of recombination between the small URA3 DNA fragment and its homologous, but rearranged, chromosomal counterpart. Naturally, if one inserts a large fragment of yeast DNA into any of the available restriction enzyme cleavage sites, the new hybrid can again transform at low frequency by integration (7). However, when a small (1.4kb) piece of DNA carrying the TRP1 gene is inserted into YIp5 it transforms the ura3 rearrangement strain at very high frequency (5,000-10,000 transformants per μg DNA). This was a fortuitous finding indeed. What caused this high frequency of transformation?

Transformation by integration can be easily distinguished from the alternate modes of transformation discussed in the introduction by analyzing the state of the foreign DNA. DNA from Trp+ transformants may be electrophoresed uncut on a low percentage agarose gel such that chromosomal DNA migrates much slower than small, extrachromosomal DNA. The transforming DNA can be detected by transferring the gel to nitrocellulose and hybridizing the paper-bound DNA to ^{32}P-labelled bacterial plasmid sequences. In all the transformants we examined, the hybrid DNA is present in the yeast cells only as supercoiled or nicked, open circles (1,5).

Therefore, the TRP1 fragment bears a second element that we have termed ars1 for autonomously replicating sequence. The properties of ars1 are as follows (1):

　　1.) it acts in cis on any colinear DNA, allowing the autonomous replication of hybrids from 1.4 to 50 kb.

　　2.) a single molecule containing several copies of ars1 is fully functional,

　　3.) ars1 can stably integrate into the yeast genome at low frequency, and

　　4.) the ars-containing plasmids are mitotically unstable in yeast cells. Yeast lacking the ars plasmids are segregated at high frequency.

What is the cellular function of a chromosomal DNA fragment that allows autonomous replication? An isolated origin of DNA replication should demonstrate all the properties enumerated above. Indeed, these same genetic properties are demonstrated by oriC, the E. coli chromosomal origin (15). However, we've yet to demonstrate either that DNA replication initiates at ars sequences or that ars replication is subject to cell cycle controls. It is possible that all plasmids replicate in yeast and ars acts in cis to permit expression of the selectable marker or to control the intracellular location of transformed DNA. Alternatively, autonomous replication could by due to mitochondrial, repair, or recombinational DNA synthesis. Biochemical experiments that test whether ars1 is a chromosomal origin of DNA replication are in progress.

II. ISOLATION OF OTHER ARSES

Replication is initiated at many sites along the eukaryotic chromosome. If origins share DNA sequences that reflect their common role in initiating DNA replication, then ars1 should hybridize to several yeast chromosomal DNA fragments. We have shown that this is not the case; ars1 shows no hybridization to other sites in the yeast genome even under conditions that favor the formation of mismatched hybrids (1).

Since other arses could not be detected by their physical similarities, we chose to use a functional selection (2). Ars1, when inserted into YIp5, allows high frequency transformation of the ura3⁻ rearrangement strain. We inserted restriction endonuclease-generated DNA fragments from each of six eukaryotes (Saccharomyces cerevisiae, Neurospora crassa, Dictyostelium discoideum, Caenorhabditis elegans, Drosophila melanogaster, and Zea mays) and from E. coli into YIp5. We then attempted to transform the ura3 yeast strain with the collections of hybrid molecules. No E. coli fragments permitted transformation of the yeast strain. Yet DNA fragments from all six of the eukaryotes tested allowed transformation by the URA3 vector. Upon transformation of yeast, these hybrid plasmids behaved like molecules containing ars1: the molecules replicated autonomously and were mitotically unstable.

Other investigators have used similar selection schemes to isolate yeast arses (16,17). They found sequences that allowed autonomous replication at a frequency consistent with the estimated total number of yeast origins of replication. Thus, arses are as abundant as the initiation sites of DNA replication in yeast chromosomes.

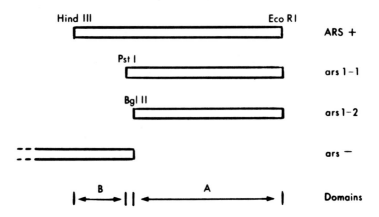

Figure 1. Structure of ars1.
The DNA fragments shown were inserted into the vector, YIp5. ars1-1 and ars1-2 denote fragments with partial function. The large BglII/EcoRI fragment extends off the figure to the left and is devoid of function.

III. ARS STRUCTURE

We have been analyzing the sequences responsible for ars1 function. The major characteristics of ars1 are diagrammed in Figure 1. ars1 is contained within an 800 base pair fragment defined by HindIII and EcoRI cleavage sites. Removal of an additional 212 base pairs from the left end (by deletion to the PstI site) reduces ars function; transformants are more unstable and have lower copy numbers. This led us to believe that ars function resides at or near the PstI endonuclease cleavage site (1). Another 23 base pairs was deleted by digestion with the enzyme BglII. If ars is an uninterrupted continuous locus, further deletion should result in further loss of function. Yet the hybrid plasmid containing the deleted fragment, YIp5-Sc4106, transformed ura3 yeast strains to Ura+ at high frequency. The transformants were very unstable. They grew under selective conditions with 12 hour doubling times, behaving exactly like the PstI deletion mentioned above. We conclude that ars1 is bipartite. One domain (A) between the BglII and the rightward EcoRI site allows high frequency transformation with only limited autonomous replication. The second domain (B) between the HindIII and PstI cleavage sites facilitates replication of domain A. Alone, domain B will not allow autonomous replication. The behavior of the BglII deletion

```
         HindIII  10            20            30            40            50
      5'-AAGCTTACAT   TTTATGTTAG   CTGGTGGACT   GACGCCAGAA   AATGTTGGTG
      3'-TTCGAATGTA   AAATACAATC   GACCACCTGA   CTGCGGTCTT   TTACAACCAG

                60            70            80            90           100
         ATGCGCTTAG   ATTAAATGGC   GTTATTGGTG   TTGATGTAAG   CGGAGGTGTG
         TACGCGAATC   TAATTTACCG   CAATAACCAC   AACTACATTC   GCCTCCACAC

               110           120           130           140           150
         GAGACAAATG   GTGTAAAAGA   CTCTAACAAA   ATAGCAAATT   TCGTCAAAAA
         CTCTGTTTAC   CACATTTTCT   GAGATTGTTT   TATCGTTTAA   AGCAGTTTTT

               160           170           180           190           200
         TGCTAAGAAA   TAGGTTATTA   CTGAGTAGTA   TTTATTTAAG   TATTGTTTGT
         ACTATTCTTT   ATCCAATAAT   GACTCATCAT   AAATAAATTC   ATAACAAACA

                       PstI       220           230          ***     ***     ********
         GCACTTGCCT   GCAGGCCTTT   TGAAAAGCAA   GCATAAAAGA   TCTAAACATA
         CGTGAACGGA   CGTCCGGAAA   ACTTTTCGTT   CGTATTTTCT   AGATTTGTAT
                                                              BglII

          ***   ****       270           280           290           300
         AAATCTGTAA   AATAACAAGA   TGTAAAGATA   ATGCTAAATC   ATTTGGCTTT
         TTTAGACATT   TTATTGTTCT   ACATTTCTAT   TACGATTTAG   TAAACCGAAA

               310           320           330           340           350
         TTGATTGATT   GTACAGGAAA   ATATACATCG   CAGGGGGTTG   ACTTTTACCA
         AACTAACTAA   CATGTCCTTT   TATATGTAGC   GTCCCCCAAC   TGAAAATGGT

               360           370           380           390           400
         TTTCACCGCA   ATGGAATCAA   ACTTGTTGAA   GAGAATGTTC   ACAGGCGCAT
         AAAGTGGCGT   TACCTTAGTT   TGAACAACTT   CTCTTACAAG   TGTCCGCGTA

               410           420           430           440           450
         ACGCACAATG   ACCCGATTCT   TGCTAGCCTT   TTCTCGGTCT   TGCAAACAAC
         TGCGTGTTAC   TGGGCTAAGA   ACGATCGGAA   AAGAGCCAGA   ACGTTTGTTG

               460           470           480           490           500
         CGCCGGCAGC   TTAGTATATA   AATACACATG   TACATACCTC   TCTTCGTATC
         GCGGCCGTCG   AATCATATAT   TTATGTGTAC   ATGTATGGAG   AGAAGCATAG

               510           520
         CTCGTAATCA   TTTTCTTGTA   TTTATCGT -3'
         GAGCATTAGT   AAAAGAACAT   AAATAGCA -5'
                  *   *** **  *   ******
```

Figure 2. Sequence of ars1.
Restriction enzyme cleavage sites mentioned in the text are labeled. Asterisks mark the nucleotides that are homologous to the Scp1 origin of replication.

constructed here is different from the similar deletion constructed by Tsumper and Carbon (18). They did not observe any ars function for the 601 bp BglII/EcoRI fragment. The primary difference between our hybrid plasmid and theirs is the orientation of ars sequences relative to bacterial plasmid DNA (see Methods and reference 18). The function of domain A may depend upon outside sequences.

We determined the sequence of nucleotides around the PstI cleavage site. Concurrently, Tschumper and Carbon sequenced the entire 1.4 kb fragment and have reported their results (18). The 528 nucleotides we sequenced are shown in Figure 2. It differs from the sequence of Tschumper and Carbon at only two nucleotides. The overall sequence is 63% AT base pairs. Two AT-rich regions of 100 base pairs lie on either side of the PstI site. There are no striking repeated sequences, either in direct or inverted orientation.

By comparing several sequences capable of autonomous replication in yeast, we hoped to determine which nucleotides could be of functional importance. Broach and Hicks have identified a region that permits replication of the endogenous yeast plasmid, Scp1 (19). They found that a section of the inverted repeat and the adjacent unique AT-rich region are required for replication upon transformation of yeast. This region corresponds to the major site of initiation of plasmid DNA replication in vivo and in vitro (20,21). Since yeast plasmid replication is controlled in a manner similar to chromosomal DNA replication (22), the Scp1 origin may share sequence homology with chromosomal origins. We compared the nucleotide sequence of ars1 to that of the Scp1 origin (23) using the DNA sequence analysis computer program developed by the MOLGEN group at Stanford University. Two different sequences from opposite strands of ars1 showed significant homology with a Scp1 sequence that abuts the plasmid inverted repeat (see Figure 3). The canonical shared sequence is TAAAPyAPyAAPu. The ars1 homologous sequences are both within domain A, the small BglII/EcoRI fragment that is deficient for replication in yeast. These sequences may be necessary but they can not be sufficient for autonomous replication.

We are also investigating the sequences responsible for the function of ars3. ars3 was found in the 1.25 kb EcoRI-generated DNA fragment that bears the yeast tyrosine tRNA suppressor, SUP11. A 300 base pair fragment, beginning at a Sau3A site between the anticodon and Ψ loops and extending 3' from the tRNA, contains ars3 (24). The region around the SUP11 tRNA has been sequenced (25), including 77 base pairs of ars3. The canonical sequence shared by ars1 and Scp ori is also found within this region of ars3 (Figure 4). Other DNA sequences that do not allow autonomous replication in yeast have been examined, including the sequences of E. coli oriC (26), the SV40 origin (27), HIS3 (28), CYC1 (29), gap63, gap491 (30), δ (31), and actin (32). None contain the ars canonical sequence.

Other investigators have made additional comparisons between ars DNA sequences. Gary Tschumper has delimited and sequenced ars2. He found only small regions of homology with ars1 (33). Likewise, Sydney Shall (34) and Ronald Pearlman (35) and their coworkers have sequenced DNA fragments from yeast and tetrahymena, respectively, that allow autonomous replication. Again, no extensive homologies have been observed. This is in striking contrast to the comparisons between ori sequences of different bacterial species: here large significant sequence homologies provide evidence for the functional domains of oriC (36).

Why are ars sequences so divergent? One must remember that the ars phenotype is loosely defined and that different ars hybrid plasmids vary in their copy number and mitotic stability. Sequence diversity may reflect this functional

---

```
Concensus:              TAAAQAQAAP

ars1:        AAGCATAAAAGATC TAAACATAAA ATCTGTA
                *   ***   ***  **********  *   ****
Scp1:        AGTAATACTAGAGA TAAACATAAA AAATGTA
                    *        ** **** * **   ***   *
ars1':       CAGCGAAAAGACGA TAAATACAAG AAAATGA

ars3:        CAGCTTTCAATATT TAAATATAAG TATTAAA
```

Figure 3. Yeast ars sequence comparisons.
Asterisks (*) designate bases matched between Scp1 and ars1 or ars1'. Underlined bases are those outside the concensus sequence that are homologous between any two of the three ars sequences. The first 10 nucleotides of the ars1' sequence shown are from Tschumper and Carbon (18). The remaining ars1 and ars1' sequences are from the top and bottom strands in Figure 2, respectively. The Scp1 sequence shown is from Hartley and Donelson (nucleotides 3688-3718, ref. 23). The inverted repeat begins at nucleotide 3714, 2 base pairs after the concensus sequence.

diversity. Furthermore, sequence diversity may be due to temporal control; DNA synthesis initiates at origins at specific times during S phase (37). The sequences shared by two different arses may be no more extensive than the common sequences of two promoters.

IV. MITOTIC STABILITY

In general, ars transformants are mitotically unstable. A Trp+ transformant containing YRp12 will continually bud off daughter cells that lack the ars hybrid plasmid, so that under selective conditions only 20-30% of the cells are Trp+. Yet the average copy number of YRp12 appears to be approximately 30 copies/cell (Table 1). The average Trp+ cell must contain some 80 copies of the ars plasmid. λ hybrid molecules containing TRP1 and ars1 transform trp1 yeast to Trp+ and autonomously replicate as covalently closed circles (8). The λ/ars1 Trp+ transformants are more unstable than YRp12 transformants: only 7.5% of the cells in a selectively grown culture are Trp+. The average copy number of such a culture is 1.6, so each Trp+ cell contains about 20 λ/ars1 hybrid molecules (Table 1). The appearance of Trp- segregants in the face of these high copy numbers indicates that ars-containing hybrid plasmids segregate nonrandomly. Autonomously replicating molecules may be compartmentalized in the nucleus during mitosis, such that transformants may accumulate the ars plasmids.

Two classes of ars-containing hybrids overcome this segregational defect. When the 1.4 kb EcoRI-generated DNA fragment bearing TRP1 and ars1 (Sc4101) is circularized and transformed into trp1 yeast cells, stable Trp+ transformants are obtained. After 10 generations of growth in nonselective media, 90-95% of the cells are Trp+. The stable transformants contain a 1.4 kb covalently closed circle at high copy number (38).

Yeast strain	ng/10^7 cells	copies/cell	% Trp+	copies/ Trp+ cell
YNN27 (Yrp12)	2.1	27	32	84
YNN27 (λ/ars1)	0.81	1.6	7.5	21
YNN27 (λ/ars1CEN4)	0.50	0.99	96	1.0

Table 1. Copy numbers of ars hybrid molecules.

Why is this small circle mitotically stable when YRp12 is not? Perhaps pBR322 sequences inhibit replication or proper segregation. Yet M13/ars1 and λ/ars1 hybrid molecules are not mitotically stable in yeast. In addition, we constructed another molecule consisting only of yeast sequences by circularizing an 8.0 kb XhoI/SalI fragment carrying TRP1 and ars1 (Sc4128). When transformed into yeast, the circular Sc4128 fragment autonomously replicates but again it is mitotically unstable. Mitotic stability of the small Sc4101 is not due merely to the absence of bacterial sequences.

Ligation of the two EcoRI ends of Sc4101 conjoins sequences that are not normally adjacent. To simulate this rearrangement, we inserted two copies of Sc4101 in the same orientation into YIp5. The double insert, termed YR^2p12, transforms the ura3 yeast strain to Ura+ at the same frequency as YRp12. The resulting transformants are mitotically unstable. Therefore, the novel joint of the Sc4101 circle does not explain its mitotic stability. The small size of the circle may permit it to escape the compartmentalization that causes asymmetric segregation of ars-containing molecules.

V. CENTROMERIC SEQUENCES

A second class of ars-containing hybrids are mitotically stable. Louise Clark and John Carbon have isolated a DNA fragment (CEN 3) from the centromeric region of chromosome III that stabilizes autonomously replicating molecules (3). When transformed into yeast, the CEN3-ars1 molecules segregate properly during mitosis and meiosis. For instance, in a diploid yeast strain, a single copy of the plasmid will replicate exactly once during premeiotic S phase, and will segregate reductionally in the first meiotic division and equationally in the second. The products of meiosis are thus two haploids with the plasmid and two haploids without it.

We have isolated another such fragment (Sc4134, see Figure 6) from the centromeric region of chromosome IV (8). Hybrid molecules carrying TRP1, ars1, and Sc4134 are mitotically stable; after 10 generations under nonselective conditions, 85-95% of the cells are TRP+. If such a TRP+ haploid is mated with a trp1⁻ haploid of opposite mating type and the resulting dipoid induced to undergo meiosis, the hybrid plasmid replicates and segrates with fidelity, producing 2 Trp+ haploids and 2 Trp- haploids. Sc4134 behaves as a centromere and the element responsible is termed CEN4.

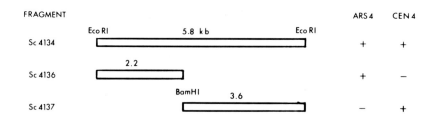

Figure 4. Structure of CEN4.
The fragments diagrammed were inserted into YIp5 to assess ars function and YRp12 (which contains ars1) to test CEN function.

Does CEN4 contain replicative as well as segregational activities? When inserted into YIp5, Sc4134 allows high frequency transformation of ura3 mutants to Ura+ and the recombinant molecules replicate autonomously. Thus, Sc 4134 contains a sequence the allows autonomous replication, ars4, as well as CEN4. The two functions are separable by a BamHI endonuclease cleavage site (see Figure 4). When inserted into YIp5, the smaller EcoRI/BamHI fragment (Sc4136) allows autonomous replication, but it does not mitotically stabilize ars1 hybrids. Conversely, YIp5 hybrids containing the larger fragment (Sc4137) can not autonomously replicate. Yet this fragment contains CEN4 since plasmids bearing both Sc4137 and ars1 are mitotically stable in yeast. Furthermore the Sc4137-ars1 plasmid meiotically segregates $2+:2^0$, indicating that ars4 is not an essential part of CEN4 function.

Ars4 transformants grow slowly, with generation times of 8.0 hours in selective media. However, CEN4 does not correct this deficiency, since CEN4-ars4 transformants also grow poorly and are mitotically unstable. Mitotic and meiotic stability of DNA in yeast requires both proper ars and CEN function.

Using yeast transformation, we have isolated sequences that allow autonomous replication and accurate segregation of extrachromosomal DNA. Meiotic analysis of these autonomous plasmids is the most sensitive test for proper replication (ars) and segregation (CEN). With this functional test we hope to characterize the sequences that control yeast chromosome behavior in more detail.

REFERENCES

1. Stinchcomb, D. T., Struhl, K., and Davis, R. W. (1979) Nature (London) 282, 39-43.
2. Stinchcomb, D. T., Thomas, M., Kelly, J., Selker, E. and Davis, R. W. (1980) Proc. Natl. Acad. Sci. USA 77, 4559-4563.
3. Clarke, L., and Carbon, J. (1980) Nature (London) 287, 504-509
4. Hinnen, A., Hicks, J. B., and Fink, G. R. (1978) Proc. Natl. Acad. Sci. USA 75, 1929-1933.
5. Struhl, K., Stinchcomb, D. T., Scherer, S., and Davis, R. W. (1979) Proc. Natl, Acad. Sci. USA 96, 1035-1039.
6. Davis, R. W., Botstein, C., and Roth, J. (1980) Advanced Bacterial Genetics : A Manual for Genetic Engineering (Cold Spring Harbor Laboratory, Cold Spring Harbor, New York 11724).
7. Scherer, S., and Davis, R. W., (1979) Proc. Natl. Acad. Sce. USA 76, 4951-4955.
8. Stinchcomb, D. T., Mann, C., and Davis, R. W., in preparation.
9. Chen, C. W., and Thomas, D. A., (1980) Anal. Biochem. 101, 339-341.
10. Elder, R. T., St. John, T. P., Sinchcomb, D. T., and Davis, R. W., (1980) Cold Spring Harbor Symp. Quant. Biol., in press.
11. Rigby, P. W., Dieckmann, M., Rhodes, C., and Berg, P. (1977) J. Mol. Biol. 113, 237-251.
12. Selker, E., and Yanofsky, C. (1979) J. Mol. Biol. 130, 135-143.
13. Maxam, A. M., and Gilbert, W. (1977) Proc. Natl. Acad. Sci. USA 74, 560-564.
14. Bennet, B.N., Schweingruber, M.E., Brown, K.D., Squires, C., and Yanofsky, C. (1978) J. Mol. Biol. 121 113-137.
15. Yasuda, S., and Hirota, Y. (1977) Proc. Natl. Acad. Sci. USA 74, 5458-5462.
16. Beach, D., Piper, M., and Shall, S. (1980) Nature (London) 284, 185-187.
17. Chan, C., and Tye, B.-K. (1980) Proc. Natl. Acad. Sci. USA 77, 6327-6333.
18. Tschumper, G., and Carbon, J. (1980) Gene 10, 157-166.
19. Broach, J. R., and Hicks, J. B. (1980) Cell 21, 501-508.
20. Newlon, C. S., and Devenish, R. J., this volume.
21. Sugino, A., Kojo, H., Greenberg, B. Q., Brown, P. O., and Kim, K. C., this volume.
22. Zakian, V. A., Brewer, B. J., and Fangman, W. L. (1979) Cell 17, 923-934.

23. Hartley, J. L., and Donelson, J. E. (1980) Nature (London) 286, 860-865.
24. Stinchcomb, D. T., and St. John, T. P., unpublished data.
25. Philippsen, P., unpublished data.
26. Sugimoto, K., Ota, A., Sugisaki, H., Takanami, M., Nishimura, A., Yasuda, S., and Hirota, Y. (1979) Proc. Natl. Acad. Sci. USA 76, 575-579.
27. Subramanian, K. N., Dhar, R., and Weissman, S. M. (1977) J. Biol. Chem. 252, 355-367.
28. Struhl, K.(1979) Ph.D. Dissertation, Stanford University.
29. Smith, M., Leung, D. W., Gillam, S., Astall, C. R., Montgomery, D.L., and Hall, B.D.(1979) Cell 16, 753-761.
30. Holland, J. P., and Holland M. J., (1980) J. Biol. Chem. 255, 2596-2605.
31. Gafner, J., and Philippsen, P. (1980) Nature (London) 286, 414-418.
32. Ng, R., and Abelson, J. (1980) Proc. Natl. Acad. Sci. USA 77, 3912-3916.
33. Tschumper, G., and Carbon, J., this volume.
34. Shall, S., personal communication.
35. Pearlman, R. E., Kiss, G. B., and Amin, A. A., this volume.
36. Zyskind, J. W., Harding, N. E., Takeda, T., Cleary, J. M., and Smith, D. W., this volume.
37. Sheinin, R., Humbert, J., and Pearlman, R. E. (1978) Annu. Rev. Biochem. 47, 277-316.
38. Scott, J. R., this volume.

SEQUENCING AND SUBCLONING ANALYSIS
OF AUTONOMOUSLY REPLICATING SEQUENCES
FROM YEAST CHROMOSOMAL DNA

Gary Tschumper and John Carbon

Department of Biological Sciences
University of California
Santa Barbara, CA 93106

ABSTRACT

Specific cloned fragments of yeast chromosomal DNA function like chromosomal origins of replication. Plasmids containing these fragments replicate autonomously in yeast and consequently transform yeast cells with high frequency. We have used DNA sequencing and subcloning analysis to characterize two specific autonomously replicating sequences (arsl and ars2) from yeast chromosomal DNA. Arsl is contained on an 838 base pair sequence which overlaps the structural portion of the TRP1 gene. Ars2 has been localized on a 100 base pair sequence near the ARG4 gene. Ars2 is also adjacent to a tRNA^{Gln} gene and a solo delta sequence, and may be contained within a second delta sequence. The sequence similarities between the arsl and ars2 regions consist of three homologous segments of 8-10 nucleotides along with adjacent nonhomologous GC- and AT-rich areas. Plasmids containing ars2 transform yeast with greater efficiency than plasmids containing arsl. Comparison of arsl and ars2 suggests they may be representatives of different classes of yeast chromosomal origins of replication.

INTRODUCTION

The haploid genome of yeast (Saccharomyces cerevisiae) contains about 13.5×10^3 kilobase pairs (kbp) of DNA distributed among 17 chromosomes (1). These chromosomes vary in size from 150 to 2300 kbp and each consists of a linear duplex DNA molecule containing an interstrand cross-link at each terminus (2). Initiation of DNA replication occurs at multiple internal sites in the chromosomes, and these sites may be activated at different times during the replication phase of the cell cycle (3). Both the location and the temporal order of activation of the replication origins are specific with respect to adjacent gene loci (4). Approximately 150 origins of replication occur in the haploid yeast genome at an average spacing of 90 kbp, and the DNA is replicated bidirectionally from these sites (5). It is estimated that yeast contains 3000-4000 genes for growth and cell division and that 40% of the DNA in the genome is utilized for these genes (6). Because of this high gene density in yeast, many of the origins of replication may be located within cloning distance (10-20 kbp) of selectable yeast genes.

During random cloning of yeast DNA, several fragments have been obtained which carry a specific yeast gene and also function like a chromosomal origin of replication (7,8). Bacterial plasmids carrying these fragments are capable of autonomous replication in yeast (7) and the segments responsible for this replication are termed ars (autonomously replicating sequences) (9). We have characterized two specific ars fragments from yeast: ars1, which is adjacent to the TRP1 gene near the centromere of chromosome IV (7); and ars2, which is located near the ARG4 gene on the long arm of chromosome VIII (8). The yeast DNA fragments carrying the TRP1 gene (7) and the ARG4 gene (10) were originally isolated by their ability to complement the corresponding mutations in Escherichia coli. Following the development of methods for yeast transformation (11), both the plasmid containing TRP1 (7) and the plasmid containing ARG4 (8) were shown to be capable of autonomous replication in yeast. The ars activity is readily assayed since it results in high frequency transformation and rapid growth of the host cell under conditions of selective pressure for the plasmid marker. The plasmids replicate within the nucleus of the yeast cell and undergo random assortment during meiosis (12). Although these ars sequences function like chromosomal origins of replication, they have not yet been shown to be identical to the origins of replication utilized by the chromosomes.

We have used DNA sequencing and subcloning analysis to characterize and compare ars1 (13) and ars2 (14). Since these two regions of yeast DNA perform a similar function, their sequences might be expected to share some common features. The subcloning analysis was directed toward obtaining the minimum sequence which still contained ars activity. We have determined the nucleotide sequence of approximately 1500 bp spanning each of these ars elements (13,14) using the chemical sequencing method of Maxam and Gilbert (15). In the case of ars2, we have localized the ars activity to a 100 bp region of DNA. We describe here the similarities and differences between ars1 and ars2, and discuss these sequences in terms of yeast chromosomal origins of replication.

RESULTS

Analysis of the ars1 region. The ars1 activity was originally isolated on a 1.4 kbp EcoRI fragment that also contained the TRP1 gene (7). The DNA sequence of this fragment has been determined (13) and some features of this region are shown in Figure 1. The fragment is 1453 bases long and is 60% A+T, close to the value of 59% A+T for total yeast nuclear DNA. The coding portion of the TRP1 gene was located by analyzing all the potential initiation and termination codons in each phase of each strand of the DNA (13). As shown in Figure 1, the initiation codon for TRP1 starts with nucleotide 103 and the termination codon starts with nucleotide 775, and transcription proceeds toward the ars1

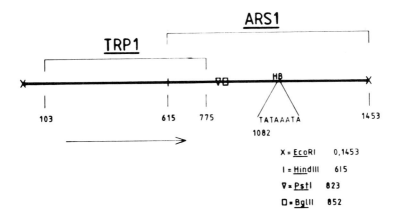

FIGURE 1. Diagram of the 1453 bp fragment containing ars1 and the TRP1 gene. The numbers refer to the nucleotide position, starting from the left. The TRP1 bracket indicates the predicted coding portion of the gene and the arrow shows the direction of transcription. The ars1 bracket indicates the maximum limits of the ars1 region. HB indicates the 'Hogness box' at nucleotide 1082.

region. The phase analysis also indicates there are no significant coding regions (greater than 100 bp) within the arsl region.

As shown in Figure 1, the 1453 bp EcoRI fragment has single restriction sites for HindIII, PstI, and BglII, and these were used in the subcloning analysis. The 823 bp EcoRI-PstI fragment contains full TRP1 activity but the 615 bp EcoRI-HindIII fragment contains no TRP1 activity. The 838 bp HindIII-EcoRI fragment contains full arsl activity (9) but the 601 bp BglII-EcoRI fragment contains no arsl activity. The 630 bp PstI-EcoRI fragment apparently has very weak arsl activity as shown by the slow growth of transformed cells under selective conditions for the plasmid marker (9). The region required for arsl activity must extend close to and possibly overlaps the 3' end of the TRP1 gene.

The subcloning results suggest that the region spanning the PstI and BglII sites of the 1453 bp fragment is essential for arsl activity. The DNA sequence in this region contains some short dyads and a direct repeat as shown in Figure 2. The two adjacent dyads of 10 and 11 bp have opposite nucleotide compositions: one is 80% G+C, the other is 80% A+T. Of the 60 bp shown in Fugure 2, 50 bp are involved in the cluster of dyads and the direct repeat. The only other obvious feature in the DNA sequence of the arsl region is a 'Hogness box' (-TATAAATA-), a possible eukaryotic promoter (16), located at nucleotide 1082 as shown in Figure 1. The location of this promoter within the arsl region and distant from any coding region may be significant if transcriptional activation or primer synthesis by RNA polymerase is required for the initiation of DNA replication in the arsl region.

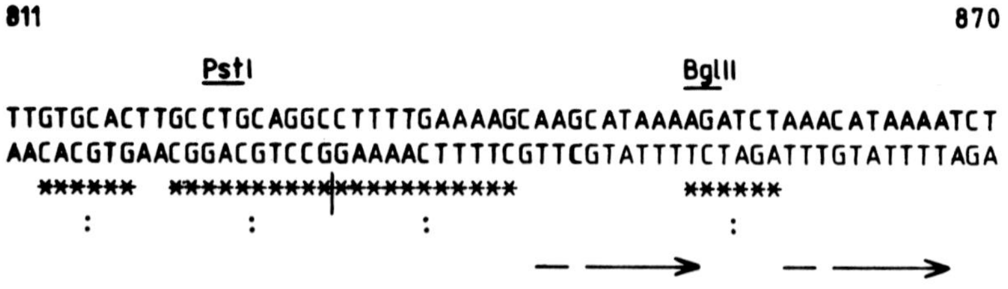

```
811                                                              870
          PstI                         BglII
TTGTGCACTTGCCTGCAGGCCTTTTGAAAAGCAAGCATAAAAGATCTAAACATAAAATCT
AACACGTGAACGGACGTCCGGAAAACTTTTCGTTCGTATTTTCTAGATTTGTATTTTAGA
******  ******************              ******
   :         :         :                    :
                                    — ——>    — ——>
```

FIGURE 2. DNA sequence of a 60 bp segment spanning the PstI and BglII sites of the arsl region. The asterisks indicate the dyad regions and the colons indicate the centers of the dyads. The arrows indicate the direct repeat and the numbers refer to the position of the sequence in Figure 1.

32 ANALYSIS OF AUTONOMOUSLY REPLICATING SEQUENCES

Analysis of the ars2 region. The region containing ars2 was originally isolated in conjuction with the ARG4 gene on a 12 kbp segment of sheared yeast DNA which had been cloned in a plasmid using poly(dA:dT) connectors (10). This plasmid was subsequently shown to transform arg4⁻ yeast strains to ARG4⁺ with high frequency and to replicate autonomously in yeast (8). A 13.5 kbp EcoRI fragment overlapping this area was obtained from a collection of recombinant phage and is shown in Figure 3. We have determined the sequence of 1517 bp of this DNA beginning at the HindIII site left of the tRNA gene and extending to the right beyond the divergent delta (14). This region is 69% A+T, considerably higher than the value of 59% A+T for total yeast nuclear DNA. The sequenced region contains three notable features in addition to the ars2 activity: a tRNA^{Gln} gene, a solo delta sequence and a divergent delta sequence. Transcription of the tRNA^{Gln} gene proceeds away from the ars2 region, the gene has no introns and the anticodon is complementary to the CAA codon.

Delta sequences are a family of 0.3 kbp repeated sequences (about 100 copies per haploid yeast genome) which exist as direct repeats at the termini of the Ty transposons or alone in the genome (17). The solo delta adjacent to ars2 in Figure 3 is the type of delta which exists alone (solo) in the genome (14) but can also function as a site for insertion of the yeast Ty1 transposon (18). Computer

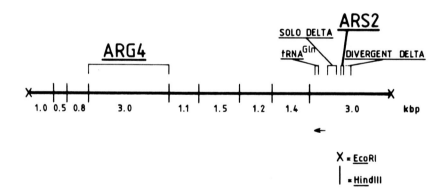

FIGURE 3. Diagram of a 13.5 kbp EcoRI fragment containing ars2 and the ARG4 gene. All the EcoRI and HindIII sites are indicated. The region starting at the HindIII site adjacent to the tRNA gene and extending 1517 bp to the right has been sequenced. The arrow indicates the direction in which the tRNA gene is transcribed.

1153

MspI

CCGGTAAAAGGGTAGTGTTGTTTTTTTGAGAGCCTTTTCTGACGGACAAATGGTC
GGCCATTTTCCCATCACAACAAAAAAACTCTCGGAAAAGACTGCCTGTTTACCAG

                                                            1254

                                                            dA:dT

GCAGGAAATATAAATTTATATTTAGTAATACGCAATCGCATACACATAAAAA
CGTCCTTTATATTTAAATATAAATCATTATGCGTTAGCGTATGTGTATTTTT
      ****

FIGURE 4. The 100 bp DNA sequence which contains <u>ars2</u>. The numbers refer to the position of the nucleotides counting from the <u>Hind</u>III site just left of the tRNA gene in Figure 3, and proceeding to the right. The sequence extends from an <u>Msp</u>I site to a point where a poly(dA) connector was added during the original isolation of the fragment. The asterisks indicate a region of dyad symmetry.

analysis of the 1517 bp DNA sequence revealed a second delta-like sequence, labeled a "divergent delta" in Figure 3. It contains 69% homology with the solo delta over a 150 bp region and is in an inverted orientation with respect to the solo delta (14). If the divergent delta has the typical delta length of about 330 bp, <u>ars2</u> would be contained within the divergent delta. This could not be determined, however, because the homology between the solo delta and the divergent delta does not persist through the <u>ars2</u> sequence.

    Ars2 has been localized by subcloning analysis to a 100 bp DNA sequence shown in Figure 4. The location of <u>ars2</u> on this 100 bp sequence is deduced from the structure of two slightly larger subclones, each containing <u>ars2</u> and having only this region in common (14). The sequence of <u>ars2</u> is distinguished by an 18 bp A+T region of perfect dyad symmetry with superimposed alphabetic symmetry as shown in Figure 5. The conjunction of this large (18 bp) region of double symmetry with the relatively small (100 bp) <u>ars2</u> region suggests a correlation. This double symmetry element would occur randomly only once in 69×10^9 bp of random DNA, a length 5000 times the size of the haploid yeast genome. It

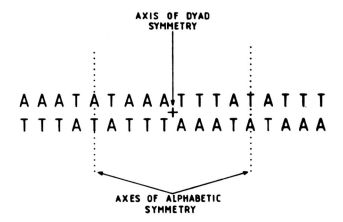

FIGURE 5. The double symmetry element that occurs in the ars2 sequence. The element consists of an 18 bp perfect dyad of A+T, each half of which has alphabetic symmetry. The dyad axis (+) projects out of the plane of the page, the alphabetic axes are the vertical dotted lines.

may be significant that the double symmetry element contains a nearly complete 'Hogness box' in each strand and is adjacent to a GC-rich region. A phase analysis of potential initiation and termination codons indicated no coding regions of significant size (greater than 100 bp) either near the ars2 region or anywhere in the 1517 bp of sequenced DNA.

Comparison of the ars1 and ars2 regions. We have used the 100 bp ars2 sequence shown in Figure 4 as the basis for comparison with the larger ars1 sequence. The best region of homology as determined by computer analysis between ars1 and the 100 bp ars2 sequence is shown in Figure 6. The portion of the ars1 region shown spans the 'Hogness box' (see Figure 1) and contains three exactly homologous sequences of 8 bp or larger in common with the 100 bp ars2 sequence. The common sequences occur in the same direction and in the same linear order, but not in the same spatial order. The comparison also shows that a GC-rich region of 9-11 bp occurs adjacent to the 'Hogness box' regions in each sequence. In addition, the 100 bp ars2 sequence has only two other regions of 8-10 bp in common with the 838 bp ars1 sequence, and these are widely separated and do not occur in the same direction. On the basis of homology, the portion of the ars1 sequence shown in Figure 6 is the best candidate to contain the ars1

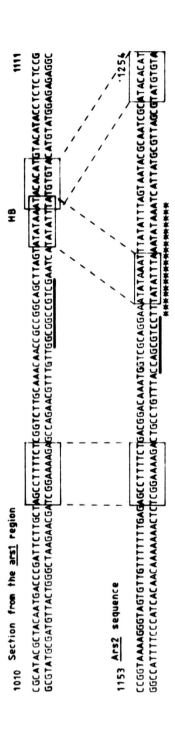

FIGURE 6. A comparison of the 100 bp ars2 sequence with a region of the ars1 sequence containing some partial homologies. The numbers refer to the location of these sequences in the fragment shown in Figure 1 (ars1) or Figure 3 (ars2). The HB indicates the 'Hogness box' in the region of DNA from ars1, and the asterisks indicate the double symmetry element (see Figure 5) from the ars2 sequence. The boxed areas enclose the three direct repeats which are common to these two sequences. The heavy underlines indicate the GC-rich areas adjacent to the AT-rich regions in each sequence.

32 ANALYSIS OF AUTONOMOUSLY REPLICATING SEQUENCES

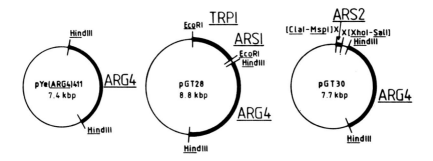

FIGURE 7. Plasmids used to compare the relative transformation efficiencies of DNAs containing arsl and ars2. The thick lines indicate the yeast DNA, the thin lines indicate pBR322 DNA. The X's (pGT30) mark sites where fragments containing complementary ends produced by the indicated enzymes were joined, thereby eliminating both sites.

activity. However this region is several hundred bp distant from the sequence shown in Figure 2, which is apparently required for autonomous replication. If both these regions are necessary for autonomous replication, arsl includes a much larger sequence of DNA than does ars2.

We have compared the ability of circular DNAs containing arsl or ars2 to transform yeast, using the plasmids shown in Figure 7. Plasmid pYe(ARG4)411 consists of pBR322 containing the 3 kbp HindIII fragment bearing the ARG4 gene (see Figure 3). This plasmid DNA was used as a control in the transformation experiments and also for construction of the other two plasmids shown in Figure 7. Plasmid pGT28 contains the entire 1453 bp fragment (described in Figure 1) containing arsl. Plasmid pGT30 contains a 220 bp region of yeast DNA beginning with the 100 bp ars2 region and extending beyond the site of the poly(dA:dT) connector (14). These plasmids were used to transform S. cerevisiae strain X3656-7D(arg4$^-$) selecting for Arg$^+$ transformants. The results of three separate transformations using the plasmids shown in Figure 7 are given in Table 1. These results show that the transformation efficiency of plasmid DNA containing ars2 is consistently greater than is observed with a plasmid of

TABLE 1. Transformation Efficiency of Plasmids

Plasmid	size(kbp)	ars	transformants/μg DNA			
			1	2	3	avg
pGT28	8.8	ars1	280	650	90	340
pGT30	7.6	ars2	1100	2420	540	1350
pYe(ARG4)411	7.3	none	0	0	0	0

S. cerevisiae strain X3656-7D (a leu1 ade6 ura1 arg4-1 thr1) was used in all transformations.

similar size containing ars1. On the average, four times more transformants were obtained with pGT30 DNA (containing ars2) than with pGT28 DNA (containing ars1). Both pGT28 and pGT30 DNA segregate with high frequency when the host cells are grown under nonselective conditions, and no differences are observed in their relative mitotic stabilities.

DISCUSSION

The limited homology between the sequences in ars1 and ars2 described in Figure 6, combined with the differences in transformation efficiency shown in Table 1, suggest that ars1 and ars2 could be representative of different classes of replication origins. This is consistent with recent findings that yeast DNA sequences containing ars fall into two categories: those that are unique in the genome and those that are repetitive (19). Hybridization results suggest that the ars1 region is unique in the yeast genome (9), whereas the delta sequences surrounding and possibly overlapping ars2 are members of a family of dispersed, repetitive sequences in yeast (17). Analysis of yeast origins of replication by fiber autoradiography (5) and electron microscopy (3) indicates that all origins are not activated at the same time, suggesting that all origins are not identical.

It is possible that one or both of the ars described here do not function as origins of replication in the genome. Measurement of the average spacing of origins of replication in yeast by electron microscopy (3) and fiber autoradiography

(5) gives average values of 60 kbp and 90 kbp, respectively. However, random cloning experiments indicate the average spacing of ars elements in the yeast genome to be 32 kbp (19) or 40 kbp (20). These results give average values of 180 origins of replication and 375 ars elements per haploid genome. One explanation for this two-fold difference might be that the origins of replication normally used by the yeast cell are a subset of the ars elements. It is possible that the requirements for initiation of DNA replication in a circular supercoiled molecule such as a plasmid may be less rigorous than in a linear chromosome.

We have compared the ars1 and ars2 sequences described here with various phage, plasmid and bacterial origins of replication without detecting significant homology. The origin of replication of SV40 (16), a virus which must rely on eukaryotic replication mechanisms, also lacks sequence homology with ars1 and ars2. The DNA regions containing ars1 and ars2 do however have three general features in common with the SV40 origin of replication. All three contain at least one dyad spanning 10 or more base pairs. Secondly, they all contain an A+T region identical to or closely resembling a 'Hogness box'. Finally, all three contain a GC-rich region of at least 10 base pairs immediately adjacent to the A+T region. As more eukaryotic origins of replication or ars are characterized, it will be interesting to see if these three features persist.

ACKNOWLEDGMENTS

This work was supported by a research grant (CA-11034) from the National Cancer Institute, National Institutes of Health.

REFERENCES

1. Lauer, G. D., Roberts, T. M., and Klotz, L. C. (1977). J. Mol. Biol. 114, 507.
2. Forte, M. A., and Fangman, W. L. (1979). Chromosoma 72, 131.
3. Newlon, C. S., Petes, T. D., Hereford, L. M., and Fangman, W. L. (1974). Nature 247, 32.
4. Burke, W., and Fangman, W. L. (1975). Cell 5, 263.
5. Petes, T. D., and Williamson, D. H. (1975). Exptl. Cell Res. 95, 103.

6. Hereford, L. M., and Rosbash, M. (1977). Cell 10, 453.
7. Struhl, K., Stinchcomb, D. T., Scherer, S., and Davis, R. W. (1979). Proc. Natl. Acad. Sci. 76, 1035.
8. Hsiao, C. L., and Carbon, J. (1979). Proc. Natl. Acad. Sci. 76, 3829.
9. Stinchcomb, D. T., Struhl, K., and Davis, R. W. (1979). Nature 282, 39.
10. Clarke, L., and Carbon, J. (1978). J. Mol. Biol. 120, 517.
11. Hinnen, A., Hicks, J. B., and Fink, G. R. (1978). Proc. Natl. Acad. Sci. 75, 1929.
12. Kingsman, A. J., Clarke, L., Mortimer, R. K., and Carbon, J. (1979). Gene 7, 141.
13. Tschumper, G., and Carbon, J. (1980). Gene 10, 157.
14. Tschumper, G., and Carbon, J., manuscript in preparation.
15. Maxam, A. M., and Gilbert, W. (1977). Proc. Natl. Acad. Sci. 74, 560.
16. Kornberg, A. (1980). "DNA Replication". W. H. Freeman, San Francisco, Ch. 7.7.
17. Cameron, J. R., Loh, E. Y., and Davis, R. W. (1979). Cell 16, 739.
18. Gafner, J., and Philippsen, P. (1980). Nature 286, 414.
19. Chan, C. S. M., and Tye, B-K. (1980). Proc. Natl. Acad. Sci. 77, 6329.
20. Beach, D., Piper, M., and Schall, S. (1980). Nature 284, 185.

REPLICATION ORIGINS USED *IN VIVO* IN YEAST[1]

Carol S. Newlon
Rodney J. Devenish
Peter A. Suci
C. J. Roffis

Department of Zoology
University of Iowa
Iowa City, Iowa

ABSTRACT

A 190 kilobase (kb) ring chromosome derived from chromosome III can be purified in good yield from the yeast *Saccharomyces cerevisiae*. This chromosome should contain 5-6 replication origins. Electron microscopic techniques are being used to map origins used *in vivo* on this molecule and autonomously replicating segments (ars's) are being identified and mapped using the yeast transformation system.
Replicating 2-μm plasmid DNA has been prepared from *cdc8* strains and from cells synchronized by the α-*factor-cdc7* method. Replication origins have been mapped with respect to known restriction sites. At least two origins of replication are required to account for the observed molecules, one of which may be the same as the mapped *ars* of the plasmid.

[1] This work was supported by an NIH Research Grant GM21510 to CSN. PAS was supported by NIH Genetics Training Grant GM07091.

INTRODUCTION

In contrast to prokaryotic chromosomes, where DNA replication initiates at a single site, eukaryotic chromosomes contain multiple replication origins. The precision with which origins of replication on eukaryotic chromosomes are spatially and temporally ordered is largely unknown. A number of studies in both higher and lower eukaryotes (1-4) suggest that DNA replication is temporally ordered. However, in none of these studies was the resolution sufficient to detect individual replication units. Studies on both *Drosophila* and *Triturus* have demonstrated that the spacing between replication origins varies with developmental stage (5, 6). These results mean that either not every origin is used in every S phase or that the temporal order of activation of origins can vary.

The questions of location and sequence specificity of replication origins have been difficult to approach because of the extremely large size of eukaryotic chromosomal DNA's. We used electron microscopy to study the replication of the smallest chromosomal DNA's from the yeast, *Saccharomyces cerevisiae* (7), and concluded that origins of replication are located nonrandomly on yeast DNA at intervals of approximately 36 kilobases (kb). A number of laboratories have recently reported the isolation of DNA sequences from *S. cerevisiae* and other organisms called *ars*'s (for autonomously replicating segments) which allow autonomous replication of DNA colinear with them when they are used to transform yeast (8-13). As a result of their ability to replicate autonomously, the transformation efficiency of *ars*-containing plasmids is about 1000-fold higher than plasmids which cannot replicate. These *ars*'s may be chromosomal replication origins; their frequency of occurrence in yeast DNA, one per 40 kb, is consistent with this notion (11, 12). However, it needs to be demonstrated that these sequences actually function as initiation sites on chromosomes *in vivo*.

We are taking two approaches to further define the location and temporal order of activation of replication origins used *in vivo* and to determine whether these origins correspond to autonomously replicating segments identified by the transformation assay. The first is based on our recent identification of a unique, *circular* derivative of yeast chromosome III, which contains approximately 190 kb of DNA (14). Based on the average spacing between origins (7), we expect that this ring chromosome should contain five or six replication origins. This molecule has several advantages for the study of DNA replication. Circularity of a replicating molecule visualized by electron microscopy provides assurance that the molecule is intact and represents only chromosome III. In addition, the ring chromosome can be purified substantially free from other DNA and used directly for constructing or screening banks of cloned DNA for chromosome III sequences. Thus, identifying and mapping chromosome III ars's will be greatly facilitated. In this paper we report progress toward these goals.

Our second approach is to map the origin(s) of replication used *in vivo* on the yeast 2-μm plasmid. Most laboratory strains of *S. cerevisiae* carry 50-100 copies of a small plasmid DNA that is approximately 2-μm in circumference. Although the function of this plasmid is unknown, a variety of studies have shown that it replicates under the same control as chromosomal DNA: it follows the same pattern as chromosomal DNA replication in cell cycle mutants, failing to replicate in mutants blocked in G1 and exactly doubling in mutants blocked in G2 (15); replicating 2-μm DNA accumulates at the nonpermissive temperature in mutants defective in DNA elongation (16); it replicates during S phase and, in contrast to bacterial plasmids, each 2-μm DNA molecule replicates only once (17). Thus, the origin(s) of replication on the 2-μm plasmid may serve as a good model for chromosomal origins. Using high frequency transformation of yeast as a criterion for origin function, Broach and coworkers have identified a single 350 base pair region of the 2-μm plasmid genome which contains the 2-μm *ars* (18, 19). We have isolated replicating 2-μm DNA from yeast and mapped the position of replication structures relative to unique restriction sites. In this paper we present evidence for at least two origins of replication on this plasmid, one of which is consistent with the *ars* identified using the transformation assay.

RESULTS AND DISCUSSION

A. *Studies on Ring Chromosome III*

The events involved in the generation of the ring chromosome are diagrammed in Fig. 1. Yeast mating type is controlled by genes that reside on transposable genetic elements, the *a* and α cassettes (20). These cassettes reside in three loci on chromosome III, *HML, HMR* and *MAT*. The cassette at *MAT* is expressed, while cassettes at *HML* and *HMR* are normally silent. By selecting for cells which have changed mating type, rearrangements of chromosome III can be isolated which lead simultaneously to the loss of the gene at *MAT* and the

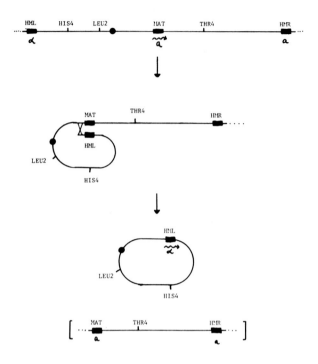

FIGURE 1. Diagram of the events involved in the generation of the MATα-lethal ring chromosome. Homologous recombination between the left ends of MATa and HMLα yields a MATα ring chromosome and an acentric fragment which is lost during mitosis (14, 21).

activation of the gene at HML or HMR. Strains in which the gene at HML is activated contain a circular chromosome which results from an intramolecular recombination between HML and MAT (14, 21). This ring chromosome is 62.6 ± 5.7 μm (~190kb) in length and can be isolated as covalently closed circular DNA in CsCl-ethidium bromide gradients (14). Diploids carrying the ring chromosome can be distinguished from normal diploids which arise in the same selection by a rare switch from MATa to MATα by simple genetic analysis. The ring chromosome is haplolethal and diploids carrying the ring yield a maximum of two viable spores per tetrad, both MATa, while the normal diploids yield four viable spores per tetrad, 2MATa and 2MATα.

The initial isolation procedures for the ring chromosome yielded only about 1% of the expected DNA. In order to isolate enough ring DNA for physical studies, we have modified a procedure developed by Casse et al. (22) for the isolation of large Ti plasmids from *Rhizobium*. Using this new procedure, we can isolate at least half the expected ring DNA (1 molecule per cell or 0.8% of the total DNA). It involves lysis of spheroplasts at pH 12.45, which denatures linear but not covalently closed circular DNA, a gentle phenol extraction to remove single-stranded DNA, and then an ethanol precipitation. The extract contains 5-10% of the total DNA of the culture. When large circular DNA's are electrophoresed in 0.7% agarose gels, the circular molecules migrate more slowly than the largest linear DNA's (23). The left hand tracks of Fig. 2 show an ethidium bromide-stained 0.7% agarose gel in which extracts of a diploid carrying the ring chromosome and an isogenic wild type diploid have been electrophoresed. A band present in the upper part of the ring extract track (arrow) is missing from the normal diploid extract. That this band is chromosome III is demonstrated by the autoradiograms shown in the same figure. Pairs of tracks from the gel shown on the left were transferred to nitrocellulose by the method of Southern (24) and hybridized to various nick-translated plasmids containing yeast DNA fragments. It can be seen that the putative ring chromosome band hybridizes to a *MATα* fragment from chromosome III but fails to hybridize to a *CYC7* fragment from chromosome V. This analysis has been extended to include the use of hybrid plasmids containing the *HIS4* and *LEU2* genes from chromosome III as well as *CYC1* from chromosome X, *HIS3* from chromosome XV and rDNA from chromosome XII. In all cases, fragments from chromosome III hybridized to the band as expected while fragments from other chromosomes failed to hybridize to any DNA in the extract. In addition to the slowly migrating band, chromosome III fragments hybridize to a second band which migrates at the same rate as linear fragments of chromosomal DNA. Since the band contains no fragments from other, linear chromosomes as evidenced by its failure to hybridize fragments from chromosomes V, X, and XV, it presumably results from broken ring chromosomes. These results provide strong evidence that the band indicated by the arrow is ring chromosome III.

In addition to the ring chromosome, the extract contains 2-μm plasmid DNA and its multimers (Fig. 2), as well as some circular ribosomal DNA (data not shown). The bulk of the material in the extract which fails to enter the gel is ribosomal DNA, presumably the network formed by chromosomal rDNA cistrons annealing out of register.

We have quantitated the amount of ring DNA in these preparations by running extracts made from cells uniformly labelled with ^{3}H-adenine on agarose gels. Fig. 3 shows the profile of radioactivity in the tracks of the gel shown in Fig. 2. In this preparation, 7.5% of the radioactivity was

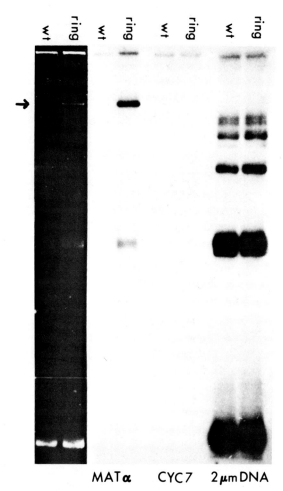

FIGURE 2. Separation of circular DNA's by agarose gel electrophoresis. Extracts were prepared by the method of Casse et al. (22) from strains XG1 #1 (labelled wt for wild type) and XG1 #24b (labelled ring, ref. 14) and subjected to electrophoresis in 0.7% agarose gels (20mA, 17 hrs). The gel was stained with ethidium bromide and photographed, and DNA from pairs of tracks was transferred to nitrocellulose (24) after partial depurination (26). The DNA on the filters was hybridized to ^{32}P-labelled, nick-translated DNA as previously described (14). Hybridization probes were as follows: MATα, the EcoRI-HindIII yeast fragment from MATα (21); CYC7, a Pst-EcoRI yeast fragment from pYeCYC7 (27); 2-μm, the 2-μm DNA monomer excised from plasmid 82-6B with Pst (28). The slot is the bright band at the top of the stained tracks.

in the ring band, 30% in supercoiled 2-μm monomers (lower band), 18% at the slot and 18% associated with the remaining visible band which consists of nicked 2-μm monomers, supercoiled 2-μm tetramers and linear chromosomal DNA (see Fig. 2). Similar results have been obtained in three other experiments. In these experiments we have recovered between .35 and .5% of the total DNA in the ring band, which is 50-70% of the theoretical yield.

It is surprising that we have found only a single band from the ring chromosome. Based on previous work with bacterial plasmids of a similar size (e.g., see ref. 22), we expected to find a band containing supercoils and a band

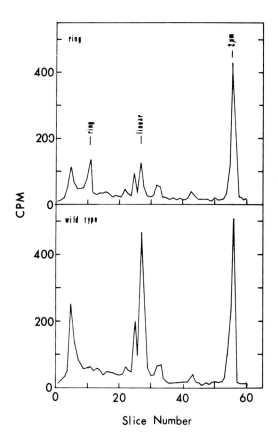

FIGURE 3. ^3H profile of DNA in agarose gel in Figure 2. Extracts were prepared from cells grown for six generations in 8-^3H-adenine (0.033μCi/μg, 15μg/ml) as described in Figure 2. The gel was dried, sliced into ~1mm slices and counted (17).

containing nicked circles. Although definitive proof is lacking, three lines of evidence suggest that the single band we obtain consists of nicked circles: 1) the migration rate is substantially the same in gels containing ethidium bromide as in gels without ethidium; 2) when the extract is spread for electron microscopy, only relaxed circles of the size expected for the ring chromosome are seen; no supercoils of the expected size have been found; 3) the rate of migration of the band is similar to that of nicked bacterial plasmids of similar size and considerably less than that of supercoils (see ref. 22). If the band we observe is, in fact, nicked circles, then it is not clear why they survive the extraction. Possibly a nicking-closing enzyme holds the ends of the nicked strands together.

We have used purified ring DNA to screen banks of yeast DNA cloned in λ Charon 4A and the integrating vector YIp5 (25). Chromosome III fragments in these vectors are being used to construct a restriction map of the ring chromosome and to identify and map *ars*'s on the ring. In addition we have constructed a diploid carrying the ring chromosome that can be synchronized (a *cdc7* homozygote) and are isolating replicating ring DNA's on sucrose gradients for mapping of origins used *in vivo*.

B. *Mapping the Origins of Replication on 2-μm DNA*

As an approach to determining whether *ars*'s identified by their ability to promote autonomous replication of plasmids are the same as replication origins, we have attempted to map the origin(s) of replication on the 2-μm plasmid which are used *in vivo*. The plasmid contains two identical 599 base-pair inverted repeats (28). Intramolecular recombinations between these repeats generate two isomeric forms (diagrammed in Fig. 4) which are present in equal amounts. Broach and Hicks (18) have localized the 2-μm plasmid *ars* to a sequence which extends from the middle of one of the inverted repeats into the large unique region (Fig. 4). Our strategy has been to prepare replicating 2-μm DNA from synchronized cells, linearize it with a restriction endonuclease that cuts the molecule only once, and compare the position of replication structures with the known restriction map (28). As indicated in Fig. 4, both PstI and BclI cleave 2-μm DNA at unique sites. If replication *in vivo* initiates within the *ars* and proceeds bidirectionally, then the position of the center of replication structures (the origin of replication) relative to restriction sites can be predicted. For PstI-cleaved molecules the origin should be 15-21% of the distance from one end and should be the same for form A and form

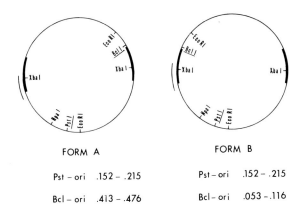

FIGURE 4. Diagram of the isomers of 2-μm plasmid DNA. The two forms are related by recombination between the inverted repeats (indicated by heavy lines). The 2-μm ars has been mapped to the region indicated by the thin lines outside the circles (18). The fractional distances from the unique restriction endonuclease sites to the ars (called ori in the figure) are indicated in the lower part of the figure.

B. Since the BclI site is in the unique region opposite the ars, two classes of molecules are expected, with form A having an origin 5-12% of the distance from one end and form B having an origin 5-12% of the distance from one end. Although bidirectional replication of 2-μm DNA has not been demonstrated in vivo, Sugino (this volume) has evidence that replication proceeds bidirectionally in vitro.

Replicating 2-μm DNA from cdc8. In the first experiments, we attempted to enrich for replicating 2-μm DNA using the temperature sensitive cell cycle mutant cdc8. This mutant is defective in DNA chain elongation (29), and has been shown to accumulate small replication bubbles at the normal spacing in chromosomal DNA (30), as well as replicating 2-μm plasmid (16) at the nonpermissive temperature. Cultures of strain 13052 (cdc-3) or 198 (cdc8-1) were treated with α-factor at 23° for the equivalent of one doubling to block cells in G1 (31). After washing cells to remove the α-factor, cultures were shifted to 36° and incubated for 3 hours before extracts enriched in 2-μm DNA were prepared by the method of Livingston and Kupfer (15). DNA from the extracts, which contained 5 to 10% of the DNA in the culture, was centrifuged to equilibrium in CsCl-ethidium bromide gradients (31) and DNA from both the covalently closed circular peak and the chromosomal DNA peak, which consisted largely of nicked 2-μm plasmid DNA, was examined by electron microscopy. The only replicating molecules observed were 2-μm plasmid DNA; about 1-2% of the plasmid molecules from the chromosomal DNA peak contained replication structures (Fig. 5a). Molecules were linearized by cutting them with the restriction endonuclease PstI, spread for electron microscopy and micrographs of replicating molecules made and measured (Fig. 5). More than one-hundred replicating molecules were analyzed. They appear to be authentic 2-μm DNA; the average length of the linearized replicating molecules was 1.96 ± 0.10 μm, to be compared with a length of 2.03 ± .07 μm obtained for uncut 2-μm DNA (14).

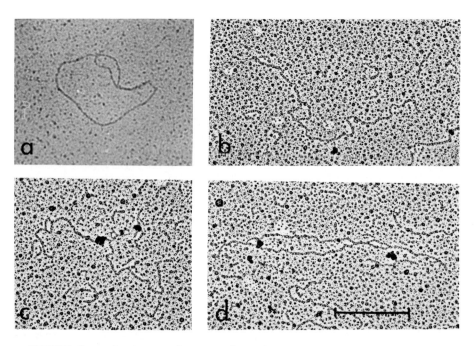

FIGURE 5. *Electron micrographs of replicating 2-μm DNA obtained from* cdc8. *DNA from CsCl-ethidium bromide gradients was prepared for electron microscopy and analyzed as described previously (14) except that in some experiments grids were shadowed with Pt-Pd. In (a) is shown a typical, relaxed θ-form replication structure characteristic of the 2-μm plasmid (16). Panels b, c and d show replicating plasmid molecules which have been linearized with PstI. The molecules in (b) and (c) contain a single replication bubble while the one in (d) contains two bubbles, the large one centered at 0.39 and the small one at 0.71. The bar represents 0.5 micron.*

33 REPLICATION ORIGINS USED *IN VIVO* IN YEAST

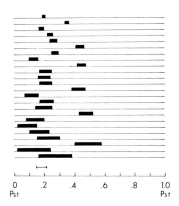

FIGURE 6. Diagram of Pst-cleaved 2-μm molecules less than 20% replicated from cdc8. The heavy lines represent the replicated region of the molecules. Molecules are oriented with the replicated region nearest the left end. The bar above the scale is the position of the mapped ars (Fig. 4), and the scale represents fractional length of the molecule with respect to the PstI cleavage site.

The fraction of the molecule replicated ranged from less than 5% to more than 90%.

Molecules with the smallest replication structures allow the most accurate determination of the position of the replication origin, since interpretation of such molecules requires no assumptions about the direction of fork movement or relative rates of movement of the two forks. This may be especially true for molecules obtained from cdc8 strains, because the elongation deficiency may affect the two forks in a single molecule unequally. In Fig. 6 are diagrammed the 24 Pst-cut molecules from these experiments that are less than 20% replicated. Inspection of the figure reveals that in 75% (18/24) of the molecules, at least some portion of the replicated region overlaps the mapped ars. The average midpoint of these replication bubbles is .19 ± .05 of the distance from the Pst site. However, 25% of the molecules have replication bubbles that cannot possibly have initiated in the ars (the 2nd, 6th, 9th, 13th, 17th, and 22nd molecules in Fig. 6). The average midpoint of the replication bubbles in these six molecules is at 0.46 ± .01 from the Pst site. We conclude that although a majority of the replicating 2-μm plasmid molecules we obtained from cdc8 could have initiated replication within the ars, at least one additional initiation site is required to account for the molecules we have observed. This site is in the small unique region of the molecule, most likely in about the middle of the small unique region (see below). This conclusion is strengthened by two observations. First, in the collection of Pst-cleaved molecules, two molecules with 2 replication bubbles were found (Fig. 5d), demonstrating that two initiations can occur in some molecules. Second, a more limited collection of EcoRI-cleaved molecules also requires an initiation site in addition to the ars to explain the observed replication structures (data not shown).

Replicating 2-μm DNA from cells synchronized by the α-factor-cdc7 method. We were concerned that the second initiation site found in 2-μm DNA isolated from *cdc8* might be an artifact induced by incubating the elongation mutant at the nonpermissive temperature. We therefore undertook to enrich for replicating 2-μm DNA by a different method. We synchronized strain 4008 (*cdc7-4*) using the α-factor-cdc7 double block method (32). *Cdc7* is a temperature-sensitive cell cycle mutant that blocks in late·G1, just prior to the initiation of DNA replication (30). When it is shifted to the permissive temperature, a highly synchronous round of DNA synthesis ensues. Zakian *et al.* (17) have used this procedure to show that 2-μm DNA replication occurs early in S phase. We therefore took samples early in S-phase (8 minutes and 12 minutes after the shift to the permissive temperature), and prepared 2-μm DNA as described above. In this experiment, replication structures were found in approximately 1% of both supercoiled and relaxed circular molecules. (This suggests that incubation of *cdc8* at the nonpermissive temperature results in nicking of supercoils; see above and ref. 16). Supercoiled 2-μm DNA was linearized with BclI and spread for electron microscopy. Figure 7 shows seventeen replicating molecules obtained from the 8 minute sample. As in the previous experiment, the molecules are of the length expected for 2-μm DNA. If initiation of replication were confined to the *ars*, we would expect to find half the molecules with replication structures near the middle of the molecule (Form A, Fig. 4) and half with replication structures 5-12% of the distance from one end (Form B, Fig. 4). In this small sample, 10 of the molecules (60%) have initiated replication at a site consistent with the Form A *ars*; only one has a bubble at a position consistent with the Form B *ars*. Six of the molecules (35%) have replication structures which do not overlap the *ars*. The average midpoint of these structures is at .26 ± .036 of the distance from the BclI site. One of the two sites at which molecules could have initiated replication is in about the middle of the small unique region (see Fig. 4), in the same region of the molecule as predicted by the *cdc8* data described above. Although we do not yet have data

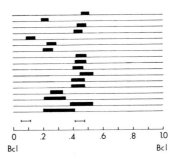

FIGURE 7. Diagram of Bcl-cleaved 2-μm DNA molecules from cultures synchronized by the α-factor-cdc7 method. The heavy lines represent the replicated region of the molecule. Molecules are oriented with the replicated region nearest the left end. The bars above the scale represent the position of the mapped ars relative to the Bcl site in Form B and Form A of the molecule (see Fig. 4). The scale represents the fractional length of the molecule with respect to the BclI cleavage site.

on Pst-cleaved molecules obtained using this synchronization procedure, it is clear that, again, at least one site in addition to the previously mapped ars is required to explain the replication structures observed in 2-μm DNA.

It is surprising that only one molecule has a replication structure that could have initiated in the Form B ars. It is possible that the paucity of molecules in this class is simply the result of having looked at only a small number of molecules. Other explanations are that Form B does not replicate in vivo or that Form B uses the alternate origin in vivo. These possibilities seem somewhat unlikely in light of the observation that the small EcoRI fragment of Form B functions as efficiently as an ars as the large EcoRI fragment of Form A (18, 19, see Fig. 4), but more data will be required to resolve the question.

In summary, at least two origins of replication on the yeast 2-μm plasmid are required to explain the replication structures we observe in 2-μm DNA prepared from cells synchronized by two different methods. One of the origins (used in 60-75% of the molecules observed) is consistent with the mapped ars, but until we have mapped the site with respect to two restriction enzyme cleavage sites in the same DNA preparation, we cannot state conclusively that the ars is used in vivo. The second site is probably near the middle of the small unique region, but the same reservation applies.

The implication of these experiments is that not all replication initiation sites function as autonomous replicators in the transformation assay. There are several possible explanations for this finding. First, the second site used on the plasmid *in vivo* may be a weak *ars*, and rate of loss of the plasmid due to inefficient replication may prohibit colony formation under selective conditions. Along the same line, efficient utilization of this site may require the 2-μm plasmid-encoded replication protein (18, 19). If this were the case, then one would have to argue that the replication protein cannot be expressed from the large EcoRI fragment of Form B because this fragment contains both the putative second origin and the entire open reading frame for the replication protein (19, 28).

The second possibility is that the situation in 2-μm DNA is similar to that in the bacterial plasmids R6K (33) and ColE1 (34), where a sequence some distance away from the site at which replication actually begins is required for replication. In R6K, replication forks usually initiate *in vivo* at $ori\alpha$ or $ori\beta$. However, a sequence, $ori\gamma$, which is 1-2 kb away from these sites is required in cis for replication to initiate at $ori\alpha$ or $ori\beta$. *In vivo*, replication initiates at low frequency at $ori\gamma$, and $ori\gamma$ is the only one of the three replication origins able to function as a replicator when cloned. In ColE1, transcription of the RNA primer used at the origin begins about 550 bp upstream from the origin. Mutations in the promoter of this transcript interfere with replication of the plasmid.

Finally, it is still possible that *ars*'s identified in the transformation assay and replication origins used *in vivo* may not be directly related. Additional studies on the yeast 2-μm plasmid as well as on ring chromosome III should clarify these issues.

ACKNOWLEDGMENTS

We thank Jeffrey Strathern, Benjamin Hall, Donna Montgomery, John Donelson, Gerry Fink, and John Carbon for yeast plasmids; Cathy Henderson for assistance; and Deena Staub for typing the manuscript.

REFERENCES

1. Braun, R., and Wili, H. (1969). *Biochim and Biophys. Acta 174,* 246.
2. Lima-de-Faria, A., and Jaworska, H. (1972). *Hereditas 70,* 39.
3. Burke, W. and Fangman, W. L. (1975). *Cell 5,* 263.
4. Kee, S. G. and Haber, J. E. (1975). *Proc. Nat. Acad. Sci. 72,* 1179.
5. Blumenthal, A. B., Kriegstein, H. J. and Hogness, D. S. *Cold Spring Harbor Symp. Quant. Biol. 38,* 205.
6. Callan, H. G. (1972). *Phil. Trans. Roy. Soc. B 181,* 19.
7. Newlon, C. S., and Burke, W. G. (1980). In "Mechanistic Studies of DNA Replication and Recombination" (B. Alberts and C. F. Fox, eds.), p 399. Academic Press, New York.
8. Stinchcomb, D. T., Struhl, K., and Davis, R. W. (1979). *Nature 282,* 39.
9. Kingsman, A. J., Clarke, L., Mortimer, R. K., and Carbon, J. (1979). *Gene 7,* 141.
10. Hsiao, C.-L., and Carbon, J. (1979). *Proc. Nat. Acad. Sci. 76,* 1035.
11. Beach, D., Piper, M., and Shall, S. (1980). *Nature 284,* 185.
12. Chan, C. M., and Tye, B.-K. (1980). *Proc. Nat. Acad. Sci. 77,* 6329.
13. Stinchcomb, D. T., Thomas, M., Kelly, J., Selker, E., and Davis, R. W. (1980). *Proc. Nat. Acad. Sci. 77,* 4559.
14. Strathern, J. N., Newlon, C. S., Herskowitz, I., and Hicks, J. (1979). *Cell 18,* 309.
15. Livingston, D., and Kupfer, D. (1977). *J. Mol. Biol. 116,* 249.
16. Petes, T. D., and Williamson, D. (1975). *Cell 4,* 249.
17. Zakian, V. A., Brewer, B. J., and Fangman, W. L. (1979). *Cell 17,* 923.
18. Broach, J. R., and Hicks, J. B. (1980). *Cell 21,* 501.
19. Broach, J. R., and Hartley, J. (1980). In "Mechanistic Studies of DNA Replication and Recombination" (B. Alberts and C. F. Fox, eds.), p. 389. Academic Press, New York.
20. Hicks, J., Strathern, J., and Herskowitz, I. (1977). In "DNA Insertion Elements, Plasmids, and Episomes" (A. Bukhari, J. Shapiro, and S. Adhya, eds.), p. 457. Cold Spring Harbor Laboratory, Cold Spring Harbor, New York.
21. Strathern, J. N., Spatola, E., McGill, C., and Hicks, J. B. (1980). *Proc. Nat. Acad. Sci. 77,* 2839.

22. Casse, F., Boucher, C., Julliot, J. S., Michel, M., and Dénarié, J. (1979). *J. Gen. Microbiol.* *113*, 229.
23. Meyers, J. A., Sanchez., D., Elwell, L. P. and Falkow, S. (1976). *J. Bacteriol.* *127*, 1529.
24. Southern, E. (1975). *J. Mol. Biol.* *98*, 503.
25. Struhl, K., Stinchcomb, D. T., Scherer, S., and Davis, R. W. (1979). *Proc. Nat. Acad. Sci. 76*, 1035.
26. Wahl, G. M., Stern, M., and Stark, G. R. (1979). *Proc. Nat. Acad. Sci. 76*, 3683.
27. Montgomery, D. L., Leung, D. W., Faye, G., Shalit, P., Smith, M., and Hall, B. D. (1980). *Proc. Nat. Acad. Sci. 77*, 541.
28. Hartley, J. L., and Donelson, J. E. (1980). *Nature 286*, 860.
29. Hereford, L. M., and Hartwell, L. H. (1971). *Nature New Biol. 234*, 171.
30. Petes, T. D., and Newlon, C. S. (1974). *Nature 251*, 637.
31. Newlon, C. S., and Fangman, W. L. (1975). *Cell 5*, 423.
32. Newlon, C. S., Petes, T. D., Hereford, L. M., and Fangman, W. L. (1974). *Nature 247*, 32.
33. Helinski, D. R., Inuzuka, M., Inuzuka, I., Kolter, R., Stalker, D. M., and Thomas, C. M. (1980). In "Mechanistic Studies of DNA Replication and Recombination" (B. Alberts and C. F. Fox, eds.), p. 293. Academic Press, New York.
34. Itoh, T., and Tomizawa, J.-I. (1980). *Proc. Nat. Acad. Sci. 77*, 2450.

REPLICATION PROPERTIES OF Trp1-R1-CIRCLE:
A HIGH COPY NUMBER YEAST CHROMOSOMAL DNA PLASMID[1]

John F. Scott
Connie M. Brajkovich

Molecular Biology Institute
University of California
Los Angeles, California 90024

I. ABSTRACT

A 1.4 kilobase restriction fragment of *Saccharomyces cerevisiae* chromosomal DNA, containing the Trp1 gene and an autonomously replicating sequence (Ars1) was excised from the yeast vector YRp7 and ligated at low DNA concentration to favor intramolecular circularization. The Ars1 presumably contains an origin of chromosomal DNA replication. When yeast cells were transformed with the ligation mixture, selecting for *trp1* complementation, colonies were obtained at high frequency (about 1000 per microgram of Trp1 fragment ligated). The transformants were found to contain a 1.4 kilobase circular plasmid (Trp1-R1-Circle) in more than 100 copies per cell. The Trp1-R1-Circle was purified from yeast and used to re-transform *trp1* strains containing or not containing normal amounts of the endogenous yeast "2μm" DNA plasmid. The Trp1 plasmid was found to be highly stable mitotically in both hosts in comparison with YRp7. The Trp1 plasmid was also found to be maintained with high

---

[1] *This work was supported in part by funds from the MBI Parvin Core Grant (USPHS CA 16163), the ACS UCLA Institutional Grant (IN-131) and an NIH Research Grant (USPHS 1 R01 GM 27000).*

stability through meiosis and frequently segregated 4 Trp$^+$: 0 Trp$^-$ to the haploid progeny. The Trp1-R1-Circle is therefore not under the segregation control of a centromere, but is maintained with high stability mitotically and meiotically, probably due only to its high copy number and small size.

II. INTRODUCTION

Our laboratory is interested in the investigation of the mechanism of initiation of chromosomal DNA replication in eukaryotic cells. We have chosen to use as model templates for this study plasmids capable of autonomous replication in yeast. Our decision to make use of yeast was based on its suitability for use in biochemical, genetic and molecular cloning approaches.

The endogenous "2µm" DNA plasmid present in most laboratory yeast strains is a potentially attractive model template. It is high-copy-number (30-50 per haploid cell) (1), stable mitotically and meiotically (2), has been completely sequenced (3), is replicated under nuclear control (4) once per cell division early in S-phase (5) and is organized in nucleosomes comparable to those of yeast nuclear chromatin (1,6,7). However, the 2µm plasmid has several disadvantages as well. In its naturally occurring form the plasmid contains no selectable gene and its presence in the cell is non-essential (2). Thus the presence of the plasmid can be verified only by analysis of extracted DNA. In addition, 2µm DNA contains a 599 base pair inverted repeat and undergoes frequent intramolecular recombination in yeast, mediated by a plasmid coded function (3,8,9). The recombination events result in the presence of two forms of the molecule in yeast and serve to complicate the topological analysis of replicative intermediates (10,11). These disadvantages have been minimized or eliminated by molecular engineering (8). Properties of these modified plasmids have pointed to one remaining, unavoidable disadvantage in the use of 2µm DNA as a model for chromosomal replication. The 2µm replicator is from an unknown source, and while it is replicated in a manner similar to chromosomal

DNA, it differs in that its replication efficiency is influenced in a positive way by a second plasmid coded function (REP). When plasmids were constructed which contained the 2μm replicator but no REP function, and the plasmids were introduced into yeast containing no 2μm DNA, they replicated, but much less efficiently than when REP was present on the plasmid or provided in "trans" by a second plasmid (8). As a result of these considerations we have chosen to concentrate on model templates known to be derived from the yeast chromosomes.

The ability of recombinant plasmids to replicate in yeast as units separate from chromosomes or endogenous plasmids has been shown by several laboratories to be dependent on the presence on the plasmid of a functional unit referred to as an autonomously replicating sequence (Ars) (8,12-15). While it has not yet been proven unequivocally, the Ars most likely contains an origin of DNA replication. The first such Ars derived from yeast chromosomal DNA, Ars1, was discovered as a component which coincidentally resided on the same 1.4 kilobase EcoR1 restriction fragment as the Trp1 gene (12,13). When the fragment was sub-cloned in the EcoR1 site of *E. coli* plasmid pBR322, the resulting plasmid, YRp7, was found to transform yeast at high frequency (500-5000 colonies per μg DNA) and to exist as an autonomous plasmid (12). The properties of YRp7 in yeast originally reported included low mitotic stability (95% loss after fifteen generations of growth without selection) and low-copy-number per cell in a population of cells grown with Trp$^+$ selection (1-5 per cell at stationary phase) (12) as well as being very unstable meitoically (15). While YRp7 has proven to be a very useful vector for molecular cloning in yeast, its low copy number and low stability represent significant disadvantages to its use as an endogenous template in or from yeast. We therefore chose to construct a new plasmid containing the known chromosomal Ars1 and selectable Trp1 gene which would possess as many advantages and as few disadvantages of the 2μm and YRp7 plasmids as possible.

III. RESULTS

The construction and preliminary characterization of such a plasmid, the Trp1-R1-Circle (Figure 1) has been previously described briefly (16) and is to be published in detail elsewhere. In summary, the Trp1 containing EcoR1 restriction fragment of yeast chromosomal DNA, the complete sequence of which has been reported (17), was excised from YRp7 and ligated at low DNA concentration to favor intramolecular circularization. When yeast cells were transformed with the ligation mixture selecting for *trp1* complementation, colonies were obtained at high frequency (about 1000 colonies per microgram of Trp1 fragment ligated). The transformants were found to contain a 1.4 kb circular plasmid (Trp1-R1-Circle) in more than 100 copies per cell. The Trp1-R1-Circle was purified from yeast and returned by transformation to *trp1* strains containing or not containing the endogenous 2μm DNA plasmid. The copy-number and mitotic stabilities were found to be comparable in the two hosts. After 15 generations of selected growth to stationary cultures, followed by 15 generations of growth in rich medium, the Trp1 plasmid was found to be present in 80-90 percent of the cells. In comparison, less than 5 percent of YRp7 transformants contain that plasmid when grown in the same manner.

Diploids homozygous for the *trp1* mutation and containing the Trp1-R1-Circle were formed by mating haploid strains. The diploids were selected Trp^+. They were then sporulated and tetrads dissected under non-selective conditions. The parents in the cross were a *trp1* strain of a mating type containing a normal complement of endogenous 2μm plasmid DNA and a second *trp1* strain of α mating type which had been transformed with Trp1-R1-Circle purified from yeast and which was thereby phenotypically Trp^+. The diploid was selected for its ability to grow on minimal media. This insured that the parent diploid used in the sporulation and segregation experiments to follow contained the Trp1 plasmid since the haploids mated were both *trp1* and the diploid was therefore homozygous *trp1* chromosomally. That the diploid chosen contained the 2μm DNA plasmid was assumed on the basis of mating and segregation

34 REPLICATION PROPERTIES OF Trp1-R1-CIRCLE

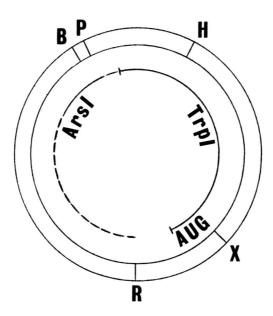

Figure 1. Partial restriction map of the 1453 base pair Trp1-R1-Circle. This map is based on the sequence of the Trp1 EcoR1 fragment published by Tschumper and Carbon (17). The sites indicated are useful unique cleavage points on the plasmid. The circle was closed at the EcoR1 (R) site at nucleotide residue [0/1453]. Reference points on the nucleotide sequence are indicated counter-clockwise from that point. Other sites shown are XbaI (X) at [186], HindIII (H) at [615], PstI (P) at [827] and BglII (B) at [852]. The Trp1 coding region is marked as a solid arc beginning with the AUG at [103] and ending with UAG at [775]. The Ars1 is marked as a partially solid and partially dashed arc. The minimum functional Ars1 has not yet been delimited but it has been shown to be completely contained in the EcoR1-HindIII fragment [615-1453] and must span the BglII site (B) based on deletion constructions published previously by Stinchcomb and Davis (2) and by Tschumper and Carbon (17), respectively.

experiments done with that plasmid by others (2), and was confirmed to be true in this experiment. The diploid was then sporulated without maintenance of Trp$^+$ selection at 23°C in the usual manner (18), and tetrads were dissected. One complete four spore tetrad derived from this cross was chosen for further analysis. All four progeny haploid clones were found to be Trp$^+$, suggesting that the Trp1-R1-Circle had segregated to all 4 spores. The genotypes and phenotypes of the parents, the diploid and the progeny strains are listed in Table 1. The Northern analysis (19) of the plasmid DNA contents of the strains is shown in Figure 2. Cultures of the haploid parents, the diploid and the haploid progeny were grown to mid-log phase in selective media and DNA preparations were made from each as described in the figure legend. The DNA samples were displayed on an agarose gel, transferred to diazotized paper and hybridized using a mixture of 2μm DNA and Trp1-R1-Circle probes. The presence of 2μm DNA but no Trp1 plasmid in one parent (lane A) and Trp1 plasmid but no 2μm DNA in the other (lane B) was confirmed. The diploid was found to contain both 2μm DNA and Trp1 plasmid DNA as expected (lane C). The presence of Trp1 plasmid in all 4 progeny haploids was also verified, as was the segregation of 2μm DNA to all four spores of the tetrad (lanes D-G).

The analysis of the chromosomally linked markers in the four haploid progeny (JSY 315-318) (Table 1) demonstrated that *mat, his*3V-1, *gal*2,10, and TS all segregated 2:2 as anticipated. The Ura$^+$ phenotype segregated 1+ : 3$^-$ due to the independent segregation of *ura*1 and *ura*3. We have analyzed a total of 5 four spore tetrads generated in the same manner and all 5 exhibited 4+ : 0$^-$ segregation of the Trp1-R1-Circle. It is of interest that when sporulation was done at 30°C the plasmid was segregated less efficiently. In 4 four spore tetrads analyzed from 30°C sporulation only one was 4 Trp$^+$: 0 Trp$^-$ with one each 3 Trp$^+$: 1 Trp$^-$, 1+ : 3-, and 0+ : 4- being observed. In addition, a total of 300 haploid progeny generated by the random spore method (18) after sporulation of several Trp1-R1-Circle containing diploids at 23°C were scored and 78% were found to be Trp$^+$. Approximately 20 of those which were Trp$^+$ have been subsequently subjected to analysis for Trp1 plasmid, and all

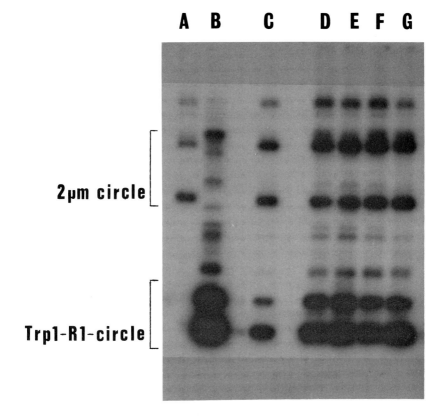

Figure 2. Meiotic segregation of Trp1-R1-Circle. Cells were grown to approximately 8×10^7 per ml in yeast minimal medium containing per liter 0.7 g Difco yeast nitrogen base, 5 g ammonium sulfate, 20 g D-glucose, and supplemented with 40 µg/ml adenine sulfate, 20 µg/ml uracil, 30 µg/ml lysine, 20 µg/ml histidine, and 30 µg/ml tyrosine. The cells were collected by centrifugation, frozen in liquid nitrogen and stored at $-20°C$ until used. DNA was prepared from each strain essentially as described in (18), p. 95-96. The concentration of DNA in each sample was estimated on the basis of ultraviolet absorbance at 260 nm. DNA samples were loaded on a 0.7 percent agarose gel in a buffer containing per liter 10.8 g Trizma base (Sigma), 5.5 g boric acid and 0.93 g disodium (Ethylenedinitrilo)-tetraacetic acid. Electrophoresis was for 5 h at approximately 5 volts per cm. The gel was acid treated and the DNA was transferred to diazotized paper prepared as described in (19,20). The resulting

blot was pretreated and hybridized (48 h) with a mixture of ^{32}P labelled probes containing 10^6 counts per minute (cpm) Trp1-R1-Circle probe and 10^6 cpm p82-6B probe [a partial dimer of 2µm DNA cloned in pMB9 (8)]. The nick-translated probes were prepared essentially as described in (21). Pretreatment, hybridization and washing of the blot was done as in (20). Autoradiography was on Kodak XRP-1 film at 23°C for 12 h. Lane contents were approximately 1 µg DNA from each strain as follows: (A) JSY222, (B) JSY12, (C) JSY310, (D) JSY315, (E) JSY316, (F) JSY317, (G) JSY318. Genotypes of the strains are listed in Table 1. Strain JSY12 is equivalent to the his3v-1 strain described in (22) into which Trp1-R1-Circle was introduced by transformation using a procedure essentially as described in (23). Variation in hybridization from lane to lane was probably due to differences in relative recovery of plasmid and chromosomal DNA among strains.

TABLE I. Genotypes and Phenotypes of Yeast Strains Constructed

Strain	Genotype	Phenotype
JSY 12	mat α, his3v-1, ura3 trp1, gal2, gal10/2µm Circle⁻, Trp1-R1-Circle⁺	His⁻, Ura⁻, Gal⁻, Trp⁺, TR [a]
JSY222	mat a, ura1, trp1/2µm Circle⁺	Ura⁻, Trp⁻, TS [a]
JSY310	mat a, ura1, trp1/2µm Circle⁺ / mat α, his3v-1, ura3, trp1, gal2, gal10, /Trp1-R1-Circle⁺	Gal⁻, Trp⁺, TR
JSY315	mat a, ura, his3v-1, trp1, gal/ 2µm Circle⁺, Trp1-R1-Circle⁺	Ura⁻, His⁻, Trp⁺, Gal⁻, TS
JSY316	mat α, trp1/2µm Circle⁺, Trp1-R1-Circle⁺	Trp⁺, Gal⁺, TR
JSY317	mat α, ura, trp1, gal/ 2µm Circle⁺, Trp1-R1-Circle⁺	Ura⁻, Trp⁺, Gal⁻, TR
JSY318	mat a, ura, his3v-1, trp1/ 2µm Circle⁺, Trp1-R1-Circle⁺	Ura⁻, His⁻, Trp⁺, Gal⁺, TS

[a] TR = Temperature Resistant; TS = Temperature Sensitive

tested thus far have been confirmed to contain the Trp1 plasmid. Thus, when a diploid is sporulated at 23°C which contains both 2μm DNA and Trp1-R1-Circle, we can generally expect to find both plasmids present in all haploid progeny which are phenotypically Trp+.

IV. DISCUSSION

We have summarized the construction and properties of a novel yeast plasmid, the Trp1-R1-Circle. The plasmid consists entirely of an unmodified segment of yeast chromosomal DNA, which has in effect simply been closed into a circle. When returned to yeast by transformation, selecting for complementation of a chromosomal mutation by the plasmid, it accumulates to a copy-number of more than one hundred per cell. This surprisingly high copy-number, probably in combination with the very small size of the plasmid (1453 base pairs), results in high mitotic and meiotic stability.

In other work we have exploited the ease with which Trp1-R1-Circle can be introduced into new *trp1* strains by standard yeast genetic crosses to construct a number of strains which are useful for the investigation of the replication properties of the plasmid in the yeast cell and in cell-free extracts. In particular we have constructed a collection of strains containing the Trp1 plasmid and one of several temperature-sensitive cell division cycle (*cdc*^{ts}) mutants which affect DNA replication (24). While such strains could also be constructed by transformation of each host (constructed previously by genetic crosses) using Trp1-R1-Circle DNA, the introduction of the plasmid genetically is much less laborious, especially when many different strains are involved. We have therefore been able to generate several strains for each *cdc* mutant used, which contain various combinations of mating type and nutritional markers, and which can be chosen to meet the needs of particular experiments, at the expense of relatively little additional work. The *cdc*^{ts}/Trp1-R1-Circle strains have been used by Dr. Virginia Zakian for the analysis of Trp1-R1-Circle DNA synthesis in the cells under conditions permissive and restrictive for the *cdc* mutant. In those experiments (to be published elsewhere), the Trp1 plasmid has been shown to replicate under the influence of the mutations in a manner indistinguishable from nuclear chromosomal DNA and 2μm DNA. The Trp1-R1-Circle has also been shown in our laboratory to be organized in nucleosomes comparable to bulk nuclear chromatin (to be published elsewhere). These properties combine to make the Trp1-R1-Circle a very attractive template for replication studies in yeast cells and cell-free extracts.

ACKNOWLEDGMENTS

We would like to thank Mary Rykowski for a careful reading of this manuscript and Dr. Virginia Zakian for many helpful conversations during the course of the work.

REFERENCES

1. Seligy, V.L., Thomas, D.Y. and Miki, B.L.A., *Nucl. Acid. Res. 8*, 3371 (1980).
2. Livingston, D.M., *Genetics 86*, 73 (1977).
3. Hartley, J.L. and Donolson, J.E., *Nature 286*, 860 (1980).
4. Livingston, D.M. and Kupfer, D.M., *J. Mol. Biol. 116*, 249 (1978).
5. Zakian, V.A., Brewer, B.J. and Fangman, W.L., *Cell 17*, 293 (1979).
6. Livingston, D.M. and Hahne, S., *Proc. Natl. Acad. Sci. U.S.A. 76*, 3727 (1979).
7. Nelson, R.G. and Fangman, W.L., *Proc. Natl. Acad. Sci. U.S.A. 76*, 6515 (1979).
8. Broach, J. and Hicks, J., *Cell 21*, 501 (1980).
9. Hindley, J. and Phear, G.A., *Nucl. Acid. Res. 7*, 361 (1979).
10. Livingston, D.M. and Klein, H.L., *J. Bact. 129*, 472 (1977).
11. Cameron, J.R., Philipsen, P. and Davis, R.W. *Nucl. Acid. Res. 4*, 1429 (1977).
12. Struhl, K., Stinchcomb, D., Scherer, S. and Davis, R., *Proc. Natl. Acad. Sci. U.S.A. 76*, 1035 (1979).

13. Stinchcomb, D., Struhl, K. and Davis, R., *Nature 282*, 39 (1979).
14. Hsia, C.-L. and Carbon, J., *Proc. Natl. Acad. Sci. U.S.A. 76*, 3829 (1979).
15. Kingsman, A.J., Clarke, L., Mortimer, R.K. and Carbon, J., *Gene 7*, 141 (1979).
16. Scott, J.F., *J. Cell Biol. 87*, A104 (1980).
17. Tschumper, G. and Carbon, J., *Gene 10*, 157 (1980).
18. Sherman, F., Fink, G.R. and Lawrence, C.W., *in* "Methods in Yeast Genetics", Cold Spring Harbor Laboratory, Cold Spring Harbor, New York (1979).
19. Alwine, J.C., Kemp, D.J., Parker, B.A., Reiser, J., Renart, J., Stark, G.R. and Wahl, G.M., *in* "Methods in Enzymology" *68*, 220 (1979).
20. Wahl, G.M., Stern, M. and Stark, G.R., *Proc. Natl. Acad. Sci. U.S.A. 76*, 3683 (1979).
21. Rigby, P.W.S., Dieckman, M., Rhodes, C. and Berg, P., *J. Mol. Biol. 113*, 237 (1977).
22. Scherer, S. and Davis, R.W., *Proc. Natl. Acad. Sci. U.S.A. 76*, 4951 (1979).
23. Hinnen, A., Hicks, J.B. and Fink, G.R., *Proc. Natl. Acad. Sci. U.S.A. 75*, 1929 (1978).
24. Hereford, L.M. and Hartwell, L.H., *J. Mol. Biol. 84*, 445 (1974).

IN VITRO REPLICATION OF YEAST 2-μm PLASMID DNA

Akio Sugino[*,†], Hitoshi Kojo[*,1], Barry D. Greenberg[*,†],
Patrick O. Brown[§], and Kwang C. Kim[*]

[*]Laboratory of Molecular Genetics, National Institute of Environmental Health Sciences, P.O.Box 12233, Research Triangle Park, N. C. 27709,[†]Curriculum of Genetics,[§]The University of North Carolina, Chapel Hill, N. C. 27514, and Department of Biochemistry, the University of Chicago, 920 East 58th Street, Chicago, IL. 60637.

ABSTRACT A soluble extract from exponentially growing yeast *Saccharomyces cerevisiae* initiates semiconservative replication of exogenously added yeast 2-μm plasmid DNA and *Escherichia coli* chimeric plasmids containing the 2-μm DNA. The replication system utilizes about 10% of the added DNA molecules and completes one round of replication. The major products in early stages of replication are θ ("eye") forms of plasmid DNA which start within one of the 599 base-pair inverted repeats (at 140 ± 50 nucleotides from the end of the inverted repeat) and DNA replication proceeds bidirectionally and discontinuously. Temperature sensitivity of *in vitro* 2-μm DNA synthesis is exhibited by the extracts prepared from the cell division cycle mutant, *cdc8*, suggesting that this system resemble *in vivo* 2-μm plasmid DNA replication.
Yeast has two DNA polymerases I and II which are immunologically distinguishable. Both DNA polymerases are quite sensitive to aphidicolin, although yeast cell growth and *in vivo* DNA replication are immune to this drug. However, aphidicolin-sensitive mutants can be isolated after mutagenesis, and one which possesses an aphidicolin-resistant DNA polymerase I has been obtained from an aphidicolin-sensitive yeast mutant.

[1]*Present address: Research Laboratories, Fujisawa Pharmaciutical Co., Ltd., Osaka 532, Japan.*

Two-μm plasmid DNA synthesis supported by obtained a crude extract from the aphidicolin-resistant DNA polymerase I mutant is relatively resistant to aphidicolin, proving that 2-μm DNA replication utilizes mainly DNA polymerase I of yeast. Hence, this system should provide a useful assay for the purification and characterization of 2-μm plasmid and nuclear DNA replication proteins.

INTRODUCTION

Knowledge of prokaryotic DNA replication mechanisms has been significantly advanced by the development of *in vitro* DNA replication systems (1). By use of small coliphage DNAs as probes, over a dozen *E. coli* DNA replication proteins have been identified and purified, as a result of such studies. Furthermore, reconstitution of the *in vitro* DNA replication systems has been accomplished using highly purified proteins (2,3), and several of those proteins were later functionally and genetically characterized (1-4).

Similar pursuits involving eukaryotic virus-specific *in vitro* DNA replication systems have been described (5-7). Some of the proteins required for activity recently have been purified (6,8) and reconstituted (9). However, these systems require virus-specific protein(s) for their initiation of DNA replication and are not exact model systems for nuclear DNA replication. Moreover, due to the complexity of eukaryotic genetics, the identification and characterization of genes encoded by the host cell have proven relatively intractable (10).

Yeast, one of the lower eukaryotes, has one of the most finely understood genetic systems and many DNA replication and cell division cycle (*cdc*) mutants have been isolated (11). Nevertheless, the only known correspondence between a gene and its encoded product is the DNA ligase gene (*cdc9*) (12).

Some of yeast strains have plasmid-like extrachromosomal DNAs and the most well known DNA is the 2-μm DNA. The 2-μm DNA, currently used as a cloning vector for specific genes via transformation of yeast cells, also provides a good model system for studying eukaryotic DNA replication (13-15). This trototypic utility is attributable to its shared characteristics with the chromosomal process. Both are under strict cell cycle control and depend on the same nuclear gene products (16-18).

An *in vitro* DNA replication system of yeast 2-μm plasmid has been described (14) which appears to mimic *in vivo* DNA replication. However, it was not well characterized and neither the site which DNA replication is initiated nor the mode of replication are known. Furthermore, the activity of the system appears prohibitively low, encumbering the isolation of DNA replication proteins by simple complementation assays.

Recently we have developed another *in vitro* 2-μm plasmid DNA replication system (19). In this paper, we summarize a system which confers activity at least an order of magnitude greater than the previously described system (14). Moreover, the origin and direction of replication are identified as those of the *in vivo* process using either native 2-μm DNA or a chimeric plasmid containing the 2-μm DNA as a template. Therefore, this system provides a reasonable assay for the purification and characterization of DNA replication proteins, and for precise studies of 2-μm plasmid and nuclear DNA replication processes.

METHODS

Preparation of crude extracts for in vitro DNA replication:
Yeast *Saccharomyces cerevisiae* was grown in 1 l of YPD medium to 5 x 10^7 cells/ml, rapidly harvested by centrifugation at 5,000xg for 5 min at room temperature and the cells were suspended in 10 ml of 10% sucrose, 50 mM Tris-HCl, pH7.5, 1 mM EDTA. The suspension was frozen dropwise in liquid N_2 and stored at -80°C until used. To make crude extract, the cell suspension was thawed at room temperature and chilled to 0°C. 0.25 M KCl, 5 mM spermidine, 10 mM 2-mercaptoethanol, 10 mM EDTA, and either 2mg/ml Glusulase (Sigma Chemicals) or 0.5 mg/ml Zymolyase 60,000 (Kirin Brewery, Co., Japan) were added simultaneously. After incubation at 0°C for 30 min, 0.1 mM phenylmethansulfonyl fluoride (PhMeSO$_2$F) was added and the spheroplasts were collected at 3,000 rpm for 5 min in a Sorvall HB4 rotor. Pellets were resuspended in 20 ml of 50 mM Tris-HCl, pH 7.5, 1 mM EDTA, 0.25 M KCl, 0.1 mM PhMeSO$_2$F, and 1 mM dithiothreitol (DTT), and incubated at 0°C for 30 min. Cell lysate were centrifuged at 40,000 rpm for 30 min (at 2-4°C) in a Spinco SW41 rotor and supernatants were then adjusted to about pH 7.5 with 2 M Trizma base (Sigma Chemicals). Solid ammonium sulfate was added to 50% saturation over 20 min at 4 °C with stirring, and following an additional 20 min at 4°C, the solutions were divided into 1 ml aliquots, transfered to 1.5 ml plastic Eppendorf tubes and centrifuged for 5 min in a Brinkman mini-centrifuge at 4°C. Supernatants were carefully and completely removed and the pellets were frozen at -80°C until used. At this stage, no significant decrease of *in vitro* DNA replication activity was observed for at least three months. Just prior to use, each pellet was suspended in 0.5 ml 50 mM Hepes buffer, pH7.8, 10% glycerol, 1 mM EDTA, 1 mM DTT, 0.1 mM PhMeSO$_2$F.

Assay for 2-μm plasmid DNA replication: The reaction mixture (100 μl) contained 35 mM Hepes buffer, pH7.8, 10 mM $MgCl_2$, 1 mM DTT, 2 mM spermidine, 5 mM ATP, each of the other three rNTPs at 0.2 mM, all four dNTPs at 10 μM (dTTP was either ^3H or ^{32}P-labeled, 200-1,000 cpm/pmole), 0.1 mg/ml bovine serum albumin, 1 mM DPN, 25% glycerol, 1-5 μg (about 0.1 pmole) of either native 2-μm plasmid DNA or the chimeric plasmid containing 2-μm DNA and 1-5 mg protein/ml from the crude extracts. Incubation was at 30°C for the indicated periods. The reaction was terminated by the addition of either 2 ml of cold trichloroacetic acid (TCA) or an equal volume of 20 mM EDTA (pH8) plus 1% sodium dodecylsulfate.

Electron microscopic analyses of the in vitro DNA replication products: In vitro replication products were treated with 0.1 mg/ml Proteinase K at 37°C for 30 min in the presence of 0.5% sodium dodecylsulfate, extracted with phenol saturated with 0.1 M Trizma base, and dialyzed against 10 mM Tris-HCl, pH8.0, 1 mM EDTA. The DNA samples were examined by the method of Kleinschmidt (20) either with or without prior cleavage with restriction endonuclease *EcoRI*. DNA lengths were measured from photograghic enlargements using a map measurer.

Isolation of aphidicolin-sensitive yeast mutants: Yeast *S. cerevisiae* A364A was grown in 100 ml YPD medium to 5 x 10^7 cells/ml, collected by centrifugation, washed twice with water, and suspended in 100 ml 50 mM Tris-HCl, pH7.5. Ethylmethane sulfonate (EMS) was added to 1% and the suspension was incubated at room temperature for 3 hrs. The cells were collected, washed with water, incubated with 6% sodium thiosulfate for 10 min to neutralized the EMS, resuspended into 100 ml YPD medium and grown overnight at 28°C to express any mutants generated by the EMS treatment. The cell culture was diluted 5-fold with fresh YPD medium and aphidicolin (provided by Dr. A. H. Todd of the Imperial Chemical Industries, Inc., England) was added to 0.1 mg/ml. Incubation was continued at 28°C for 10 min and Nystatin (Sigma Chemicals) was added to 0.1 mg/ml to enrich aphidicolin-sensitive mutants. Incubation was resumed for an additional 60 min at which time the cells were collected washed twice with fresh YPD medium and suspended to the original volume. Aphidicolin-sensitive cells were selected using replica agar plates containing YPD medium ± 100 μg/ml aphidicolin. Of 200 colonies tested which failed to grow on drug-containing plates, 15 aphidicolin-sensitive mutants were detected.

Isolation of aphidicolin-resistant DNA polymerase I mutants: *S. cerevisiae* A364A aphidicolin-sensitive mutant, aph^s-15, was treated with 1% EMS as described above. After neutralization of EMS with 6% sodium thiosulfate, cells were plated on YPD-plates containing 150 µg/ml aphidicolin, and grown at 30°C for 2 days. The colonies which appeared were grown in YPD medium containing 100 µg/ml aphidicolin or 10 µg/ml ethidium bromide. Those which grew in the presence of aphidicolin but not in the presence of ethidium bromide were further analyzed by measuring the aphidicolin-sensitivity of DNA polymerase activities in crude extracts. Aph^R1-1 and aph^R2-6 were isolated in this manner and identified as aphidicolin-resistant DNA polymerase I mutants following partial purification of DNA polymerases I and II by phosphocellulose and DEAE-sephacel (Pharmacia Fine Chemicals) column chromatographies.

DNA polymerase assay: DNA polymerase activity was measured by the incorporation of [^{3}H]dTTP into acid-insoluble material. The standard reaction mixture (0.1 ml) contained 42 mM Hepes buffer, pH 7.8, 12 mM $MgCl_2$, 1 mM DTT, 0.125 mg/ml bovine serum albumin, 0.18 mg/ml activated calf thymus DNA, and 2.5 µM each four dNTPs (dTTP was ^{3}H-labeled, sp. act. 600 cpm/pmol). After a 20-min incubation, the reaction was terminated with 2 ml of 5% TCA. TCA-insoluble radioactivity was collected on a Whatman GF/C glass filter and counted in Beckman liquid scintillation spectrometer.

Other methods: Alkaline and neutral sucrose density gradient centrifugations were performed as published (21) using a Spinco SW27 rotor. One percent agarose gel electrophoresis of DNAs was as previously described (22), DNA bands were visualized after staining with 0.5 µg/ml ethidium bromide. DNA transfer from agarose gels to nitrocellulose filters and DNA-DNA hybridization were performed as published (23) with slight modifications.

RESULTS

Properties of in vitro 2-µm plasmid DNA replication. Labeled dNTPs are incorporated into acid-insoluble materials when either purified supercoiled 2-µm plasmid DNA or chimeric plasmids containing the 2-µm yeast plasmid DNA moety were incubated with crude extracts from yeast *S. cerevisiae*. As shown in Fig. 1A, this activity was totally dependent on exogenously added DNA. Detailed requirement are shown in Table 1. The synthesis reaction was essentially completed within 30 min at 30°C (Fig. 1A). Further incubation for up to 2 hrs resulted in no significant increase or decrease in incorporated radioactivity due to nucleolytic breakdown of synthesized products (data not shown).

The optimal concentration of crude extract in this reac-

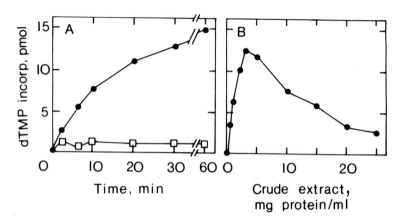

FIG. 1. *Kinetics and optimal condition of yeast 2-μm DNA synthesis in vitro (19). A) The kinetics of DNA synthesis was measured by using the standard reaction conditions described in Methods. The reaction mixtures contained 3 mg/ml of protein from S. cerevisiae A364A (cir^+) crude extract and either no added (☐), or 0.2 pmole (●) of native 2-μm plasmid DNA. After indicated incubation at 30°C, acid insoluble radioactivity was measured. B) The optimal concentration of crude extract was determined in the standard reaction as in A) by changing the concentration of the crude extract. The reaction was carried out at 30°C for 30 min.*

tion was about 3 mg/ml of protein. Higher concentrations caused a gradual decrease of incorporation (Fig. 1B), but this might not be simple degradation of DNA template, since no significant increase of incorporation was observed by further addition of new DNA template (data not shown).

Table 2 shows the template specificity for *in vitro* DNA replication using crude extracts from isogenic cells either containing (cir^+) or lacking (cir^o) the 2-μm plasmid. Native 2-μm DNA and the chimeric plasmids containing either the entire 2-μm DNA (pJDB36 and 41) (13 and see Fig. 2) or the *EcoRI* fragment of 2-μm DNA thought to contain the *in vivo* replication origin (CV17 and 7) (25 and see Fig. 2) were the most efficient templates. Conversely, the chimeric plasmids containing the other *EcoRI* fragments (CV18 and 3) (25, and see Fig. 2) support synthesis at levels no higher than parental plasmid (CV03) (Fig. 2). Interestingly, both pMB9 and YRp7,which contains pBR322 and one of the yeast replicons (27,28), also support appreciable synthesis. CV03, which contains pBR322 plus the yeast *LEU-2* gene without any replicon sequence also supports synthesis at levels significantly above background, but not as

TABLE 1. *Requirement for DNA replication in vitro (19)*

Omissions and additions	Activity dTMP incorporation (pmoles)	%
Complete	10.4	100
-ATP	1.1	10
-CTP, GTP, and UTP	9.3	89
-ATP, CTP, GTP, and UTP	0.9	9
-dCTP or dATP	2.1	20
-DPN	10.7	100
-Spermidine	10.4	100
-DTT	10.3	100
+KCl, 40 mM	7.5	72
+KCl, 100 mM	1.2	12
+Glycerol, 25%	15.4	148
+DNA polymerase I, 0.5 unit	12.5	120
+DNA polymerase II, 0.5 unit	12.0	116

DNA synthesis activity of the crude extract from S. cerevisiae A364A was measured in the standard assay described in Methods. 0.12 pmole of native 2-μm DNA and 3 mg/ml of protein from crude extract were used and the incubation was 30 min at 30°C. Yeast DNA polymerases I and II were the same as before (24).

well as pMB9. Further implications will be discussed below.

There are obvious differences in DNA synthesis between cir^+ and cir^o extracts. Concequently, whereas no 2-μm DNA encoded functions are reqired for *in vitro* DNA synthesis, extracts from cir^+ cells may contain undefined functions which enhance this capability.

The observed DNA synthesis was not DNA repair synthesis as determined by experiments replacing dTTP with BrdUTP. Reaction products using native 2-μm DNA were subjected to CsCl equilibrium density gradient centrifugation and all the newly synthesized DNA banded at the heavy position (data not shown). Furthermore, virtually all incorporated radioactivity after short reaction times sedimented at 4-5 S in alkaline sucrose density gradients (Fig. 3). Since the template DNA was not significantly degraded (Fig. 3), these data are consistent with discontinuous synthesis in contrast with continuous strand extension expected in repair DNA synthesis.

Table 3 shows the effects of various antibiotics and inhibitors on the *in vitro* DNA replication of 2-μm DNA.

TABLE 2. Template Specificity of DNA Synthesis in vitro

DNA source	DNA synthesis (pmoles of dTMP incorporated)	
	cir^+	cir^0
NONE	0.4	0.3
Native 2-μm plasmid	10.5	5.1
CV03	2.2	<0.5
CV18	1.2	<0.5
CV17	10.2	8.2
CV3	1.3	<0.5
CV7	8.4	6.0
pMB9	7.2	4.6
pBR322	2.1	<0.5
pJDB36	25.9	15.6
pJDB41	24.2	14.2
YRp7	5.1	4.8

DNA synthesis activity of the extract from either S. cerevisiae NCYC74-CB11' (cir^+) or NCYC74-11 (cir^0) cells was measured as in Table 1, except that 0.12 pmole of the indicated plasmid DNAs was added. CV03 DNA consists of E. coli plasmid pBR322 and yeast LEU2 DNA CV18, 17, 3 and 7 DNAs consist of CV03 and either the small EcoRI fragment of 2-μm Form A DNA, the large EcoRI fragment of 2-μm Form A DNA, the large EcoRI fragment of 2-μm Form B, or the small EcoRI Fragment of 2-μm Form B DNA, respectively. pJDB36 and 41 consist of pMB9 and either 2-μm Form A or Form B DNA, respectively. Finally, YRp7 DNA consists of pBR322 and yeast TRP-1 DNA fragment and possibly one of the yeast nuclear DNA replicons.

Among the inhibitors, only N-ethylmaleimide (NEM) and aphidicolin exhibit significant potency. Aphidicolin is a specific inhibitor of DNA polymerase α in higher eukaryotes (29) and yeast DNA polymerases I and II (24,30). As shown in Fig. 4, the sensitivity of *in vitro* 2-μm DNA replication to aphidicolin was equivalent to those of the purified yeast DNA polymerases I and II, suggesting that either enzyme, or both, participate in 2-μm DNA synthesis *in vitro*.

Of the other antibiotics, only novobiocin and coumermycin A_1 appreciably inhibited DNA synthesis, and then only at relatively high concentrations. The importance of such a result is questionable, however, since various enzymatic activities are nonspecifically inhibited by elevated levels of those drugs (31).

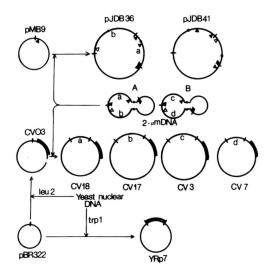

FIG. 2. Schematic diagram of the two forms of 2-μm plasmid DNA and the chimeric plasmids containing either the entire 2-μm DNA or the portion of 2-μm DNA. Native 2-μm DNA preparations contain equimolar quantities of the two isomeric forms, A and B (center of this figure). Included in the figure are chimeric plasmid DNAs used in this study and recognition sites for the restriction endonucleases EcoRI (■), HindIII (▲), XbaI (●), and HpaI (△).

TABLE 3. Sensitivity of in vitro DNA replication to various inhibitors (19)

Inhibitors	Activity dTMP incorporated (pmoles)	%
control (no inhibitor)	10.4	100
NEM, 1 mM	0.3	3
araCTP[a], 50 μM	7.0	67
ddTTP[b], 50 μM	8.7	84
aphidicolin, 1 μg/ml	6.7	64
aphidicolin, 5 μg/ml	1.1	3
novobiocin, 0.5 mg/ml	3.3	32
coumermycin A_1, 50 μg/ml	2.1	20
nalidixic acid, 1 mg/ml	8.3	80
α-amanitin, 0.1 mg/ml	10.7	103

DNA synthesis activity in the presence of the various inhibitors and antibiotics indicated was measured as Table 1 using native 2-μm plasmid DNA as template.
[a] araCTP: 1-β-D-arabinofuranosylcytosine 5'-triphosphate.
[b] ddTTP: 2',3'-dideoxythymidine 5'-triphosphate.

Analyses of in vitro products. As mentioned previously (Fig. 3), reaction products after short incubation times sedimented at about 4 S in alkaline sucrose gradients. Upon longer incubations, the S value of the slowly sedimenting materials increased slightly to about 5-6 S region after 20 min incubation, hence the ligation of small DNA species appears inefficient in this system. Accordingly, the crude extracts contained little DNA ligase activity, whereas the fraction removed from these extracts (50-75% saturation of ammonium sulfate) contained the majority of this activity (data not shown).

In neutral sucrose gradients, the newly synthesized DNA labeled with ^{32}P either cosedimented with the ^{3}H-labeled template DNA or much more slowly (Fig. 3d-f). These slowly sedimenting molecules were not degradation products since no template DNA was degraded to slowly sedimenting forms. Nor were they of chromosomal DNA origin since they hybridized exclusively to either 2-μm DNA or chimeric plasmids containing 2-μm DNA (data not shown). Hence they were probably replication intermediates which were separated from the template DNA during analysis (33).

35 IN VITRO REPLICATION OF YEAST 2-μm PLASMID DNA

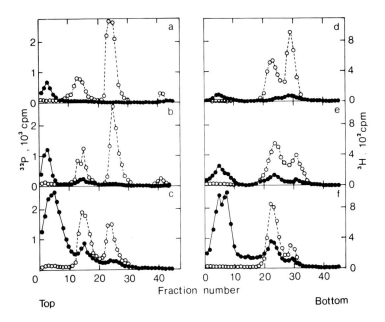

Fig. 3. Sucrose density gradient sedimentation of 2-μm DNA synthesized in vitro (19). The products were formed with crude extract from S. cerevisiae A364A in the presence of ^{3}H-labeled 2-μm supercoiled DNA and [α-^{32}P]dTTP as described in Methods. Reactions were incubated at 30°C for 1, 5, and 20 min. After the reactions were terminated with 0.1% sodium dodecylsulfate and 10 mM EDTA, the products were analyzed either on alkaline (a-c) or on neutral (d-f) sucrose density gradients. After 10 hrs at 25,000 rpm and 4°C in a Spinco SW 27 rotor, the gradients were fractionated and acid insoluble radioactivity was measured. -●-; newly synthesized ^{32}P-labeled DNA, -O-; ^{3}H-labeled parental 2-μm DNA template. The rapidly and slowly sedimenting [^{3}H]2-μm DNAs in alkaline sucrose gradients are the denatured supercoiled 2-μm DNA and a combination of denatured linear and circular single-stranded forms, respectively. The rapidly and slowly sedimenting ^{3}H-labeled 2-μm DNAs in neutral sucrose gradients are supercoiled and relaxed (both nicked and covalentlyclosed) DNA, respectively.

Electron microscopic analysis of in vitro products (Fig. 5) showed both supercoiled and open circular DNA molecules containing an "eye" form (theta-form) structure, presumably a replication intermediate.

Table 4 summarizes the analysis of products obtained from

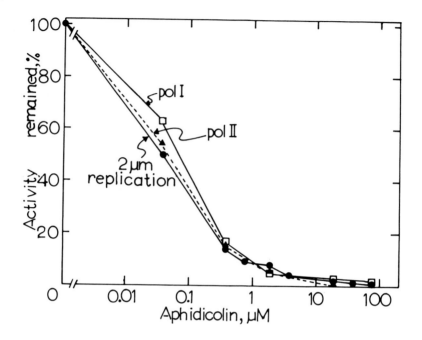

Fig. 4. Aphidicolin sensitivity of in vitro 2-μm plasmid DNA replication system and the purified DNA polymerases I and II of yeast. The aphidicolin sensitivity of in vitro 2-μm DNA replication was measured in the presence of various concentrations of aphidicolin as in Table 1. The aphidicolin sensitivity of the purified DNA polymerases I and II was measured in the standard condition as in Methods using activated calf thymus DNA as template-primer and 0.1 unit of each DNA polymerase. The reaction time was 20 min at 30°C.

5 and 20 min incubations. Within 5 min, 7% of the total DNA molecules contained replication bubbles, most of which were between 0.05 and 0.3 μm, or roughly 150-1,000 nucleotides long, whereas some molecules were replicated over 50% (about 5,000 nucleotides) of the genome. The percentage of replicating molecules was unchanged by 20 min-incubation, however, there was a decrease in the percentage of early replication forms (less than 50% replication) with a commensurate increase in late replication forms (over 50% replication). There was also a simaltaneous increase in open circular and linear forms.

The origin and direction of replication. Native 2-μm plasmid DNA contains two identical 599 base-pair inverted repeats (34).

Recombination can occur within the inverted repeats resulting in two equimolar isomeric forms (Fig. 2), differing in the orientation of one unique region (13). Since this was expected to introduce some complication in the following assay, the chimeric plasmid pJDB36 was used as template. This plasmid consists of pMB9 plus Form A of the native 2-μm DNA (13 and Fig. 2).

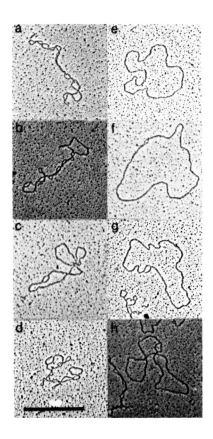

Fig. 5. Electron micrographs of the in vitro replicating 2-μm DNA molecules (19). Representative molecules observed in the 5-min synthesis reactions, described in Table 4, are shown. a and e are non-replicating supercoiled and relaxed DNA molecules, respectively. b, c, and d show replicating supercoiled molecules and f, g, and h are replicating relaxed DNA molecules. The bar represents 0.5 μm.

*Eco*RI digestion of pJDB36 generates three fragments, the largest of which is pMB9. The second fragment corresponds to the larger *Eco*RI fragment of the Form A 2-μm DNA which presumably contains the *in vivo* origin (25), and the third fragment corresponds to the smaller native 2-μm Form A *Eco*RI fragment (Fig. 2). *In vitro* replication products were analyzed by electron microscopy following *Eco*RI digestion and the results are depicted in Fig. 6. Replication bubbles were observed in all three fragments, however, the frequency was highest in fragment B. Moreover, the locations of these bubbles were nonrandom. Their centers were the same in each cleavage product, and expanded bidirectionally. The site in the B fragment corresponded to the identical site expected to be the *in vivo* origin (25), within the inverted repeat region. The low frequency of replication bubbles in the C fragment, which contains the other inverted repeat region, rendered these results statistically insignificant. This will be discussed below. Interestingly, the replication bubbles in the pMB9 moety were equally specific, but not coincident with the *in vivo* pMB9 origin in *E. coli*. Furthermore, when pMB9 DNA was incubat-

TABLE 4. *Electron microscopic analysis of the products from in vitro 2-μm DNA replication (19).*

Reaction time (min)	Number(%) of molecules					
	Super-twisted	Open circle	Replicating DNA		Linear DNA	Total
			Early	Late		
0	503 (90)	49 (9)	1 (0.2)	2 (0.4)	3 (0.5)	558
5	978 (73)	169 (13)	86 (6)	7 (0.5)	103 (8)	1343
20	922 (54)	466 (27)	21 (1)	88 (5)	212 (12)	1719

The reaction products of in vitro 2-μm DNA synthesis were purified and analyzed by electron microscopy. After taking the pictures from randomly selected fields, the various DNA species were scored as indicated. Early replicating molecules were those in which less than 50% of the 2-μm DNA were replicated, while the late replicating molecules were those in which greater than 50% were replicated. Parentheses show the percentage of each type.

ed with crude extracts, the DNA synthesis was initiated at the same site as in Fig. 6C, *in vitro* DNA replication is principally initiated at either the inverted repeat region of the larger *EcoRI* fragment of Form A 2-μm DNA or near one end of the pMB9 moety linearized with *EcoRI*.

Specific initiation of *in vitro* DNA replication was also supported by pulse-labeling experiments (Fig. 7). Pulse-labeled DNA was digested with *EcoRI* and analyzed by autoradiography of agarose gel electropherograms. After 1-min incubation, the region between bands A (pMB9) and B (*EcoRI* large fragment of Form A 2-μm DNA) was primarily labeled. With increasing incubation times, the region between bands B and C (*EcoRI* small fragment of Form A 2-μm DNA) became labeled. Since replication forms migrate more slowly than the native DNA fragments, these data are indicative of preferential initiation within the B fragment. By 20-30 min-pulse, radioactivity was detected in all three fragments as expected for completely replicated forms. However, this evidence was not conclusive since, as previously mentioned, the ligation of short fragments into longer forms was inefficient and such fragments were easily separable from the template (Fig. 3). We also used the Southern blotting method to identify the replication origin. *EcoRI*-digested plasmid DNA was transferred to nitrocellulose sheets from agarose gels by the method of Southern (23) and labeled *in vitro* replication products were used as hybridization probes to this DNA. Preferential hybridization to the B fragment was observed with replication products from short incubation times, in support of the above conclusions.

Effect of cdc mutants on in vitro DNA replication. Cell division cycle mutants *cdc*4, 7, and 28 are thought to affect the ability of the cell to enter S phase (35). The *cdc*8 gene is required for nuclear DNA replication, presumably for DNA strand elongation, but not for initiation (35,36). Both chromosomal and 2-μm DNA replication initiation are dependent on the completion of each stage in the *cdc*28 → 4 → 7 → 8 passway (16,17). Fig. 8 shows the *in vitro* 2-μm DNA synthesizing activity in crude extracts from these *cdc* mutants, at various reaction temperatures. Contrary to the *in vivo* observations, the activity decreased only in *cdc*8 extract at higher temperatures whereas the other extracts paralleled the wild-type control. However, if these cells were heated to the restrictive temperature (37°C) prior to harvesting, DNA synthesis activity in their subsequent crude extracts was considerably lower even at permissive temperatures (such as 25°C), as in the published system (14). The only exception was the *cdc*9 mutant which showed no decrease in activity in either case (data not shown). Implications of these experiments will be

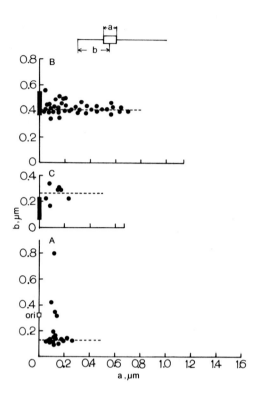

Fig. 6. Relation between the center of the replication bubble and the size of the bubble in the in vitro replicating pJDB36 plasmid molecules treated with restriction endonuclease EcoRI (19). The length of the EcoRI treated pJDB36 products in the 5-min synthesis reactions were measured and classified into three distinctive groups, A, B, and C. These are the pMB9 moety (1.61 ± 0.15 μm), the EcoRI large fragment of 2-μm Form A (1.16 ± 0.06 μm), and the EcoRI small fragment of 2-μm Form A (0.67 ± 0.07 μm) respectively. The center (b) and the width (a) of the replication bubbles were measured as shown above, simultaneously. The dashed lines represent the average distance of the replication bubbles from one end of the molecule and correspond to 0.13 ± 0.07 μm, 0.41 ± 0.15 μm, and 0.27 ± 0.10 μm in the pMB9, the EcoRI large fragment, and the small EcoRI fragment, respectively. In the case of pMB9, four molecules had the replication bubbles significantly further from the others and were not included in the average; and ⊑⊐ in the ordinate axis indicates the locations of the 599-base-pair inverted repeats of 2-μm DNA and the in vivo origin of pMB9 DNA replication in E. coli.

35 IN VITRO REPLICATION OF YEAST 2-μm PLASMID DNA

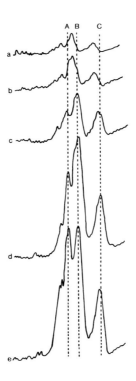

Fig. 7. Identification of initiation site of in vitro DNA replication. In vitro DNA replication products using pJDB36 chimeric plasmid DNA as template were labeled with [α-^{32}P]dTTP for 20 sec (a), 1 min (b), 5 min (c), 10 min (d), and 20 min (e). The products were purified as in Methods and analyzed by 1% agarose gel electrophoresis followed by the digestion with EcoRI. The gel was dried and the autoradiogram was taken. The figure shows the scanning of the autoradiogram by E.-C. densitometor. DNAs migrated from left to right and the position of each DNA bands were identified after staining with 0.5 μg/ml ethidium bromide.

presented below.

DNA polymerase I is a DNA replicase of 2-μm plasmid. As previously mentioned, aphidicolin is a specific inhibitor of yeast DNA polymerases I and II. It also severely inhibits *in vitro* 2-μm DNA replication as well as DNA polymerase activity in crude extracts (24 and Fig. 4). However, as much as 1 mg/ml aphidicolin has little effect on cell growth or *in vivo* DNA

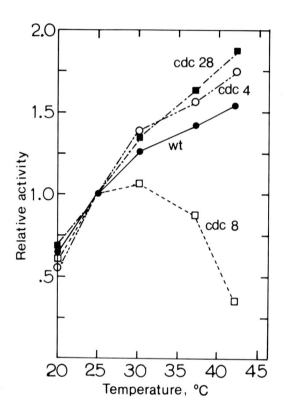

Fig. 8. Temperature sensitivity of in vitro 2-µm DNA synthesis in the crude extracts from the various cdc mutants (19).

Crude extracts were prepared from the various cdc mutants as described in Methods, except that the cells were grown at 23°C. In vitro DNA synthesis was carried out for 30 min at various temperatures in the presence of pJDB36 chimeric plasmid DNA as template. The activities are shown the relative value to the activity at 25°C. The DNA synthesis activities in wild-type S. cerevisiae A364A, cdc4, cdc8, and cdc28 at 25°C were 19.5, 14.1, 10.1, and 12.0 pmoles of dTTP incorporated, respectively.

replication (Fig. 9) presumably due to a permeability barrier. Thus, in order to investigate the effects of aphidicolin on DNA replication of yeast, a permeability mutant was required, followed by a second mutant resistant to the effects of this drug. This could possibly enable us to identify the true DNA replicase of 2-μm plasmid. To this end we isolated several aphidicolin-sensitive cells following EMS mutagenesis, as described in Methods. One such mutant, aph^s-15, exhibits significantly more sensitive cell growth and *in vivo* DNA synthesis than wild-type cells. Moreover, 10 μg/ml ethidium bromide killed these cells whereas the lethal dose of wild-type cells is over 0.1 mg/ml. These two phenotypes were coordinately expressed, as selection for either aphidicolin or ethidium bromide-resistant revertants yielded cells with both resistant phenotypes. This implies that both phenotypes were caused by

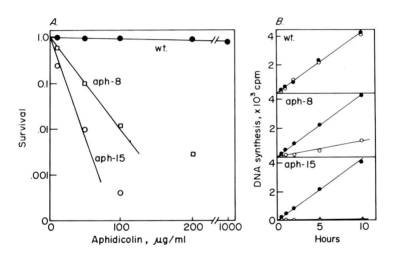

Fig. 9. Aphidicolin sensitivity of cell growth and DNA synthesis in wild-type and aphidicolin-sensitive mutants of yeast (24). (A) Aphidicolin sensitivity of cell growth was measured as cell growth on YPD medium plates containing the indicated concentrations of aphidicolin. After two days at 30°C, the visible colonies were counted. (B) The cells were grown in 10 ml of YPD medium to 1×10^7 cells/ml and 50 μg/ml aphidicolin and 10 μCi of ^{3}H-uridine were added. After incubation at 30°C, 0.5 ml aliquots were removed to assay the extent of alkaline-resistant and acid-insoluble radioactivity, as published (21). -●-; without aphidicolin, -○-; with aphidicolin.

a single mutation breaching the natural cellular permeability barrier.

We then attempted to isolate aphidicolin-resistant mutants from aph^S-15, maintaining the permeability mutation by mutual selection for ethidium bromide sensitivity. Of over 200 such colonies, two were identified as possessing an aphidicolin-resistant DNA polymerase I ($aph^R 1$-1 and $aph^R 2$-6), while retaining a normal DNA polymerase II. As shown in Fig. 10, the partially purified DNA polymerase I from both strains was at least 20 times more resistant to aphidicolin than the same enzyme from either aph^S-15 or its parent strain A364A. The cell growth and in vivo DNA synthesis were also much more resistant to aphidicolin than aph^S-15, but not as resistant as strain A364A.

Crude extracts prepared from either $aph^R 1$-1 or $aph^R 2$-6 supported in vitro 2-μm plasmid DNA synthesis at levels 10-20 times greater than their parent strains in the presence of aphidicolin. We therefore conclude that DNA polymerase I is the major DNA replicase in yeast, as previously speculated (37), and it is responsible for 2-μm plasmid DNA synthesis in vivo and in vitro.

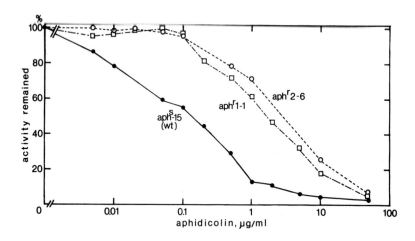

Fig. 10. Aphidicolin sensitivities of DNA polymerase I from aph^S-15 (wt), $aph^R 1$-1, and $aph^R 2$-6 cells. DNA polymerase I was partially purified from the crude extracts by successive column chromatographies of phosphocellulose and DEAE-Sephacel (A. S., H. K., and J. Arendes, in preparation). Aphidicolin-sensitivity of the DNA polymerase I reaction was the same as in Fig. 4.

DISCUSSION

We have described another *in vitro* replication system of yeast 2-μm plasmid DNA. The activity of this system is at least an order of magnitude higher than the system previously described (14) when a chimeric plasmid consisting of pMB9 DNA plus the 2-μm DNA is used as template. It may therefore represent a better system for the purification of replication proteins.

This system initiates DNA synthesis at the same locus used *in vivo* (25) and proceeds bidirectionally. It also exhibits similar requirements for various gene products as the *in vivo* process. The only eukaryotic templates which exhibit initiation and *de novo* DNA synthesis *in vitro* are adenovirus DNA and the yeast 2-μm plasmid. However these two systems are very different in that adenovirus DNA requires a specific protein attachment for initiation, and it exhibits displacement synthesis (38). Conversely, no such protein requirement has been observed with 2-μm DNA replication and the mode of its synthesis is bidirectional as is yeast nuclear DNA synthesis. Since the processes of yeast 2-μm and nuclear DNA synthesis manifest similar requirements, our *in vitro* system may provide a useful model toward the understanding of the *in vivo* initiation of chromosomal DNA replication.

The location of the *in vitro* 2-μm DNA replication origin is contained within one of the inverted repeats, 140 ± 50 nucleotides from the end of the inverted repeat proximal to the *XbaI* site (34) (Fig. 2). The nucleotide sequence near the site shows very interesting features (34) including similarities to published sequences of prokaryotic and eukaryotic replication origins (39-41). However any particular structural features based on the sequence data are insufficient to explain this specific initiation since only one of the two perfect inverted repeats is used as an origin (Fig. 6). Hence some portion of the unique sequence must confer the necessary characteristics for preferential initiation at this site. Broach and Hicks recently showed that efficient transformation of 2-μm DNA into yeast cells requires about 100 base pairs adjacent to the inverted repeat region, in addition to the inverted repeat allegedly containing the origin (25).

In addition to the 2-μm DNA and chimeric plasmids containing the *in vivo* 2-μm origin, *E. coli* plasmid DNAs, particularly pMB9, were active templates for *in vitro* synthesis (Table 3). Several observations argue against these results being artefactual. Electron micrographs of replication products showed the same type of "eye" structure observed in native 2-μm DNA. This structure appeared at one specific region of the molecule (Fig. 6C) which is not the *in vivo* pMB9 origin. Strathern, et al. (42) showed some homology between 2-μm DNA and bacterial plasmids ColEI, pBR313 and pBR322, although these observations are not supported by sequence data (34). Nevertheless we have performed similar experiments to compare regions of pMB9 DNA to the 2-μm plasmid. Particularly, the *HaeIII* fragment of pMB9 DNA which contains the observed *in vitro* origin (Fig. 6C) hybridizes to the small *EcoRI* fragment of Form B 2-μm DNA, indicating some homology. We are consequently searching for similarities with either 2-μm DNA replication or another type of replication complex in yeast, such as a nuclear DNA complex which recognizes a different sequence than in 2-μm DNA. Nucleotide sequence determination of this fragment is currently underway to resolve this point.

Our crude extracts also supported DNA synthesis in the presence of YRp7 plasmid DNA, which consists of pBR322, the yeast *trp1* gene and possibly a yeast nuclear DNA replicon (26, 27). Electron micrographs show an "eye" form structure after brief DNA synthesis but the *in vivo* origin of the plasmid is not known. Since pBR322 DNA is only poorly replicated by our crude extracts (Table 2), it cannot be assumed that this plasmid replicates from the pBR322 origin. Hence *in vitro* product analysis may provide useful information regarding a natural chromosomal origin.

Chan and Tye (43) have recently cloned a number of DNA segments from yeast DNA which permit the autonomous replication of pBR322 containing the yeast *leu2* gene. These inserts represent both unique and repetitive segments of the yeast genome. Due to the large number of yeast chromosomal replicons, it is unlikely that each should be recognized by a unique replication complex, hence some sequence ambiguity is likely at origins recognized by specific replication complexes. This may explain why our crude extracts replicate 2-μm, YRp7 and pMB9 DNA.

The *in vitro* 2-μm DNA synthesis described here is very sensitivie to aphidicolin, a specific inhibitor of yeast DNA polymerases I and II. We have successfully isolated aphidicolin-resistant DNA polymerase I mutants from the aphidicolin sensitive yeast mutant aph^s-15(Fig. 9), and in crude extracts from this resistant mutant, 2-μm DNA replication was 10-20 times more resistant to aphidicolin than in wild type extracts. Furthermore, *in vitro* replication of 2-μm DNA in crude extracts from these resistant mutants occurred. At levels of aphidicolin which do not effect the resistant polymerase I but almost completely inhibit polymerase II, and replication products were the same as those in crude extracts from wild type cells in the absence of aphidicolin. The 2-μm DNA replicase therefore appears to be DNA polymerase I.

Although *in vivo* DNA replication is temperature sensitive in all the *cdc* mutants used in this study, only the *cdc* 8 crudy extract exhibited temperature sensitivity *in vitro* 2-μm DNA replication. Several explanations are possible. Residual *cdc* gene product may be sufficient for the relatively low level of *in vitro* synthesis observed in our system or the mutants may be leaky. Alternatively if the temperature sensitivity is manifested by proteolytic breakdown of required gene products which are unused at nonpermissive temperatures, and hence still available after a downward temperature shift, the necessary proteases may be purified away by our crude extract preparative procedure. Nevertheless, the possibility that our *in vitro* replication system does not totally resemble *in vivo* replication cannot be eliminated. When wild type cell extracts were fractionated on DNA cellulose columns and fractions were added to crude extracts from *cdc*8 mutants pretreated at the restrictive temperature, DNA synthesis could be restored to levels present in *cdc*8 extracts without pretreatment (data not shown). This indicates that at least the *cdc*8 gene product may be purified by complementation assays using this *in vitro* system.

Several enzymatic activities have been identified and purified from yeast, which may participate in both 2-μm plasmid and nuclear DNA replication. Among these are DNA ligase, DNA-dependent ATPases, DNA binding protein(s), topeisomerases, RNase H, RNA polymerase(s), and DNA polymerases I and II. Presently, *cdc*4, 7, 8, 9, 13, 14, 23 and 28 have been identified as DNA replication mutants, however the only determined correspondence between these mutations and a specific gene is *cdc*9, the DNA ligase. We should therefore expect that new mutants can be generated to further elucidate the process of DNA replication.

ACKNOWLEDGMENTS

We are indebted to Drs. J. Broach, J. D. Beggs and J. Carbon for various chimeric plasmid DNAs, to Dr. D. M. Livingston and the Yeast Stock Center, University of California, Berkeley for various yeast strains.

A. S. especially thanks Dr. J. Broach for his communication before publication.

REFERENCES

1. Kornberg, A. (1980) *DNA Replication* (W. H. Freeman and Company, San Francisco).

2. Meyer, R. R., Shlomai, J., Kobori, J., Bates, D. L., Rowen, L., McMacKen, R., Ueda, K. and Kornberg, A. (1978) *Cold Spring Harbor Symp. Quant. Biol. 43*, 283-293.
3. Sumida-Yasumoto, C., Ikeda, J. -E., Benz, E., Marians, K. J., Vicuna, R., Sugrue, S., Zipursky, S. L. and Hurwitz, J. (1978) *Cold Spring Harbor Symp. Quant. Biol. 43*, 311-329.
4. Hubscher, U. and Kornberg, A. (1980) *J. Biol. Chem. 255*, 11698-11703.
5. Challberg, M. and Kelly, T. J., Jr. (1979) *Proc. Natl. Acad. Sci. USA 76*, 655-659.
6. Kaplan, L. M., Ariga, H., Hurwitz, J. and Horwitz, M. S. (1979) *Proc. Natl. Acad. Sci. USA 76*, 5534-5538.
7. Su, R. T. and DePamphilis, M. L. (1978) *J. Virol. 28*, 53-65.
8. Krokan, H., Schaffer, P. and DePamphilis, M. L. (1979) *Biochemistry 18*, 4431-4443.
9. Ikeda, J. -E., Enomoto, K. and Hurwitz, J. (1981) *Proc. Natl. Acad. Sci. USA 78*, 884-888.
10. Sheinin, R., Humbert, J., and Pearlman, R. E. (1978) *Ann. Rev. Biochem. 47*, 277-316.
11. Mortimer, R. K. and Schild, D. (1980) *Microbiol. Rev. 44*, 519-571.
12. Johnston, L. H. and Nasmyth, K. A. (1978) *Nature 274*, 891-893.
13. Beggs, J. D.(1978) *Nature 275*, 104-109.
14. Jazwinski, S. M. and Edelman, G. M. (1979) *Proc. Natl. Acad. Sci. USA 76*, 1223-1227.
15. Broach, J. R., Strathern, J. N. and Ficks, J. B. (1979) *Gene 8*, 121-133.
16. Petes, T. D. and Williamson, D. H. (1975) *Cell 4*, 249-253.
17. Livingston, D. M. and Kupfer, D. (1977) *J. Mol. Biol. 116*, 249-260.
18. Zakian, V. A., Brewer, B. J. and Fangman, W. L. (1979) *Cell 17*, 923-934.
19. Kojo, H., Greenberg, B. D., and Sugino, A. (1981) *Proc. Natl. Acad. Sci. USA in press*.
20. Davis, R. W., Simon, M. and Davidson, N. (1971) *Methods Enzymol. 21*, 413-428.
21. Okazaki, R. (1971) *Methods in Enzymology 21*, 296-304.
22. Sugino, A., Higgins, N. P., Brown, P. O., Peebles, C. L. and Cozzarelli, N. R. (1978) *Proc. Natl. Acad. Sci. USA 75*, 4838-4842.
23. Davis, R. W., Botstein, D. and Roth, J. R. (1980) *Advanced Bacterial Genetics* (Cold Spring Harbor Laboratory, Cold Spring Harbor, New York).
24. Kojo, H. and Sugino, A. (1981) *Nucleic Acids Res. in press*.

25. Broach, J. R. and Hicks, J. B. (1980) *Cell, 21,* 501-508.
26. Stinchcomb, D. T., Struhl, K. and Davis, R. W. (1979) *Nature 282,* 39-43.
27. Tschumper, G. and Carbon, J. (1980) *Gene 10,* 157-166.
28. Livingston, D. M. (1977) *Genetics 86,* 73-84.
29. Huberman, J. A. (1981) *Cell in press.*
30. Plevani, P., Badaracca, G., Ginelli, E., and Sora, S. (1980) *Antimicrb. Ag. Chemother. 18,* 50-57.
31. Nakayama, K. and Sugino, A. (1980) *Biochem. Biophys. Res. Commun. 96,* 306-312.
32. Ogawa, T. and Okazaki, T.(1980) *Ann. Rev. Biochem. 49,* 421-457.
33. Okazaki, R., Okazaki, T., Sakabe, K., Sugimoto, K., Kainuma, R., Sugino, A. and Iwatsuki, N.(1968) *Cold Spring Harbor Symp. Quant. Biol. 33,* 129-143.
34. Hartley, J. L. and Donelson, J. E. (1980) *Nature 286,* 860-864.
35. Hereford, L. M. and Hartwell, L. H. (1974) *J. Mol. Biol. 84,* 445-461.
36. Hartwell, L. H. (1976) *J. Mol. Biol. 104,* 803-817.
37. Chang, L. M. S. (1977) *J. Biol. Chem. 252,* 1873-1880.
38. Winnacker, E. L. (1978) *Cell 14,* 761-773.
39. Sims, J., Capon, D. and Dressler, D. (1979) *J. Biol. Chem. 254,* 12615-12628.
40. Sugimoto, K., Oka, A., Sugisaki, H., Takanami, M., Nishimura, A., Yasuda, S. and Hirota, Y. (1979) *Proc. Natl. Acad. Sci. USA 76,* 575-579.
41. Subramanian, K. W., Dhar, R. and Weissman, S. M. (1977) *J. Biol. Chem. 252,* 355-367.
42. Strathern, J. N., Newlon, C. S., Herskowitz, I. and Hicks, J. B. (1979) *Cell 18,* 309-319.
43. Chan, C. S. M. and Tye, B. -K. (1980) *Proc. Natl. Acad. Sci. USA 77,* 6329-6333.

REPLICATION IN MONKEY CELLS
OF PLASMID DNA CONTAINING
THE MINIMAL SV40 ORIGIN

R. Marc Learned
Richard M. Myers and
Robert Tjian

Department of Biochemistry
University of California
Berkeley, California 94720

ABSTRACT We have cloned the SV40 origin of DNA replication into a bacterial plasmid, generated deletion mutations of the T antigen binding sites, and measured the replication of these plasmid DNAs in a monkey cell line, COS7, that synthesizes SV40 large T antigen. Our results indicate that a sequence of approximately 80-85 bp, containing T antigen binding sites I and II but lacking site III, is sufficient to serve as the origin of viral DNA replication. A filter binding assay was used to measure the binding of a purified T antigen to wild type and mutant origin templates. Our findings indicate that the removal of binding sites II and III has no effect on the binding of the protein to site I.

We have also constructed plasmids, containing the wild type SV40 origin of DNA synthesis, that replicate as efficiently as SV40 DNA in COS7 monkey cells. Because these plasmids are small, contain several restriction sites, and have high replication efficiencies, they may be useful cloning vectors for introducing eukaryotic genes into monkey cells in order to study gene expression in vivo.

INTRODUCTION

Simian virus 40 (SV40), a small oncogenic papovavirus, is a useful model system for studying eukaryotic DNA replication and gene expression (1). One of the viral gene products, large tumor antigen (T antigen), is required for the initiation of viral DNA synthesis (2). T antigen, a 96,000 dalton phosphoprotein (3), binds sequentially (4) and cooperatively (5) to three sites encompassing the viral origin of DNA synthesis. There is good evidence that a direct interaction between T antigen and the viral origin is necessary for the initiation of viral DNA synthesis (6,7,8).

Gluzman recently reported the isolation of a monkey cell line, COS7, that is transformed by SV40 and produces large amounts of T antigen (9). We have found that a bacterial plasmid (pSV01) containing 311 bp of SV40 DNA, including the T antigen binding sites and the origin of replication, replicates when transfected into COS7 cells (7). Thus, it is possible to generate mutations in vitro in cloned origin sequences and introduce these mutant DNAs into monkey cells. Earlier we used this approach to study the effects of deletion mutations in T antigen binding sites I and II on DNA replication and T antigen binding (7). Here we use the same procedure to study the effects of deletion mutations in T antigen binding site III in order to determine the minimum amount of SV40 DNA that is required for DNA replication.

Because cloned origin plasmids replicate in both bacterial and in monkey cells, it should also be possible to clone a eukaryotic gene into these plasmids and study the expression of the gene in monkey cells. However, we (7) and others (10) have found that the replication of cloned origin templates in COS7 cells is 5-to-20 fold less efficient than the replication of SV40 DNA. This reduced replication efficiency limits the use of plasmids such as pSV01 for expression studies where a high number of DNA molecules per cell is desirable. Recently, Lusky and Botchan determined that the efficiency of replication of cloned origin plasmid DNAs in monkey cells can be increased to the level of SV40 DNA replication by removal of specific bacterial sequences from the plasmids (10). Here, we report the construction of plasmids lacking these "poison sequences", that replicate in COS7 cells as efficiently as SV40 DNA.

RESULTS AND DISCUSSION

Construction of site III deletion mutants. In an effort to determine the SV40 sequences that are necessary

and sufficient for viral DNA replication, we have measured
the ability of various mutant DNAs to replicate in monkey
cells. Earlier we used this procedure to determine the
boundary on the "early side" (Figure 1b) of the SV40 origin
of replication (7). Here we report the mapping of the boundary on the "late side". Using the procedure outlined in
Figure 1a, we introduced deletion mutations into T antigen
binding sites II and III. The plasmid pSVOR, which contains
the 311 bp Eco RII G fragment of SV40 inserted into the Eco
RI site of pBR322 with Eco RI linkers (7), was partially
digested with the restriction endonuclease Eco RI. The
resulting linear molecules were treated with nuclease BAL-31
to remove approximately 100 bp of DNA from each end. After
digesting with Pvu II to provide one common endpoint for all
of the deletions, the mixture was treated with T4 DNA ligase
and used to transform E. coli cells. Colonies containing
SV40 sequences were selected by hybridization to radioactively labeled SV40 DNA (11). Six deletion mutants analyzed
by restriction analysis and direct nucleotide sequencing
(12) were used for replication experiments (Figure 1b).

Replication of mutant site III plasmids in COS7 cells.
The replication of cloned wild type and mutant origin
plasmids was measured as previously reported (7). Monolayers of COS7 monkey cells (9) were transfected with each
DNA in the presence of DEAE-dextran. Small DNA molecules
were isolated from the cells at various times following
transfection by the Hirt lysis procedure (13), and subjected
to agarose gel electrophoresis. The DNA was transferred to
nitrocellulose membranes (14). The filters were hybridized
to pBR322 DNA labeled with ^{32}P by the nick translation procedure (15,16) and analyzed by autoradiography.

Immediately after transfection of COS7 cells with
plasmid DNA containing a wild type SV40 origin, pSVOR, forms
I, II and III of plasmid DNA were present in the Hirt
lysates (Figure 2, lane 1). At twelve hours after transfection, supercoiled molecules were no longer detected,
although forms II and III persist (Figure 2, lane 2). A
marked increase in the amount of supercoiled plasmid
molecules was observed at 50 hours after transfection (Figure 2, lane 3), indicating that replication of the plasmid
DNA occurred. This increase in supercoiled molecules is
dependent upon SV40 origin sequences, as no replication of
pBR322 DNA was observed in the assay (Figure 2, lanes 4-6).
When DNA from dl 161, a mutant plasmid containing binding
site I and II but lacking all of site III, was transfected

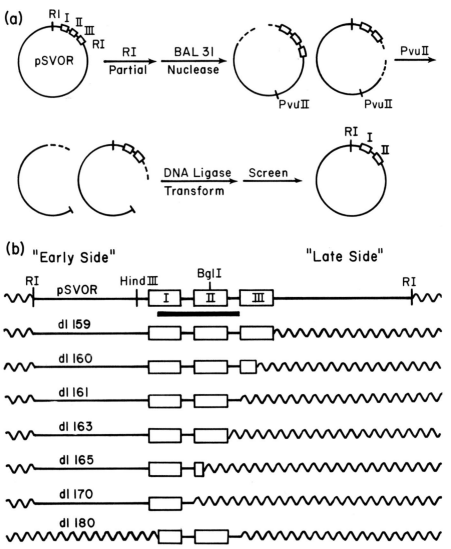

Figure 1. Construction of deletion mutants in the T antigen binding sites. (a) Scheme for constructing plasmids with deletions in sites II and III, from the "late side" of the origin. The procedure is described in Results and Discussion. (b) Map of wild type and mutant origins. SV40 sequences are indicated by bold lines, with T antigen binding sites depicted as boxes. Wavy lines represent bacterial plasmid sequences. The minimum amount of SV40 sequence that will support replication of a plasmid in COS7 cells is shown by the black bar (see dl 180 replication experiment, Figure 2).

into COS7 cells, a wild type level of replication was
observed (Figure 2, lanes 7-9). However, dl 163 DNA, which
lacks the sequences between sites II and III in addition to
site III, failed to replicate in COS7 cells (Figure 2, lanes
10-12). As expected, no replication of plasmid DNAs
occurred when COS7 cells were transfected with mutant DNAs
lacking both sites II and III, (dl 166 and dl 170), whereas
dl 158 and dl 159 DNA, which contain all of sites I and II,
replicated as efficiently as wild type DNA in the assay
(data not shown). We often observed several bands migrating
slower than forms I, II and III in the 50 hr Hirt lysates of
replicating DNAs, particularly with the smaller plasmids.
We believe that these extra forms are supercoiled, linear
and nicked circular dimeric forms of the plasmids because
treatment of the mixture with Eco RI leads to a single form
III band (data not shown).

Thus, in agreement with results obtained by Subramanian
and Shenk (17) with viable SV40 deletion mutants, we have
found that T antigen binding site III is not required for
the efficient replication of cloned origin plasmids in monkey cells. Additionally, our results confirm that sequences
in site II are required for viral DNA replication (18,19,7).
We do not yet know whether the inability of dl 163 to replicate is due to the absence of a small portion of binding
site II, the A-T rich sequence between sites II and III, or
both.

Construction of a plasmid containing the minimum SV40
sequence required for replication. The boundary for each
side of the origin of replication has thus been determined
by this laboratory and others (20,17). These results suggest that a sequence of SV40 DNA approximately 85 bp in
length corresponds to the functional origin of replication
in the viral genome. To test whether this sequence is sufficient to allow DNA replication, we constructed a bacterial
plasmid, dl 180, that contains only 85 bp of SV40 DNA
corresponding to the origin sequences. Indeed, when dl 180
DNA was introduced into COS7 cells as above, the plasmid
replicated at the same level as PSVOR DNA (Figure 2, lanes
13-15). Thus, these 85 bp of SV40 sequences, which include
T antigen binding sites I and II and the sequences between
sites II and III, must include all the information required
for a functional origin of replication.

Binding of a purified T antigen to wild type and mutant
origin sequences. We have used the quantitative filter

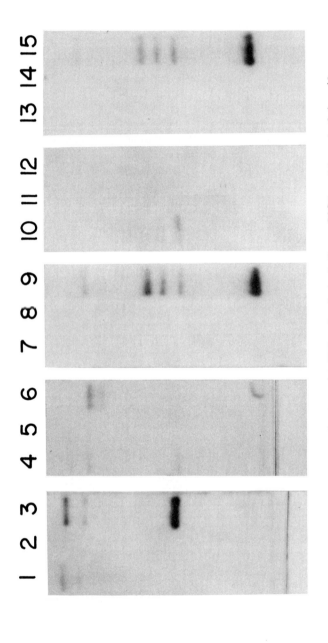

Figure 2. Analysis of plasmid DNAs isolated from COS7 cells at 0 hr (lanes 1,4,7,10,13), 12 hr (lanes 2,5,8,11,14), and 50 hr (lanes 3,6,9,12,15) after transfection. pSVOR DNA (lanes 1-3), pBR322 DNA (lanes 4-6), dl 161 DNA (lanes 7-9), dl 163 DNA (lanes 10-12) and dl 180 DNA (lanes 13-15). Supercoiled forms for each DNA are the most rapidly migrating bands in this gel system.

binding assay previously described (7) to measure the specific interaction between a T antigen-related protein, D2 protein (21), and the wild type and mutant plasmid DNAs described above. Various amounts of purified D2 protein were incubated with end-labeled DNA and filtered through nitrocellulose membranes. Because both D2 protein and authentic SV40 T antigen show the greatest affinity for binding site I (4,5), it was predicted that any mutant DNA containing this sequence would be retained on nitrocellulose membranes to the same extent as the wild type DNA. Indeed, DNA from mutant dl 170, which lacks binding sites II and III, was retained on the filter as efficiently as intact SV40 origin sequences (Figure 3). As expected, DNAs lacking site III and the sequences between sites II and III (dl 163) were also retained on the filter at wild type levels (Figure 3). These results suggest that the absence of T antigen binding sites II and III has no effect on the binding of D2 protein to site I.

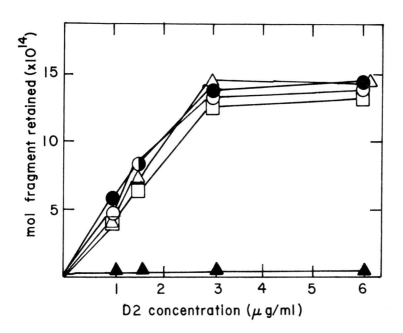

Figure 3. Binding of D2 protein to plasmid DNAs. End-labeled plasmid molecules were mixed with various amounts of D2 protein, filtered through nitrocellulose and analyzed as described in Results and Discussion. pSVOR DNA (○), pBR322 DNA (▲), dl 161 DNA (●), dl 163 DNA (△), and dl 170 DNA (□).

Construction of plasmids with increased replication efficiencies. We have reported that the plasmid pSV01, which contains the Eco RII G fragment of SV40 inserted into the Eco RI site of pBR322 with Eco RI linkers, replicates in COS7 cells less efficiently than SV40 DNA (7). Several groups (10,19) have shown that the SV40 DNA is not modified in bacteria; SV40 DNA excised from a bacterial plasmid replicates as efficiently as viral DNA. This result suggested that the lower level of replication of recombinant plasmid is due to a cis-acting pBR322 sequence. If such is the case, removing some plasmid sequences may increase the replication efficiency of cloned origin vectors in monkey cells. In order to test this hypothesis, we constructed several plasmids that lack a significant portion of the bacterial plasmid sequences. Restriction maps for two of the new plasmids are shown in Figure 4. The plasmid pSV08 was constructed by first isolating a 229 bp Eco RI/Hind III restriction fragment from pSV01 that contains the three T antigen binding sites. The Hind III site corresponds to map position 0.649 of the SV40 genome (see Appendix A in Reference 1), and the Eco RI site replaces the Eco RII site occurring at map position 0.633 on the SV40 DNA. This restriction segment was inserted into a fragment of pBR322 (a gift from C. Thummel) containing plasmid sequences from pBR322 map positions 2440 to 4361/0 (Figure 4; reference 22). One end of this 1910 bp pBR322 fragment corresponds to the Eco RI site at nucleotide number 0/4361 and the other end corresponds to a Hind III site constructed at nucleotide number 2440. The resulting 2140 bp plasmid thus contains 229 bp SV40 DNA (including the viral origin of replication), the bacterial β-lactamase gene, and the plasmid origin of replication. The plasmid pSV010 was constructed by inserting a 100 bp "polylinker" sequence into the Eco RI site of pSV08. This 100 bp insertion, kindly provided by B. Seed and T. Maniatis, contains recognition sites for several restriction enzymes.

We compared the replication of SV40, pSV01, pSV08 and pSV010 DNA in COS7 cells as described above, except that nitrocellulose membranes were hybridized to uniformly labeled pSV50 DNA, a plasmid that contains the entire SV40 genome inserted into pBR322. The replication of pSV01 DNA was approximately 5-to-10 fold lower than SV40 DNA (Figure 5, compare lanes 1-3 and 4-6). However the efficiencies of replication of pSV08 and pSV010 DNA were much higher than pSV01 (Figure 5, lanes 7-9 and 10-12). The amounts of hybridization of labeled DNA to the 50 hr time points of

Figure 4. Schematic representation of plasmids pSV08 and pSV010. T antigen binding sites I, II and III are depicted as boxes. Numbers at Eco RI, HindIII and PstI correspond to nucleotide numbers of the parent plasmid pBR322. The β-lactamase gene is indicated as Amp^R. The polylinker sequence in pSV010 is a 100 bp Eco RI fragment, constructed by Brian Seed and Tom Maniatis, that contains several single-cut restriction sites. The restriction sites present in the polylinker are, clockwise, Eco RI, Cla I, Hind III, Xba I, Bgl II, Mbo II, Pst I, Bam HI, Eco RI.

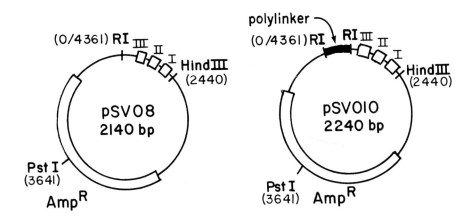

pSV08 and pSV010 are about 2 times lower than for SV40 (Figure 5). Because the plasmids are approximately 2.5 times smaller than the SV40 genome, this result indicates that the actual number of molecules of each DNA accumulating at 50 hr is about the same. Thus, the removal of pBR322 sequences between nucleotides 0 and 2440 is sufficient to overcome the decreased replication efficiency of cloned SV40 origin plasmids in COS7 monkey cells. M. Lusky and M. Botchan recently obtained similar results and mapped the pBR322 sequences responsible for the reduced replication efficiency in COS7 cells to a 1370 bp fragment between nucleotide numbers 1120 and 2490 (10). At present, it is not known how these plasmid sequences interfere with DNA replication in monkey cells.

Because the plasmids pSV08 and pSV010 replicate efficiently in COS7 cells, they will be more useful as

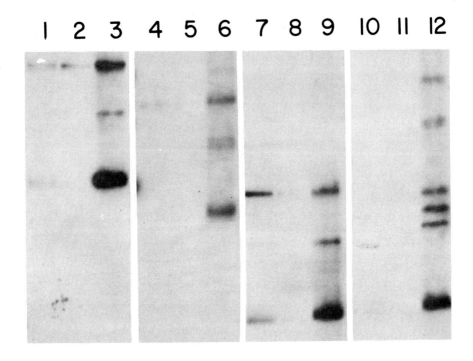

Figure 5. Analysis of SV40, pSV01, pSV08 and pSV010 DNA replication in COS7 cells. DNAs isolated at 0 hr (lanes 1,4,7,10), 12 hr (lanes 2,5,8,11), and 50 hr (lanes 3,6, 9,12) after transfection. SV40 DNA (lanes 1-3), pSV01 DNA (lanes 4-6), pSV08 DNA (lanes 7-9) and pSV010 DNA (lanes 10-12). Supercoiled forms for each DNA are the most rapidly migrating bands in this gel system.

eukaryotic vectors than the recombinant plasmid pSVO1 that we previously described (7). These new hybrid vectors have been designed to be rather small in order to accommodate relatively large inserts without interfering with the ability to propagate the DNA in bacterial cells. In addition, the plasmid pSVO10 may be particularly useful because of the polylinker sequence that was inserted in order to introduce several single-cut restriction sites into the vector. As an additional refinement, vectors derived from dl 180 DNA, which contains the minimum (85 bp) SV40 sequence required for replication in COS7 cells but lacks the viral transcriptional promoters, may be especially useful for quantitative in vivo transcription studies. Thus, these types of hybrid vectors should facilitate the amplification and expression of cloned mammalian genes and their regulatory elements in monkey cells.

ACKNOWLEDGMENTS

We wish to thank Yasha Gluzman for his generous gift of COS7 cells. We are also grateful to Monika Lusky and Mike Botchan for making their results available to us prior to publication, Taffy Mullenbach for tissue culture preparations, Don Rio for help with construction of deletion mutants, and Karen Erdley for typing the manuscript. This work was supported by research grants from the NIH and NSF.

REFERENCES

1. Tooze, J. (1980) in DNA Tumor Viruses, Molecular Biology of Tumor Viruses, ed. Tooze, J. (Cold Spring Harbor Laboratory, Cold Spring Harbor, New York), Part 2, pp. 125-204.

2. Tegtmeyer, P. (1972) J. Virol. 10, 591-598.

3. Tegtmeyer, P. (1975) J. Virol. 15, 613-618.

4. Tjian, R. (1978) Cell 13, 165-179.

5. Myers, R.M., Rio, D.C., Robbins, A.K., and Tjian, R.(1981), submitted to Cell.

6. Shortle, D.R., Margolskee, R.F., and Nathans, D. (1979) Proc. Natl. Acad. Sci. USA 76, 6128-6131.

7. Myers, R.M., and Tjian, R. (1980) Proc. Natl. Acad. Sci. USA 77, 6491-6495.

8. McKay, R., and DiMaio, D. (1981) Nature 289, 810-813.

9. Gluzman, Y. (1981) Cell 23, 175-182.

10. Lusky, M. and Botchan, M. (1981), submitted to Nature.

11. Grunstein, M., and Hogness, D.S. (1975) Proc. Natl. Acad. Sci. 72, 3961-3965.

12. Maxam, A. and Gilbert, W. (1980) in Methods in Enzymology, eds. L. Grossman and K. Moldave. (Academic Press) 65, 499-580.

13. Hirt, B. (1967) J. Mol. Biol. 26, 365-369.

14. Southern, E.M. (1975) J. Mol. Biol. 98, 503-518.

15. Maniatis, T., Jeffrey, A., and Kleid, D.G. (1975) Proc. Natl. Acad. Sci. 72, 1184-1188.

16. Rigby, P., Rhodes, M., Dieckmann, M. and Berg, P. (1977) J. Mol. Biol. 113, 237-251.

17. Subramanian, K.N., and Shenk, T. (1978) Nucleic Acids Research 5, 3635-3642.

18. Shortle, D., and Nathans, D. (1979) J. Mol. Biol 131, 801-817.

19. Gluzman, Y., Sambrook, J., and Frisque, R.J. (1980) Proc. Natl. Acad. Sci. 77, 3898-3902.

20. Lai, C.J., and Nathans, D. (1974) J. Mol. Biol. 89, 179-193.

21. Hassell, J.A., Lukanidin, E., Fey, G., and Sambrook, J. (1978) J. Mol. Biol. 120, 209-247.

22. Sutcliffe, J.G. (1979) Cold Spring Harbor Symp. Quant. Biol. 43, 77-90.

LOCATIONS OF 5'-ENDS OF NASCENT DNA AT THE ORIGIN
OF SIMIAN VIRUS 40 DNA REPLICATION

Ronald T. Hay and Melvin L. DePamphilis[1]

Department of Biological Chemistry
Harvard Medical School
Boston, Massachusetts

ABSTRACT. Initiation of the SV40 replicon has been shown to occur at a genetically defined locus (ori). Our objective has been to determine the nucleotide sequence at which the first deoxyribonucleotides are inserted within this region. By cloning SV40 DNA fragments containing ori into bacteriophage M13 and utilizing the viral DNA as a strand-specific probe it has been possible to locate, at single nucleotide resolution, the 5'-ends of nascent DNA at ori. On one strand of the SV40 ori 5'-ends of nascent DNA with the same sense as early mRNA were found at two major and a small number of minor locations. However, in the region of the complementary strand examined 5'-ends of nascent DNA were not detected. Thus preferred DNA sequences are utilized for initiation of DNA synthesis within the SV40 ori region. In addition, the data indicate that initiation of SV40 DNA replication is an asymmetric and strand-specific event.

INTRODUCTION

Perspective

Three basic questions are generally posed concerning the mechanism by which DNA replication is initiated. What DNA sequences define an origin of replication? How are these

[1]This work was supported by National Institutes of Health grant CA 15579. R.T.H. was supported by a Damon Runyon-Walter Winchell Cancer Fund postdoctoral fellowship.

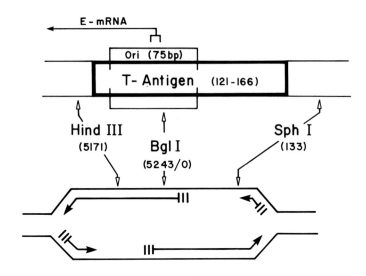

FIGURE 1. The SV40 origin of replication showing the region of T-Antigen binding, sequences essential for initiation (ori) and the cap sites for early mRNA. Also shown is a diagrammatic representation of an early replicating molecule and the nucleotide locations of relevant restriction enzyme cleavage sites (numbering system of Van Heuverswyn & Fiers, 1979).

DNA sequences recognized by the replicative machinery? Finally, how does this machinery activate the process of DNA replication? To accurately answer these questions, it is axiomatic that the locations and structures of 5'-ends of nascent DNA chains within the origin region be identified.

In addressing this problem, we have utilized Simian virus 40 (SV40) as a well-defined DNA-containing virus that provides a relatively simple, but appropriate model for eukaryotic chromosome replication (Cremisi, 1979; DePamphilis & Wassarman, 1980).

Initiation of the covalently closed circular SV40 genome occurs at a fixed point and proceeds bidirectionally until termination occurs at about 180° from the origin (Danna & Nathans, 1972; Fareed et al., 1972; Martin & Setlow, 1980; Tapper & DePamphilis, 1980). The genetic locus required for initiation (ori) has been defined as a cis acting element consisting of 75 base pairs (bp) of DNA (Fig. 1; see Nathans, this volume). Plasmids containing the ori region of SV40 DNA are capable of autonomous replication when introduced into cells producing T-Ag (Myers & Tjian,

1981). SV40 DNA synthesis is initiated within the ori region close to the single Bgl 1 site (Fareed et al., 1972; Martin & Setlow, 1980; Tapper & DePamphilis, 1980), and single base pair changes in this region alter the efficiency of replication (Shortle & Nathans, 1979). The SV40 ori region is recognised by T-Ag, the product of the SV40 A gene which is required for initiation of replication (Tegtmeyer, 1972). T-Ag binds specifically to three tandem DNA sites that span the ori region (see Tjian, this volume). SV40 revertants of ori defective mutants were found to contain second site mutations within the A gene, suggesting that the interaction of T-Ag with the ori sequence is required for initiation (Shortle et al., 1979).

We have begun to map the precise genomic positions of 5'-ends of nascent SV40 DNA chains in an effort to elucidate the requirements for initiation of DNA synthesis in eukaryotic chromosomes. One objective is to relate the synthesis of the first Okazaki fragment(s) at the origin of replication to the subsequent initiation of Okazaki fragments throughout the genome.

Experimental Approach

By measuring the distance from the 5'-ends of long nascent SV40 DNA strands to a unique restriction endonuclease site, these strands can be shown to originate at many preferred locations (Tapper & DePamphilis, 1980). This suggested that Okazaki fragments are initiated at specific DNA sequences. The validity of this hypothesis was subsequently demonstrated by mapping the precise nucleotide locations of 5'-ends on both long nascent DNA chains and Okazaki fragments (DePamphilis et al., 1980). In these experiments, the problem of determining the polarity of nascent DNA relative to the restriction site was solved by deduction, using the facts that DNA replication was bidirectional from a unique origin (Danna & Nathans, 1972; Fareed et al., 1972; Martin & Setlow, 1980; Tapper & DePamphilis, 1980), Okazaki fragments originated predominantly, if not exclusively, from retrograde DNA templates (Perlman & Huberman, 1977; Kaufmann et al., 1978; Kaufmann, 1981; Cusick et al., 1981), and only 5'-ends within 250 nucleotides of the restriction site were resolved by gel electrophoresis under conditions used for DNA sequencing (Maxam & Gilbert, 1977). However, this approach cannot be used at the origin of replication where nascent DNA on both sides of the fork will be small and similar in size. Therefore, in order to unambiguously locate the 5'-ends of nascent DNA at the origin, as well as in other genomic locations, unique sections of the SV40 genome were cloned into the DNA of E. coli phage M13. As the virion DNA of this phage is a single-stranded circle of unique polarity, recombinants containing SV40 DNA in either orientation could be prepared in large quantities. This permitted isolation of nascent SV40 DNA chains from specific genomic regions that annealed exclusively with SV40 DNA representing the template on only one arm of replication forks (Figs. 2, 3).

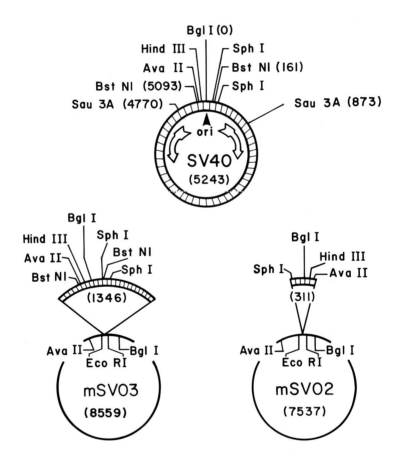

FIGURE 2. Structure of mSV clones. SV40 DNA is represented by the hatched area. mSV01 contains SV40 DNA in the opposite orientation to mSV02, while mSV04 contains SV40 DNA inserted in the opposite orientation to mSV03.

RESULTS

Cloning the SV40 ori Region Into M13 DNA.

Initial recombinants between M13 and SV40 DNA (denoted as mSV) were constructed as described by Messing et al. (1981) by ligating the 1346 bp Sau 3A fragment B, containing the ori region, into the complementary Bam Hl site of M13mp6 (Fig. 2). More highly selective probes were constructed using the shorter 311 bp Bst Nl G fragment containing the ori region (Fig. 2). In this case, the recessed 3'-ends of the fragment were extended by AMV reverse transcriptase in the presence of dTTP to permit ligation onto the flush ends of the Hinc II site in M13mp7 using T4 DNA ligase. The size of the SV40 insert was determined by gel electrophoresis following cleavage of the replicative form of mSV DNA with Eco RI (Fig. 2). The identity and orientation of each insert was established by a similar analysis following digestion with either Ava II or Bgl I, both of which provide single asymmetric cuts within the insert (Fig. 2). Hind III digestion yielded the expected single cut linear molecules, and Sph I cleaved in mSV02 once. Our conclusions were confirmed by direct DNA sequence analysis of about 150 nucleotides on either side of the single Hind III site (Maxam & Gilbert, 1977).

Large quantities of purified single-stranded DNA containing the SV40 ori region were prepared by precipitating phage with polyethylene glycol (Messing et al., 1981) and then purifying the virion DNA by phenol extraction, followed by chromatography on benzoylated-naphthylated DEAE (BND)-cellulose (Tapper & DePamphilis, 1980) and exclusion chromatography in Sepharose 4B (10 mM Tris-HCl, pH7.6, 1 mM EDTA and 100 mM NaCl). In this way, clones representing each DNA strand orientation of both Sau 3A (mSV03, mSV04) and Bst Nl (mSV01, mSV02) DNA fragments containing the SV40 ori region were purified for use in locating the 5'-ends of nascent DNA.

Locating the 5'-Ends of Nascent DNA.

The procedure used to locate the 5'-ends of nascent SV40 DNA at single-nucleotide resolution is outlined in Figure 3. Replicating SV40 DNA was labeled with [^{3}H]dThd in intact virus infected cells, extracted as described by Hirt (1967), and then purified by equilibrium centrifugation in CsCl, chromatography on BND cellulose, and sedimentation in a neutral sucrose gradient (Tapper & DePamphilis, 1980). Electron microscopic analysis revealed that about 80% of the molecules were SV40(RI) DNA and the remainder were circular SV40 DNA. The 5'-ends of nascent DNA chains in 2.5-5 µg of purified SV40(RI) DNA were dephosphorylated with bacterial alkaline phosphatase and the DNA denatured at 100°C (2 min) before labeling with [γ-^{32}P]ATP in the presence of polynucleotide kinase (Maxam & Gilbert, 1977). Single-stranded [5'-^{32}P]DNA was fractionated by electrophoresis in 10% polyacrylamide gels containing 8 M urea (Maxam & Gilbert, 1977), and DNA in the 100-200 nucleotide range was recovered by electroelution in the presence of 10 µg of the appropriate mSV DNA, 10 mM Tris-borate (pH 8.3) and 0.2 mM EDTA. [5'-^{32}P]DNA was hybridized to the cloned ori region by heating the mixture for 2 minutes at 100°C and then incubating at 65°C for 1 hour in 0.3 M NaCl. The large circular DNA hybrids were

LOCATING THE 5'-ENDS OF NASCENT SV40(RI) DNA CHAINS

1. PURIFY SV40(RI) DNA FROM INTACT CELLS
2. DENATURE DNA
3. BACTERIAL ALKALINE PHOSPHATASE
4. POLYNUCLEOTIDE KINASE + (γ-^{32}P)ATP
5. ISOLATE (5'-^{32}P)DNA, 100 - 200 NUCLEOTIDES
6. ANNEAL (5'-^{32}P)DNA TO EXCESS M13SV$_{ORI}$ VIRION DNA

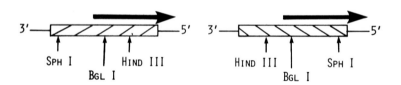

7. ISOLATE DNA HYBRIDS
8. DIGEST WITH RESTRICTION ENDONUCLEASE
9. FRACTIONATE SINGLE - STRANDED (5'-^{32}P)DNA AND COMPARE WITH APPROPRIATE (3'-^{32}P)DNA SEQUENCE

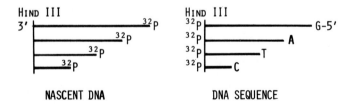

FIGURE 3. Procedure used to locate the 5'-ends of nascent DNA in ori. See the text for details.

37 LOCATIONS OF 5'-ENDS OF NASCENT DNA

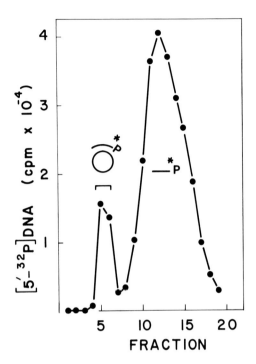

FIGURE 4. Isolation of [5'-^{32}P] DNA annealed to mSV single-stranded DNA by exclusion chromatography on Sepharose 4B. The bracket denotes those fractions appearing in the void volume which were subsequently used for 5'-end mapping.

isolated by exclusion chromatography on Sepharose 4B (Fig. 4). To locate the positions of ^{32}P-labeled 5'-ends, DNA hybrids containing mSV01 and mSV03 (Fig. 2) were digested with Hind III, while DNA hybrids containing mSV02 (Fig. 2) were cleaved by Sph I. This allowed 5'-ends originating from one strand to be analyzed independently from those of opposite polarity (Fig. 3).

In order to evaluate the ability of this mapping procedure to locate 5'-termini, a unique DNA restriction fragment was substituted for nascent DNA in replicating molecules. SV40(I) DNA was digested with Bst Nl, labeled with ^{32}P at the 5'-termini, fractionated by gel electrophoresis, and the 311 bp fragment containing the ori region isolated as described above. This [5'-^{32}P]DNA was then denatured, annealed to mSV04 and digested with Hind III. Only the expected 78 nucleotide end-labeled fragment was released, demonstrating the validity of this mapping procedure (Fig. 5).

FIGURE 5. Control experiment to map the 5'-end of a unique restriction enzyme fragment. The [5'-^{32}P] labeled 311 bp Bst N1 <u>ori</u> fragment was isolated as described for nascent DNA (text), hybridized to mSV04 and digested with Hind III. Denatured DNA fragments were separated by electrophoresis on 6% polyacrylamide containing 8M urea gels and [^{32}P] DNA visualized by autoradiography. DNA fragment size in nucleotides was determined from [5'-^{32}P] labeled, SV40 digested Bst N1 fragments.

<u>Nucleotide Sequence of 5'-Ends of Nascent DNA Within the Ori Region.</u>

To determine the precise nucleotide locations of 5'-ends of nascent DNA chains that were cleaved by Hind III, the DNA digestion products were purified, denatured in 98% formamide at 90°C, and fractionated by polyacrylamide gel electrophoresis under conditions that resolve single nucleotide differences (Fig. 6). In the absence of Hind III, all of the [5'-^{32}P]DNA migrated as a broad band in the 100-200 nucleotide region (Fig. 7), while those strands that were cleaved by Hind III migrated as shorter, discrete bands (Figs. 6,7). The 5'-^{32}P-labeled terminal nucleotide of each band was determined by direct comparison with the sequence of these nascent DNA strands. DNA was prepared for sequencing by digesting mSV04(RF I) DNA with Hind III, and the recessed 3'-OH termini labeled with [γ-^{32}P]dATP in the presence of AMV reverse transcriptase (DePamphilis <u>et al.</u>, 1980). This DNA was digested with Eco RI and the larger fragment, containing one 3'-^{32}P-labeled terminus, was isolated by gel electrophoresis. Application of the Maxam and Gilbert (1977) chemical degradation technique provided the DNA sequence of chains growing in the same direction as early mRNA (Fig. 1). Thus, the terminal

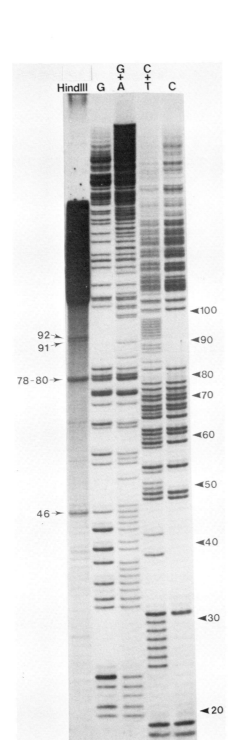

FIGURE 6. Locations of 5'-ends of nascent DNA with the same sense as early mRNA. Autoradiograph of an 8% polyacrylamide gel containing 8M urea following electrophoresis of [32P] labeled DNA. [5'-32P] labeled nascent DNA hybridized to MSV03 was cleaved with Hind III. The chemical degradation pattern of [3'-32P] labeled DNA from the same region of the genome is also shown. DNA fragment size is given in nucleotides.

5'-nucleotide on nascent DNA chains with the same polarity as early mRNA was determined up to 105 nucleotides from the Hind III site, limited only by the size class of undigested [5'-^{32}P]DNA (Fig. 6,7). A similar analysis was also carried out with [5'-^{32}P]DNA annealed to mSV02 and digested with Sph I (Fig. 7) in order to locate the 5'-ends of nascent DNA chains with the opposite polarity (Fig. 1).

The experiments using Hind III demonstrated that the 5'-ends of DNA chains from purified SV40(RI) DNA with the same orientation as early mRNA originated at two major and three minor locations within ori (Figs. 6,7). All four nucleotides are represented at the 5'-ends, although a compression effect (Van Heuverswyn & Fiers, 1979) from the

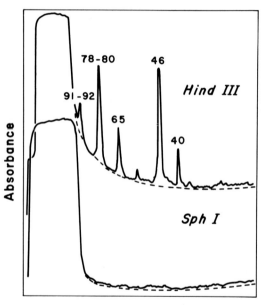

FIGURE 7. 5'-ends of nascent DNA from both ori strands which were treated with RNase T$_2$ prior to 5'-end labeling. Nascent DNA annealed to mSV01 was cleaved with Hind III while DNA annealed to mSV02 was cleaved with Sph 1. Microdensitometer tracing of an autoradiogram of an 8% polyacrylamide 8 M urea gel. The broken line indicates the tracing obtained when the DNA was not cleaved with the restriction endonuclease. DNA fragment size is given in nucleotides.

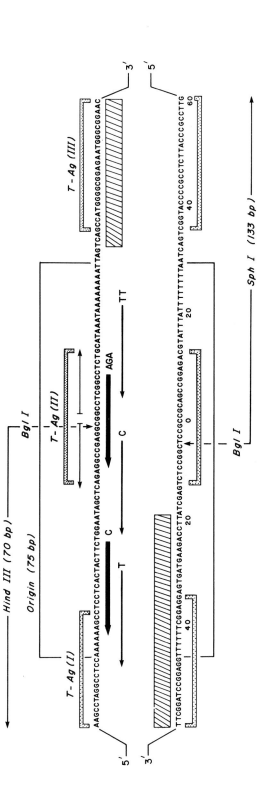

FIGURE 8. DNA sequence of SV40 ori (Van Heuverswyn & Fiers, 1979). Solid arrows indicate the location and orientation of 5'-ends of nascent DNA while the shaded area represents regions of the genome which have not yet been examined. T-Ag binding sites, the sequences required for ori function and restriction endonuclease sites used in this study, are also indicated.

27bp palindrome limited resolution of the 78-80 nucleotide band (Figs. 6,7) to GGA (Fig. 8). These data represent only 5'-ends of DNA chains because, in addition to treatment of SV40(RI) DNA with RNase A during purification, one sample (Fig. 7) was denatured and treated with RNase T2 prior to the 5'-end labeling procedure with no change in the pattern of bands (Fig. 3).

Digestion of [5'-^{32}P]DNA annealed to mSV02 with Sph 1 did not release any DNA fragments that were shorter than 155 nucleotides (size class of [5'-^{32}p] DNA examined was 155-200 nucleotides, Fig. 7). Control experiments indicated that Sph 1 was capable of digesting mSV02 double-stranded replicative form DNA to completion in the presence of a ten-fold excess of mSV02 single-stranded DNA. This demonstrates that Sph 1 was active under conditions identical to those used when nascent DNA was refractory to Sph 1 cleavage. Thus we concluded that 5'-ends of nascent DNA, with the same sense as late mRNA, are not detectable in the region stretching from nucleotide 5220 to nucleotide 80 (Fig. 8).

DISCUSSION

By cloning SV40 DNA fragments into the E. coli phage M13 and utilizing the viral DNA as a strand specific probe, it has been possible to isolate nascent SV40 DNA from a defined area of the genome which is complementary to only one of the two DNA strands. This capability has allowed the locations of 5'-ends of nascent DNA at the SV40 origin of replication to be determined. On one strand of the SV40 ori, 5'-ends of nascent DNA with the same sense as early mRNA were found at two major, and a number of less prominent, minor sites (Fig. 8). However, in the region of the complementary strand examined, 5'-ends of nascent DNA were not detected (Fig. 8). Presumably the initiation events which prime DNA synthesis on this strand take place at some point downstream (toward the Hind III site) and possibly even outwith the ori region. The locations of 5'-ends of nascent DNA in this region are currently being determined. Initiation of SV40 DNA synthesis at the origin of replication could therefore be interpreted as being initially unidirectional. In this model nascent DNA chains of the same sense as early mRNA, serve as the leading strand, while Okazaki fragments initiated on the retrograde template are extended and eventually becoming the leading strand growing in the opposite direction. A similar situation has recently been reported at the E. coli origin of replication, ori C, where 5'-ends of nascent DNA were found in two locations, 37 nucleotides apart on one strand, but none were detected on the other strand (Okazaki et al., 1980).

These data also indicate that in the SV40 ori, preferred sequences, rather than a completely unique sequence, are utilized for the initiation of the first deoxynucleotides. Previously we have shown that Okazaki fragments are initiated at preferred DNA sequences throughout the SV40 genome (DePamphilis et al., 1980), and it is possible that initiation of replication at ori represents synthesis of the first Okazaki fragments by a process which is similar to that utilized for the synthesis of Okazaki fragments throughout the genome. Thus it is possible that the observed specificity of initiation of SV40 at the unique origin of replication is a consequence of recognition of the ori region by SV40 T-antigen and not to a unique start point for DNA synthesis.

Digestion of nascent DNA with RNase T2 prior to 5'-end labeling did not significantly alter the positions or intensity of the fragments produced by Hind III cleavage. It was therefore concluded that RNA primers were not present in detectable amounts at the 5'-ends of nascent DNA, possibly due to their rapid excision in vivo. The observed 5'-ends may result from multiple initiation events or from the synthesis of a unique RNA primer which is subsequently processed and the 3'-ends generated used as primers for initiation of DNA synthesis.

It is of interest to compare initiation of transcription in SV40 to the initiation of replication, since in both cases a number of preferred, rather than unique DNA sequences, are utilized for initiation (Benoist & Chambon, 1981). Recently it has been shown that the heterogeneity of cap structures in late SV40 RNA is a consequence of independent initiation events and not of processing of a primary transcript (Contreras & Fiers, 1981).

ACKNOWLEDGMENTS

We thank David Weaver and Eric Hendrickson for their participation in cloning SV40 DNA segments into M13. M13 cloning vehicles and the appropriate E. coli strains were generously provided by Dr. Donald Oliver. All cloning experiments were carried out according to NIH guidelines.

REFERENCES

Benoist, C., and Chambon, P. (1981). Nature 290, 304.
Contreras, R., and Fiers, W. (1981). Nucl. Acids Res. 9, 215.
Cremisi, C. (1979). Microbiol. Revs. 43, 297.
Cusick, M.E., Herman, T.M., DePamphilis, M.L., and Wassarman, P.M. (1981). Biochemistry, in press.
Danna, K., and Nathans, D. (1972). Proc. Nat. Acad. Sci. USA 69, 3097.
DePamphilis, M.L., and Wassarman, P.M. (1980). Ann. Rev. Biochem. 49, 627.
DePamphilis, M.L., Anderson, S., Cusick, M., Hay, R., Herman, T., Krokan, H., Shelton, E., Tack, L., Tapper, D., Weaver, D., and Wassarman, P.M. (1980). In "Mechanistic Studies of DNA Replication and Genetic Recombination" (B. Alberts, ed.), p. 55. Academic Press, New York.
Fareed, G.C., Garon, C., and Salzman, N. (1972). J. Virol. 10, 484.
Hirt, B. (1967). J. Mol. Biol. 26, 365.
Kaufmann, G., Bar-Shavit, R., and DePamphilis, M.L. (1978). Nucl. Acids Res. 5, 2535.
Kaufmann, G. (1981). J. Mol. Biol. 147, 25.
Martin, R.G., and Setlow, V.P. (1980). Cell 20, 381.
Maxam, A.M., and Gilbert, W. (1977). Proc. Nat. Acad. Sci., USA 74, 560.
Messing, J., Crea, R., and Seeburg, P.H. (1981). Nucl. Acids Res. 9, 309.
Myers, R.M., and Tjian, R. (1980). Proc. Nat. Acad. Sci. USA 77, 6491.
Okazaki, T., Hirose, S., Fujiyama, A., and Kohara, Y. (1980). In "Mechanistic Studies of DNA Replication and Genetic Recombination" (B. Alberts, ed.), p. 429. Academic Press, New York.
Perlman, D., and Huberman, J.A. (1977). Cell 12, 1029.
Shortle, D., and Nathans, D. (1979). J. Mol. Biol. 131, 801.
Shortle, D., Margolskee, R.F., and Nathans, D. (1979). Proc. Nat. Acad. Sci, USA 76, 6128.
Tapper, D.P., and DePamphilis, M.L. (1978). J. Mol. Biol. 120, 401.
Tapper, D.P., and DePamphilis, M.L. (1980). Cell 22, 97.
Tegtmeyer, P. (1972). J. Virol. 10, 591.
Van Heuverswyn, H., and Fiers, W. (1979). Eur. J. Biochem. 100, 51.

REPLICATION DIRECTED BY A CLONED ADENOVIRUS ORIGIN[1]

George D. Pearson
Kuan-Chih Chow
Jeffry L. Corden[2]
Jerry A. Harpst[3]

Department of Biochemistry and Biophysics
Oregon State University
Corvallis, Oregon

ABSTRACT

We develop a model for the role of terminal protein in the initiation of adenovirus DNA replication based on the role of cisA protein in ϕX174 DNA replication. The two essential features of the model are: (i) adenovirus DNA is a covalently closed circular molecule, and (ii) terminal protein is an origin-specific topoisomerase. We use a cell-free system that initiates adenovirus DNA replication and specially constructed, cloned terminal sequences of adenovirus DNA to test the model.

[1] Supported by National Cancer Institute grant CA17699.

[2] Present address: Institut de Chimie Biologique, Strasbourg, France.

[3] Permanent address: Department of Biochemistry, Case-Western Reserve University School of Medicine, Cleveland, Ohio

INTRODUCTION

The linear, double-stranded DNA molecule isolated from the adenovirus particle has a molecular weight of 23×10^6. Adenovirus DNA molecules from all serotypes examined to date have two unusual structural features: an inverted terminal repetition roughly 100 to 150 base pairs long (1,2), and a 55,000-dalton protein, called terminal protein, covalently linked to the 5' end of each strand (3-5). Terminal protein is encoded by an adenovirus early gene (6), and is synthesized as a larger, presumably precursor, molecule (6,7).

The properties of replicating adenovirus DNA molecules have recently been reviewed by Winnacker (8). Lechner and Kelly (9) established that initiation occurs equally frequently from either molecular end. Replication proceeds by a strand displacement mechanism producing branched (type I) and unbranched (type II) linear molecules. Terminal protein is also covalently attached to the ends of replicating adenovirus DNA molecules (10,11). Since adenovirus cannot form circular or concatemeric replicating intermediates (12), various priming mechanisms using hairpins have been advanced to explain initiation of adenovirus DNA replication (13,14). In fact, the sequence of the inverted terminal repetition (15-18) excludes such priming mechanisms. Rekosh et al. (4) have instead proposed that the terminal protein itself may serve as a primer for replication, a notion without precedent.

In this paper, we develop a model for the role of terminal protein in the initiation of adenovirus DNA replication based on the role of cisA protein in ϕX174 DNA replication (19,20). Two essential features of the model are: (i) adenovirus DNA is a covalently closed circular molecule, and (ii) terminal protein is an origin-specific topoisomerase. We use a cell-free system that initiates adenovirus DNA replication (21,22) and specially constructed, cloned terminal sequences of adenovirus DNA to test the model.

MODEL

We base our mechanism for adenovirus replication on the extensively documented mechanism of φX174 replication, probably the most thoroughly understood system using strand displacement synthesis (23-25). The essential features of the model are:

(i) adenovirus DNA is a covalently closed circular molecule (conformation unspecified, but probably supercoiled),

(ii) terminal protein binds tightly, but not covalently, at a specific recognition sequence,

(iii) terminal protein catalyzes the breakage and reunion of a single polydeoxyribonucleotide chain at a fixed distance from the recognition site,

(iv) cleavage results in the covalent attachment of terminal protein to the 5' phosphate, and the 3' hydroxyl now acts as the primer for strand-displacement replication, and

(v) conventional methods for isolating nucleic acids (detergents, phenol extraction, etc.) uncouple breakage from reunion and result in covalent joining of terminal protein to DNA.

Fig. 1 diagrams the cycle of adenovirus DNA replication. The cycle can be divided into two separate reactions: displacement synthesis represented as DS → DS + SS, and complementary synthesis represented as SS → DS. The net reaction is then DS → 2 DS. Terminal protein initiates displacement synthesis by specifically nicking covalently closed circular adenovirus DNA. The 3' hydroxyl acts as the primer. As has been proposed for φX174 replication (20), the displaced strand is locked into the replication fork by an interaction between terminal protein and the replication machinery. The exact events terminating displacement replication have not been specified, but by analogy to φX174 replication (20), the displaced single strand has been joined into a circle by the action of terminal protein. We postulate that complementary synthesis is initiated by a cellular mechanism, perhaps involving an RNA priming event. Regardless of the specific steps, it is clear that initiation of complementary synthesis is <u>not</u> rate-limiting (26).

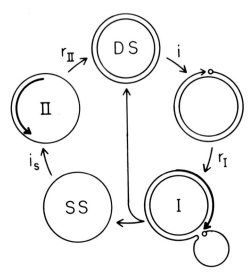

FIGURE 1. Cycle of adenovirus DNA replication. Starting at the top and proceeding clockwise: terminal protein is non-covalently bound to covalently closed circular duplex DNA (DS); initiation (i) occurs when DNA is nicked by terminal protein; terminal protein (small circle) remains covalently attached to the 5' phosphate and the 3' hydroxyl (arrow) acts as the primer for displacement synthesis (r_I); during displacement synthesis (thick line) on a type I molecule (I), the displaced strand is locked into the replication fork; termination of displacement replication yields a covalently closed circular duplex molecule (DS) and a covalently closed circular single strand (SS); initiation (i_s) of complementary synthesis (r_{II}) requires a cellular priming event and results in the formation of type II molecules (II); termination of complementary replication yields covalently closed circular duplex DNA (DS). Although not shown in this diagram, both adenovirus strands participate in displacement and complementary replication.

No <u>protein-free</u> circular adenovirus molecules, replicating or mature, have ever been found. We explain this fact by points (ii) and (v) above. There are clear precedents for both points. Plasmid relaxation proteins (27) and gyrase (28) can be isolated as stable non-covalent complexes with DNA,

38 REPLICATION DIRECTED BY A CLONED ADENOVIRUS ORIGIN

but under appropriate conditions nicking or breakage of the DNA occurs with concomitant covalent linkage of the protein. This immediately suggests a way to test our hypothesis: design and construct a specific substrate molecule for the cell-free adenovirus DNA replication system. An alternative approach, namely to show directly that adenovirus DNA is circular, would depend on establishing the right conditions by chance.

In order to design a substrate, we first need to decide what adenovirus sequences might be important for replication. Fig. 2 shows that a block of 14 base pairs is conserved absolutely in every adenovirus serotype, human or simian, sequenced to date. The block is positioned in each case exactly 8 base pairs from the ends of the linear molecule. Although the 5' terminal nucleotide is apparently always C, nucleotides at positions 2 through 7 can vary considerably. We propose that the non-covalent binding site for terminal protein includes at least the 14 base pair conserved sequence. We also propose that the binding of terminal protein is oriented since the nucleotide sequence is asymmetric. Furthermore, we stipulate that the covalent attachment of terminal protein to the 5' terminal C of each strand defines the cleavage site.

If we imagine that the ends of the linear adenovirus DNA are covalently joined to each other,

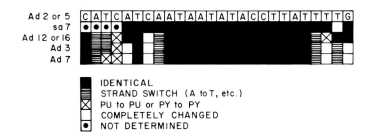

■ IDENTICAL
≡ STRAND SWITCH (A to T, etc.)
⊠ PU to PU or PY to PY
☐ COMPLETELY CHANGED
• NOT DETERMINED

FIGURE 2. Terminal nucleotide sequences in adenovirus DNA from different serotypes. From left to right the sequences read from 5' to 3'. References for individual sequences are: types 2 and 5 (15), types 3 and sa7 (16), types 12 and 16 (18), and type 7 (17).

the sequence through the ends would read as shown in Fig. 3A. Terminal protein bound at the recognition site to the right is positioned and oriented such that the β-hydroxyl group of a serine residue can initiate a nucleophilic attack on the phosphodiester bond between G and C on the top strand. As a result, terminal protein forms a phosphodiester bond with C (Fig. 3B), and the 3'-hydroxyl of G at the nick site becomes a primer for displacement synthesis (Fig. 3C). Since the first nucleotide incorporated during polymerization is C (Fig. 3C), the first phosphodiester bond contains ^{32}P donated by [α-^{32}P]dCTP. ^{32}P radioactivity is transferred to terminal protein only after a subsequent cycle of cleavage (Fig. 3D). Fig. 3E shows that ^{32}P transfer can be detected after deoxyribonuclease digestion. Challberg et al. (7) have already shown that during in vitro replication [α-^{32}P]dCTP transfers ^{32}P radioactivity to terminal protein, but α-^{32}P-labeled dGTP, dATP, and dTTP do not. In our model deoxyribonucleoside triphosphate-

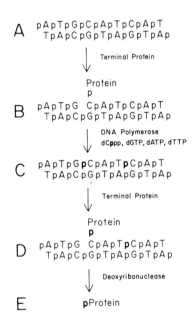

FIGURE 3. Mechanism for the transfer of ^{32}P radioactivity from [α-^{32}P]dCTP to terminal protein. The mechanism is detailed in the text. Boldface, lower case p's represent ^{32}P atoms.

specific transfer of α-^{32}P depends on the nucleotide sequence at the nick site as well as DNA synthesis. A simple expectation is that an alteration of the sequence at the nick site ought to change the specificity of ^{32}P transfer.

PLASMID

The entire 1344 base pair XbaI-E fragment from the left end of type 2 adenovirus DNA was cloned into the EcoRI site of pBR322 (J.L. Corden, unpublished results). Fig. 4 gives the structure and restriction endonuclease maps of the recombinant plasmid, called XD-7. XD-7 contains the adenovirus early 1A promoter and nearly all of the 1A transcriptional unit (29). XD-7 DNA has been used as

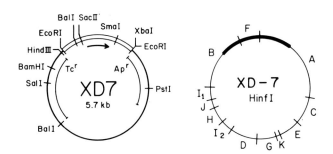

FIGURE 4. Restriction endonuclease maps of XD-7 DNA. Left: Single-cut endonucleases. The open bar represents adenovirus sequences. XD-7 contains the entire 1344 base pair XbaI-E fragment from the left end of type 2 adenovirus DNA inserted into the EcoRI site of pBR322. The arrow indicates the orientation of the adenovirus fragment (running from left to right) in the hybrid plasmid. Tc^r = tetracycline resistance. Ap^r = ampicillin resistance. Right: HinfI map of XD-7 DNA. The thick line represents adenovirus sequences. Fragment B contains an EcoRI site at the junction between pBR322 and adenovirus sequences. Fragment A contains overlapping EcoRI and XbaI sites at the junction between the two sequences.

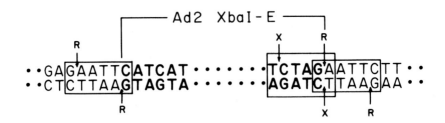

FIGURE 5. DNA sequences at the junctions between pBR322 DNA and the inserted XbaI-E fragment of type 2 adenovirus DNA in XD-7. Adenovirus DNA sequences are indicated by boldface letters and pBR322 DNA sequences by regular letters. The boxes enclose the restriction endonuclease recognition sequences for EcoRI and XbaI. R = cleavage sites for EcoRI. X = cleavage sites for XbaI. The top strand represents the l strand of adenovirus, and the bottom strand represents the r strand of adenovirus.

a template to study the initiation of transcription at the early 1A promoter in a cell-free system (30). Fig. 5 presents the DNA sequences at the junctions in XD-7 between pBR322 DNA and the inserted XbaI-E fragment of adenovirus DNA (J.L. Corden, unpublished results). It is important to note that even the adenovirus 5' terminal C residue is present in XD-7.

RESULTS

The simple expectation is that XD-7 DNA, but not pBR322 DNA, will incorporate radioactive deoxyribonucleoside triphosphates using the cell-free adenovirus DNA replication system of Challberg and Kelly (21,22). Fig. 6 shows that XD-7 DNA incorporated [α-^{32}P]dCTP into an acid-insoluble form linearly for 2.5 hr. The incorporation of [α-^{32}P]-dCTP directed by pBR322 DNA reached a plateau within 1 hr, and by 2.5 hr constituted 1/6th the incorporation directed by XD-7 DNA. This experiment has been repeated several times with similar results:

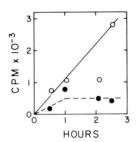

FIGURE 6. Cell-free incorporation of [α-^{32}P]-dCTP directed by XD-7 DNA (open circles) and pBR322 DNA (closed circles). Reaction conditions were essentially as described by Challberg and Kelly (21, 22). Reaction mixtures (100 μl) contained 25 μl extract, 150 ng plasmid DNA, dNTPs at 50 μM, and [α-^{32}P]dCTP at 3 Ci/mmol. Incubation was at 37°C. Samples (20 μl) were removed at the indicated times and the reactions stopped by adjusting the samples to 0.1% Sarkosyl, 10 mM EDTA, and 100 μg yeast tRNA. After precipitation with ice-cold 10% trichloroacetic acid, precipitates were collected on glass-fiber filters, dried, and counted.

the incorporation of [α-^{32}P]dCTP by XD-7 DNA always exceeded by 3 to 6 times the incorporation directed by pBR322 DNA. Moreover, we have evidence that the cloned adenovirus sequences directed the replication of XD-7 DNA. HinfI restriction mapping of XD-7 DNA labeled with [α-^{32}P]dCTP in the cell-free reaction indicated that the origin of replication was in HinfI fragment B, the fragment containing the left molecular end of adenovirus DNA. This analysis, however, lacked sufficient resolution to locate the origin precisely or to determine the direction of replication. HinfI restriction mapping of replicating pBR322 DNA showed no obvious origin, a result consistent with the notion that pBR322 DNA engages in limited repair replication.

We have begun to analyze the reaction products of the cell-free adenovirus DNA replication system by electron microscopy. Electron microscopy of the XD-7 reaction showed (195 molecules scored): 82%

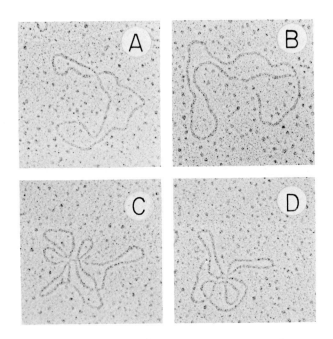

FIGURE 7. Analysis by electron microscopy of the reaction products of the cell-free replication system. Reaction conditions were as described in Fig. 6 except that all deoxyribonucleoside triphosphates were non-radioactive. Incubation was for 2 hr. DNA molecules were digested with pancreatic ribonuclease, T1 ribonuclease, and Pronase, and extracted with phenol and ether (see Fig. 8). Molecules were mounted for electron microscopy using the formamide variation of the basic protein film technique (31). Panel A shows a monomer unreplicated XD-7 DNA molecule. Panels B through D show examples of XD-7 monomer rolling circles. The bar represents 0.5 μm.

monomer circles, 7% multimer circles, 4% rolling circles, 2% theta structures, and 5% forked linear molecules. Fig. 7A shows an example of a monomer unreplicated XD-7 DNA molecule. The contour length was 2.01 ± 0.11 μm (n=5), the length expected for a 5.7 kb molecule. Figs. 7B through 7D display examples of XD-7 monomer rolling circles. The contour length of the circular portion of the replicating molecules was 1.95 ± 0.07 μm (n=5). Tails ranged from 0.1 unit length to 0.4 unit length in these molecules. In contrast, the pBR322 reaction showed (170 molecules scored): 98% monomer circles, 0.5% multimer circles, 1% rolling circles, and 0.5% forked linear molecules. Reactions lacking plasmid DNA contained no circular forms, but had about the same frequency of forked linear molecules as in the pBR322 reaction. Extracts from uninfected cells did not support the replication of either XD-7 DNA or pBR322 DNA.

A specific nick is introduced into XD-7 DNA molecules replicating in the cell-free adenovirus replication system. XD-7 DNA molecules were labeled in vitro with [α-^{32}P]dCTP, cleaved with HinfI endonuclease, and the fragments were analyzed by alkaline agarose gel electrophoresis. Fig. 8 (lane 3) shows the appearance of a new fragment, called X, with a size of 640 ± 30 bases. Since fragment X was not detected on neutral agarose gels, it clearly is not produced by HinfI cleavage alone, and cannot therefore arise as a partial digestion product (in fact, no two contiguous HinfI fragments sum to this size). Fragment X must be generated by a single-strand cleavage within HinfI fragments A or B, the only larger fragments. Cleavage of one of the strands of HinfI-A to produce fragment X would also yield a companion single-stranded fragment migrating on the alkaline gel between HinfI-B and X. Since no fragment this size has been detected, it is likely that fragment X is contained in HinfI-B, the fragment containing the origin of XD-7 DNA replication. Interestingly, the junction between pBR322 and adenovirus sequences in HinfI-B lies 636 base pairs from one HinfI site and 382 base pairs from the other. We therefore conclude that replicating XD-7 DNA molecules are nicked precisely at the junction in HinfI-B between pBR322 and adenovirus sequences, and that the nick is the origin of replication. Fig. 8 (lane 2) shows that replicating pBR322 DNA molecules were not nicked in this region. Fig. 8 (lane 1) shows that fragment X was not produced when reactions lacking plasmid DNA were analyzed. We still do not know which strand is nicked, the exact sequence at the nick, or that terminal protein causes the nick. Current experiments are designed to answer these questions.

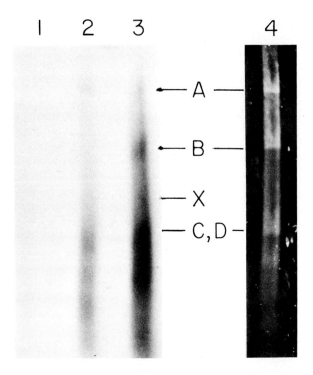

FIGURE 8. A specific nick is introduced into XD-7 DNA molecules replicating in the cell-free replication system. Reaction mixtures were as described in Fig. 6 except that dATP, dTTP, and dGTP were at 10 µM, and [α-^{32}P]dCTP was at 8.2 µM with a specific activity of 73 Ci/mmol. The reactions (100 µl) were stopped by adjusting the solutions to 0.01% Sarkosyl and 10 mM EDTA. The solutions were incubated for 1 hr at 37°C with pancreatic ribonuclease (100 µg/ml) and T1 ribonuclease (5 units/ml; Sankyo), and then incubated for 1 hr at 37°C with Pronase (100 µg/ml). The solutions were extracted twice with phenol and three times with ether, and precipitated with ethanol. DNA was taken up in HinFI buffer (50 mM NaCl, 6 mM Tris, pH 7.4, 6 mM MgCl$_2$, and 6 mM β-mercaptoethanol), and digested with 4.5 units of HinfI endonuclease in a total volume of 15 µl for 100 min at 37°C. The reaction was stopped by adjusting the solution to 6 mM EDTA. DNA was denatured by heating the solution to 100°C for 2 min, quenching on ice, and adjusting the solution to 30 mM NaOH. Samples were electrophoresed on a 1.2% agarose gel (9 cm wide, 14 cm long, 1 mm thick) in buffer containing 30 mM NaOH

and 2 mM EDTA for 2 hr at 100 V (50 ma). The gel was stained with ethidium bromide and photographed under ultraviolet illumination. After drying, the gel was autoradiographed for 10 hr using an intensifier screen. (1) No added DNA. (2) pBR322 DNA. 3) XD-7 DNA. (4) Ethidium bromide-stained XD-7 DNA. HinfI fragments E through K of XD-7 DNA are not identified in this figure.

DISCUSSION

XD-7 DNA replicates in a cell-free extract from adenovirus-infected, but not uninfected cells. pBR322 DNA does not replicate in either extract. Restriction mapping locates the XD-7 replication origin in HinfI-B, the fragment containing the left molecular end of adenovirus DNA. A specific nick is introduced into replicating XD-7 DNA molecules in HinfI-B at the junction between pBR322 and adenovirus sequences. The introduction of a specific nick may well serve as the origin of replication. The frequency of initiation of replication (6% to 11% of input XD-7 DNA) compares favorably with the results of Challberg and Kelly (21,22). Unlike the system of Challberg and Kelly (21,22) which requires adenovirus DNA-terminal protein complex, our system initiates replication on protein-free XD-7 DNA molecules. Terminal protein is presumably supplied by the cell-free extract, and becomes covalently linked to XD-7 DNA molecules at the nick site. Our replication system provides a functional assay to determine sequences required for adenovirus-specific replication. In particular, it is already clear that sequences spanning the nick site are not as important as adenovirus sequences located internally with respect to the left molecular end. Therefore, experiments with biochemically constructed deletions spanning the nick site in XD-7 DNA ought to yield valuable information on sequences required for terminal protein function. A simple expectation is that an alteration of the sequence at the nick site ought to change the specificity of ^{32}P transfer from α-^{32}P-labeled deoxyribonucleoside triphosphates to terminal protein (see Fig. 3). Our preliminary results are consistent with the notion that adenovirus DNA molecules are covalently closed circles, and that linear DNA-terminal protein complexes arise as a result of cleavage of circular molecules by terminal protein. Circularity of the adenovirus DNA molecule has profound implications for the mechanism of integration during transformation, for the control of viral transcription, and for the nucleoprotein structure of the viral chromosome (early during infection adenovirus DNA is assembled into nucleosomes (32; J.L. Corden and G.D. Pearson, manuscript in preparation)).

ACKNOWLEDGMENTS

We thank Kate Mathews for excellent technical assistance, Forest Ziemer for technical advice and assistance with certain experiments, and Dr. Andrew Taylor for instruction in electron microscopy techniques.

REFERENCES

1. Garon, C.F., Berry, K.W., and Rose, J.A.(1972). Proc. Natl. Acad. Sci. USA 69, 2391.
2. Wolfson, J., and Dressler, D. (1972). Proc. Natl. Acad. Sci. USA 69, 3054.
3. Robinson, A.J., Younghusband, H.B.,and Bellett, A.J.D. (1973). Virology 56, 54.
4. Rekosh, D.M.K., Russell, W.C., Bellett, A.J.D., and Robinson, A.J. (1977). Cell 11, 283.
5. Carusi, E.A. (1977). Virology 76, 380.
6. Stillman, B.W., Lewis, J.B., Chow, L.T., Mathews, M.B., and Smart, J.E. (1981). Cell 23, 497.
7. Challberg, M.D., Desiderio, S.V., and Kelly, T.J., Jr. (1980). Proc. Natl. Acad. Sci. USA 77, 5105.
8. Winnacker, E.-L. (1978). Cell 14, 761.
9. Lechner, R.L., and Kelly, T.J., Jr. (1977). Cell 12, 1007.
10. Robinson, A.J., Bodnar, J.W., Coombs, D.H., and Pearson, G.D. (1979). Virology 96, 143.
11. Girard, M., Bouché, J.-P., Marty, L., Revet, B., and Berthelot, B. (1977). Virology 83, 34.
12. Watson, J.D. (1972). Nature New Biol. 239, 197.
13. Cavalier-Smith, T. (1974). Nature 250, 467.
14. Bateman, A.J. (1975). Nature 253, 379.

15. Steenbergh, P.H., Maat, J., van Ormondt, H., and Sussenbach, J.S. (1977). Nucleic Acids Res. 4, 4371.
16. Tolun, A., Aleström, P., and Pettersson, U. (1979). Cell 17, 705.
17. Dijkema, R., and Dekker, B.M.M. (1979). Gene 8, 7.
18. Sugisaki, H., Sugimoto, K., Takanami, M., Fujinaga, K., and Sawada, Y., Uemizu, Y., Uesugi, S., Shimojo, H., and Shiroki, K., personal communication.
19. Langevelt, S.A., van Mansfield, A.D.M., Baas, P.D., Jansz, H.S., van Arkel, G.A., and Weisbeek, P.J. (1978). Nature 271, 417.
20. Eisenberg, S., Griffith, J., and Kornberg, A. (1977). Proc. Natl. Acad. Sci. USA 74, 3198.
21. Challberg, M.D., and Kelly, T.J., Jr. (1979). Proc. Natl. Acad. Sci. USA 76, 655.
22. Challberg, M.D., and Kelly, T.J., Jr. (1979). J. Mol. Biol. 135, 999.
23. Koths, K., and Dressler, D. (1978). Proc. Natl. Acad. Sci. USA 75, 605.
24. Keegstra, W., Baas, P.D., and Jansz, H.S. (1979). J. Mol. Biol. 135, 69.
25. Arai, K.-I., Arai, N., Schlomai, J., and Kornberg, A. (1980). Proc. Natl. Acad. Sci. USA 77, 3322.
26. Bodnar, J.W., and Pearson, G.D. (1980). Virology 105, 357.
27. Lovett, M.A., and Helinski, D.R. (1975). J. Biol. Chem. 250, 8790.
28. Cozzarelli, N. (1980). Science 207, 953.
29. Berk, A.J., and Sharp, P.A. (1977). Cell 12, 721.
30. Wasylyk, B., Kedinger, C., Corden, J., Brison, O., and Chambon, P. (1980). Nature 285, 367.
31. Chow, L.T., Roberts, J.M., Lewis, J.B., and Broker, T.R. (1977). Cell 11, 819.
32. Corden, J., Engelking, H.M., and Pearson, G.D. (1976). Proc. Natl. Acad. Sci. USA 73, 401.

EUKARYOTIC ORIGINS: STUDIES OF A CLONED SEGMENT FROM XENOPUS LAEVIS AND COMPARISONS WITH HUMAN BLUR CLONES

J. Herbert Taylor[1]
Shinichi Watanabe[1]

Institute of Molecular Biophysics
Florida State University
Tallahassee, Florida

We have estimated the number of origins functioning at the beginning of S phase in cultured cells of Chinese hamster (CHO) to be about 2000 per diploid set of chromosomes. The number of functional origin can, however, be increased in synchronized cells by blocking DNA synthesis before S phase with an inhibitor of thymidylate synthetase in a medium free of thymidine (1, 2). In the presence of such a block in DNA synthesis, the cells reach S phase and the number of potential origins increases nearly linearly with time for 10-15 hours. At 15 hours the number which will function within seconds after release of the block with ^{3}H-thymidine is about 6000 per diploid cell. We have estimated that the amount of DNA in which the functional origins are distributed is about 3% of the total to be replicated and that amount does not change much, if any, during the block. The increase in functional origins results, not from the activation of new domains of DNA, but from the preparation of many origins which would perhaps not function in a cell where DNA synthesis was not inhibited. Origins in these CHO cells are usually distributed at average intervals of 75 µm (3), but after holding cells in a block of the type described above, the functional origins can be as close together as 4-8 µm (12-24 thousand bp) with some indication of a periodic distribution at intervals which are multiples of 4 µm, at least in the first DNA replicated (4). Since the early replicating DNA in CHO cells includes a high buoyant density fraction (5,6) the periodicity may apply only to a fraction of the DNA. However, when the number of origins per genome is estimated on the assumption that potential origins are distributed at intervals of 12000 bp along all of the DNA the total origins per genome is about 250,000.

[1]*Present address: McArdle Institute, University of Wisconsin, Madison, Wisconsin.*

With these studies in mind we decided that origins might be easy to isolate and clone into a bacterial plasmid. We initiated studies of EcoRI fragments of genomic DNA. For our initial assay we decided to use the unfertilized egg of Xenopus laevis. For that reason we also decided to clone random EcoRI segments of Xenopus DNA into an E. coli plasmid, pACY189. When these had been used to transform E. coli SF8 cells and clones were grown to harvest several micrograms of DNA, we screened the plasmids first for size and selected a number that were about two times the size of the vector. Larger quantities of the selected clones were harvested, the DNA isolated and the supercoiled (form I) DNA isolated after banding in CsCl-ethidium bromide. Each plasmid along with ^{3}H-thymidine was then injected into Xenopus eggs. Those which were the best templates for DNA replication as indicated by the highest incorporation rate were retained for possible analysis for a specific sequence associated with an origin of replication. Of a number that were assayed one that was a good template was designated pSW14 and used for further studies. The recombinant plasmid selected was one of a number selected from transformed E. coli cell lines that yielded plasmids about 2 times the size of the vector, pACYC189. However, pSW14 turned out to be a head to head dimer of the vector with only a small segment of Xenopus DNA about 500 nucleotides in length. The segment has now been sequenced and found to have 507 nucleotides including the AATT sequences from the EcoRI site at each end. The sequence has two Sau3A sites ($^↓$GATC) and by using this restriction endonuclease, three subclones were produced by inserting the Sau3A fragments into pBR322. These subclones have been compared with the original pSW14 for template activity in unfertilized eggs of Xenopus. In addition, several Blur clones of human DNA were compared with pSW14 for template activity in a similar way.

MATERIALS AND METHODS

A. *Plasmid Vectors*

pACYC189 was obtained from Stanley N. Cohen who synthesized it from a PstI generated fragment of R6-5 (7) cloned into pVH51 [a col El plasmid that contains the kanamycin resistance fragment from R6-5 that Cohen et al. (8) had originally used in constructing pSC101]. It has in addition to the kanamycin gene, a chloramphenical resistance factor. The single EcoRI site is in the kanamycin gene. pBR322 (9) was obtained from H.W. Boyer (Univ. of Calif., San Francisco). The human "Blur" clones (10) were a gift from Theodore Friedman and Prescott Deininger (University of California, San Diego).

39 EUKARYOTIC ORIGINS STUDIES OF A CLONED SEGMENT

B. Injection of Xenopus Eggs

Unfertilized eggs obtained by injection of Xenopus laevis females with 350-500 units of chorionic gonadotropin (Antuitrin "S", Parke-Davis) into the dorsal lymph sac 15 hours before eggs were required. The eggs were separated in well or spring water and placed on glass fiber filters floated on a 1 x 3 microscope slide. The eggs were irradiated with ultraviolet light for 4-5 min from a Mineral Lamp (Ultraviolet Products, Inc., San Gabriel, CA) which emits 1300 $\mu\omega/cm^2$ at 15 cm. About 5-10 min after irradiation, the jelly coats were soft enough to inject the eggs with a glass needle pulled to a outside diameter of about 50 µm with a Livingston micropipette puller, mounted in a de Fonbrune micromanipulator (11) and connected to a microsyringe driven by a syringe pump set to deliver 50 µl in 11 sec. Each egg was injected with 50 µl of buffer containing 15 ng of DNA and 1-2 µCi of ^{3}H-thymidine (60 Ci/mM). After injection the eggs, which are activated by the needle puncture, were incubated in sterile buffer for 1-4 hrs as previously described (12).

C. Extraction and Assay of Injected DNA

The eggs were crushed in lysing buffer (13) containing 1% sarkosyl at 0-2°C and 50 µg/ml Pronase B (Cal Biochem). If the radioactivity was to be determined the lysate of each egg was precipitated on a glass filter with trichloroacetic acid and counted in a liquid scintillation spectrometer.

If further analysis was necessary the DNA was extracted from lysates of 20-30 eggs by shaking over phenol and precipitation with ethanol (12). The DNA was separated by agarose gel electrophoresis without further treatment or after separation of supercoiled plasmids from linear and relaxed circles in a CsCl-ethidium bromide isopycnic gradient. Restriction enzyme digestion was carried out when desired before electrophoreses to demonstrate other properties of the DNA.

RESULTS

A. Subcloning Three Fragments of the Xenopus DNA in pSW14

As previously reported the plasmid (pSW14) was labeled in vivo by methylation of adenine in the EcoRI site. Its replication in the Xenopus egg could be followed by the dilution of the methylation pattern in a semiconservative manner. If DNA is fully methylated on both chains at each EcoRI site, the products of a first replication will be hemimethylated but still fully protected from digestion by EcoRI; however after two rounds of replication without methylation, one-half of the plasmids should be free of methylation and susceptible to digestion by EcoRI. Of the ^{3}H-thymidine labeled DNA isolated by CsCl banding nearly one half was digestable by EcoRI after incubating eggs for 4 hours, a time sufficient for two rounds of replication. Some other plasmids with much larger segments of Xenopus DNA than the 507 bp of pSW14 replicated very slowly as did the vector, pACYC189. On the assumption that the 507 bp of pSW14 might contain a replication origin, it was sequenced (the sequence will be published elsewhere).

Since the structural features of pSW14 did not reveal where the origin might be, we decided to subclone it. It has two Sau3A sites which divide it into unequal sized segments. The center segment was cloned into the BamHI site of pBR322 by taking advantage of the fact that the 5' single strand extensions produced by Sau3A and BamHI are identical (GATC). The other two segments which have a Sau3A end and an EcoRI end were cloned in pBR322 which had been cleaved with both BamHI and EcoRI. This releases a short segment of pBR322 but provides a site for the insertion of the two end segments of the Xenopus fragment (Figure 1). The fragments were subcloned as follows. The Xenopus fragment was cut with Sau3A and ligated into either pBR322 cut with BamHI or pBR322 cut with both BamHI and EcoRI. Several colonies resistant to ampicillin and susceptible to tetracycline (gene cut by BamHI) were isolated from E. coli SF8 (r_k^-, m_k^-, lop-11, rec BC-) transformed with both the whole vector as well as the one with a gap (Figure 1). DNA was isolated from each clone and digested with either HpaII or HaeIII to identify the fragment cloned into pBR322. Recombinant plasmids containing 66 bp and 174 bp of the Sau3A fragments between the EcoRI site and the BamHI site were designated pSW141 and pSW142, respectively. The plasmid carrying the center 263 bp Sau3A fragment at the BamHI site of pBR322 was designated pSW143.

B. *Template Efficiency of the Subclones of pSW14*

Plasmids were harvested from each clone, purified by isopycnic banding in CsCl-ethidium bromide and the supercoiled form of the plasmids injected into unfertilized eggs of Xenopus. The results are presented in Table 1. None of the three subclones replicated nearly as well as the original pSW14 and were not much better than the vector (pBR322) alone.

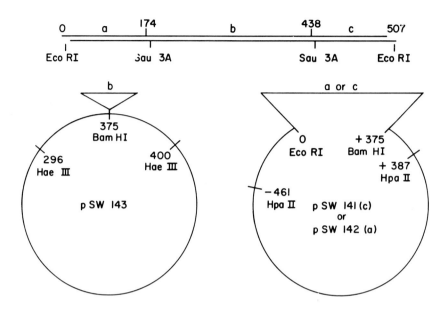

FIGURE 1. Construction of three subclones with the Xenopus segment from pSW14 cloned into pBR322. Two Sau3A sites are present in the 507 bp segment. The fragment was isolated, digested with Sau3A and the three sub-fragments ligated into pBR322. Since the single chain extensions produced by BamHI are the same as those for Sau3A, the center sub-fragment (b) was spliced into the BamHI site. The other two sub-fragments have one EcoRI end and one Sau3A end, and were spliced into the gap produced by digestion of pBR322 with both EcoRI and Bam HI. The 3 recombinant plasmids were identified by digestion with HaeIII or HpaII and separation of the fragments on 8% polyacrylamide gel.

Our tentative conclusion is that Sau3A cleaves a sequence which is necessary for efficient initiation of replication in the 507 bp of Xenopus DNA.

C. *Template Efficiency of the Hman Blur Clones in Xenopus Eggs*

Some plasmids derived by recombining about 300 bp human repetitive sequences into the BamHI site of pBR322, the Blur clones (14, 10), were obtained from Prescott L. Deininger and tested for template activity by injection into Xenopus eggs

TABLE 1. *Relative Activity of pSW14, pBR322 and Subclones of pSW14 Xenopus Fragment for Promoting DNA Replication in Xenopus eggs*

Plasmid	Number of eggs injected	^{3}H-thymidine incorporated mean ($\bar{x}$) per egg	Standard deviation
pSW14	17	18,218	7605
pSW141	17	5,583	2193
pSW142	18	4,250	1431
pSW143	19	1,030	562
pBR322	26	431	270

along with an equal amount of pSW14 and ^{3}H-thymidine. A mixture of 7.5 ng of pSW14 and 7.5 ng of each the Blur plasmids, Blur 2, Blur 6, Blur 8, Blur 11 and Blur 19 have been tested. All of these plasmids, except Blur 19, have an AluI site which divides the sequence into segments of about 170 and 130 bp, and are known as the Alu family of sequences. Some sequence variations exist among the member of the Alu family but they share extensive homology and make up at least 3% of the human genome.

The template activity is indicated in Figure 2 and Table 2. Figure 2 shows the distribution of radioactivity in a gel run with the plasmid isolated from 25-30 eggs four hours after injection. Carrier plasmids were mixed with the ng amounts of ^{3}H-labeled DNA isolated from the pooled lysate of 25-30 eggs. After electrophoresis for a time required to separate four peaks of radioactivity representing, in order from the top of the gel, form II of pSW14, form I of pSW14, form II of Blur 6, and form I of Blur 6 the gel was sliced in equal sized segments and the acid precipitable counts determined. The efficiency of replication is probably about the same for the two plasmids although the radioactivity in Blur 6 DNA is a little more than in pSW14. Plasmid SW14 is more than 2X the size of Blur 6, which means that there are about two times as many origins in 7.5 ng of Blur 6 as in the same amount of pSW14. On a per origin basis then, the Blur clone is probably a less efficient template than the Xenopus segment. More than 50% of the radioactivity remained at the top of the gel and was counted in fraction 1. This material was consistently seen when the DNA prepared by phenol extraction was applied directly to the gel. However, the fraction was lost when only supercoiled DNA was analyzed in gels after banding in CsCl-ethidium bromide (12); therefore this material may not be supercoiled and probably

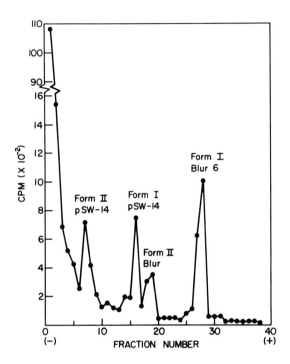

FIGURE 2. Graph showing the incorporation of ^3H-thymidine into pSW14 and plasmid Blur 6 injected together in equal amounts into <u>Xenopus</u> eggs. The eggs were incubated for 4 hours after injection, lysed and the lysate extracted with phenol. After ethanol precipitation the DNA was run into a 2% agarose gel along with unlabeled form I and form II pSW14 and Blur 6. The gel was sliced into equal sized disks and counted by liquid scintillation.

represents catenated plasmids. When the DNA was first digested with EcoRI before placing on the gel, the material ran as a single band in a 1% agar gel (data not shown). The linear pBR322and the linear pACYC189 produced by digesting Blur and pSW14 DNA, respectively were not easily resolved in a 1% agarose gel sliced for counting radioactivity. Therefore the broad band was assumed to represent linear DNA (pSW14 is cut to the size of the original vector, pACYC189) of both plasmids running together. This result indicates that the DNA at the top is either a very large linear polymer of the plasmid

TABLE 2. *Relative Activity of Blur Plasmids and pSW14 in Promoting DNA Replication in* Xenopus *Eggs*

Blur plasmid	Relative activity to pSW14	Amount of radioactivity remaining at top of gel
Blur 2	0.49	43.7%
Blur 6	1.09	54.4%
Blur 8	0.85	29.9%
Blur 11	1.04	34.4%
Blur 19	0.32	40.8%

genomes or an interlocked network of plasmids. The fraction has been consistently found when plasmid DNA from Xenopus eggs is separated on a gel without banding in CsCl-ethidium bromide. Other tests (data not shown) in our laboratory indicate that if only one of the two plasmids is cut some circular forms of the other are released. This indicates that the large material that fails to enter the gel are catenanes of the two plasmids, probably produced by topoisomerase II which has been reported to be present in the germinal vesicles of Xenopus oocytes (15) and is very likely in the egg also.

A comparison of all of the Blur clones tested along with pSW14 is given in Table 2. The least efficient template is Blur 19 which does not contain an Alu I site. It probably replicated only one time since the gel from which the data in Table 2 was taken showed that very little of the Blur 19 plasmid appeared in a supercoil form, while most of pSW14 in the same gel was supercoiled. The Alu family of repetitive sequences appears to function reasonably well in the Xenopus egg to initiate replication of pBR322 which replicates very poorly without a eukaryotic segment spliced into it.

DISCUSSION

Inclusion of an appropriate segment of eukaryotic DNA in a bacterial plasmid enhances its efficiency as a template for DNA replication 20-30 fold although some replication may occur on plasmids with only bacterial origins. Whether the bacterial origin is of any significance in such replication can not be ascertained from our data. The efficiency of the bacterial plasmid is too low for good comparative studies in Xenopus eggs. Harland and Laskey (16) have reported the replication of a variety of templates which were injected into Xenopus eggs. They did not report a comparison of two templates injected at concentrations approaching saturation for the egg's enzyme systems as we have done in our comparisons. However, they cut SV40 into four segments with Hind III and recyclized the fragments by ligation. These circles of different sizes were injected into Xenopus eggs at a concentration of 40 µg/ml (compared to 300 µg/ml in this report). At least four of the circles could be demonstrated to have replicated two times in trace amounts.

It seems clear that circles of DNA without an origin can replicate in the Xenopus egg, but I do not reach the conclusion, as Harland and Laskey (16) did, that a specialized DNA sequence is not involved in replication of eukaryotic DNA's. The replication system in eggs is probably operating in a relaxed mode, because we know that the usual controls that operate to regulate replication over the cell cycle do not come into operation until late blastula or even later in mammals where the behavior of the late replicating X-chromosome becomes fixed or differentiated. Nevertheless some regulatory systems must be operating to prevent DNA segments from replicating more than once in each cell cycle even in the early cleavage stages. Functional origins which become inactive after one round of replication in a cell cycle would appear to be part of such a control system (17).

ACKNOWLEDGMENTS

This research was supported by grants from the U.S. Department of Energy (EY-78-S-05-5854) and from the National Institute on Aging (1 R01 AGo1807-01).

REFERENCES

1. Taylor, J. H., *Chromosoma 62*, 291 (1977).
2. Laughlin, T. J., and Taylor, J. H., *Chromosoma 75*, 19 (1979)
3. Ockey, C. H., and Saffhill, R., *Exptl. Cell Res. 103*, 361 (1976)
4. Taylor, J. H., and Hozier, J. C., *Chromosoma 57*, 341 (1976)
5. Tobia, A. M., and Schildkraut, C. L., and Maio, J. J., *J. Mol. Biol. 54*, 499 (1970)

6. Bostock, C. J., and Prescott, D. M., *Exptl. Cell Res. 64*, 481 (1971)
7. Cabello, F., Timmis, K., and Cohen, S. N., *Mol. Gen. Genetics 162*, 121 (1978)
8. Cohen, S. N., Chang., A. C. Y., Boyer, H. W., and Helling, R. B., *Proc. Natl. Acad. Sci. USA 70*, 3240 (1973)
9. Sutcliffe, J. G., *Cold Spr. Harb. Symp. Quant. Biol. 43*, 77 (1979)
10. Jelinek, W. F., Toomey, T. P., Leinwand, L., Duncan, C. H., Biro, P. A., Choudary, P. V., Weisman, S. M., Rubin, C. M., Houck, C. M., Deininger, P. L., and Schmid, C. W., *Proc. Natl. Acad. Sci. USA 77*, 1398 (1980)
11. McKinnel, R. G., *in* "Cloning: Nuclear Transplantation in Amphibia" p. 230, University of Minnesota Press, Minneapolis, Minnesota (1978)
12. Watanabe, S., and Taylor, J. H., *Proc. Natl. Acad. Sci. USA 77*, 5292 (1980)
13. Guy, A. L., and Taylor, J. H., *Proc. Natl. Acad. Sci. USA 75*, 6088 (1978)
14. Rubin, C. M., Houck, C. M., Deininger, P. L., Friedmann, T., and Schmid, C. W., *Nature 284*, 372 (1980)
15. Baldi, M. I., Benedetti, P., Mattoccia, E., and Tocchini-Valentine, G. P., *Cell 20*, 461 (1980)
16. Harland, R. M., and Laskey, R. A., *Cell 21*, 761 (1980)
17. Taylor, J. H., *in* "DNA Synthesis: Present and Future" (Molineux, I., and Kohiyoma, M., eds.) p. 143, Plenum Press, New York and London (1978)

REGIONS OF TETRAHYMENA rDNA ALLOWING AUTONOMOUS REPLICATION OF PLASMIDS IN YEAST[1]

Gyorgy B. Kiss[2]
Anthony A. Amin
Ronald E. Pearlman

Department of Biology
York University
Toronto, Ontario
Canada

ABSTRACT

Regions near the center and the termini of the extrachromosomal rDNA of Tetrahymena thermophila allow autonomous replication of plasmids in yeast. Both these regions contain greater than 80% A-T base pairs. The region near the center of the rDNA includes the in vivo origin of replication of the molecule. Aside from short homooligomeric stretches, no extensive homology is evident between these regions of Tetrahymena DNA or between Tetrahymena DNA and regions of yeast DNA containing putative origins of replication.

INTRODUCTION

Combining a selectable genetic marker with a DNA fragment and assaying for autonomous replication has been a procedure much used in prokaryotes to isolate and study origins of DNA replication (see this volume). To extend these studies to eukaryotes, Saccharomyces cerevisiae can be used as a host to assay for autonomously replicating elements. Transformation of yeast at low frequency is indicative of integration of the transforming DNA into

[1] Supported by grants from NCI and NSERC
[2] Present address: Institute of Genetics, Biological Research Center, Szeged, Hungary

the host chromosome (1) while a high frequency of transformation reflects the autonomous replication of the transforming DNA (2, 3, 4). Other criteria characterizing autonomous replication of a genetic element are mitotic instability in the absence of selection and the ability to re-isolate and characterize the autonomously replicating element from the transformed yeast.

Transformation of yeast has been used to study regions on the yeast 2 µm circle plasmid implicated as origins of replication (5) as well as yeast chromosomal segments which are putative origins of replication (6, 7, 8, 9). A fraction of DNA from a number of eukaryotes, but no E. coli sequences, also allow autonomous replication of plasmids in yeast (10). However it is not known if these sequences represent origins of replication from these eukaryotes, although a 2.2 kb fragment from Xenopus laevis mitochondrial DNA which includes the in vivo origin of replication does function as an autonomously replicating segment (ars (8)) in yeast (11).

In our continuing studies on the replication of the extrachromosomal palindromic rDNA from the ciliate protozoan Tetrahymena thermophila (12, 13) we have looked for regions of this molecule capable of autonomous replication in yeast. Non-transcribed sequences near the center and the termini, but not the transcribed coding region allow autonomous replication of plasmids in yeast. The ars element from the center of rDNA includes the in vivo origin of replication which from electron microscopy experiments, is found to be near the center of the molecule (T. Cech, personal communication). The nucleotide sequences of these ars elements are known (14, 15) and thus the studies described here represent the most detailed characterization at present of non-yeast sequences which allow autonomous replication of plasmids in yeast. These studies will be very important in defining sequences recognized by the yeast DNA replication apparatus as origins of replication and also in studying specific interactions of Tetrahymena proteins with putative replication origins.

RESULTS

The extrachromosomal, macronuclear rDNA of the ciliate protozoan T. thermophila is a palindromic molecule containing two genes coding for the primary transcript of rRNA (16, 17). Restriction maps and a detailed transcription map of this molecule have been obtained (18). In order to ask if regions from this molecule, particularly a DNA segment which includes the in vivo origin of replication, can function as ars elements in yeast, we have cloned the non-transcribed central and terminal regions and the coding region of the molecule onto plasmids. In Figure 1 is presented a map of the rDNA of T. thermophila including restriction sites used in cloning. Figure 1 also shows the regions of rDNA cloned for this study.

TABLE I. Transformation of Saccharomyces cerevisiae by plasmids containing Tetrahymena rDNA sequences

Plasmid used	Transformants per μg DNA	LL 20 His$^+$ Transformants	Segregation % His$^+$	Re-isolated plasmid
pGY 24 (pGY 19Ω his$^+$)	2.8-10.0 x 10^3	GY 764	0.1	pGY 24
pGY 28 (pGY 17ζ his$^+$)	9-12	GY 768	99.0	-
pRP 35 (pRP 14ζ his$^+$)	20	RP 44	99.9	-
pRP 38 (pRP 7ζ his$^+$)	14	RP 45	99.9	-
pGY 43 (pGY 39Ω his$^+$)	5.8 x 10^3	RP 46	0.1	pGY 43
pYF 92 (3)	1.0- 1.3 x 10^3	GY 761	66.8	pYF 92
pYF 177 (5)	4.6 x 10^3	RP 1177	60.5	pYF 177
pGY 25 (pACYC 184ζ his$^+$)	1- 7	GY 758	99.6	-
pGY 35 (pACYC 177ζ his$^+$)	2-20	GY 759	99.9	-

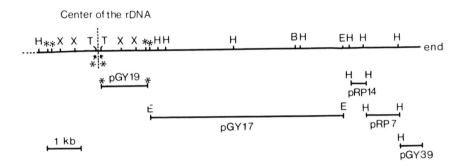

FIGURE 1   Restriction map of the rDNA of T. thermophila illustrating regions cloned for this study. Details of the mapping and of the cloning strategy are presented elsewhere (15, 18). pGY 17 and pGY 19 were cloned into the Eco RI site of pACYC 184 (19). pRP 7 and pRP 14 were cloned into the Hind III site of pBR 322 (20). pGY 39 contains the terminal 650 bp Hind III fragment of rDNA and 180 bp of pGA 22 (15) cloned into the Hind III site of pACYC 184 (19). The end point of pGY 19 near the center of rDNA was determined by nucleotide sequence analysis (14). E - Eco RI; H - Hind III; * - Eco RI*; X - Xba I; T - Taq I; B - Bam HI.

The His 3 gene of yeast, present on a Bam HI fragment was cloned into the unique Bam HI sites of pGY 19, pGY 39, pRP 7 and pRP 14. Because the rDNA fragment in pGY 17 contains a Bam HI site, the Eco RI rDNA containing fragment from pGY 17 was cloned into the unique Eco RI site of pYF 111, a plasmid containing the yeast His 3 gene inserted into pBR 322. His 3 is expressed in both yeast and E. coli and the above plasmids were maintained and propagated in E. coli.

Covalently closed circular plasmid DNA was used to transform His$^-$ yeast selecting for His$^+$ transformants. Besides the DNA containing rDNA sequences, plasmid DNA containing yeast 2 μm circle DNA and plasmid DNA containing no yeast replicator sequences were used to transform yeast. Data from the yeast transformation experiments are presented in Table I.

Saccharomyces cerevisiae strain LL20 (α, leu2-3, leu2-112, his3-11, his3-15) obtained from G. Fink, Cornell University, was transformed to His$^+$ as described by Hinnen et al (1). Plasmids used were as described in Fig. 1 and the text. Segregation was determined by growth of transformed yeast for 10 generations in the absence of selection and is expressed as the % His⁺ colonies of the total colonies tested. -, no plasmids could be reisolated.

Plasmids capable of autonomous replication in yeast should transform yeast at high frequency. Transformants should be unstable in the absence of selection and plasmid DNA should be isolatable from yeast and following propagation in E. coli, shown to be identical to the original transforming DNA. Of the non-Tetrahymena containing DNA, only plasmids including the yeast 2 μm circle origin of replication were capable of autonomous replication in yeast. Of the rDNA containing plasmids, only those including 1.9 kb at the center of the molecule (pGY 19) or 650 bp at the termini (pGY 39) were capable of autonomous replication in yeast. Other regions of the rDNA of Tetrahymena did not allow autonomous replication of plasmids in yeast.

DISCUSSION

A 1.9 kb fragment from the central nontranscribed region of Tetrahymena rDNA functions as an ars element in yeast. This fragment extends to within 38 bp of the center of the rDNA (14). In vivo, rDNA replication in T. thermophila is bidirectional from an initiation site near the center of the molecule. The best estimates suggest the origin is approximately 600 bp from the center of the molecule (T. Cech, personal communication). Thus pGY 19 contains the origin of replication of rDNA. We are sub-cloning the rDNA in pGY 19 to localize the ars element more precisely.

A 650 bp fragment of Tetrahymena rDNA from the termini of the molecule also allows autonomous replication of plasmids in yeast. This region does not contain an origin of replication for extrachromosomal rDNA but may contain a chromosomal origin of replication. Alternatively, ars function of this region may reflect aspects of the replication of the ends of linear extrachromosomal rDNA, may reflect amplification of rDNA during development, or it may be fortuitous that these Tetrahymena sequences are recognized by the yeast replication apparatus. We are presently subcloning the rDNA from pGY 39 and site specific mutagenesis of these smaller plasmids should allow us to identify sequences recognized by yeast as origins of replication.

The nucleotide sequence of most of an approximately 700 bp region from the center of rDNA has been determined (14). This region is included in pGY 19 and includes the in vivo origin of replication of the molecule. The entire rDNA contained in pGy 39 has also been sequenced (15). This 650 bp fragment contains approximately 50 repeats of a C_4A_2 sequence present at the termini of native rDNA (21). This C_4A_2 repeat is unlikely to function as an origin of replication and thus the ars element in pGY 39 is probably contained in a 355 bp DNA fragment. Both of these Tetrahymena fragments contain 85% A-T base pairs but other than homooligomeric stretches, no extensive sequence homology is seen in these regions. A perfect "Hogness box" (TATAAATA) is found approximately 535 bp from the center of rDNA and a number of other near perfect "Hogness boxes" are found in both pGY 19 and pGY 39. This is not unexpected in an A-T rich sequence but the TATAAATA sequence is the only extensive homology between the Tetrahymena sequences and the ars-1 (22) and ars-2 (J. Carbon, personal communication) yeast chromosomal sequences. It is not yet known however if the TATAAATA sequence is important in defining a yeast chromosomal origin of replication. The ars-1 and ars-2 sequences are not A-T rich but do contain some short stretches of high A-T content. Furthermore, no extensive sequence homology exists between the Tetrahymena sequences and the region of the yeast 2 μm circle plasmid containing the origin of replication (23).

Of other non-yeast sequences allowing autonomous replication of plasmids in yeast (10, 11), none except a 2.2 kb fragment from Xenopus laevis mitochondrial DNA (11) are known to include in vivo replication origins. In the case of the Xenopus mitochondrial DNA, this fragment is known to include the in vivo origin of replication but it is not known what sequences are recognized by yeast. No DNA sequence data are available for these heterologous ars elements.

Non yeast DNA can allow autonomous replication of plasmids in yeast. Two non-yeast sequences, the central region of Tetrahymena rDNA and a 2.2 kb fragment from Xenopus mitochondrial DNA include known in vivo origins of replication. From the limited sequence data available, it appears that the yeast replication apparatus does not recognize a single sequence as an origin of replication but may recognize at least a class of sequences. Alternatively, secondary structure, which has not as yet been analyzed in any ars sequences, may be important in defining origins of replication. Whether yeast recognizes the same sequences used as origins of replication in Tetrahymena is also not yet known.

Despite the limitations, assaying in yeast for high frequency of transformation of a selectable marker with a plasmid containing non-yeast sequences is a useful way of isolating and characterizing sequences which may be replication origins. We are at present using the cloned Tetrahymena ars elements to isolate and characterize Tetrahymena DNA binding proteins and in in vitro assays which reflect initiation of replication. These studies should be useful in correlating ars sequences with in vivo origins of replication.

ACKNOWLEDGMENTS

We thank Anna Bankuti, Brian Morrisson and Nora Tsao for expert technical assistance, Dr. Reg Storms for helpful discussion and Christina Sutherland for secretarial assistance.

REFERENCES

1. Hinnen, A., Hicks, J.B. and Fink, G.R. (1978). Transformation of yeast. Proc. Natl. Acad. of Sci. 75: 1929-1933.
2. Beggs, J.D. (1978). Transformation of yeast by a replicating hybrid plasmid. Nature 275: 104-109.
3. Storms, R.K., McNeil, J.B., Khandekar, P.S., An, G., Parker, J. and Friesen, J.D. (1979). Chimeric plasmids for the cloning of deoxyribonucleic acid sequences in Saccharomyces cerevisiae. J. Bact. 140: 73-82.
4. Struhl, K., Stinchcomb, D.T., Scherer, S. and Davis, R.W. (1979). High-frequency transformation of yeast: autonomous replication of hybrid DNA molecules. Proc. Natl. Acad. Sci. USA 76: 1035-1039.
5. McNeil, J.B., Storms, R.K. and Friesen, J.D. (1980). High frequency recombination and the expression of genes cloned on chimeric yeast plasmids: identification of a fragment of a 2-μm circle essential for transformation. Current Genetics. 2: 17-25.
6. Beach, D., Piper, M. and Shall, S. (1980). Isolation of chromosomal origins of replication in yeast. Nature 284: 185-187.
7. Chan, C.S.M. and Tye, B-K. (1980). Autonomous replicating sequences in Saccharomyces cerevisiae. Proc. Natl. Acad. Sci. USA 77: 6329-6333.
8. Stinchcomb, D.T., Struhl, K. and Davis, R.W. (1979). Isolation and characterization of a yeast chromosomal replicator. Nature (London) 282: 39-43.
9. Szostak, J.W. and Wu, R. (1980). Unequal crossing over in the ribosomal DNA of Saccharomyces cerevisiae. Nature 284: 426-430.
10. Stinchcomb, D.T., Thomas, M., Kelly, J., Selker, E. and Davis R.W. (1980). Eukaryotic DNA segments capable of autonomous replication in yeast. Proc. Natl. Acad. Sci. USA 77: 4559-4563.
11. Zakian, V.A. (1981). The origin of replication from Xenopus laevis mitochondrial DNA promotes high frequency transformation of yeast. Proc. Natl. Acad. Sci. USA. In Press.
12. Ganz, P.R. and Pearlman, R.E. (1980). Purification from Tetrahymena thermophila of DNA polymerase and a protein which modifies its activity. Eur. J. Biochem. 113: 159-173.
13. Ganz, P.R., Kiss, G.B. and Pearlman, R.E. (1981). Oligo (A) stimulated Tetrahymena rDNA synthesis in vitro. Can. J. Biochem. In Press.
14. Kiss, G.B. and Pearlman, R.E. (1981). Extrachromosomal rDNA of T. thermophila is not a perfect palindrome. Gene. In Press.

15. Kiss, G.B., Amin, A.A. and Pearlman, R.E. (1981). Two separate regions of the extrachromosomal rDNA of Tetrahymena thermophila enable autonomous replication of plasmids in yeast. Molecular and Cellular Biology. In Press.
16. Engberg, J., Andersson, P., Leick, V. and Collins, J. (1976). Free ribosomal DNA molecules from Tetrahymena pyriformis GL are giant palindromes J. Mol. Biol. 104: 455-470.
17. Karrer, K.M. and Gall, J.G. (1976). The macronuclear ribosomal DNA of Tetrahymena pyriformis is a palindrome. J. Mol. Biol. 104: 421-453.
18. Engberg, J., Din, N., Eckert, W.A., Kaffenberger, W. and Pearlman, R.E. (1980). Detailed transcription map of the extrachromosomal ribosomal RNA genes in Tetrahymena thermophila. J. Mol. Biol. 142: 289-313.
19. Chang, A.C.Y. and Cohen, S.N. (1978). Construction and characterization of amplifiable multicopy DNA cloning vehicles derived from the P15A cryptic miniplasmid. J. Bacteriol. 134: 1141-1156.
20. Bolivar, F., Rodriguez, R.L., Green, P.J., Betlach, M.C., Heyneker, H.L., Boyer, H.W., Crosa, J.H. and Falkow, S. (1977). The circular restriction map of pBR 322. In DNA Insertion Elements, Plasmids and Episomes, A.I. Bukhari, J.A. Shapiro and S.L. Adhya, ed. Cold Spring Harbor Laboratory pp. 686-687.
21. Blackburn, E.H. and Gall, J.G. (1978). A tandemly repeated sequence at the termini of the extrachromosomal ribosomal RNA genes in Tetrahymena. J. Mol. Biol. 120: 33-53.
22. Tschumper, G. and Carbon, J. (1980). Sequence of a yeast DNA fragment containing a chromosomal replicator and the TRP 1 gene. Gene 10: 157-166.
23. Hartley, J.L. and Donelson, J.E. (1980). Nucleotide sequence of the yeast plasmid. Nature (London) 286: 860-864.

HIGH-FREQUENCY TRANSFORMATION OF MOUSE L CELLS

Friedrich Grummt[1], Erika Dinkl[1,2],
Ursula Wolf[1] and Werner Goebel[2]

ABSTRACT A vector consisting of pBR322-, HSV-tk- and yeast 2 μ circle origin DNA transforms mouse L-tk⁻cells at high frequency (approximately 4000 transformants/μg DNA x 10^6 cells). The transforming DNA is replicated autonomously in mouse cells without being integrated into chromosomal DNA. The extrachromosomal vector DNA extracted from transformants by the Hirt procedure is identical with the original transforming DNA as determined by restriction analysis and blot hybridization. A copy number of at least 600 per transformed cell was determined by this techniques. No segregation of the transformed state was observed in the absence of selection pressure. Thymidine kinase activity can be measured in extracts from transformed cells. This activity is significantly higher in Ltk⁺ transformants than in wild-type L cells.

INTRODUCTION

Mammalian cells can be genetically transformed by the introduction of foreign DNA. A favorite system to study the events of biochemical transformation represent mouse L cells lacking the enzyme thymidine kinase. These Ltk⁻ cells have been shown to be susceptible to in vitro transformation by thymidine kinase gene containing Herpes simplex virus DNA (1-3).

The resulting tk⁺ phenotype can be isolated in selective medium (hypoxanthine, aminopterin, thymidine-HAT) which prevents growth of the Ltk⁻ parental phenotype. The transformation frequency observed in experiments using HSV-tk⁺ gene DNA is relatively low, e.g. about 1-10 colonies per μg of donor DNA and 10^6 cells (4). The tk⁺ gene becomes

[1]Present address: Institut für Biochemie der Universität Würzburg, Röntgenring 11, D-87 Würzburg F. R. G.
[2]Present address: Institut für Mikrobiologie und Genetik der Universität Würzburg, Röntgenring 11 D-8700 Würzburg, F. R. G.

integrated into the chromosomal DNA of all transformants. The event of integration seems to be rate limiting for the transformation process. In contrast to this commonly observed type of low-frequency transformation Struhl et al. (5) reported recently high-frequency transformation in yeast cells. The vectors at high frequencies (5000 - 20,000 colonies per µg donor DNA) replicate autonomously as closed circles or as chromosomally integrated structures in the recipient yeast cells. Other vectors also transforming cells at high frequency contain centromeric DNA and behave as minichromosomes because they always replicate autonomously and do not integrate into the genome.

In the studies described here, a hybrid bacterial plasmid containing Herpes simplex virus and yeast 2 µ plasmid DNA was used as a transformation vector. The plasmid pBSG1 is a hybrid composed of Escherichia coli plasmid pBR322, an origin of replication containing Hind III fragment of the yeast 2 µ plasmid and the tk gene containing Kpn I fragment of Herpes simplex virus type 1. Using this plasmid we can transform a stable tk^- mutant of mouse L cells to a tk^+ phenotype at high frequency. During the transformation event the transforming vector DNA is not integrated into the genome of the recipient cells but replicates autonomously.

RESULTS AND DISCUSSION

<u>Transformation Vectors</u>: Transformation of mouse Ltk^- cells was carried out using two different vectors, pLE578 and pBSG1. pLE578 (kindly provided by Lynn Enquist) contains the HSV-1 tk gene as a 3.4kb Bam H1 fragment cloned into the Bam H1 cleavage site of pBR322. pBSG1 consists of the tk gene-containing KpnI fragment of HSV-1, the origin of replication-containing Hind III fragment of the yeast 2 µ circle and pBR322 (Fig. 1). These two tk gene-containing transfection vectors were applied to Ltk^- recipient cells by the calcium phosphate procedure together with cellular DNA as carrier.

<u>Transformation to the tk^+ phenotype</u>: Mouse L cells lacking the thymidine kinase gene can be transformed both by the vector pLE578 and by pBSG1 to the Ltk^+ phenotype. The results of a typical set of experiments are shown in Table 1. With the vector pLE578 biochemical transformation occurs at low

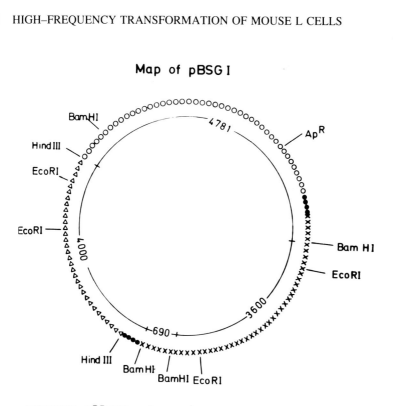

Fig. 1. Structure of the vector pBSG1 used for high-frequency transformation of mouse Ltk⁻cells.

frequency and is usually detected by the ability of the rare transformed cell to grow under selective conditions in HAT medium. However, when the chimeric vector pBSG1 is used a much higher transformation frequency can be observed. This high-frequency type of transformation results in about 4000 transformants per 10^6 cells per µg DNA. The transfection event is strictly dose-dependent with regard to the transforming vector DNA (Table 1). No transformants resulted from treatment with carrier DNA isolated from mouse Ltk⁻cells and no tk⁺ colonies arose in untreated cultures. The reversion frequency of the tk⁻mutation in Ltk⁻cells is extremely low. We have never observed a spontaneous revertant of Ltk⁻cells to Ltk⁺.

Table 1. Transformation of Ltk⁻ cells by HSV-1 tk gene-containing hybrid plasmids

tk hybrid plasmid	ng DNA/dish	Colonies per dish	Transformants per µg hybrid DNA per dish	
pBSG 1	20	84	4 200	
	20	49	2 450	
	20	64	3 200	
	20	41	2 050	
	20	62	3 100	
	20	108	5 400	M̄ ∿ 4 000
	20	99	4 950	
	2	10	5 000	
	2	11	5 500	
	2	6	3 000	
	0.2	0	0	
	0.2	1	5 000	
	0.02	0	0	
	0.02	0	0	
	0.02	0	0	
LE 578	200	1	5	
	200	0	0	
	200	7	35	
	200	4	20	
	200	3	15	M̄ ∿ 10
	200	3	15	
	200	3	15	
	200	1	5	
	200	0	0	

Table 1. Experiments were carried out using a modification of the method developed by Wigler et al. (4). One day prior to transformation cells were seeded at 3 x 10⁵ per dish. Carrier DNA (40 µg/ml) and plasmid DNA were solved in 1 mM Tris (pH 7.9) 0.1 mM EDTA/0.25 M CaCl$_2$ DNA/CaCl$_2$ solution was added dropwise to an equal volume of 280 mM NaCl/ 50 mM Hepes (pH 7.1) 1.5 mM sodium phosphate while introducing air bubbles at the bottom of the tube through a pipette. After 30 min, 0.5 ml of the

suspension was added to cultures in 6 cm dishes containing 5 ml of growth medium. After 16 hr the medium was replaced and the cells were allowed to incubate for an additional 24 hr. At that time, selective pressure was applied by addition of HAT medium.

Recovery of transfecting vector DNA from Ltk^+ cells and copy number: Individual transformant colonies were picked from separate plates to ensure that they represent independent transformation events. These colonies were grown into mass cultures under selective conditions. DNA was extracted from individual clones by the Hirt procedure (6). Both the Hirt supernatant and pellets were analyzed for the presence of vector DNA. Most of the transfecting DNA was found in the Hirt supernatant. The re-extracted DNA was analyzed by digestion with the restriction endonuclease Bam H1, agarose gel electrophoresis and Southern blot hybridization with nick-translated pBR322 or tk-gene DNA.
Fig. 2 shows that the Bam H1 fragment of the re-extracted DNA hybridizes against nick-translated pBR322 DNA and is identical in size with that of the vector DNA used originally in the transfection experiment. From this result it is concluded that at least the main portion of re-extracted vector DNA was not incorporated into the genome of the recipient cells but replicated autonomously. The best evidence that the re-extracted vector DNA is identical with the original transformation vector comes from the behavior of these plasmids when they were re-introduced into E. coli, amplified in and re-isolated from the bacteria and used again for the transformation of L-tk⁻cells. No differences in structure and transformation efficiency were observed between the originally constructed and re-extracted recombinant plasmid after these procedures (not shown).

The copy number of the vector DNA per transformed cells was also calculated from the blot hybridization data. DNA was extracted from a known number of transfected cells, the DNA of the Hirt supernatant fraction blot hybridized against nick-translated pBR322 DNA. In parallel increasing amounts of cold pBR322 DNA were hybridized against the radioactive probe to estimate the amount of the pBR322 sequences present in the re-extracted vector

DNA fraction (Fig. 2). The data demonstrated that the copy number in one clone is at least 600 and in a second one higher than 450.

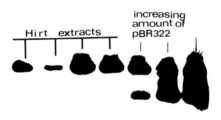

Fig. 2. Southern blot hybridization of transformation vector pBSG1 re-extracted from transformed mouse L-cells after digestion with Bam H1.

Stability of the transformant: The presence of a high amount of autonomously replicating vector DNA in the transformed cells leads to the question as to whether the transformed phenotype is stable or not. To answer this question a transformed clone was grown in nonselective medium. After 24 and 38 days 720 and 320 cells, respectively, were recloned and the number of colonies arising under selective pressure scored. Table 2 shows that all colonies survived in HAT medium. This high stability of the transformed state might be due to the high copy number of tk genes in the transformants.

TABLE 2

Studies on segregation rate in transformed mouse L-cells

time of culture without HAT (days)	number of initial clones	number of surviving clones after re-addition of HAT
24	720	720
38	320	320

Thymidine kinase activity in transformed L-cells. The appearance of thymidine kinase activity was studied in cell extracts from mouse L-cells. Fig. 3 shows that significant enzyme activity can be observed in extracts of Ltk^+ transformants, whereas the original Ltk^- phenotype is completely lacking thymidine kinase activity. Extracts of the transformed Ltk^+ phenotype show only a 4-5 fold higher kinase activity than wild type extracts, although the genetic information is present in at least 600 copies. This relatively slight elevation of the enzyme activity in comparison to the wild type could hint to a down regulation either of the thymidine kinase gene expression or enzymic activity.

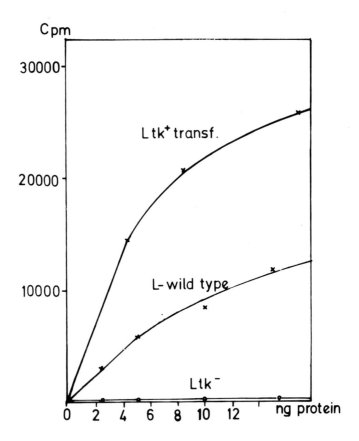

Fig. 3. Thymidine kinase activity in mouse L cell extracts in vitro.
Cell extracts were prepared and thymidine kinase activity assayed according to Lee and Cheng (7).

REFERENCES

1. Davis, D.B., Munyon, W., Buchsbaum, R. and Chawda, R. (1974). J. Virol. 13, 140-145.
2. Maitland, N.J. and McDougall, J.K. (1977). Cell 11, 233-241.
3. Pellicer, A., Wigler, M., Axel, R. and Silverstein, S. (1978). Cell 14, 133-141.
4. Wigler, M., Pellicer, A., Silverstein, S., Axel, R., Urlaub, G. and Chasin, L. (1979). Proc. Natl. Acad. Sci. USA 76, 1373-1376.
5. Struhl, K., Stinchcomb, D.T., Scherer, S., and Davis, R.W. (1979). Proc. Natl. Acad. Sci. USA 76, 1035-1039.
6. Hirt, B. (1967). J. Mol. Biol. 26, 365-369.
7. Lee, L-S. and Cheng, Y.-C. (1976). J. Biol. Chem. 251, 2600-2604.

INDEX

A

Adenovirus, 38
Aphidicolin, 35
  DNA pol I resistance, 35
  Yeast mutants, 35
ars, 31
asnA, 2
  Enterobacter aerogenes, 2
  Erwinia carotovora, 2
  E. coli, 2
  Salmonella typhimurium, 2
ATPase, DNA dependent, 26, 27
Attenuation, 20
Autonomously replicating sequence, 31, 32, 34, 40

C

cdc Mutant, 35
Centromeric sequence, 30
col E 1, 11, 12
  lagging strand synthesis, 12
Complementation assay, 25
Consensus sequence, 2, 3
Copy number (plasmid), 10
  control, 8, 11
  R plasmid NR1, 10
cos 7, monkey cells, 36

D

dam Methylase, 2
das Loci, 20
Direct repeats of nucleotide sequences, 9

DNA binding proteins, 7
DNA polymerase I, 35
DNA polymerase II, 35
DNA polymerase III
  β subunit, 28
  chemical inhibition, 28
  Escherichia coli, 28
  replication initiation, 28
  subunit structure, 27
DNA–protein interactions, 26
dna genes
  plaque complementation assay, 25
dnaA gene, 20, 21, 23, 25
  amber mutations, 21
  rpoB suppression, 20, 21
  supressor mutations, 20 21
dnaB gene, 12
dnaB protein, 12, 27
dnaC gene, 25
dnaC protein, 27
dnaE gene
  intergeneric complementation, 25
dnaG, 12
dnaN gene
  protein, 27
dnaT gene, 23
  suppression, 23
dnaZ protein, 22

E

Electron microscopy
  displacement synthesis, 19
  G4 secondary structure, 16

2μm DNA replication, 35
φ29 replication, 29
RNA polymerase binding sites, 6
T7 replication, 27
Elongation, 28
Enterobacter aerogenes, 2, 3
  asnA, 2
  origin of replication, 2, 3
  uncB, 2
Erwinia carotovora, 2, 3
  asnA, 2
  origin of replication, 2, 3
Escherichia coli
  asnA, 2
  ATPases, 27
  chromosomal replication origin, see ori C
  DNA polymerase III, 27, 28
  origin of DNA replication, see ori C
  primosome, 27
  promoters, 4
  uncB, 2
Eukaryotic cloning vectors, 36
Exonuclease III
  protection assay, 18

F

FII Incompatibility Group, 8
Footprinting, 18

G

G4
  secondary structure, 16
  replication origins, 13, 16
Gene 4 protein, 26
Gene 32 protein, 19

H

Human Blur clones, replication in *Xenopus* eggs, 39

I

IS5 insertion element, 17
Inceptors, 17, 18
Incompatibility, 8, 9, 10
  direct repeat nucleotide sequences, 9
  genetic mapping, 8
  Inc FII R Plasmid NR1, 10
  models, 8, 10
Initiation (of DNA replication)
  complementary strand, 13
  determinants, 13

determinant cloning, 12, 13
DNA polymerase III (*E. coli*), 28
dnaA protein, 20, 21
dnaA mutations, 20, 21
in vitro, 26, 27
lagging strand synthesis, 17, 26
minichromosomes, 5, 6
mutants, 6
p3 terminal protein, 29
φ29, 29
primary, 19
protein synthesis requirement, 24
Initiation
  rec A dependent, 23, 24
  rpo B suppression, 20, 21
  recombination, 19
  rifampicin-resistant, 12, 13
  SV40, 37
  secondary, 19
  strand separation, 8
  T4, 19
  T7, 26
  topoisomerase, 19
  transcriptional activation, 5, 26
  viral strand, 13

K

Klebsiella pneumoniae, 2, 3
  origin of replication, 2, 3

L

Lagging strand DNA synthesis, 26
λ Origin of replication
  deletions, 18
  growth interference, 18
  O protein interaction, 17, 18
  repeated sequence, 18
  sequence homology, 17
λ Replication control, 18
Leading strand DNA synthesis, 17, 26
Lethal gene, 2

M

M13, 12, 13
Maturation genes, 19
Membrane attachments, 7
Minichromosomes, 5, 6
  cloning in pBR322, 5
  dna mutants, 6
  initiation mutants, 6
  nucleotide sequence, 6
Mismatch repair, 2

INDEX 627

N

N-ethylmaleimide, 28
n' Protein, 27

O

o-Protein of λ, 17, 18
Okazaki fragments
  DNA polymerase III, 28
  initiation, 13, 26, 28
  T7, 26
ori C
  cloning, 2, 4, 5
  consensus sequences, 2, 3
  defined, 1, 4, 6
  functional sites, 6
  membrane binding, 7
  minichromosomes, 6
  mutations, 1
  nearby genes, 5, 6
  plasmid chimeras, 2, 4, 5
  promoters, 2, 4, 5, 6
  Tn 10 insertion mutants, 5
  terminators, 5
  transcriptional activation, 6
Origins (of DNA replication), see also ori C
  adenovirus, 38
  Chinese hamster cells, 39
  col E1, 12
  consensus sequence, 1, 2, 3
  direct repeat nucleotide sequences, 9
  Enterobacter aerogenes, 2
  Erwinia carotovora, 2
  Escherichia coli see also ori C, 2
  eukaryotic, 31
  evolution, 2
  F plasmid, 9
  G4, 13, 16
  incompatability regions, 8, 9, 10
  Klebsiella pneumoniae, 2
  λ, 17, 18
  M13, 13
  2-μm DNA, 33, 35
  mutations, 3
  φX174, 13, 14, 15, 16
  R plasmid NR1, 8, 10
  R1 plasmid, 8
  RGK plasmid, 9
  R100 plasmid, 8, 10
  RK2 plasmid, 9
  repeats, 2
  SV-40, 36, 37

Saccharomyces cerevisiae, 30, 31, 32, 33
Salmonella typhimurium, 2, 25
secondary structure, 2, 16
T7, 26
Tetrahymena thermophila, 40
Vibrio (Beneckea) harveyi, 2
Xenopus laevis, 39

P

Protein of λ, 18
Phasmid, 11
φ29 Phage, 29
φX174
  complementary strand origin, 27
  gene A protein, 14, 15, 16
  host DNA replication proteins, 27
  interference, 15
  origin (of DNA replication), 14
  origin mutants, 14
  prepriming, 27
  primosome, 27
  replication determinants, 14, 15
  replication *in vitro*, 27
  SS→RF inhibition, 15
  secondary structure, 16
  viral strand origin, 16
Plaque complementation assay, 25
Plasmids
  col E1, 11, 12, 13
  copy number mutants, 8, 10
  evolution, 8
  incompatibility, 10
  initiation protein, 9
  origin structure, 9
  pBR322 5, 13
  pGA46, 4
  pMB1, 11
  pSW14, 39
  R plasmid NR1, 10
  R1, 8
  R100, 8
  replication, 8, 9, 10, 11, 24
polA, 2
Pribnow box, 2
Primase
  T7, 26
  promoter sequence, 17
Primosome, 17, 27
Promoters
  conserved sequences, 2
  insertion activation analysis, 4, 5
  mapping, 5
  minichromosomes, 6

ori C, 2, 4, 6
primase, 17
relative strength, 5
T7, 26

R

rec A, 23, 24
rpoB mutations, 20
Recombination, 19
Replication (see also origins, initiation)
  complementary strand, 27
  discontinuous, 26
  inceptor, 17
  intermediates, 35
  in vitro, 26, 29, 35
  mutants, 11
  priming, 29
  rec A dependent, 23, 24
  rin, recA–independent, 24
  regulation, 8, 23
  rolling circle, 15, 38
  strand displacement, 38
  T7, 26
Resistance plasmids, 8
RNA polymerase
  binding sites, 6
  E. coli, 6
  signal sequence, 26
  T7, 26
rpo B, 20, 21

S

SV40
  initiation of DNA replication, 37
  T antigen, 36
Saccharomyces cerevisiae
  aphidicolin mutants, 35
  autonomously replicating sequences, 31, 32, 33, 34
  centromere, 31
  DNA polymerases I and II, 35
  2-μm DNA, 33, 35
  plasmid replication, 40
  replication origins, 30, 31, 32, 33
  ring chromosome III, 33
  transformation, 31, 40
Salmonella typhimurium
  asnA, 2
  dna genes, 25
  dna mutants, 25

ori sequence, 2, 3
plaque complementation assay, 25
uncB, 2
Scatchard binding analysis, 7
sdr, 23, 24
sdr T, 23
Secondary replication origins, 17, 26
Shine-Delgarno sequence, 8
SOS repair, 23
Stable DNA replication, 24

T

T4 phage, 19
T7 phage, 26
T7 DNA Polymerase, 26
Target site duplication, 17
5'-terminal protein, 29, 38
Termination (of DNA replication), 15
tet Insertion activation, 4, 5
Tetrahymena thermophila, 40
Tn 10, 5, 25
Topoisomerase, 19, 38
Transcription
  incompatability mechanism control, 10
  in vitro, 6, 8, 10
  minichromosomes, 6
  pause site, 17
  R plasmid NR1, 10
  replication control, 8
  role in initiation, 6
  T7, 26
  terminators, 5, 17
  translation requirement, 5
Transcriptional activation, 5, 6, 17, 26
Transduction mapping, 25
Transformation vector, 41
Transposon, 25
Trp1, 34
Trp1-R1-Circle, 34
tyrT tRNA promoter, 2

U

uncB, 2

V

Vibrio (Beneckea) harveyi, 2

Y

Yeast see *Saccharomyces cerevisiae*